The Gurob Ship-Cart Model and Its Mediterranean Context

Ed Rachal Foundation Nautical Archaeology Series
In association with the
Institute of Nautical Archaeology

THE GUROB SHIP-CART MODEL AND ITS MEDITERRANEAN CONTEXT

Shelley Wachsmann

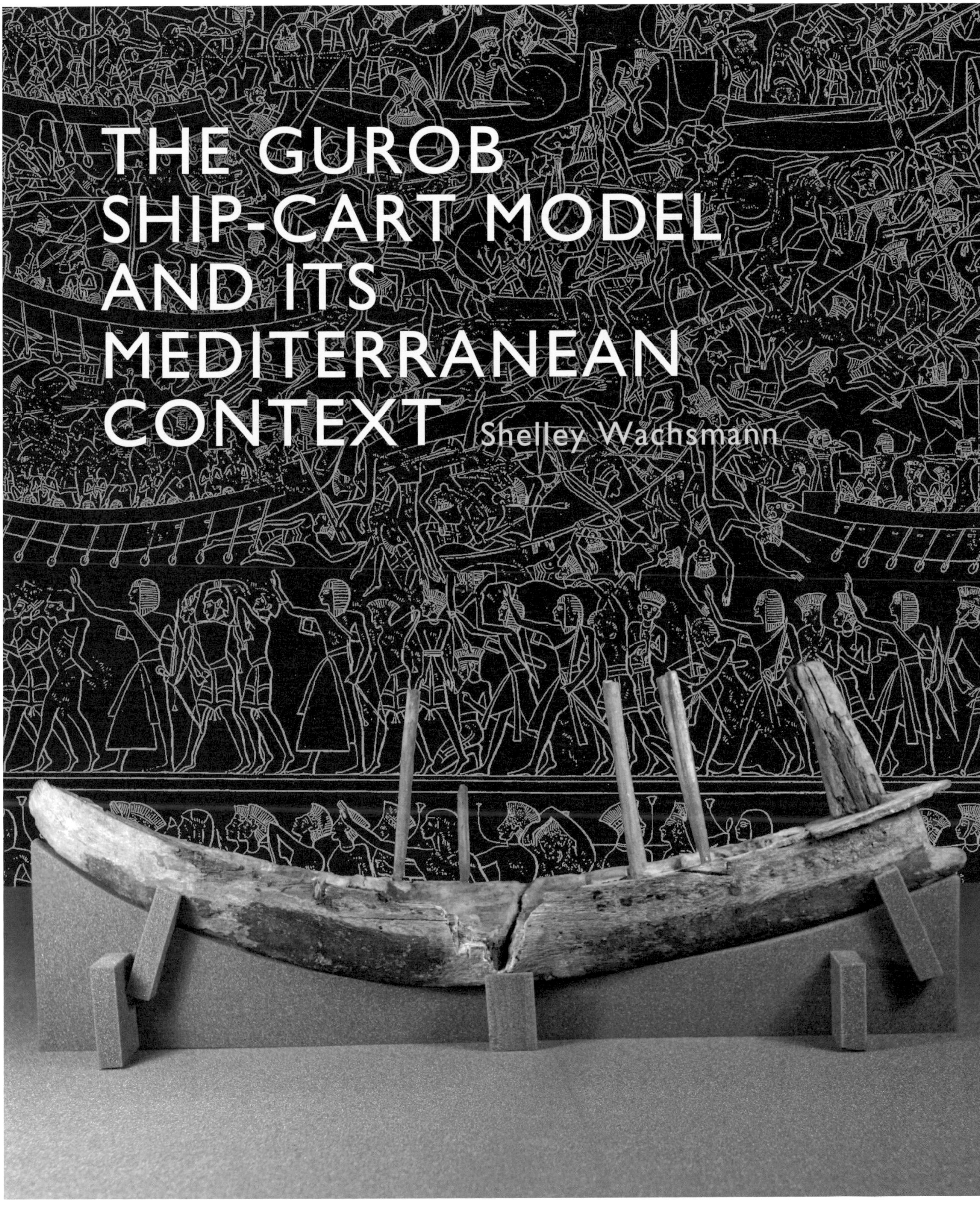

TEXAS A&M UNIVERSITY PRESS • COLLEGE STATION

First edition

This paper meets the requirements of ANSI/NISO Z39.48-1992 (Permanence of Paper).
Binding materials have been chosen for durability.
∞

Library of Congress Cataloging-in-Publication Data

Wachsmann, Shelley.
The Gurob ship-cart model and its Mediterranean context / Shelley Wachsmann. — 1st ed.
p. cm. — (Ed Rachal Foundation nautical archaeology series)
Includes bibliographical references and index.
ISBN 978-1-60344-429-3 (cloth : alk. paper) —
ISBN 1-60344-429-7 (cloth : alk. paper) —
ISBN 978-1-60344-746-1 (e-book) —
ISBN 1-60344-746-6 (e-book)
1. Galleys—Models—Egypt—Gurob (Extinct city) 2. Carriages and carts—Models—Egypt—Gurob (Extinct city) 3. Egypt—Antiquities. 4. Gurob (Extinct city)—Relations—Sources. 5. Shipbuilding—Mediterranean Region—History—To 1500—Sources. 6. Sea Peoples—Sources. I. Title. II. Series: Ed Rachal Foundation nautical archaeology series.
DT73.G85W33 2012
932'.2—dc23
2012004368

Frontispiece: Gurob ship model seen against the background of Ramses III's nautical battle against the Sea Peoples as depicted on his mortuary temple at Medinet Habu. Photo of model: S. Wachsmann, courtesy of the Petrie Museum of Archaeology. Background, after The Epigraphic Survey, 1930. *Medinet Habu I: Earlier Historical Records of Ramses III,* pl. 37, courtesy of the Oriental Institute of the University of Chicago.

Publication of this book has been aided by a grant from the von Bothmer Publication Fund of the Archaeological Institute of America.

The virtual reality images of the Gurob ship-cart model and the associated digital resources available at: http://www.vizin.org/Gurob/Gurob.html were produced by Donald H. Sanders and the Institute for the Visualization of History, Inc. (VIZIN).

For J. Richard "Dick" Steffy

May 1, 1924–November 29, 2007

Contents

The rabbit-hole went straight on like a tunnel for some way, and then dipped suddenly down, so suddenly that Alice had not a moment to think about stopping herself before she found herself falling down what seemed to be a very deep well.

From *Alice's Adventures in Wonderland*
by Lewis Carroll

Map 1: Overview. Map courtesy D. Davis.

Mare
aja
Sea of Azov
Black Sea
Caspian Sea
Aral Sea
Troy
Aegean Sea
ANATOLIA (ASIA MINOR)
Kültepe
Çatalhüyük
Tarsus
Ugarit
Til Barsip
Knossos
CYPRUS (ALASHIA)
CRETE (CAPHTOR)
Hama
Tigris
Euphrates
Byblos
Ashkelon
Alexandria
Map 3
Siwa Oasis
EGYPT
Nile
Thebes
Map 2
Persian Gulf
Red Sea
Napata
© Dan Davis

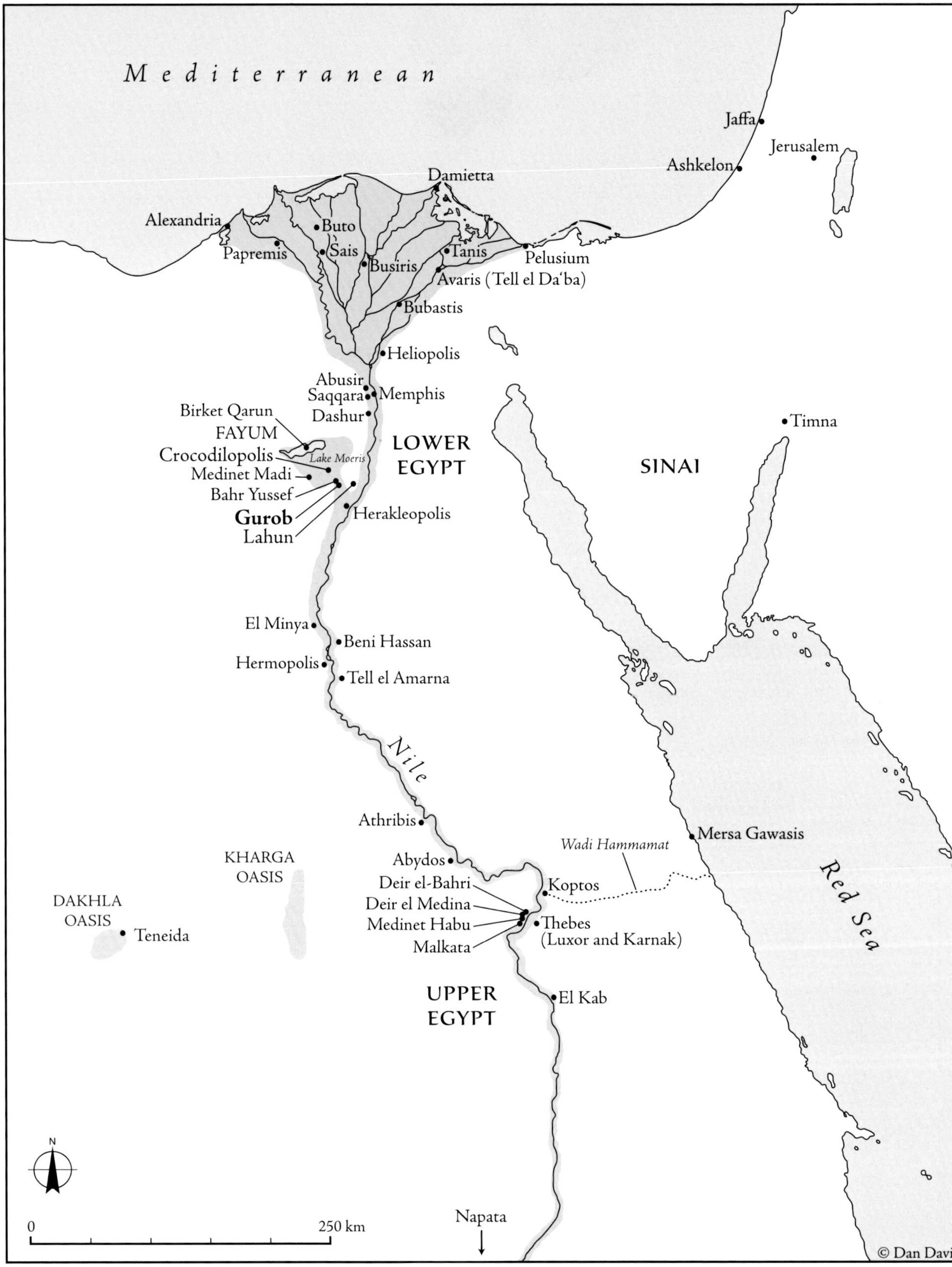

Map 2: Egypt. Map courtesy D. Davis.

Map 3: The Levant. Map courtesy D. Davis.

Map 4: The Aegean. Map courtesy D. Davis.

Map 5: The Corinthiad. Map courtesy D. Davis.

Preface

This book is about the one that got away.

I "discovered" it one clear dry afternoon in spring 1998. I was sitting in the small, specialized library of Texas A&M University's Nautical Archaeology Program, where I teach. I had been thumbing through various titles that I had noticed but had previously not found time to peruse. It was one of those days when I allowed myself to do something I *wanted* to do rather than what I *should have been doing.* I was feeling good. My latest book, *Seagoing Ships and Seamanship in the Bronze Age Levant,* was about to be published by Texas A&M University Press.[1] In that book I covered the waterfront for the Bronze Age in the eastern Mediterranean. A major portion of the book dealt with ships of the Aegean.

Among the pile of books that I had pulled off the shelves sat a bound copy of Chris Monroe's expanded 1990 master's thesis, titled "The Boatbuilding Industry of New Kingdom Egypt." As I flipped through Chris's illustrations one figure stopped me cold in my tracks.[2]

This was the first time I remember seeing a representation of the Gurob ship-cart model (or at least the first time that it registered for me). Somehow, surprisingly, the model simply had never made it into the list of "usual suspects" so familiar to those of us dealing with ancient Mediterranean ship iconography. Looking at the photograph of the model, I immediately realized that, despite the fact that it had been found buried in the sands of Egypt, the model actually represented a Helladic galley of a type used by the Mycenaeans and adopted by the Sea Peoples. To date no one has ever found even a splinter of one of these vessels. Trust me. I have been one of those searching.[3]

Looking at the photo, I got the impression that one of the Pyrgos Livanaton (Kynos) galley depictions—among the most detailed known representations of this ship type—had sailed off its sherds into a three-dimensional reality.[4] The Gurob model's details were unequivocal: the shape of the hull, the stempost ending in the head of a waterbird with a strange vertical beak, the ramlike bow extension, and particularly the stanchions.

All the other images of this ship type that I had studied were either two dimensional or crudely made models. Here, for the first time, was what appeared to be a relatively well-made model, which appeared to be painted and showed clearly details in three dimensions, unavailable on any other representation of which I was aware. Admittedly, some aspects of the model's reconstruction seemed strange to me: Why was the quarter rudder leaning against the bow? In addition, the "awnings" made no sense to me. I assumed that a closer examination would resolve most, if not all, of the outstanding issues.

I suddenly realized that I had left this representation out of *Seagoing Ships.* I slapped my forehead and asked myself aloud, "How could you have missed this one?" (Fortunately, no one else was in the library to hear me.) There was nothing I could do about it at that late stage of the book's production.

How embarrassing!

At the same time, I felt remarkably elated. Here, in front of me—and better late than never I quickly realized—was an incredible resource for our understanding of a complex, fascinating, and pivotal ship type about which so little is known. I immediately contacted Chris and asked him whether he had any intentions of publishing the model. He wrote back that he did not and encouraged me to study it. I then wrote to the Petrie Egyptological Museum, which owns the model, requesting permission to do so. To my dismay, the museum's response was that the model had recently been sent away for conservation, for an indefinite but most definitely a very long period,

Fig. P.1 "Ceci n'est pas une pipe" ["This is not a pipe"]. 1928–29. Oil on canvas. William N. Copley Collection, New York. (From Torczyner 1979: 71 fig. 87).

and until it returned to the museum I would be unable to examine it. Several years later I wrote again and received basically the same reply.

Time passed, and my research interests took other paths: the model receded into the back of my memory. Then, in 2005, realizing that it would be relatively easy for me to schedule a stopover in London while doing my summer work in the Mediterranean, I wrote again to the museum. This time I addressed my request to Stephen Quirke, the museum's curator. He informed me that the model indeed had returned to the Petrie Museum from conservation and graciously invited me to publish my observations. That summer I spent a week in London studying, recording, and photographing the model.

When I began work on the model, I expected that this research would result in a medium-length, peer-reviewed article. I initially thought that the model would be a quick and easy study for me. In fact, I was so convinced of this that I stopped working on another book manuscript that dealt with archery in biblical times in order to devote my attention to the Gurob model, thinking I could complete it quickly. After all, the model represented only one more iconographic representation of a Helladic ship type with which I had become quite familiar. How difficult could it be to write a comprehensive report on the model?

As it turns out, *seven years later,* it was difficult. Very difficult. I had been incredibly naïve. I soon discovered just how much so when, at every turn, the model pointed me in the direction of new and unexpected avenues of research until I sometimes felt like a latter-day Alice free-falling down a rabbit hole. I began to see the model as a type of key to the past.

For example, the model was found with four wheels. At the beginning of my research I confess that I did not pay much attention to them. I assumed that the wheels probably simply indicated that the model had been constructed as a child's plaything. I was eventually disabused of that assumption, however.

A major breakthrough in understanding the Gurob model came several years into my research, when I read a 1917 publication by Georges Legrain, titled "Le logement et transport des barques sacrées et des statues des dieux dans quelques temples égyptiens."[5] In this work the Egyptologist discusses the support device that he terms a *pavois,* which the Egyptians employed with their terrestrial cult ships to allow for the attachment of long poles, which would enable porters to carry the vessels. This article was truly a game changer as I soon realized that one of the loose wood pieces found with the model and which fitted into a hole drilled into the bottom of the hull amidships was actually a miniature *pavois.*[6] *This, then, meant that the Gurob model, far from being a child's toy, actually replicated a syncretic cult ship on wheels that mixed both European and Egyptian cultic features.* My first reaction to this realization was one of disappointment, as this meant that the model might be, and indeed probably is, at least once removed from the actual Helladic galley that it represents: In other words, it is a copy of a copy. My negative view was again premature, however, for this new understanding led me down a long and fascinating gallery of ancient and arcane Mediterranean boat-related cultic customs.[7]

A note of warning. This book deals to a large degree with the iconography of ancient ships. When working with this type of evidence, we must fully comprehend the character of the objects of our inquiry. In this regard it is worth reflecting on the meaning behind the iconic image of a smoker's pipe under which appears the phrase "Ceci n'est pas une pipe," which the Belgian surrealist painter René Magritte depicted in a series of paintings titled *La trahison des images, Les deux mystères, L'air et le chanson,* and so on (Fig. P.1).[8] Of course,

Magritte is correct. We do not see an actual smoker's pipe but rather an image of one. To put it another way, *a representation of an object is not the object itself.*

When studying ship iconography, we must keep this concept firmly in mind. We must always be aware that these are not the ships themselves under our scrutiny but rather representations of them. Thus, although these images of ships, be they two dimensional or in the round, represent the actual water transport that they copy, they do so refracted through the thought processes, the artistic abilities, and the limitations of their creators. The type of medium employed, as well as problems of space constraints placed on the artist, can also cause the image to deviate from the original. Thus, when, for convenience, I refer to "ships" or "galleys" without qualifying them as depictions, this should be understood. I also refer to the Gurob ship-cart model simply as a ship model, although here, too, the larger meaning is to be assumed.

Acknowledgments

I have had much welcome assistance during the research and writing of this book. My study of the physical remains of the Gurob ship-cart model, which took place during the summer of 2005, would not have been possible without the gracious and the generous assistance of the staff of the Petrie Museum of Egyptian Archaeology, who did everything possible to make my study of the model pleasant and productive. I am particularly indebted to Stephen Quirke, the museum's curator, who gave me permission to study the model and to publish my research, as well as to the museum's staff for their support: Suzy MacDonald and Hugh Kilmister for administrative assistance, Ivore Pridder and Suzi Pancaldo for help in aspects of the model's conservation, and Tracey Golding and Judy Joseph for facilitating my work.

Readers will realize the crucial contribution of the Dakhla Oasis Helladic ship graffito for this study.[1] The only photograph of the graffito resides in the H. A. Winkler photograph collections of the Egypt Exploration Society. I am grateful to Andrew Bednarski and Patricia Spencer for making available for publication here Winkler's references and images from Dakhla Oasis Site 69. I also thank Mike Morrow for allowing me to use excerpts from the Winkler diaries, translated and transposed by Ursula Murphy.[2]

When I first began work on the model, I envisioned a digital component to the publication that would allow the reader to interact with a computerized three-dimensional virtual artifact, in addition to the standard methods of presenting such materials. Donald Sanders, president of the Institute for the Visualization of Ancient History, and his team have been outstanding partners in generating these digital components (available for online viewing at www.vizin.org/Gurob/Gurob.html). I am also grateful to John A. Stewart for helping ensure that the virtual component functions in the now Intel-based Apple Macintosh environment.

In research such as this it is inevitable that one would seek advice from scholars from a variety of disciplines, and, indeed, colleagues have been unstinting in helping me better understand aspects of the research related to their own areas of expertise. I owe them all an enormous debt of thanks. I was exceptionally privileged to discuss aspects of hull construction as it relates to the model with the late J. Richard ("Dick") Steffy. On other aspects of ancient ships I received valuable advice from Lucien Basch, Olaf Höckmann, and William M. Murray. On Egyptological matters I turned to Lanny Bell, Manfred Bietak, Noreen Doyle, David Jeffreys, Peter Lacovara, Christine Lilyquist, David O'Conner, and Anson Rainey. On topics related to the Aegean Bronze Age I received valuable guidance from Richard Janks, Vassos Karageorghis, Daniel M. Master, Penelope Mountjoy, and Joseph W. Shaw. On issues pertaining to Classical archaeology I learned from Bridget Buxton, David Blackman, Dan Davis, Kevin Glowacki, John Hale, Robert L. Hohlfelder, Jeffrey Hurwit, Demi Kaminou, Harry Kritzas, John Lenz, Mary B. Moore, William M. Murray, Naomi J. Norman, Harry Tzalas, and Peter van Alfen. For insights into subjects related to central European antiquity, I am beholden to Jan Bouzek and Kristian Kristiansan. For clarifications on paleoethnobotanical issues I thank Vaughn Bryant and Jennifer Ramsay. It goes without saying, however, that I am responsible for any errors herein, whether of omission or commission.

Graduate students are the lifeblood of scholarly research. I am grateful to my student assistants, past and present, who worked diligently on various stages of this volume: (now Dr.) Dante Bartoli, Heather Hatch, Megan Lickliter-Mundon, Veronica Morriss, Megan Collier, and Meko Kofahl. I also thank Alexis Catsambis, Sarah Kampbell, Troy Nowak, and Mark Polzer.

Texas A&M University's Sterling C. Evans Library is, simply put, the best archaeological library in which I have had the pleasure to work. The library has a truly formidable collection of Egyptian, Near Eastern, and Classical holdings. Over the years the university has aggressively added to the two major library purchases that form the heart of the collection, organized by George F. Bass soon after the creation of the Nautical Archaeology Program in the 1970s. It also has an excellent and helpful staff and an indispensible Interlibrary Loan System, so that any item not available from the library itself normally arrives within a few days of its request. Furthermore, the library's outstanding E-Docs system, which permits readers to order Adobe Acrobat PDF files of articles unavailable in the library catalogue (or up to fifty pages of any work owned by the library), truly changes the manner in which one carries out research today. I am particularly grateful to Stephen Bales, Boyd Conerway, Larry Reynolds, and Lan Yang for their help. Many of the more unique and valuable volumes owned by Texas A&M University reside in the adjacent noncirculating Cushing Memorial Library. I thank its librarians, Valerie Coleman and Lindsey Raney, for their assistance.

For the many kindnesses they have showered upon me throughout the research and writing of this book I am grateful to the administrative staffs, past and present, of the Department of Anthropology and the Institute of Nautical Archaeology (INA): Michelle Chmelar, Tamara Hebert, Chasity Hedlund, Cindy Hurt, Claudia LeDoux, Rebekah Luza, Andrasia Prejean, and Monica Sommerfield.

I am grateful to the staff of Texas A&M University Press for their unstinting efforts in seeing this book through all the steps to publication. The text has been much improved through the copyediting skills of Carol Hoke.

I wrote this book while holding the Meadows Professorship in Biblical Archaeology in Texas A&M University's Nautical Archaeology Program. I consider myself blessed to have received this honor: It has dramatically aided in allowing me to pursue my research goals. For this privilege I am grateful to the Meadows Foundation, Texas A&M University, INA, and George F. Bass. I also thank INA board member Lucy Darden for her support of my research. I am also indebted to the Archaeological Institute of America for a grant from its von Bothmer Publication Fund supporting the publication of this book.

Finally, I thank Karen, my soul mate, for her endless patience and support.

Shelley Wachsmann
College Station, Texas
February 2012

Note Regarding Online Resources

The virtual reality model and other associated materials related to the Gurob ship-cart model were specially created to accompany this book, and they are vital to making it both more understandable and more useful. These digital resources, created by Donald H. Sanders and the Institute for the Visualization of History, Inc. (VIZIN), are available online at http://www.vizin.org/Gurob/Gurob.html.

The reader is urged to make full use of them and also to note appendix 2 in this volume, in which their development, applicability, and implications are discussed more fully, along with complete instructions for use.

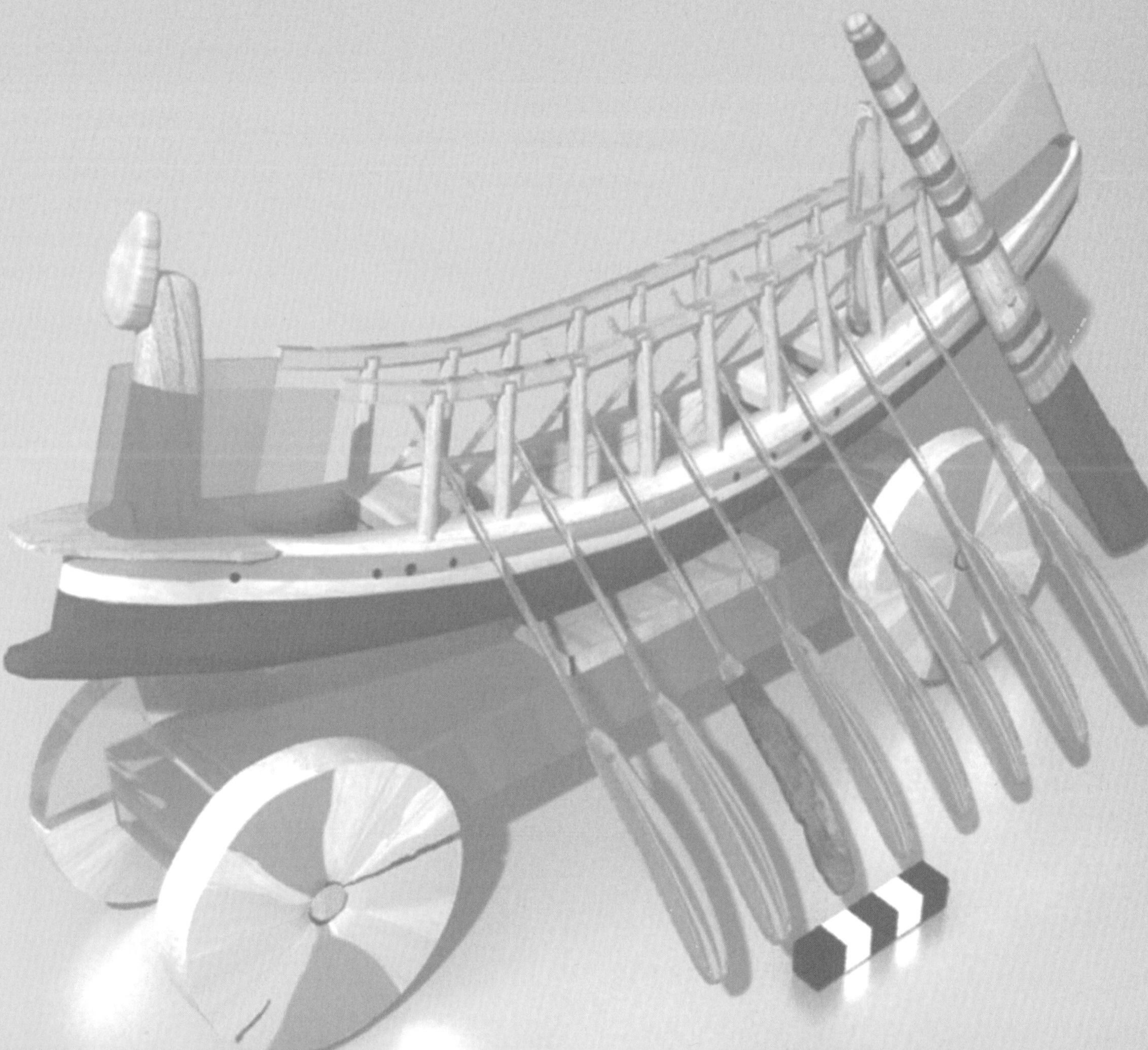

The Gurob Ship-Cart Model and Its Mediterranean Context

Fig. 1.1: Gurob is located at the entrance to the Fayum. This satellite image places it in its geographical context of the eastern Mediterranean. Courtesy Google.

1 The Gurob Ship-Cart Model

In 1920 William Mathews Flinders Petrie assigned two of his assistants, Guy Brunton and Reginald Engleback, to excavate Gurob, a site that he had first examined three decades earlier (Fig. 1.1).[1] Petrie's renewed interest in the ancient settlement resulted from his concern for the site's destruction and the loss of antiquities to illicit excavations carried out by various antiquity dealers, as well as the collection of *sabbâkhin* (settlement deposit used as fertilizer) by the local fellahin, which had lowered the level of the site to that of the surrounding desert.[2] The expedition's primary goal was the examination of the site's tombs, which were still believed to hold promise.

The 1920 season yielded a remarkable wooden model of a ship, found in Tomb 611 and now located in the Petrie Museum of Egyptian Archaeology, London (Fig. 1.2).[3] No photographs exist of the artifact in situ. The terse tomb registration card, apparently filled out by Petrie himself, notes only "Frags of painted wooden boat on wheels" (Fig. 1.3). No other documentation of the tomb is known to exist.[4] The excavation report describes the tomb simply as a "shallow shaft with chamber on west."[5]

The model was the only artifact found in Tomb 611. Along with a reconstruction drawing by Petrie showing the model in port sheer plan, as well as plan views, the excavation report included the following laconic description of the model (Fig. 1.4):[6]

> Professor Petrie has reconstructed this in his drawing which also shows the colouring in red, blue and yellow. The short pegs, of which eight of the twelve remain, are loose and fit the holes in the deck. The tall poles, six in number, are broken off, leaving their stumps in the deck. The awnings are cut out to pass between the tall poles. The gap between the awnings is to allow entering over the side. The figure of a man has been drawn in to show the relative size. It would seem to be the model of a war galley with a ram(?) and projection at the prow for boarding other vessels. The circles along the sides are doubtless meant for oars.

Six years later, in the second of a two-part article dealing with Egyptian watercraft, Petrie supplied another, somewhat different, drawing—this time in an elevated starboard view along with the following description (Fig. 1.5):[7]

A

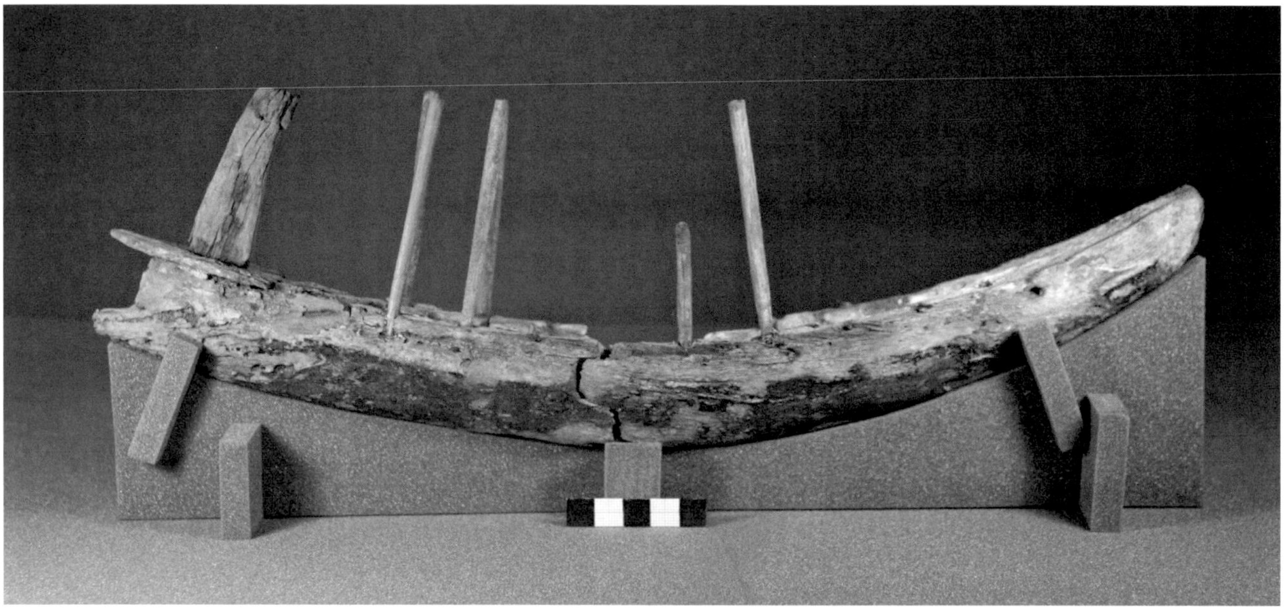

B

Fig. 1.2: The bow and stern sections reconstructed in port (A) and starboard (B) views.

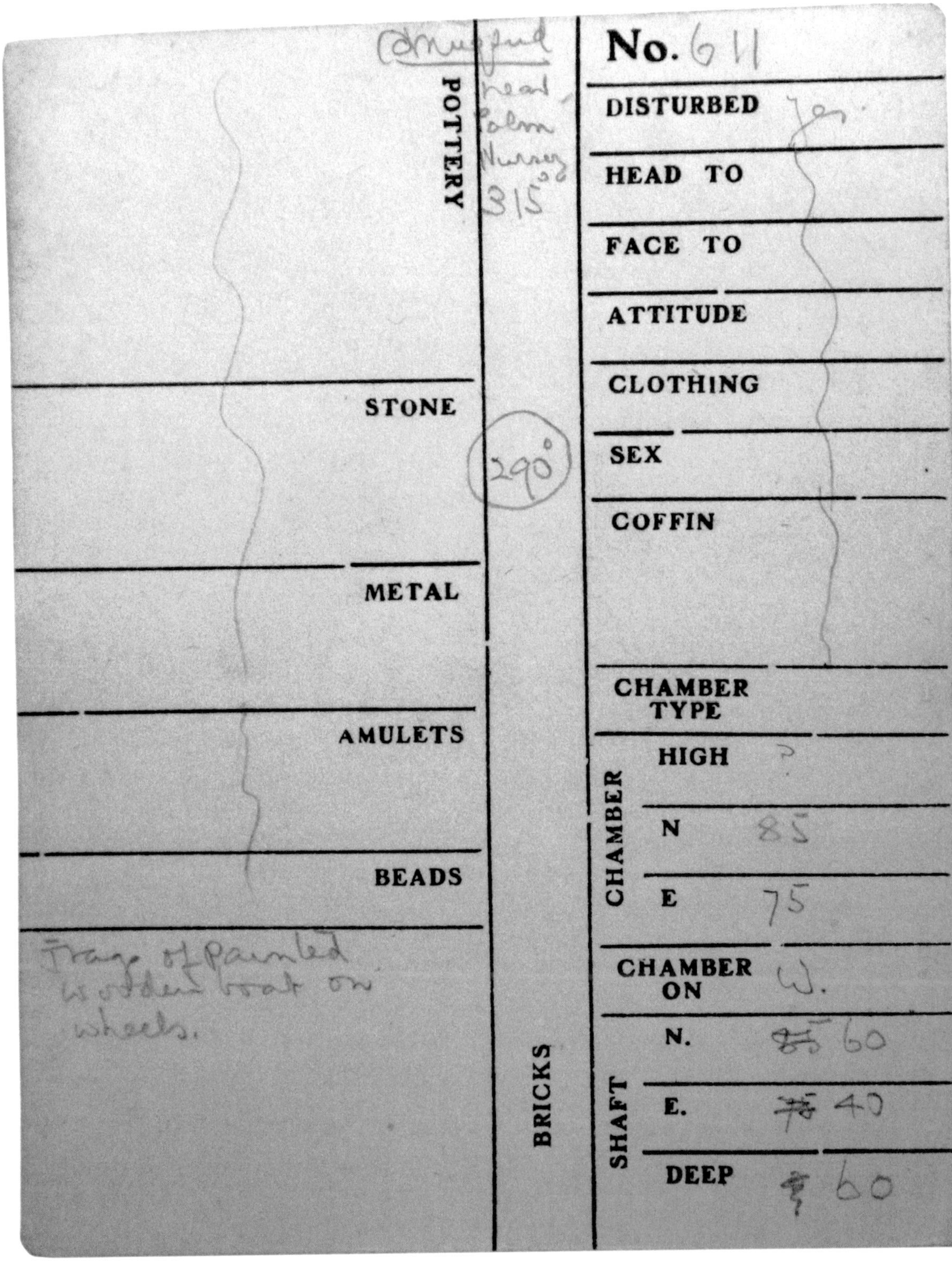

Fig. 1.3: W. M. F. Petrie's tomb registration card for Tomb 611 at Gurob. Courtesy Petrie Museum of Egyptian Archaeology.

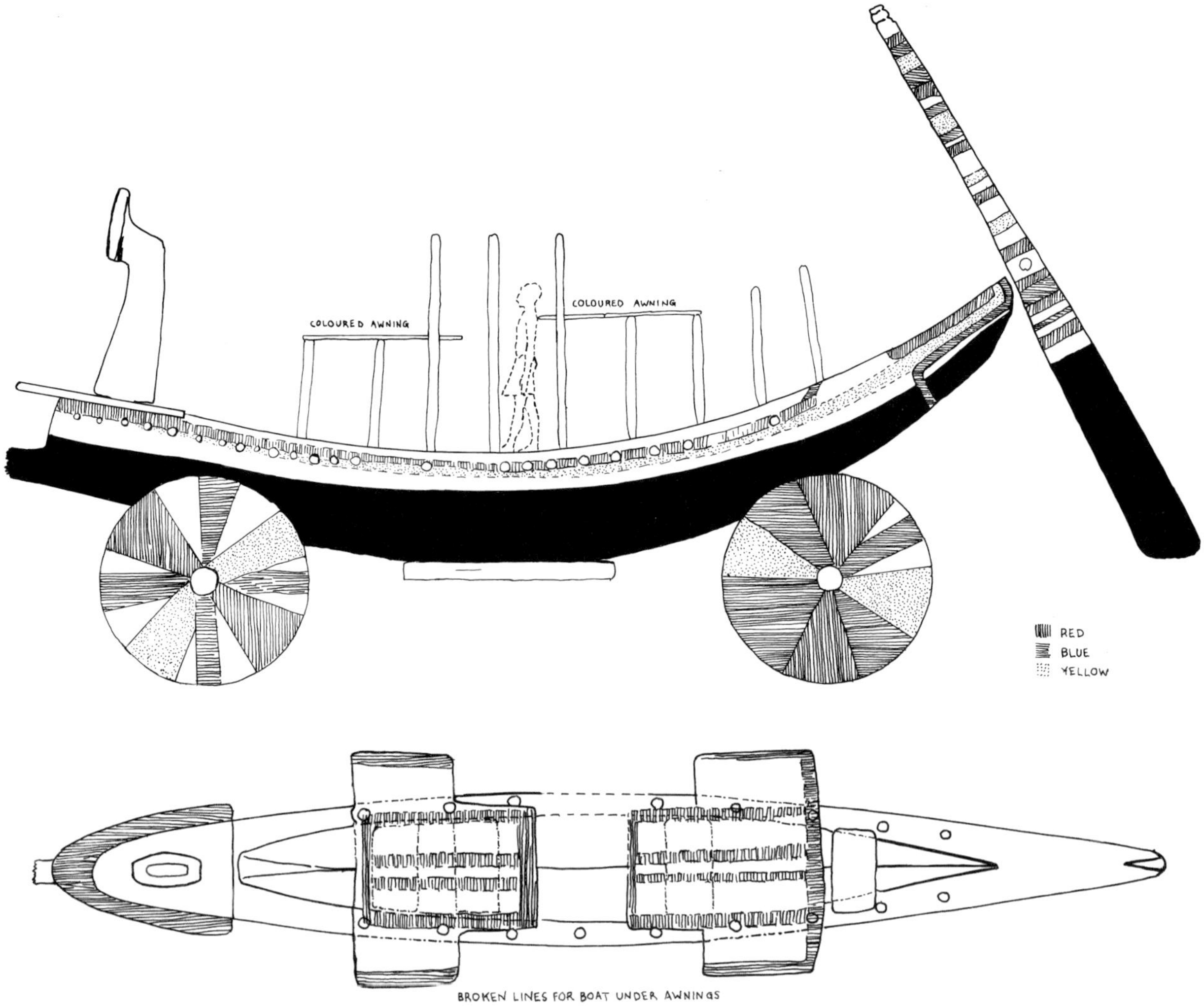

Fig. 1.4: Port (A) and plan (B) views of the model as originally reconstructed by Petrie. After Brunton and Engelbach 1927: pl. LII.

> [This] is another vessel of much the same date [as the ships depicted at Medinet Habu], a little pirate boat, designed for rapid attack and boarding. There are four thwarts, seating eight rowers. A light wood roof shades part, and would suffice to bear a cloth thrown over it. In the bows is a platform for landing, and below it a deadly little ram. The action would be for the pirates to lie to by some island, and when a small trading vessel passed them row out swiftly, ram it so that the crew would be occupied in saving the vessel; the pirates would then swarm up, catch the top of the stern [*sic*] post to swing on to the boarding platform and so on to the trading vessel. Seizing what they could secure by force, they could return quickly and be off while the trader was saving himself from sinking. The model was mounted on wheels as a child's toy, and was found broken up in a late tomb, probably dating from *c.* 1,000 B.C.

In her monograph on Gurob, Angela Thomas published a photo of the model in port sheer view (Fig. 1.6: A).[8] In this photo the model is arranged in a manner identical to Petrie's second drawing with two important changes: The quarter rudder is incongruously propped up at the bow and the hull rests on a modern carriage fitted with the four wheels found with the model. A second, until now, unpublished photograph in the Petrie Museum collection, apparently taken at the same time, shows a top view of the

Fig. 1.5: Petrie's second reconstruction of the model. After Petrie 1933B: 74 fig. 85 (NTS).

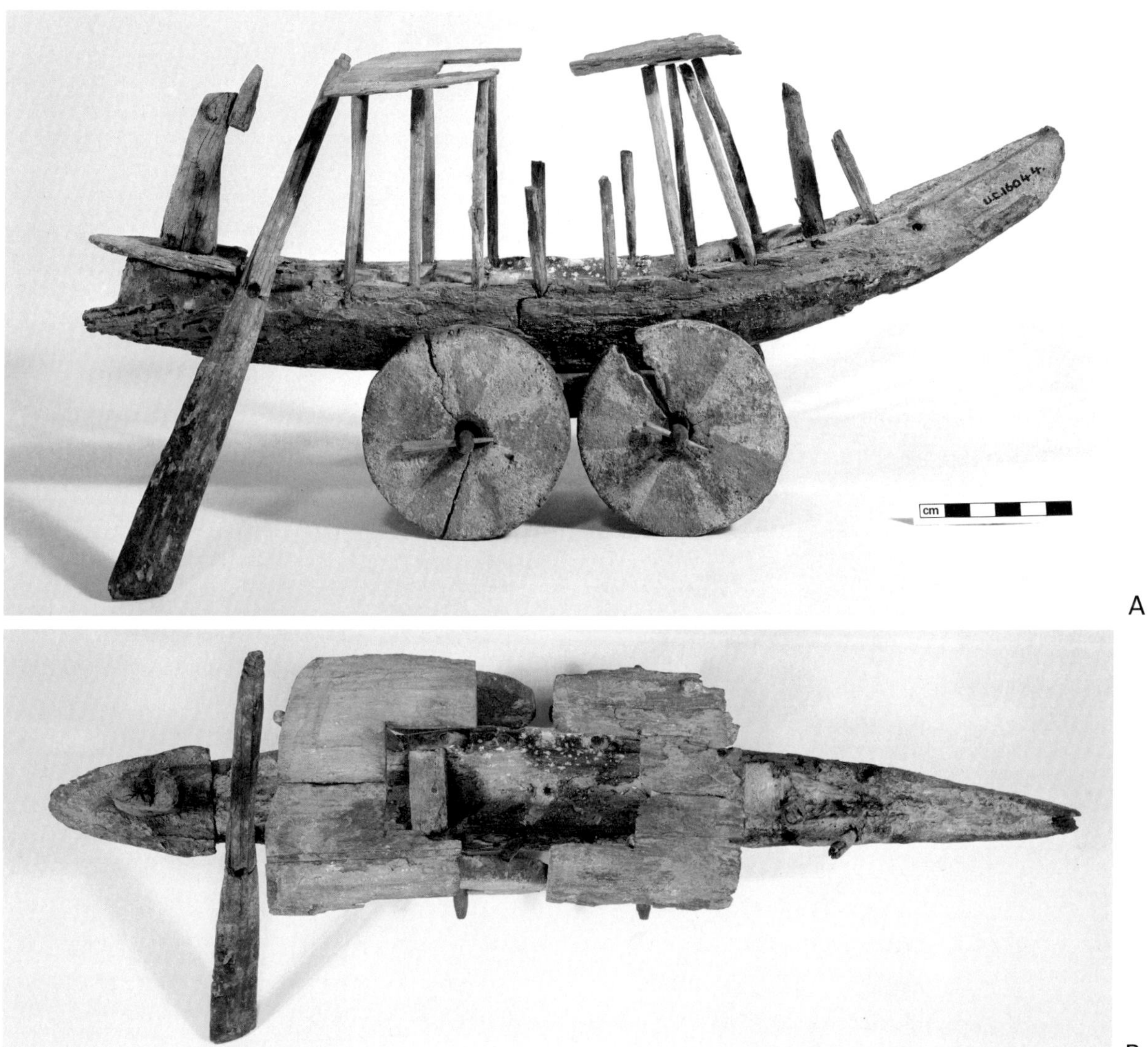

Fig. 1.6: Port (A) and top (B) views of the model. Date of photographs unknown. Petrie Museum negatives 904–905. Courtesy Petrie Museum of Egyptian Archaeology.

A

B

Fig. 1.7: Forward section of the model in port (A) and starboard (B) views.

model (Fig. 1.6: B). Thomas describes the model in the accompanying catalogue thus:[9]

> 747. A model boat. (UC 16044) . . .
>
> Wood, painted blue, red and yellow. The boat has a ramming device in the bows, with a platform above. There are four thwarts seating eight rowers. Six tall poles forward and four aft bear light awnings. There is a large steering oar. The boat is mounted on wheels as a child's toy, and was found in pieces. Tomb 611.
>
> N.K.
>
> Very fragile. Not quite complete, with some unplaced fragments.
>
> As the boat is extremely fragile and is tied together in a display case in the Petrie Collection, it could not be removed for close examination. It may have been a funerary boat adapted for use as a child's toy.
>
> L. 405mm

The model's deteriorating condition led the Petrie Museum to have it conserved in 1998. An unnamed conservator submitted the following report:[10]

> Has beee [*sic*] previously joined and badly gapfilled using plaster. Old joins failing, plaster detaching and breaking the wood. Some parts have been made out of modern wood (eg balsa wood) to replaced [*sic*] missing parts. Some pieces have bee [*sic*] glued in the wrong place Dirty.

Commenting on the painted designs decorating the wheels found with the model, M. A. Littauer and J. H. Crouwel note that "the wheels, which are solid, are painted in an unintelligible manner."[11] Martha Bell mentions the model in her discussion of nearby Tomb 605, C. Monroe includes a photo from Thomas's publication in a revised version of his master's thesis, and P. P. Creasman and Noreen Doyle discuss the model in relation to evidence for overland transport of ships in ancient Egypt.[12] Most recently, Ann Merriman references the model, noting that it "is unique since the boat has a non-standard bow finial on a platform and an odd projection under the waterline that has been interpreted as a 'ram' by some scholars."[13] This appears to be the sum total of extant documentation regarding the model—a stunning lack of discussion given the artifact's importance.

The hull.—The model was broken in half in antiquity, apparently intentionally (Figs. 1.7–8).[14] A section of the starboard side is missing at this break (Fig. 1.2: B). A white substance, presumably the remains of the plaster referred to in the conservation report, covers portions of the break. When the two parts of the hull are aligned, the model is 38.5 cm long in a straight line from stem to stern, approximately 43 cm long following the bottom line of the hull, and 40.3 cm long along the caprail (Fig. 1.9). The hull's maximum breadth is 5.5 cm, giving the hull a 1:7 beam-to-length ratio.

The hull is rockered (Fig. 1.2: A–B).[15] The wood used in its construction, which is light in weight and porous, has been identified as sycomore fig *(Ficus sycomorus)*.[16]

The bow section is 20.2 cm long (Fig. 1.7). The starboard side of this section is well preserved, while the port side has been severely damaged by wood-boring insects (Figs. 1.7: A, 10: A).

The stem is nearly vertical, sloping slightly inward. A robust waterline projection extends 1.5 cm beyond the end of the lower extremity of the bow (Figs. 1.7, 10: A, 11: B, 12: C–D). A flat, roughly triangular, forecastle deck extends over the model's bow (Figs. 1.12, 13: A). The waterline projection's extremity is now in line with the forward edge of the forecastle deck above it, but as the end has been abraded and damaged by woodborers, it could have been longer originally.

The forecastle deck is 2–3 mm thick and extends by 1.3 cm over the hull at the bow, by 6 mm to starboard, and by 5 mm to port. Remnants of light blue paint survive along the upper perimeter. Inboard of this is a dark horseshoe-shaped stain that indicates the location of a now-missing forecastle screen. The stain continues onto the caprail to starboard, suggesting that the screen also may have continued this far (Fig. 1.12: B–C).

Figure 1.12: C: C represents a tentative reconstruction of the missing screen. The blue paint of the port side shows a clear edge where the screen piece had sided to it (Fig. 1.12:

A

B

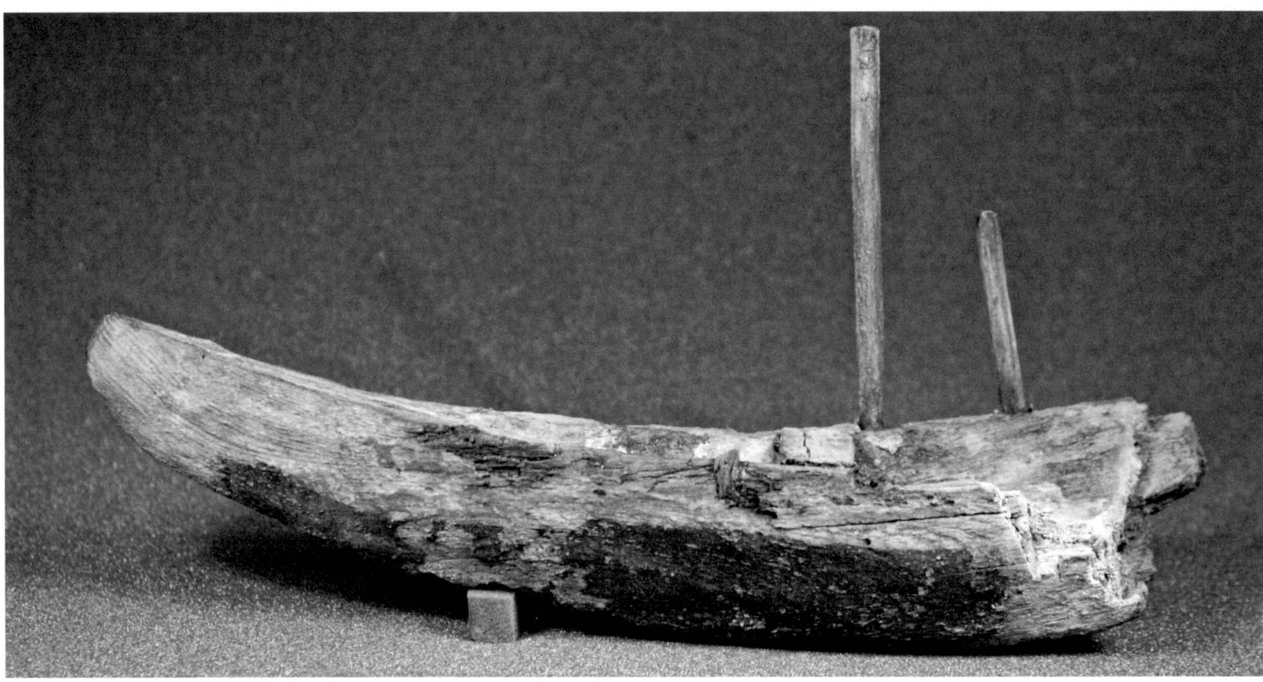

Fig. 1.8: Stern section of the model in port (A) and starboard (B) views. The arrow indicates the hole for attaching the quarter rudder.

A: A–B, B: A–B). This border is also evident on the starboard side but to a lesser degree. These details indicate that at least part of the screen had been painted blue.

The model carries a prominent stempost in the bow (Fig. 1.11: A–C). The stempost does not continue the line of the stem, however. Rather it rises, oddly, in the center of the forecastle. In this case, the model maker sacrificed accuracy for convenience, as the stempost serves to lock the model's forecastle in place.[17]

The stempost originally had an amygdaloidal-shaped vertical continuation at its upper extremity, visible in Petrie's drawings and the museum's photographs. In the 1927

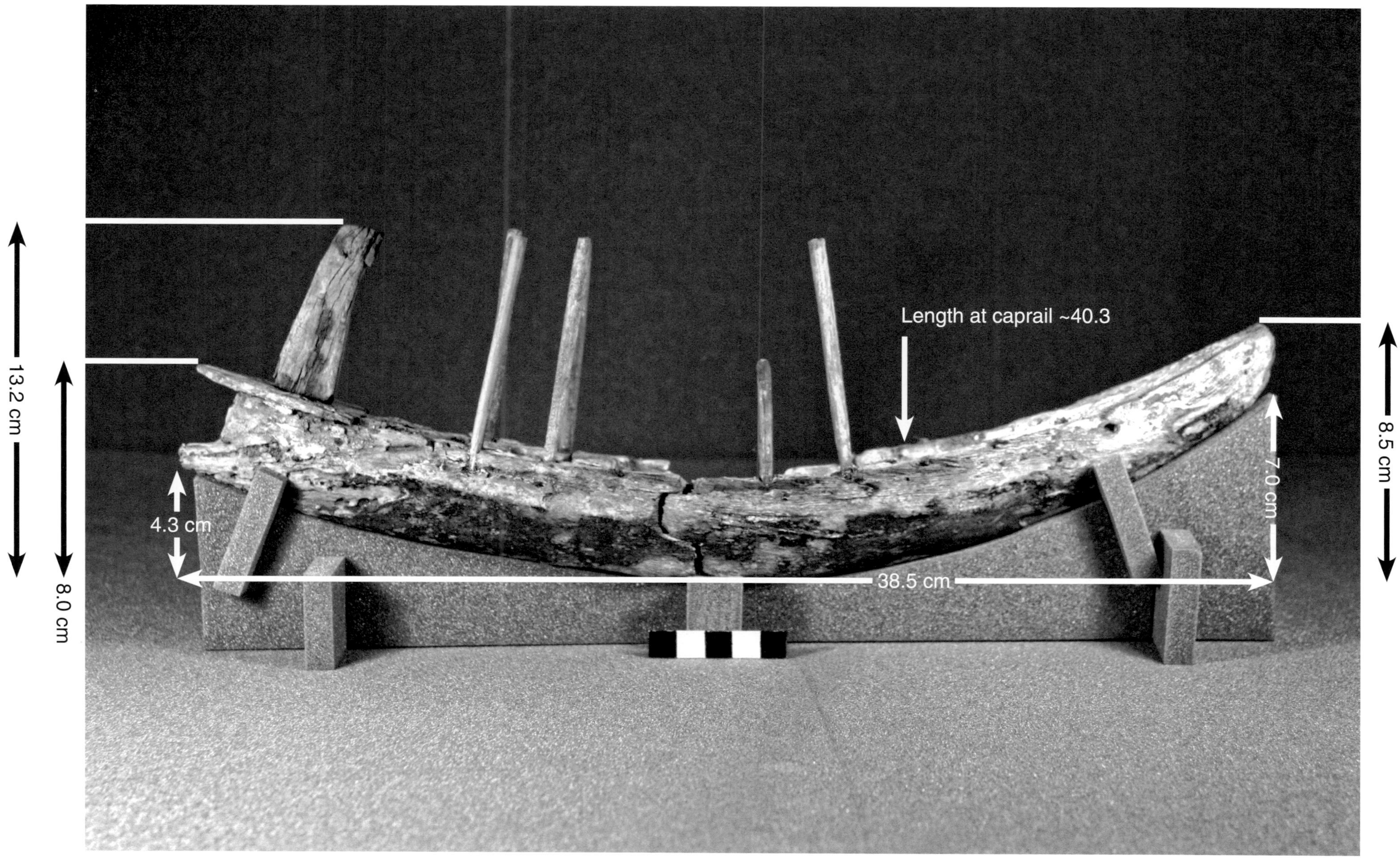

Fig. 1.9: Port view of the model with basic measurements.

A

B

Fig. 1.10: Bottom view of the model, forward (A) and stern (B) sections.

publication this element faced the bow (Fig. 1.4: A). By 1933, however, the stem had been reversed, presumably by Petrie, so that the protuberance faced the stern (Figs. 1.5, 6: A). The stempost is now glued solidly into the hull, and the vertical continuation has broken off (Fig. 1.11: D).[18] Table 1 presents the dimensions of these pieces.

The base of the hull is flattened amidships to permit the model to sit upright on a flat surface (Fig. 1.10). A knot is visible at the bottom of the hull on the bow section adjacent to the midship break (Fig. 1.10: A).

On the underside of the model at the break is a hole, 1 cm in diameter by 1.3 cm deep (Fig. 1.14: B). This hole served to receive a peg that would have attached the hull to its support structure. Remains of blue paint are evident at the hole's lower edge (arrow).

The model is hollowed to signify a planked watercraft (Figs. 1.13, 14: A).[19] The caprail, where it is preserved, averages 8 mm sided. The hull groove begins immediately aft of the forecastle deck and continues to 7 cm from the stern. The interior of the hull is carved out to a depth of 2 cm below the caprails. The upper surface of the hull is covered with a layer of white to yellow gesso, which, in areas in the bow and stern, is applied in globs.[20]

The stern section's port side is better preserved than the starboard side (Fig. 1.8). On the port quarter is a hole for attaching the quarter rudder (Fig. 1.8: A: arrow).

A

B

C

D

Fig. 1.11: Views of the stempost: (A) port, (B) forward, (C) starboard, and (D) detached bird beak (item no. 6).

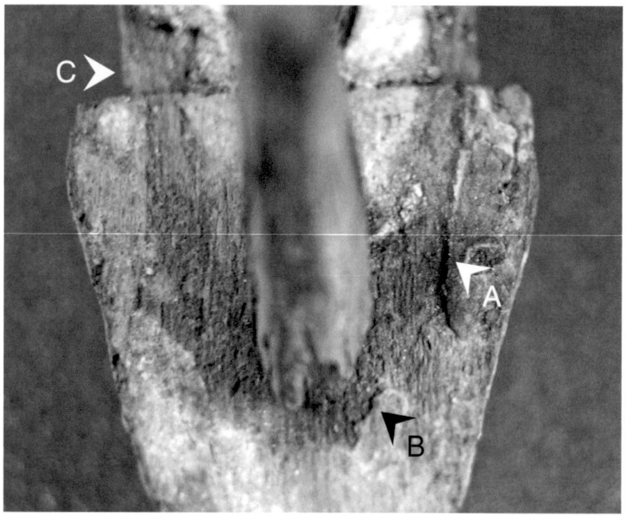

A

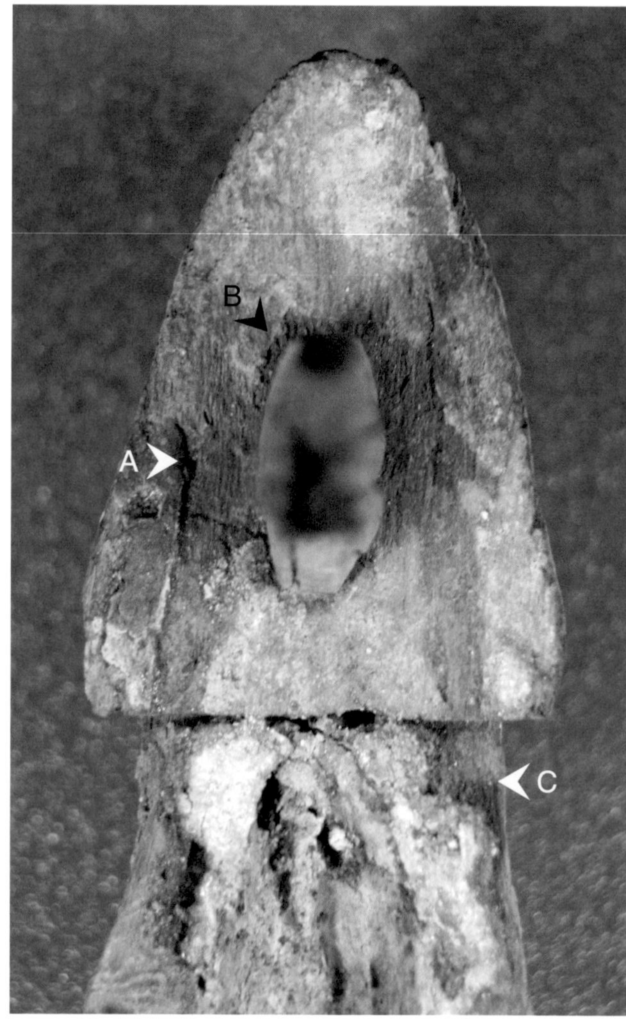

B

C

D

Fig. 1.12: (A) Forecastle deck, top view facing the stern. (B) Forecastle deck, top view. (C) The forecastle in starboard view with tentative reconstructed fencing. (D) Detail of bow below the forecastle deck.

A

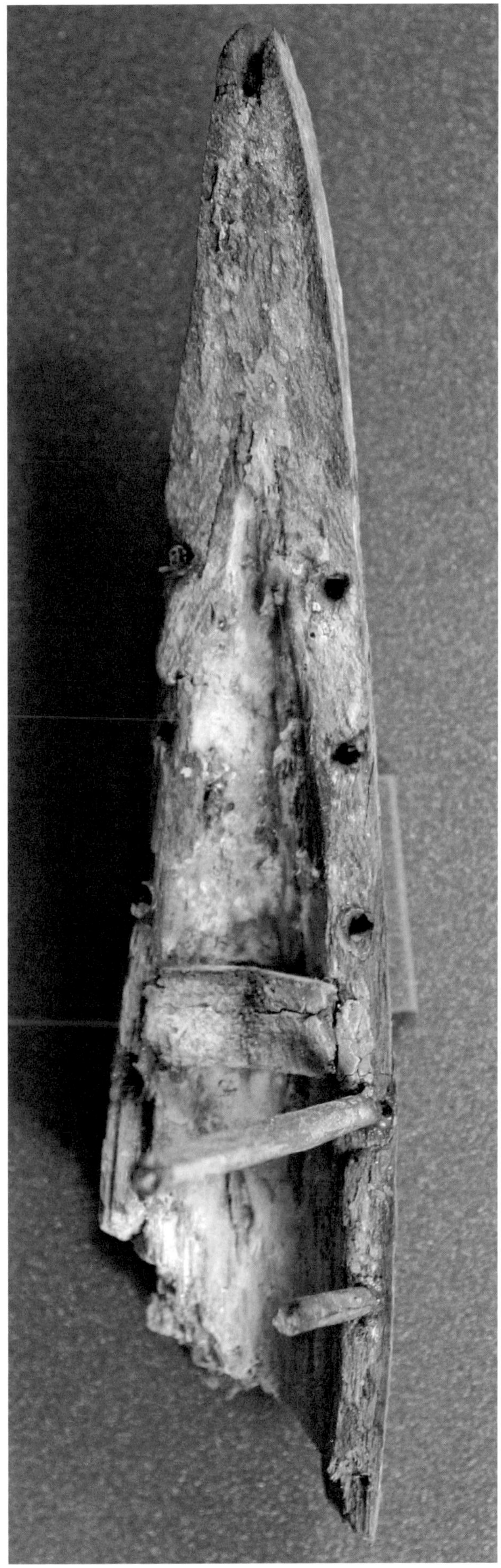

B

Fig. 1.13: Top view of the model, forward (A) and stern (B) sections.

A

B

Fig. 1.14: (A) View of the V-section shape of the hull with thwart in the model's forward section viewed from the break. (B) Sectional view of the stern section break amidships showing part of the hole for a peg that would have supported the model. At the bottom of the hole are traces of blue paint (arrow).

Piece	Dimension	Measurement (cms)
Stempost	height above deck piece	6.5
Stempost	molded dimension at base	2.3
Stempost	molded dimension at base of protrusion	1.5
Stempost	molded dimension at top	1.4
Stempost	sided dimension	1.1
Vertical element	height	2.7
Vertical element	molded dimension	1.4
Vertical element	sided dimension	0.7

Table 1. Measurements of the Stempost and Its Protuberance

Port Side	
Between Hole Numbers	Distance (cm)
1–2	2.7
2–3	2.6
3–4	2.5 (at break)
4–5	2.8
5–6	2.9
6–7	2.8
7–8	2.7
8–9	2.5
Starboard Side	
1–2	2.9
2–3	2.7
4	missing
5	missing
6–7	2.8
7–8	2.5
8–9	2.8

Table 2. Distances between the Centers of the Stanchion Holes

It is located 6.5 cm from the stern and 5 mm below the caprail. Astern of it is a blue line in an "L" shape turned 90 degrees clockwise. Also, just below the caprail, forward of the quarter-rudder hole, is a small patch of blue paint. The stern ends in a shallow 5 (deep) × 4 (wide) mm notch (Figs. 1.13: B, 15: A–B). A stanchion (no. 15) experimentally placed in the notch canted at a 70-degree angle above the stern (Fig. 1.15: C).

Originally, the model had nine pairs of stanchion holes on either side (Fig. 1.13). The first pair is set back 5.5 cm, measured from the aft end of the stem platform to the centers of the holes. The average distance between the holes is 2.7 cm.

Fig. 1.15: (A) Top view of the stern section. Note the notch and the narrow spaces along the sheer strakes and on the inner side of the hull (arrows) where gesso is missing. (B) The stern notch, view facing bow. (C) Stanchion no.15 placed in the notch demonstrates the notch's angle (70 degrees) of incline.

The bow section of the model has three holes on either side. The first two of these on either side now have long pegs solidly glued into them. The third hole on the starboard side is destroyed at the midship break; its port companion is 5 (diameter) × 4.5 (depth) mm.

The port side of the stern section is 22.5 cm long and contains six stanchion holes (Fig. 1.8: A, 13: B). The forwardmost of these holes split in half when the model was broken. The hole is 7 (depth) × 4 (width) mm. A stanchion is solidly glued into the second hole from the bow, and a long peg is glued into the third hole.

Table 2 lists distances between the centers of the stanchion holes. Table 3 lists the depths and diameters of the stanchion holes on the port caprail.

The gesso covering the stern is absent in a clear pattern along the upper edges of the caprail (Fig. 1.13: B, 15: A: white arrows). These lines of missing gesso are 2–3 mm wide and run for 7 cm on the starboard quarter and 6 cm on the port quarter. In the center of this area is an elongated

Hole Number	Depth (mm)	Diameter (mm)	Comments
1	7	-.-	tall peg in hole; external side of hole missing
2	-.-	-.-	unknown; peg in hole
3	4.5	5	
4	8.5	7	at break
5–6	-.-	-.-	unknown; pegs in holes
7	6	5	
8	7.5	4	
9	7.5	5	

Table 3. Depths and Diameters of the Port Stanchion Holes

Thwart	Dimension	Measurement (in cms)
thwart no. 1*	length	2.5 (forward)
thwart no. 1*	length	2.8 (astern)
thwart no. 1*	molded dimension	0.4
thwart no. 1*	sided dimension	1.4
thwart no. 4†	length	3.5
thwart no. 4†	molded dimension	0.7
thwart no. 4†	sided dimension	1.5

*Thwart nearest bow.

† Thwart nearest stern, now detached from the hull and glued to the quarter rudder stanchion (no. 26:B, Fig. 1.19:A–B).

Table 4. Thwart Dimensions

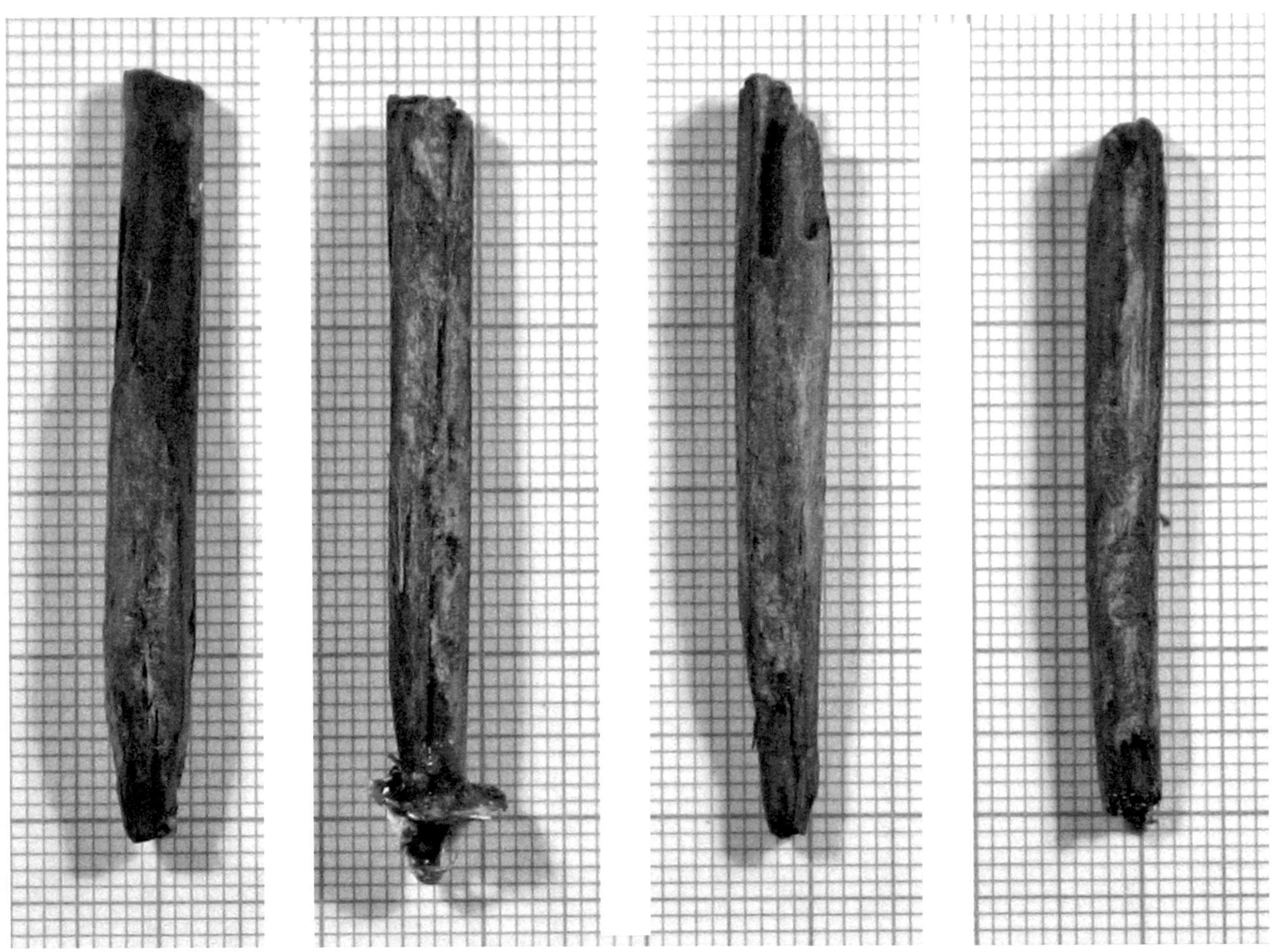

#13 #14 #15 #16

Fig. 1.16: The disconnected stanchions (item nos. 13–16).

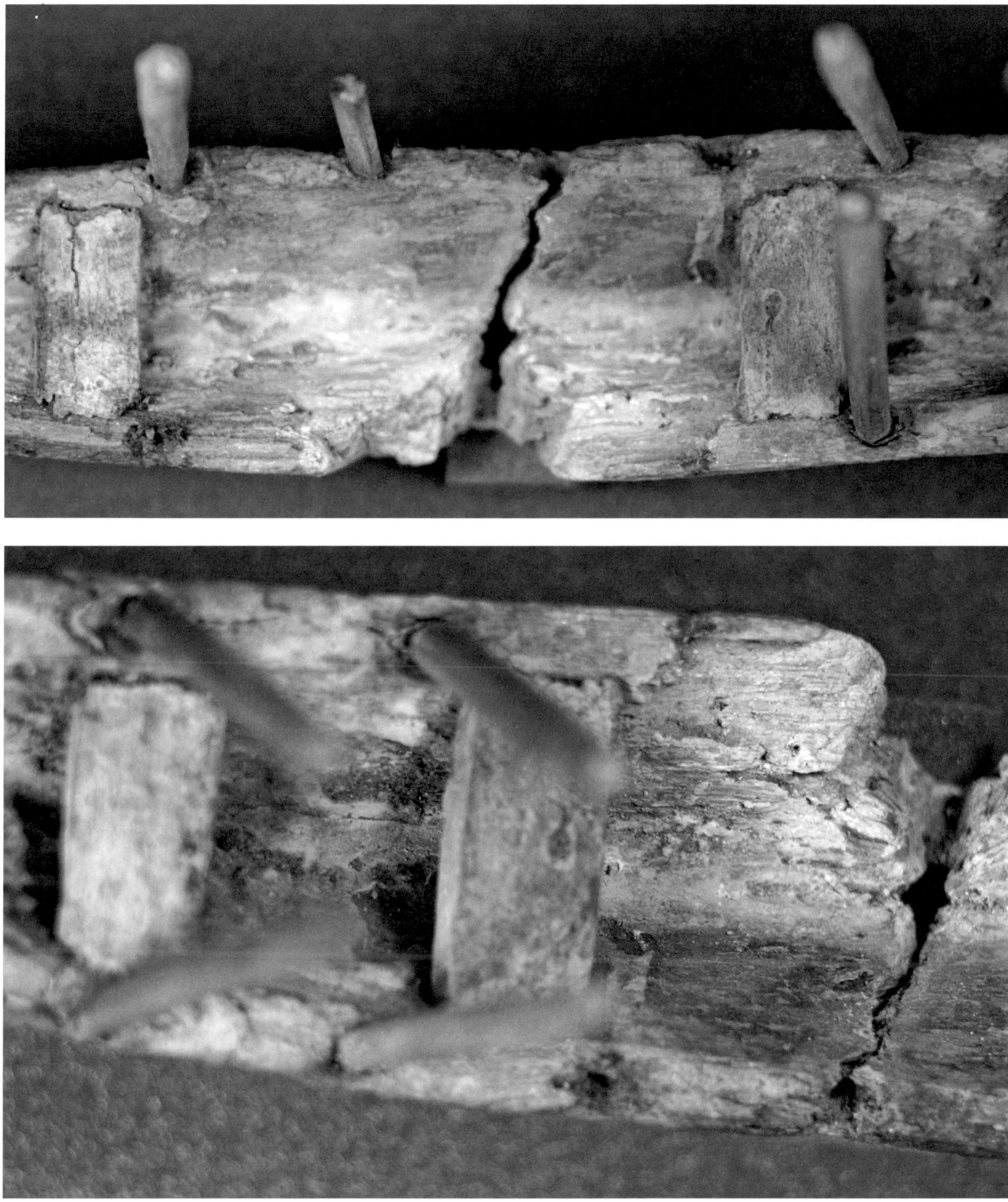

A

B

Fig. 1.17: Missing gesso on the inner sides of the hull amidships suggests the existence of a now-missing element, perhaps a maststep. (A) Amidships, port at top. (B) Amidships, starboard side at top.

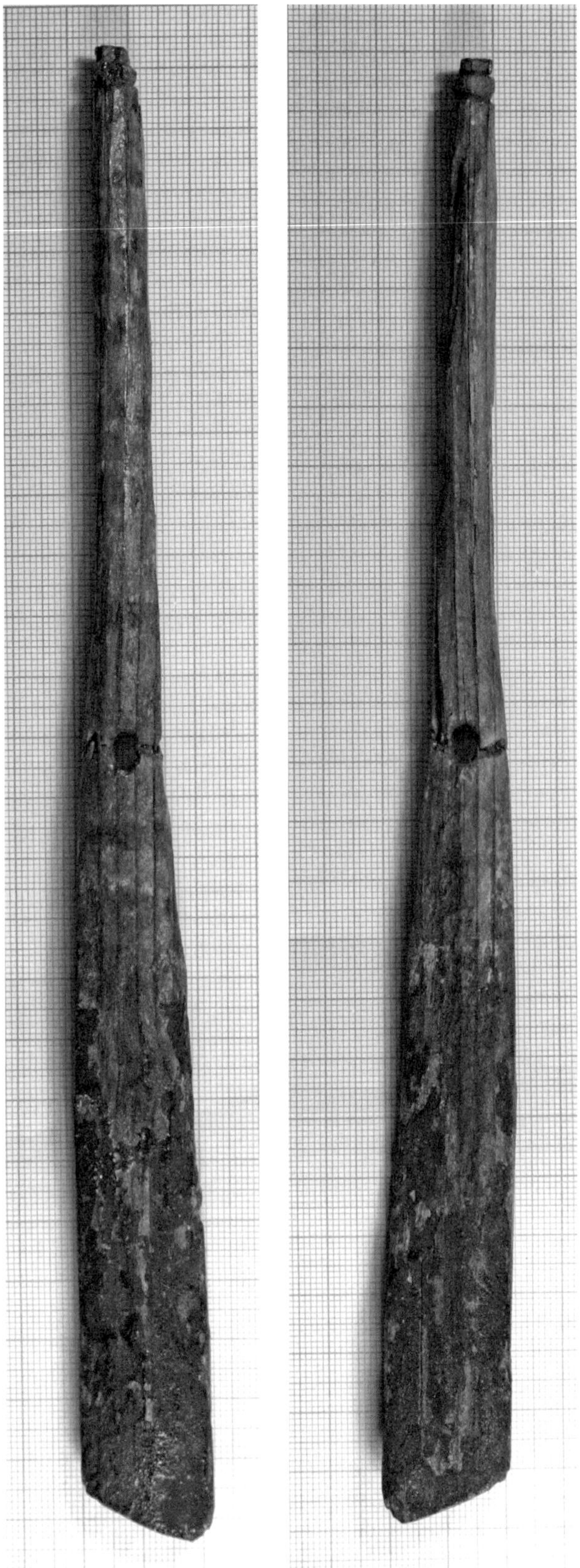
A B

Fig. 1.18: The quarter rudder. (A) Outer (visible) side. (B) Inner (hidden) side.

oval lacking gesso (Fig. 1.15: A: black arrows). Together, these areas of missing gesso strongly suggest that some form of sterncastle and screen existed there.[21]

The model has three attached broad thwarts (Figs. 1.4: B, 5, 6: B, 13). Their upper surfaces are covered with gesso. A fourth sternmost thwart has come off the model and is now glued to a broad peg that has been reconstructed as a quarter rudder stanchion (Figs. 1.5, 6: A, 19).[22] Its original location on the model is indicated by patches of missing gesso inside the hull next to stanchion hole station no. 8 (Fig. 1.13: B). Judging from this dislocated thwart, no pegs or other attachments served to attach the others to the hull. Rather, they were wedged into place inside the hull (Fig. 1.14: A). This caused the thwart attached to the stern to crack down its center (Fig. 1.13: B). Table 4 gives the measurements of two of the thwarts.

There is no sign of a hole to step a mast.[23] The break itself seems to have caused the small opening at its top center. On the other hand, gesso is missing amidships along the inner side of the hull (Figs. 1.13: A, 17). This is particularly noticeable on the port side, where the area lacking gesso extends for 8 cm. Some evidence appears on the starboard side also, but much of this area is lost in the break. These considerations suggest that a now-missing thwart or maststep/partner piece had been in place when the hull's interior received its coating of gesso.

Some of the hull's details are rendered in paint. The polychromatic decoration is best preserved on the starboard side of the bow section, which is probably indicative of the original appearance of the rest of the model (Fig. 1.7: B).

The bottom of the hull (from 0.0 to a height of 2.5 cm) is painted black. Above the black paint and about half of the distance between its upper edge and the caprail runs a sloppy line of irregularly spaced black dots (0.5 to 1.3 cm apart, from center to center) painted over a light wash of gesso. The dots begin at the stern edge of the forecastle deck. In Petrie's 1927 reconstruction, however, the dots begin almost at the stem (Fig. 1.4: A). These are not visible today.

The starboard side of the model's forward section bears nine irregular dots along a length of 9 cm (Fig. 1.7: B). The dots seem to have been applied haphazardly, and not all are equidistant. The distance between centers varies

A

B

Fig. 1.19: (A) "Quarter-rudder stanchion" now detached from the model and glued to a thwart. (B) Detail of the thwart.

from 0.7 to 1.5 cm, with an average distance of ~1.0 cm. On the starboard side the foremost dot appears behind the sternward termination of the forecastle deck, and the last visible dot is located beneath the last stanchion hole on the stern section. The distance from the stern end of the stem deck platform to the last stanchion is about 25 cm, and if the well-preserved starboard section is representative of the entire model, 25 dots per side seems a reasonable number.

Above the line of dots, at a height of 3.5 cm from the bottom of the hull, and situated just below the caprail, runs a continuous, faded horizontal line of red paint, which continues under the forecastle deck but disappears beneath the edge of the stempost. Where it is best preserved, the line is 4 mm wide. As the stem section has suffered the loss of its painted surface, the red line may well have continued around the stem.

Detached Pieces

Thirty unattached pieces are associated with the model. While the purpose of some items, such as the stanchions, quarter rudder, and wheels, is self-evident, the fragmentary condition of some pieces makes it difficult, if not impossible, to determine their purpose on the model. Indeed, some items may not even be related to the model. In some cases items were added in the modern era. No human or animal figures are associated with the model.

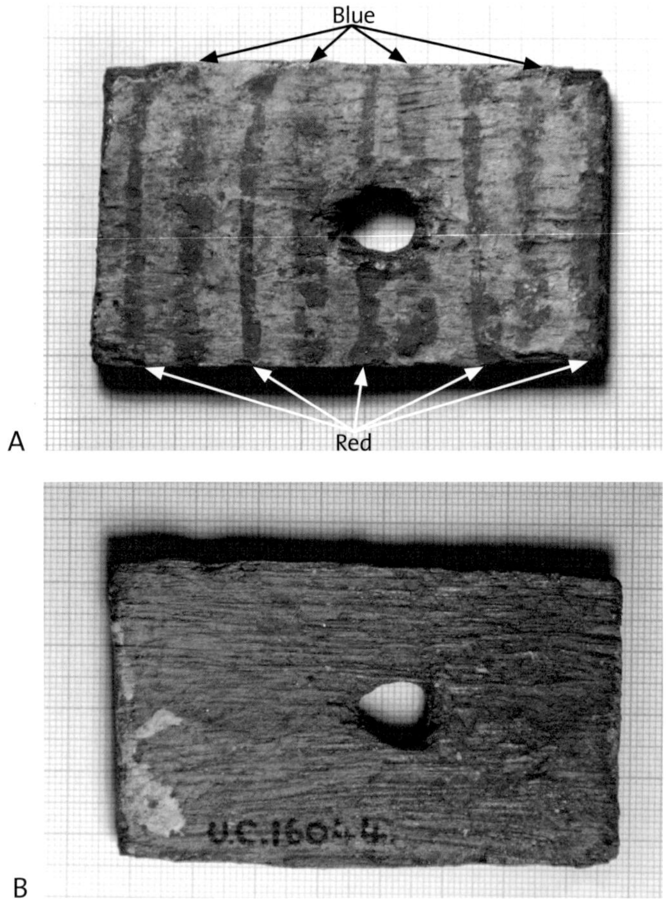

Fig. 1.20. Wooden rectangle *(pavois)* decorated with roughly parallel red and blue lines painted on a coating of gesso.

Stanchions.—Brunton mentions eight "short pegs" found with the model.[24] Today only five of these have survived: One is glued into a stanchion hole on the stern starboard quarter, and four (nos. 13–16) are loose (Figs. 1.8, 16). Clearly, these items served as stanchions.

Brunton assumes an original total of twelve short pegs. His calculation apparently refers to Petrie's accompanying reconstruction (Fig. 1.4). The model had a total of eighteen stanchion holes. Petrie inserted the six long pegs into the midship stanchion holes, thus leaving twelve holes for the stanchions.

The stanchions probably slotted into pieces representing wales that ran over them. None of the surviving fragments appear to have served this function, however.

Quarter rudder.—The model carried a rudder (no. 17) on its port quarter attached by a now-missing peg through holes at the center of the rudder and on the model (Figs. 1.18, 8: A: arrow). The rudder is constructed from a single piece of wood: The loom and blade merge gently. The exterior (port) side of the quarter rudder is decorated with alternating black and red stripes. The lower 9 cm of the object are painted black, matching the appearance of the hull. The top of the loom has a groove. There is no evidence of a tiller. Presumably, the quarter rudder would have been positioned on the model so that the blade's shoe rode horizontal.

Thick peg.—In his initial reconstruction Petrie's axial, or stern-mounted, steering oar did not feature a supporting stanchion (Fig. 1.4: A), but by 1933 he had moved the quarter rudder to the port quarter and added a stanchion for it (Fig. 1.5). The stanchion is now attached with modern glue to the sternmost thwart, which has detached from the model (nos. 26 A–B) (Fig. 1.19). There is no hole for the attachment of a quarter rudder stanchion, however. Thus, this model appears to have lacked one in its original state.

Pavois.—Based on its position and shape, a rectangular plaque decorated with parallel blue and red lines and with a hole pierced through its center represents the structure—termed a *pavois* by G. Legrain—which served as a base for Egyptian cult and funerary ships when transported overland (Fig. 1.20).[25] In 1927 Petrie positioned this piece amidships below the hull (Fig. 1.4: A). In support of this reconstruction, note the similar diameters of the hole in this piece and the one amidships at the bottom of the model, 1.0 and 1.2 cm respectively, as well as the blue paint still evident inside the model's hole, which may have resulted from color rubbing off the plaque when a peg was pushed through it and into the model (Fig. 1.14: B: arrow). This artifact indicates that the model depicts not a ship but rather a cultic ship-cart.

Wheels.—Four wheels were found with the model (Figs. 1.21–22). They are decorated with radial wedges—and in some cases squiggles—of white, red, blue, and brown on their external faces. As noted by M. A. Littauer and J. H. Crouwel, the painted decorations cannot represent spokes.[26]

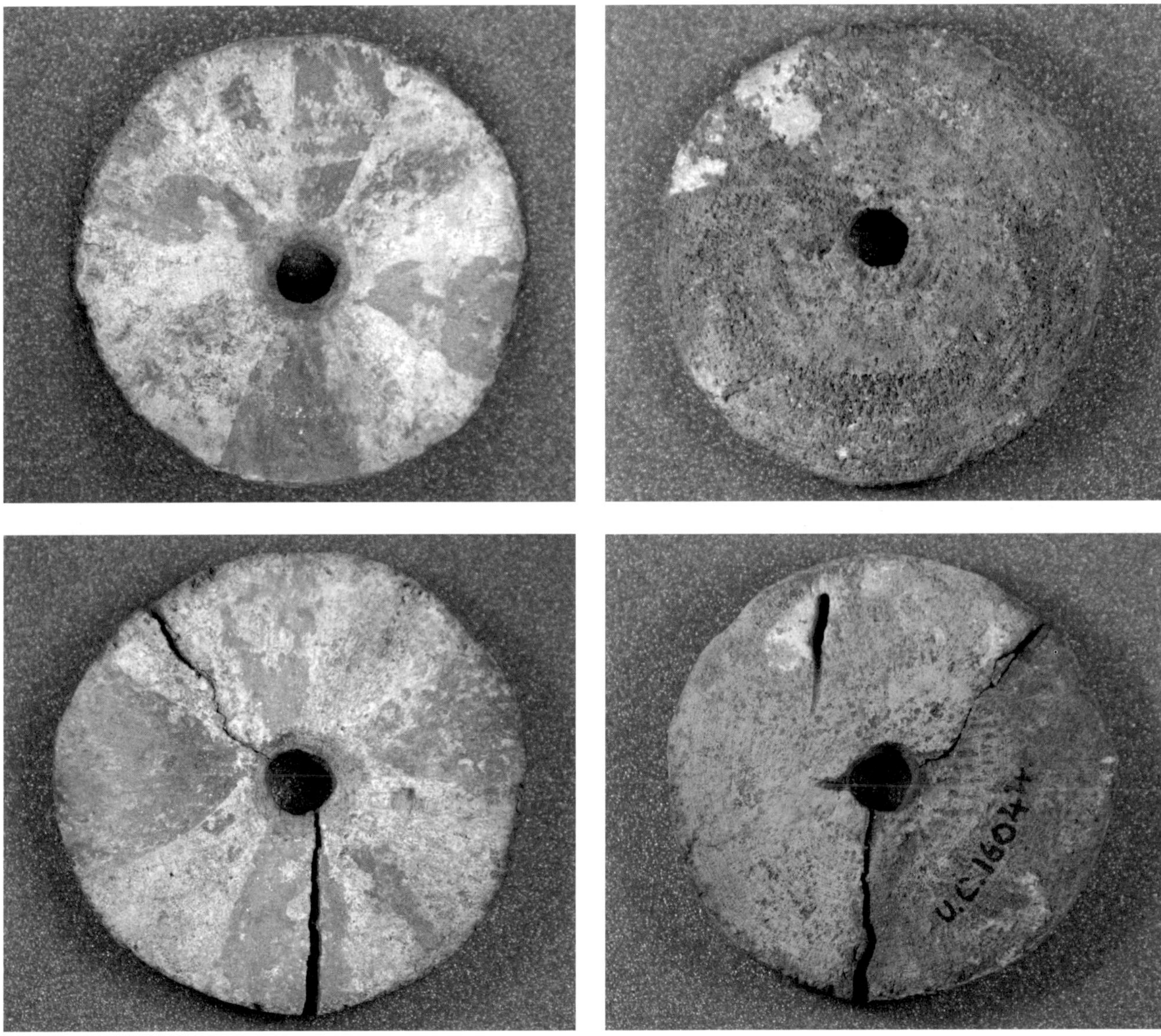

A

B

Fig. 1.21: Wheels 1–2. (A) Item no. 27. (B) Item no. 28.

When I examined the model, its four wheels were attached to a clearly modern axle-and-chassis system partially visible in Fig. 1.6: A. Lynchpins held the wheels to the axles. I removed the wheels from the carriage and recorded them individually. The preservation of the wheels varies: One (no. 27) is intact, two are cracked (nos. 28–29), and one (no. 30) has a wedge missing. Abrasions at the outer edges of the axle holes indicate that the wheels had actually seen use.

There are clearly two *sets* of wheels. The larger wheels (nos. 28 and 29) have a diameter of 8.1 cm: The smaller wheels (nos. 27 and 30) have a diameter of 7.6–7.7 cm. This discrepancy in the diameter of the wheels was dramatically demonstrated when, while developing the virtual reality model, the two larger wheels were inadvertently placed on one side of the model and the two smaller ones on the other, causing the model to lean distinctly to one side (Fig. App. 2.10).[27] The larger wheels were likely placed at the stern to help the quarter rudder clear the base.

The wheels presume a chassis or carriage. While some loose pieces may have belonged to such an item, these fragments are too incomplete to indicate its size and shape. Presumably the chassis was a rectangular block of wood of sufficient thickness to allow the quarter rudder to clear the base on which the wheels stood (Fig. App 2.4) while having a peg in its center, which penetrated the *pavois* and the hole at the model's base.[28]

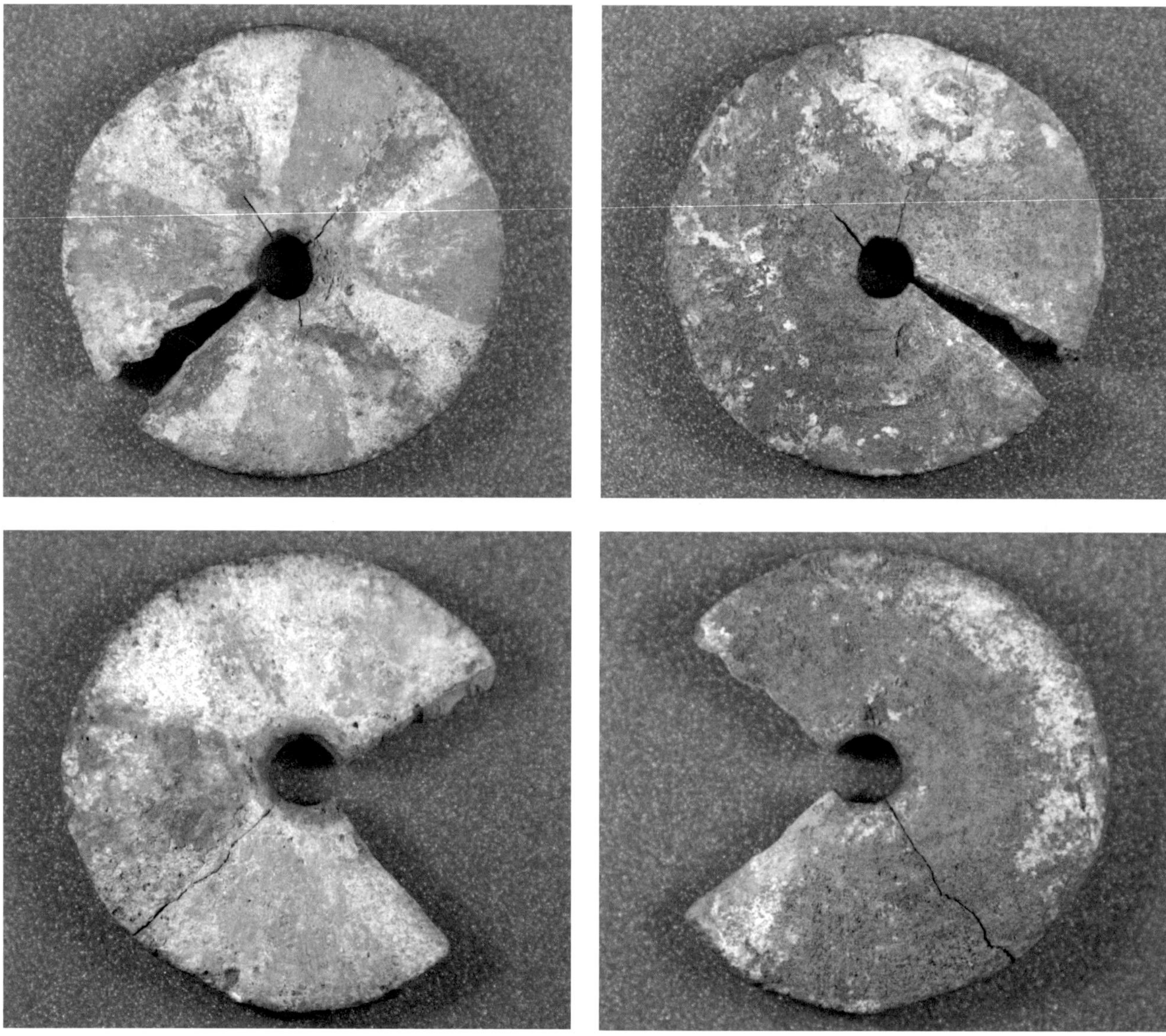

A

B

Fig. 1.22: Wheels 3–4. (A) Item no. 29. (B) Item no. 30.

Stanchion Hole	Total Length (cms)	Length above Caprail (cms)	Thickness (cms)	Depth of Stanchion Hole (cms)
Port bow	8.6	7.7	0.5	0.7
Port center*	–.–	7.9	0.5	–.–
Starboard bow	–.–	7.9	0.5	–.–
Starboard center	–.–	8.0	0.6	–.–

–.– Not available/relevant.
* Light white encrustation.

Table 5. Measurements of the Long Pegs Now Attached to the Model's Bow Section

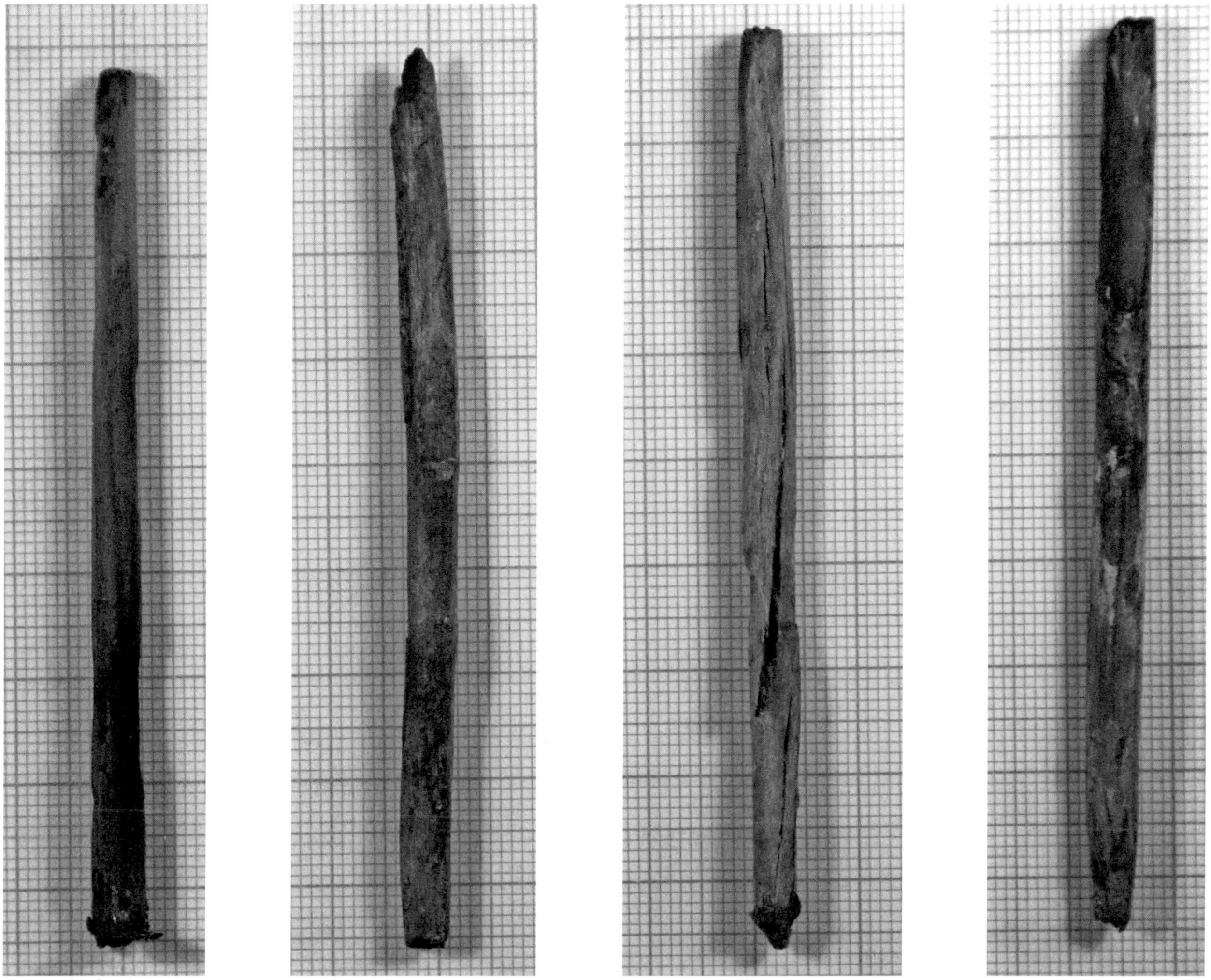

#18 #19 #20 #21

Fig. 1.23: The four loose long pegs (item nos. 18–21).

Long pegs.—In addition to the five long pegs now glued into the model's stanchion holes, the collection includes four loose pegs (nos. 18–21; Fig. 1.23). At least three of the long pegs must be modern additions as Brunton mentions only six "tall poles" found with the model.[29] Table 5 lists the relevant measurements of the long pegs now glued into the model's bow-section stanchion holes.

"Awnings."—Petrie reconstructed the model to include a pair of what he termed "awnings." In the Gurob report drawing/reconstruction these are T shaped and rest on short stanchions (Fig. 1.4). In his 1933 article, however, the awnings have become U shaped and have been raised so that they are supported on the long stanchions (Fig. 1.5). In fact, Petrie's awnings consist of four separate wood fragments that show no evidence of having been configured in antiquity in either of the manners he proposed (nos. 22–25; Fig. 1.24).

The shape of these items resembles the silhouette of a smoker's pipe bowl. They do not attach to the model or to each other in any logical manner. At first glance the awnings seem to be ideal to reconstruct as castle screens.[30] They, however, do not fit the imprint left on the hull, indicating that they did not serve this purpose. We experimented with the possibility that the awnings may have served as "fenders" for the chassis on both a Styrofoam mockup and the virtual model but finally concluded that insufficient evidence exists to support this hypothesis (Figs. 1.25, App. 2.9). At present their purpose remains an enigma.

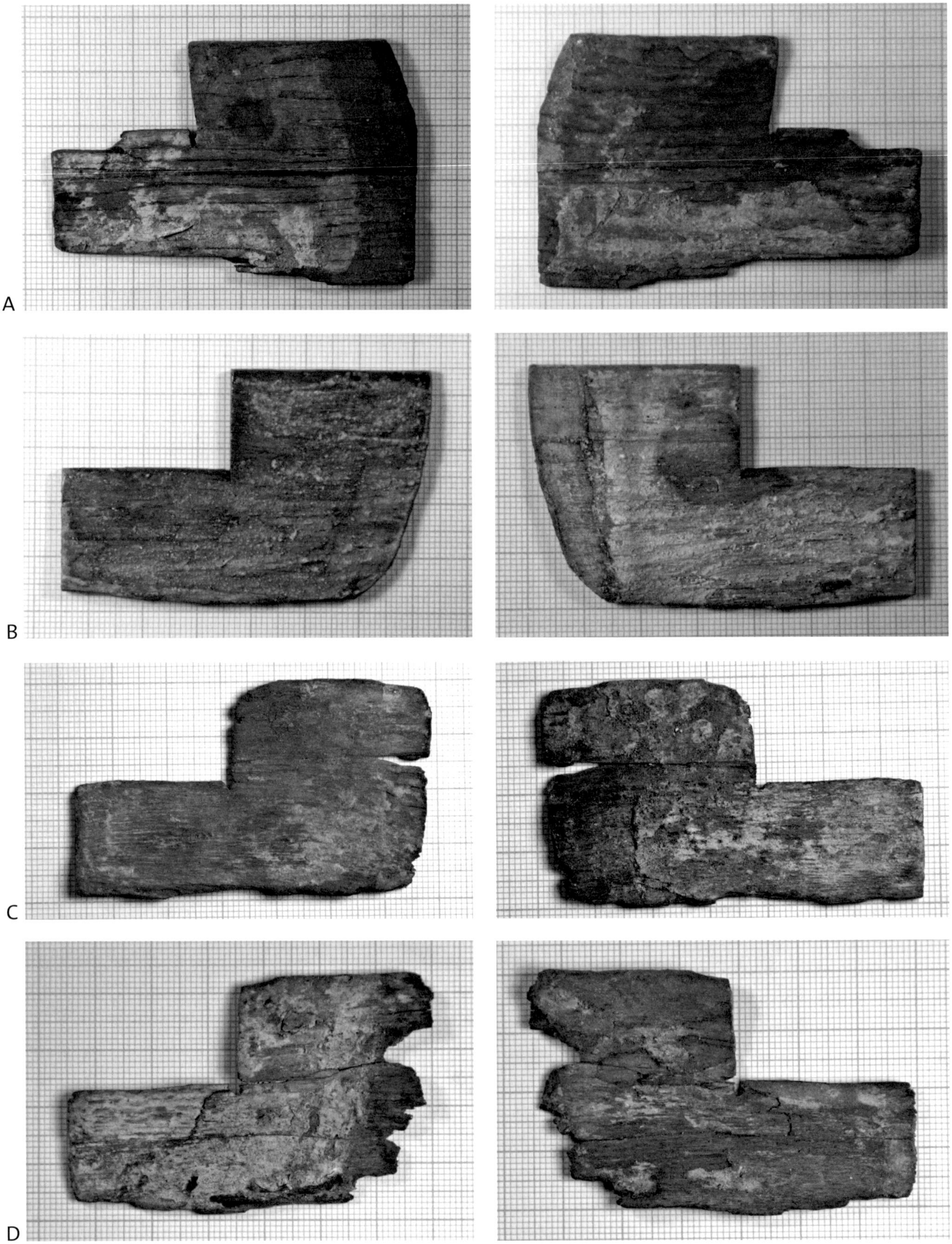

Fig. 1.24: "Awnings." (A) Item no. 22. (B) Item no. 23. (C) Item no. 24. (D) Item no. 25.

Discussion

New Kingdom wooden ship models.—The Gurob model is exceptional in that it is a nonroyal New Kingdom wooden ship model. Wooden models of watercraft are common from the Old to the Middle Kingdoms, but their popularity appears to have waned afterward. Only two ship models are known from the Second Intermediate period, and they are made of metal.[31] Apart from the Gurob model and two other exceptions, New Kingdom wooden ship models are limited to the ones found in the royal tombs of Thutmose III, Amenhotep II, and Tutankhamun.[32] Regarding this phenomenon Reisner notes the following:[33]

> The only New Empire boat models, aside from the boat found in the Tomb of the Priests of Amon (n° 4929), were the models found in the royal tombs of Amenophis II and Thotmes III at Thebes. These are all shown by their decorations to be more or less ceremonial in character but are scarcely to be called either funerary or solar barques. They were fitted with long lines of rowers, cabins, landing planks and other implements inconsistent with their identification as funerary or solar barques. Nor are they of the form of such barques shown in the papyri and the reliefs of the period. The form is manifestly derived from that of Type IV and the function is no doubt the same—that of pleasure boats in this case for the use of the king.[34]
>
> The practical boats and ships of the period are to be found in abundance in the reliefs of the New Empire. But models are not to be found; and a discussion of the boats and ships of later times, however interesting, lies outside the present field, that of boat-models.

A sterncastle deck on the Gurob model?—As noted, gesso absent at the stern indicates a now-missing castle. The wooden ship models and model fragments from the royal tombs of the XVIIIth Dynasty may help in understanding the pattern of missing gesso. The Gurob model's forecastle deck is conceptually similar to those on, or detached from, models from the tombs of Thutmose III, Amenhotep II, and Tutankhamun (Fig. 1.26).[35] These decks are constructed of roughly triangular slips of wood that overhang their hulls and carry (or once carried) superstructures.

Glue or pegs secured the castle decks onto the royal ship models.[36] On the Gurob model, the stempost secures the forecastle deck in place. The pattern of missing gesso at its stern can best be explained by a similar slip of wood representing the sterncastle deck glued to the center of the hull and resting on the caprails (Fig. 1.15: A). If so, the stern notch could have served to receive a connecting peg to further secure the sterncastle deck (Fig. 1.15: C).

Wood.—Samples of wood from one of the wheels and a loose fragment (item #12) were identified as sycomore fig (*Ficus sycomorus*).[37]

Fig. 1.25: The "awnings" experimentally reconstructed as fenders on a Styrofoam mockup of the missing chassis. Photo: D. Bartoli.

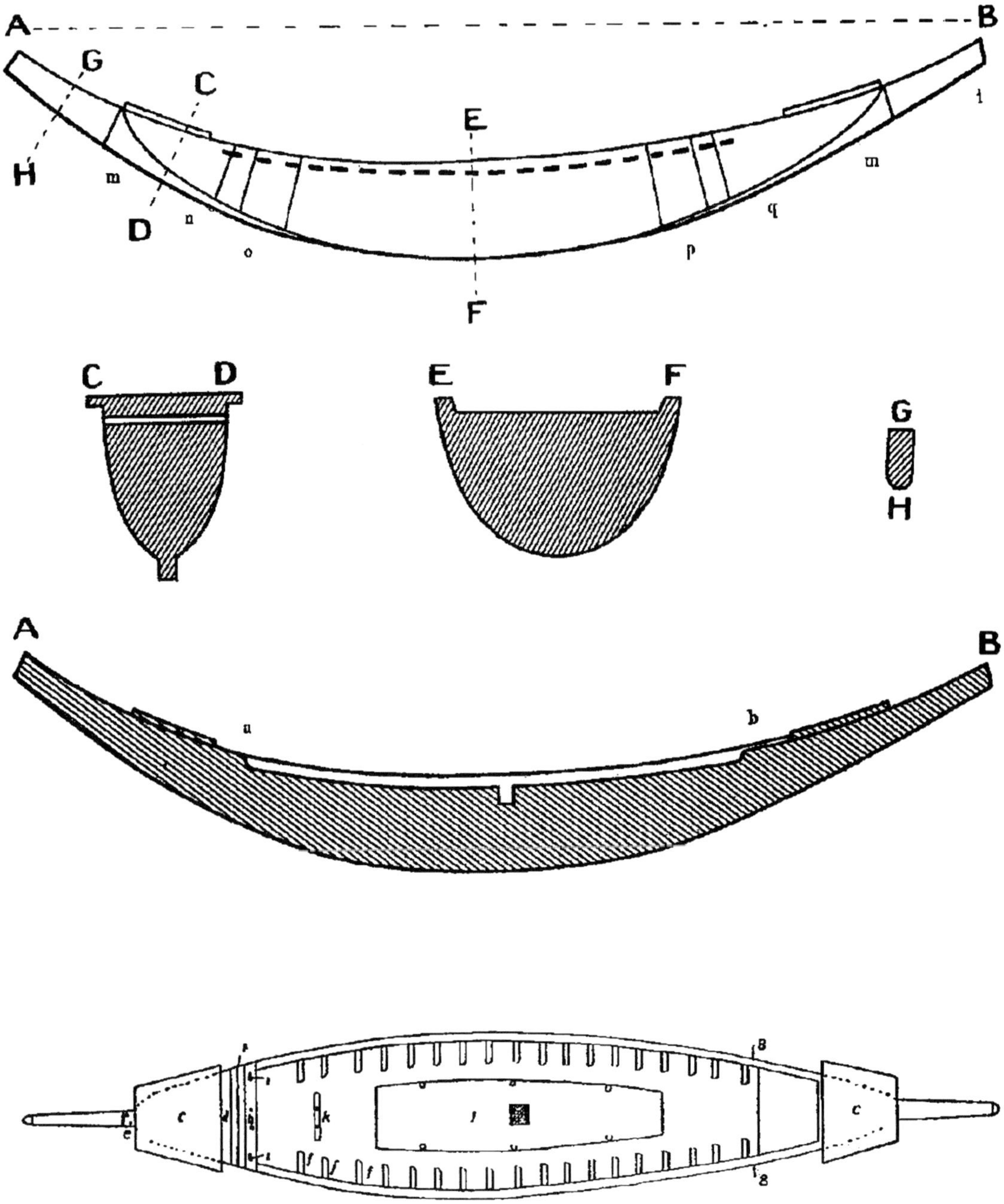

Fig. 1.26: Wooden ship model from the tomb of Amenhotep II (NTS). From Reisner 1913: 96 figs. 348–349 (no. 4944).

Color.—Although found in Egypt, the Gurob ship model replicates a Helladic galley type.[38] Depictions of ships on ceramics remain the primary source of Greek ship imagery from the Late Bronze Age to the Classical Period and beyond. The monochromatic character of these images from the Mycenaean to the Geometric periods largely precludes them, by their very nature, from elucidating questions regarding these vessels' coloration.[39] These images depict silhouetted shapes of the vessels with varying degrees of linear detail. The representations of a Sea Peoples ship, also of Helladic origin, appearing in the painted and carved tableau of Ramses III's naval battle relief at Medinet Habu, lost their paint and much of their carved detail long ago.[40] Thus, the Gurob model supplies a rare, although certainly not a unique, view of how the pattern of colored decoration of Greek ships *might* have appeared before the sixth century B.C., when numerous depictions of colored ships appear on black-figure vases.

Note that a Cypriot terracotta model with bichrome decoration of unknown provenance now in the Israel National Maritime Museum and dated to the mid-eleventh century B.C. is a nearly contemporaneous parallel to the Gurob model (Fig. 1.27).[41] The stem and sternposts are decorated with three pairs of colored bands, alternating red, black, and red. A line of red paint on each caprail connects the lower pairs of red stripes on the posts. The bottom of the hull is painted black. The Cypriot counterpart thus replicates, albeit crudely, the color scheme of the Gurob model.

The value of the colors on the Gurob model must be measured against several considerations. First, *the model is itself a representation, not a ship,* and as such it could have a color scheme different from its prototype for many possible reasons. Also, it is not clear what chemical changes have occurred in the various paints used on the model and its related artifacts.[42] Thus, the colors visible today may, and probably do, diverge from their original hues.[43]

Homer includes color epithets in his descriptions of ships. Most commonly he describes them as *black,* although he on occasion terms them *red-cheeked, purple-cheeked,* and *blue-* or *dark-prowed.*[44] While Homer composed his epics during the eighth or early seventh century B.C., he wrote of events that took place at the end of the Late Bronze Age: His descriptions of ships fit equally well with what we know of vessels that appear in Mycenaean art, as well as with ships depicted in Geometric art, which are approximately contemporaneous with the date of writing.[45]

Egyptian artists employed a wide palette of colors consisting of black, blue, brown, green, gray, orange, pink, red, white, and yellow.[46] Most closely contemporaneous with the Gurob model are those from the XVIIIth Dynasty royal tombs.[47] These models are decorated with the following colors: black, blue, blue green, brown, green, red, white, yellow, yellow brown, and gold leaf.[48]

We have no way of knowing how many of these hues were available to the artisan who painted the Gurob model, but his palette included at least the four colors that he used on the ship, the wheels, the *pavois,* and other parts. Today, although degraded and dimmed, apart from a white (huntite) undercoating these colors appear as black, blue, red, and yellow -brown. In decorating the model, however, the painter limited his decoration to only three colors—black, blue, and red. Homer specifically mentions two of these colors in connection with ships—black and red. What, then, can the Gurob model's painted decoration teach us about Homer's descriptions of ships' colors?

The *black* epithet Homer used to describe his ships presumably derives from the pitch smeared on the outside of hulls as an antifouling and waterproofing agent.[49] The Gurob model seems to show what Homer intended: The pitch covered only the bottom of the hull up to about the waterline.

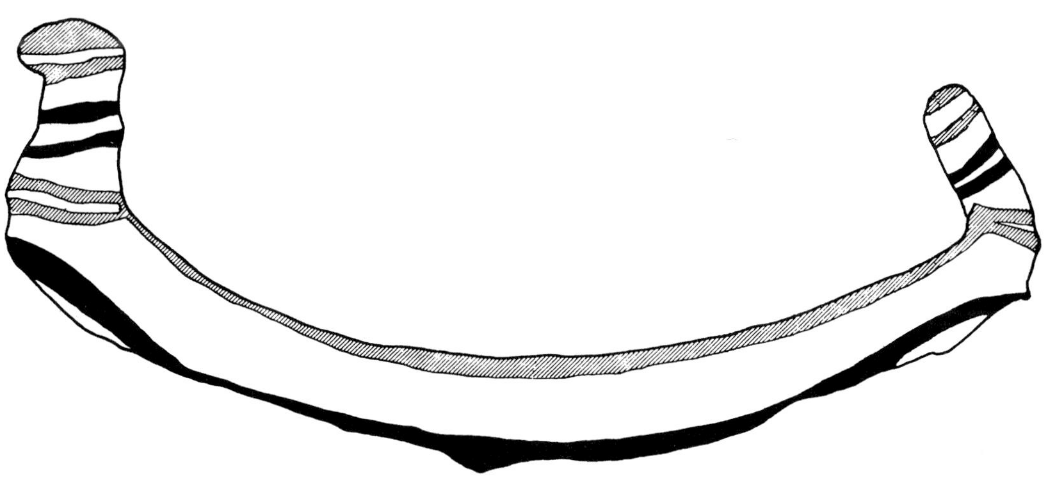

Fig. 1.27: Unprovenienced terracotta ship model from Cyprus with red and black paint, now in the Israel National Maritime Museum (NTS). After Göttlicher 1978: 7: 107.

The remaining color epithets generally have been taken to indicate that the ships were brilliantly painted in these colors.[50] Vessels appearing on black-figure vases and terracotta models of the sixth century B.C. have red lines along the sides of their hulls, reminiscent of those on the Gurob model.[51] Somewhat later, red paint has been found in a mid-fifth-century B.C. context in a shipshed at the Sicilian Naxos.[52] The paint has been tentatively identified as *minium/miltos* (lead oxide). All this strongly implies that select portions of ships' sides were painted red during the interim period for which we lack evidence.

One might be tempted to identify this decoration with Homer's *red-cheeked* epithet. Traces of red paint also survive on the rim of the Gurob model's forecastle deck and on the Cypriot model.[53] It is possible that the upper parts of the former's missing castle screens might also have been painted red, as, for example, is the forecastle screen of a galley depicted on an Attic black-figure amphora.[54]

The lower part of the Gurob model's forecastle screen piece was apparently painted blue. Might this be what Homer intended in terming ships *blue-* or *dark*-prowed (*kyaneos*)?[55] Apparently not. As D. Davis notes, Homer's use of this term probably means simply "dark" or "black."[56]

Finally, although Minoan/Cycladic vessels followed a tradition different from that of Helladic ships, one cannot discuss colors on ships in the Aegean without mentioning, if only in passing, the coloration of the craft depicted in the Miniature Frieze on Thera.[57]

The larger ships taking part in the procession are highly decorated with paint. Most of the hulls are white and blue and include decoration in the form of lions and dolphins. The processional vessel nearest the arrival town in the upper row, however, has a yellow-brown hull. The same color appears on the sheer strakes of the other processional ships.[58] The creator(s) of the Miniature Frieze also chose yellow-brown to represent timbers appearing in the construction of buildings and uprights in the ships' *ikria*.[59] Thus, the yellow-brown parts of the ships may represent unpainted planking rather than paint. If this interpretation is correct, then the hulls may have been constructed of pine, which is of similar color, grew indigenously on Thera in the Bronze Age, and was commonly used for shipbuilding in the ancient Mediterranean.[60]

Only one processional ship bears a red line, which separates its blue spiral decoration from the white hull below it.[61] Perhaps of greatest interest is the red color of the rowed ship, however (Fig. 3.8: A).[62] The smaller vessels in the cove before the procession are rendered in the same color.[63] This might indicate that the smaller vessels are either constructed of a type of wood different from that of the processional ships or coated in a red covering of paint or resin.[64]

The rowed ship's sheer strake is painted black, and the sternpost is blue.[65] The sailed ship in the same scene also has a black sheer strake.[66] The two fragmentary ships on the north wall below and to the left of the soldiers marching on shore have hulls painted white but with a black sheer strake.[67] In their sterns, however, is a red horizontal element, the identity of which remains unclear. Thus, little direct correlation exists between the coloring on the Gurob model and that on the ships portrayed in the Miniature Frieze.

Date of the Model

Radiocarbon tests of the ship model and one of its wheels indicate a two-sigma calendar age horizon of 1256–1054 BC.[68] Typologically, the ship model finds its closest iconographic parallels to representations of Helladic galleys of the Late Helladic IIIB–C periods, ca. 1300–1125 BC, although the upturned beak of the Gurob model's bird-head stem ornament is best compared to those on Late Helladic IIIC ship depictions.[69] Gurob Tomb 611, which contained the boat model, belongs to a graveyard dating primarily to the XIXth Dynasty.[70] Given these considerations, a date in the late thirteenth or early twelfth centuries BC seems most likely for the Gurob ship-cart model.

Catalogue of Loose Pieces

1) *Pavois.* Wooden rectangle decorated with blue and red painted bands on its upper surface (Fig. 1.20). Hole through the center. Notation "U.C. 16044" on the back.[71]

L. = 8.0 cm; W. = 4.8 cm; Th.[72] = 0.6 cm. Hole: D. (max.) = 1.2 cm.

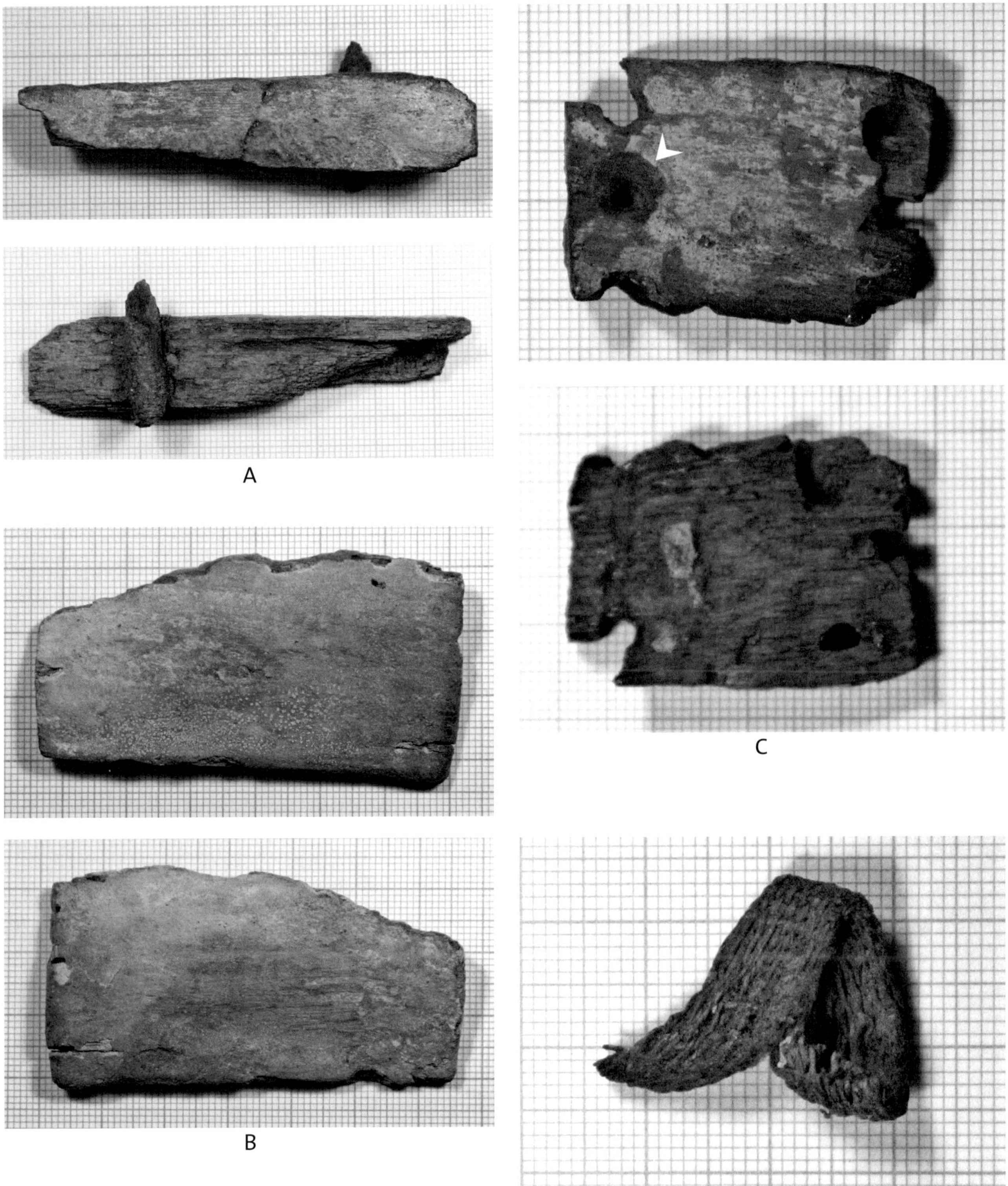

Fig. 1.28: Loose fragments (item nos. 2–4) found with the model. (A) Item no. 2. (B) Item no. 3. (C) Item no. 4. (D) Item no. 7. Curved piece of webbing.

A B

Fig. 1.29: Loose fragments (item nos. 5 and 8) found with the model. (A) Item no. 5. (B) Item no. 8.

2) Wood fragment with portion of a peg attached (Fig. 1.28: A). Cream-colored gesso on the side away from the peg. The fragment is so badly abraded and so diminished that its actual use is difficult to determine.

L. = 7.5 cm; H. = 1.6 cm. Peg: L. = 2.4 cm; D. = 0.5 cm.

3) Wood fragment (Fig. 1.28: B). One edge is curved.

L. = 6.4 cm; W. (max) = 3.3 cm; Th. = 6 mm.

4) Wood fragment with piercings (Fig. 1.28: C). The piece has three holes on either edge. One of the center holes (arrow) is blind. Traces of red paint survive on the narrow sides, and along one of the long sides traces of blue paint remain.

L. = 3.5 cm; W. (max) = 2.5 cm; Th. = 5 mm.

5) Wood fragment (Fig. 1.29: A). Remains of blue and red paint. Beveled for 4 mm on the side with paint (arrow).

L. = 4 cm; W. (max) = 2.4 cm; Th. = 2 mm.

6) Amygdaloidal-shaped beak of bird-head stem ornament, now broken off from the model's stempost (Figs. 1.11: D and also 1.4: A, 1.5, 1.6: A).

L. = 2.7 cm; W. (max) = 1.4 cm; Th. = 7 mm.

7) Woven material (Fig. 1.28: D). Small, brown, rough, curled piece of woven material.

L. = 2.5 cm (curled); W. (max) = 1.6 cm.

8) Semicircular wood fragment (Fig. 1.29: B).

L. = 3.5 cm; W. (max) = 2.4 cm; Th. = 3.5 mm.

9) Wood fragment (Fig. 1.30: A). Painted with gesso on one side with drops of blue paint (arrows).

L. = 6.7 cm; W. (max) = 1.8 cm; Th. = 7.6 mm.

10) Wood fragment (Fig. 1.30: B). Blue paint (arrow) on one side. Oar blade fragment?

L. = 6.4 cm; W. (max) = 1.35 cm; Th. = 5 mm.

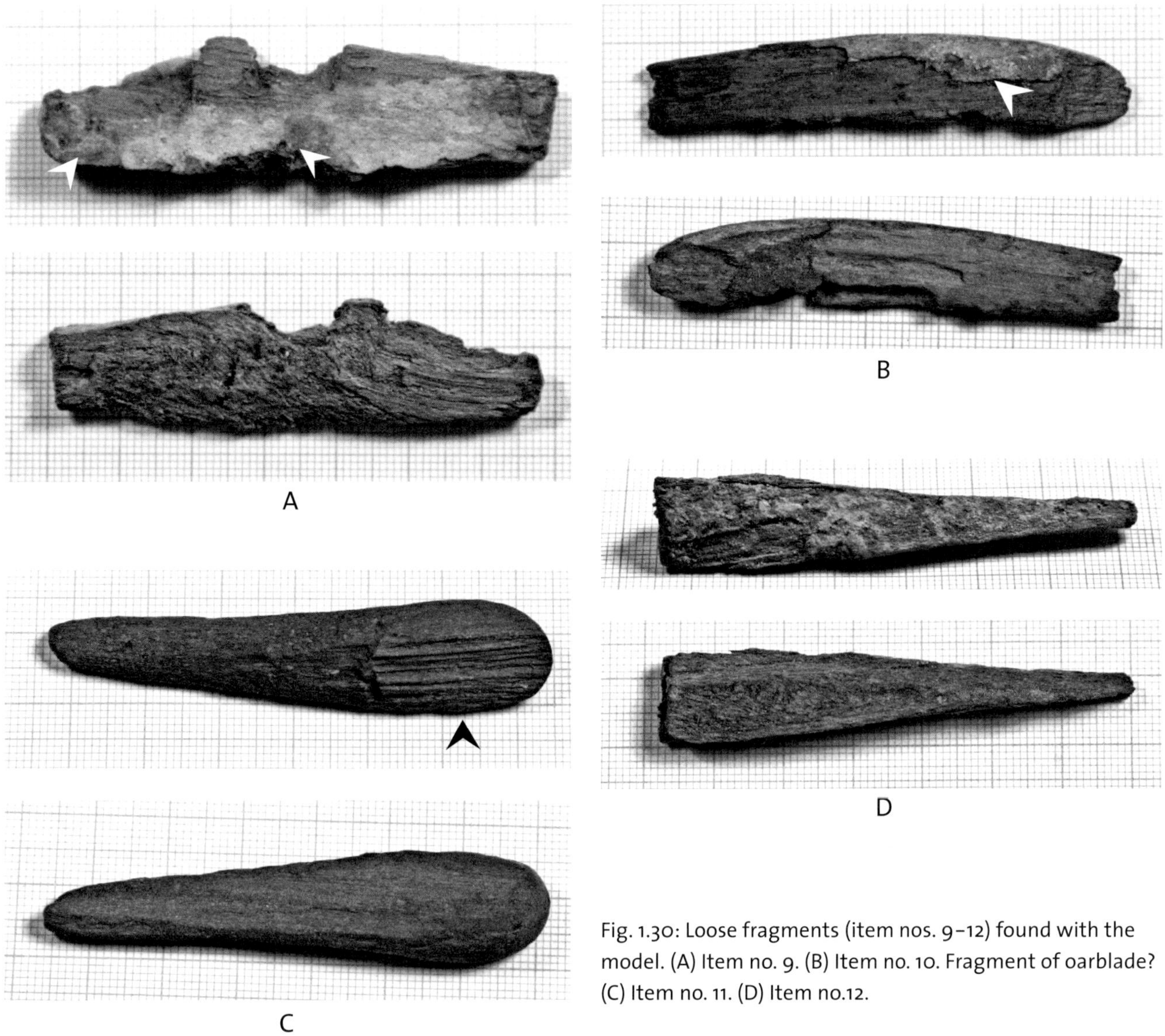

Fig. 1.30: Loose fragments (item nos. 9–12) found with the model. (A) Item no. 9. (B) Item no. 10. Fragment of oarblade? (C) Item no. 11. (D) Item no.12.

11) Wood fragment (Fig. 1.30: C). Abraded and partially sheered off (arrow).

L. = 6.8 cm; W. (max) = 1.5 cm; Th. = 8 mm.

12) Wood fragment (Fig. 1.30: D). Gesso on one side with traces of blue paint.

L. = 6.75 cm; W. (max) = 1.1 cm; Th. = 7 mm.

13) Stanchion (Fig. 1.16). Evidence of glue on the sharpened end.

L. = 5.7 cm; Th. = 5.5 mm; Th. at sharpened end = 3.1 mm.

14) Stanchion (Fig. 1.16). Broken at top. Modern glue adhering to base precludes determining thickness at lower end.

L. = 4.75 cm; Th. = 6 mm.

15) Stanchion (Fig. 1.16). Broken at top.

L. = 4.6 cm; Th. = 6 mm; Th. at sharpened end = 3 mm.

16) Stanchion (Fig. 1.16).

L. = 4.25 cm; Th. = 5.5 mm; Th. at sharpened end = 4 mm.

17) Quarter rudder (Fig. 1.18). External side decorated with black and red bands. The bottom is painted black. A hole in the piece at midlength was used to secure the quarter rudder to a hole in the starboard quarter by means of a now-missing peg (Fig. 1.8: arrow). The blade's

tip slants, suggesting the quarter rudder's angle of incline when attached to the model. The top of the loom bears a groove.

L. = 24.4 cm; W. (max) = 2.2 cm; Th. = 8 mm. Hole: D. 6 mm.

18) Long peg (Fig. 1.23).

L. = 8.35 cm; Th. = 4 mm.

19) Long peg (Fig. 1.23).

L. = 8.5 cm; Th. = 5 mm.

20) Long peg (Fig. 1.23). Traces of red and blue paint.

L. = 8.7 cm; Th. = 5.5 mm.

21) Long peg (Fig. 1.23). Traces of red and white paint.

L. = 8.55 cm; Th. = 6 mm.

22) "Awning" (Fig. 1.24: A). Light wash of white gesso on both sides. Pair of right-angled red lines. Daub of blue. Chamfering at one edge.

L. = 6.75 cm; W. (max) = 4.5 cm; Th. = 2 mm.

23) "Awning" (Fig. 1.24: B). Light wash of white gesso. Chamfering on one edge.

L. = 6.8 cm; W. (max) = 4.3 cm; Th. = 2 mm.

24) "Awning" (Fig. 1.24: C). Light wash of white gesso. Red stain. Chamfering on one edge.

L. = 6.5 cm; W. (max) = 4 cm; Th. = 2 mm.

25) "Awning" (Fig. 1.24: D). Light wash of white gesso. Signs of red paint. Chamfering on one edge.

L. = 6.6 cm; W. (max) = 4.2 cm; Th. = 1.5 mm.

26A–B) Large, thick peg (A) reconstructed in modern times (by Petrie?) as the quarter rudder stanchion and glued for that purpose to a thwart (B), now loose from the model's stern (Fig. 1.19). There is no ancient evidence to support this reconstruction.

Peg: L. = 7 cm; Th. = 1 cm.

Thwart: L. = 2.6 cm; W. (max) = 1.6 cm; Th. = 0.5 cm.

27) Wheel (Fig. 1.21: A). Excellent condition. Light white gesso wash on the outer side covered with eight radial wedges of which two are red, two are brown, and four are blue. Evidence of wear surrounding the hole for 1–5 mm. Remains of gesso and evidence of lath marks on inner side. The wheel's diameter indicates that it was paired with no. 30.

D. = 7.6 cm; D. of hole = 1.1 cm; Th. = 8 mm.

28) Wheel (Fig. 1.21: B). Light white gesso wash on the outer side with an eight-wedged radial decoration covering it in red (two), brown (two), and blue (four) paint. Most of the blue paint has flaked off. The wheel has cracked—perhaps the result of shrinkage—into two unequal parts, which are now glued together. The notation "U.C. 16044" appears on the back. The inner side of the larger section of the wheel bears a 2.1-cm-long groove, perhaps caused by a wood-boring insect. Splashes of gesso and lath marks are visible on the interior side. The wheel's diameter indicates that it was paired with no. 29.

D. = 8.1 cm; D. of hole = 1.1 cm; Th. = 1.5 cm.

29) Wheel (Fig. 1.22: A). Light white gesso wash on the outer side with an eight-wedged radial decoration, of which two wedges are red, two are brown, and four are blue. Signs of wear at the external side of the axle hole. Cracked due to shrinkage but still in one piece.

D. = 8.1 cm; D. of hole = 1.2 cm; Th. = 1.1 cm.

30) Wheel (Fig. 1.22: B). A wedge of the wheel equivalent to 80 degrees of its circumference is missing. The surviving portion consists of two sections glued together. Seven of the radial wedges survive: One is red, two are brown, and four are blue. Much of the blue and brown paint has flaked off. Signs of wear at the external side of the axle hole. Blue rectangle on interior side, as well as some white coloring, apparently remains of gesso or perhaps plaster from modern reconstructions. Lath marks visible on the interior surface.

D. = 7.7 cm; D. of hole = 1.2 cm; Th. = 1.0 cm.

2 The Iconographic Evidence

Despite its Egyptian provenience, as we shall see, the ultimate prototype of the Gurob ship model is clearly a Helladic-style galley of the Late Bronze/Early Iron Ages.[1] All later Greek galleys of the Protogeometric, Geometric, Archaic, Classical, and Hellenistic periods essentially evolved from this original Mycenaean galley type.[2] The model is remarkable in being *the* most detailed known representation of this galley type, supplying structural details in a unique, multihued, three-dimensional manner, which contemporaneous ship depictions either ignore or, at best, illuminate in two dimensions only. The model takes on added significance considering that the hull of such a ship has yet to be found, and, even if one were to be discovered in the future, it would be unlikely for its superstructure to have survived.

The Helladic galley also saw service with the Sea Peoples, who adopted and adapted it to their needs.[3] As we shall see, this may explain the appearance in Egypt of a model of this ship type.[4] Its use by the Sea Peoples also supports the evidence of significant Mycenaean links recorded in the material culture of the Philistines, who remain the best archaeologically and historically documented of the various ethnic groups that made up the Sea Peoples.[5]

Helladic Ship Representations in Egypt

Representations of Helladic ships appear at two other Egyptian sites: Medinet Habu and Dakhla Oasis.

Medinet Habu.—Ramses III reports that in his Year 8 (ca. 1176 B.C.) Egypt came under attack by a coalition of Sea Peoples, whom he defeated in two clashes, one on land, the other on water. Tableaus depicting these events decorate the central part of the northern exterior wall of his mortuary temple at Medinet Habu.[6]

The naval encounter was not a sea battle per se as it must have taken place inside the Delta, possibly on the Pelusiac branch of the Nile.[7] The combat pitted the invading Sea Peoples' fleet against Egyptian ships and land-based archers (Fig. 2.1). In the tableau, the invading galleys float immobile in the water, with their sails furled and their oars shipped. Apparently, Ramses surprised the enemy while they lay at anchor or were moored alongshore.[8]

Regrettably, the relief is a far cry today from how it would have appeared to a visitor in the twelfth century B.C. The wall had been both carved and painted. No traces of

Fig. 2.1: Ramses III's naval battle scene at Medinet Habu (NTS). From *MH* I: pl. 37. Courtesy of the Oriental Institute of the University of Chicago.

the paint remain now, and the layers of plaster employed to smooth the surface and to make corrections have long since crumbled away. Notes H. H. Nelson:[9]

> Another point not always appreciated in dealing with these Medinet Habu reliefs is the extensive ancient use of plaster to cover up defects in the masonry and to eliminate lines or whole figures which, although already carved into the stone, were nevertheless expunged from the final composition. The surfaces of the temple walls, when the masons had done with them, presented many hollows or depressions where certain stones did not come out quite to the plane of their fellows. These hollows, as well as the interstices between the stones, were filled with plaster so as to present an even surface to the artists who were to adorn the building. This plaster, especially in the exposed portions of the wall, has for the most part fallen away, carrying with it practically all of the design once carved into or painted upon it. Though this method of work was slovenly, the result produced when the building was new was not, for the time being, unpleasing; but it has had disastrous effects on the permanency of the records.

In this art form, the Egyptians did not differentiate between carved and painted detail. If anything, carving was apparently subordinate to the painted detail. The staggering loss of detail that may have disappeared with the paint may be inferred from Nelson's description of another battle scene at Medinet Habu where the paint survived:[10]

> Here, in the upper portions of the relief, even the water-color paint is unusually well preserved, and we find that the bare sculpture has been extensively supplemented by painted details distinctly enriching the composition. The colors of the garments worn by the Tjemhu stand out clearly. Between the bodies of the slain as they lie upon the battlefield appear pools of blood. The painter has suggested the presence of the open country by painting in wild flowers which spring up among the dead. Moreover, it is apparent that the action takes place in a hilly region, for streams of blood run down between the bodies as the enemy attempt to escape across the hills from the Pharaoh's pursuing shafts. The details of the monarch's accouterments are indicated in color, relieving him of the almost naked appearance often presented by his sculptured figure when divested of its paint. It is not infrequent to find such details as bow strings or lance shafts partly carved and partly represented in paint. The characteristic tattoo marks on the bodies of the Tjemhu are also painted in pigment only. When all these painted details have disappeared, though the sculptured design may remain in fairly good condition, much of the life of the original scene is gone and many aids to its interpretation are lost.

The nine ships—four Egyptian and five of the Sea Peoples—that define the naval battle are arranged in three columns.[11] A single ship type represents each fleet. This is an obvious oversimplification, however, as several considerations indicate that the Sea Peoples' fleet must have been far more heterogeneous in appearance than the scene implies.

First, the accompanying text refers to three types of vessels that took part in the battle on the Egyptian side.[12] Thus, *the scene depicts five representations of a single prototype Sea Peoples' ship in various degrees of detail.* Second, a considerable corpus of Helladic/Sea Peoples ships exists, and while they all are similar, no two are virtually identical, as is the case with the Sea Peoples' ships at Medinet Habu.[13] Third, the Egyptian artists peopled the invading ships with groups of warriors from at least two cultures. Sherden warriors wearing horned helmets crew two of the ships—Nelson's N.2 and N.4 (Figs. 2.3–4). Feather(?)-helmeted warriors crew ships N.1, N.3, and N.5 (Figs. 2.5–7).[14] Feather helmets are most commonly associated with the Peleset/Philistines, but the Sekels (Sikila/Tjeker) and the Denyen wore similar helmets.[15] Regarding the Sekels, this is perhaps best illustrated on the right façade of the Migdol high gate at the entrance to the Medinet Habu temple complex, where their bearded leader wears a similar helmet (Figs. 2.8, 4.16).[16]

Although at first impression the naval scene has a nearly photographic quality, seeming to capture a critical moment during the battle, in fact Nelson demonstrates that the relief is based on spatial, ideological, and temporal

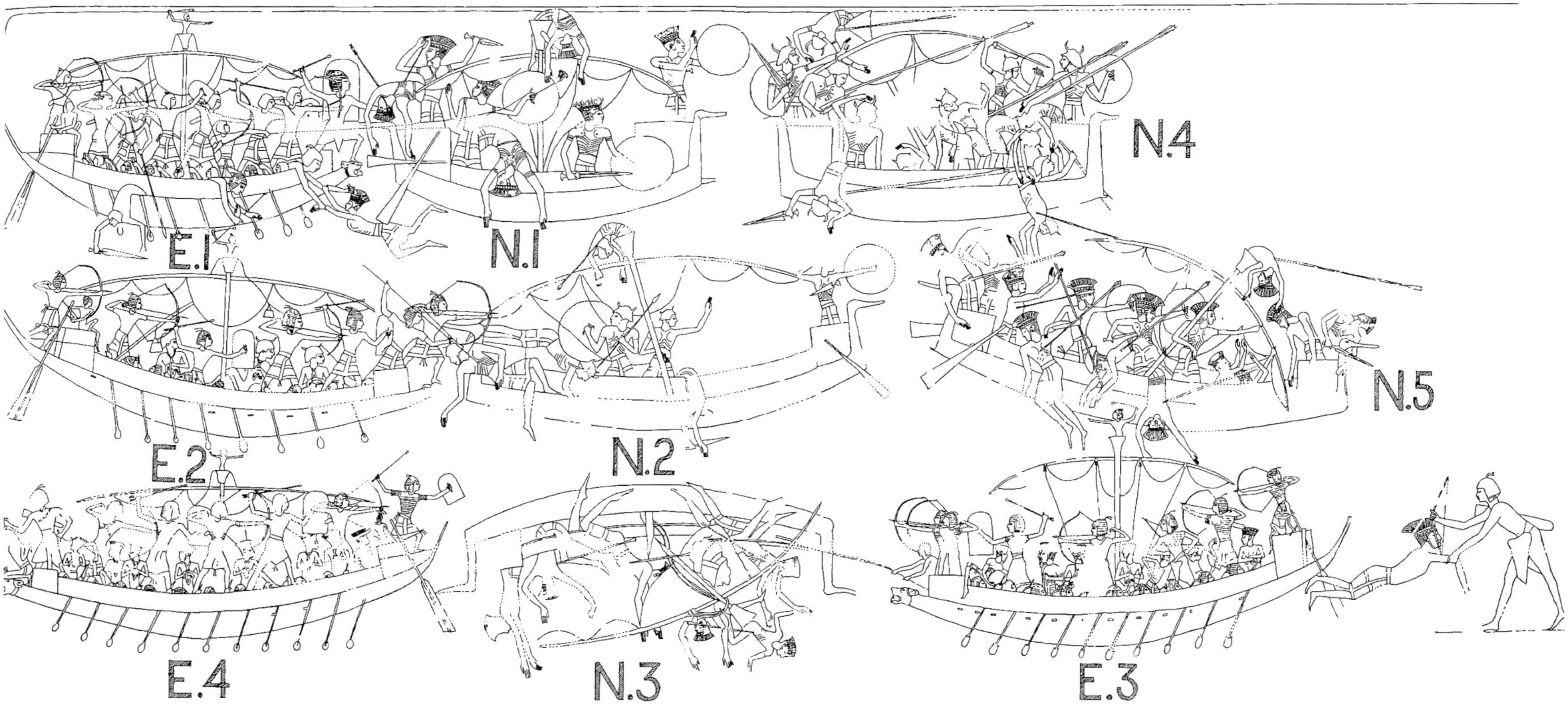

Fig. 2.2: The naval battle with floating dead bodies removed. Ramses III (NTS). From Nelson 1943: fig. 4. Courtesy of the University of Chicago Press.

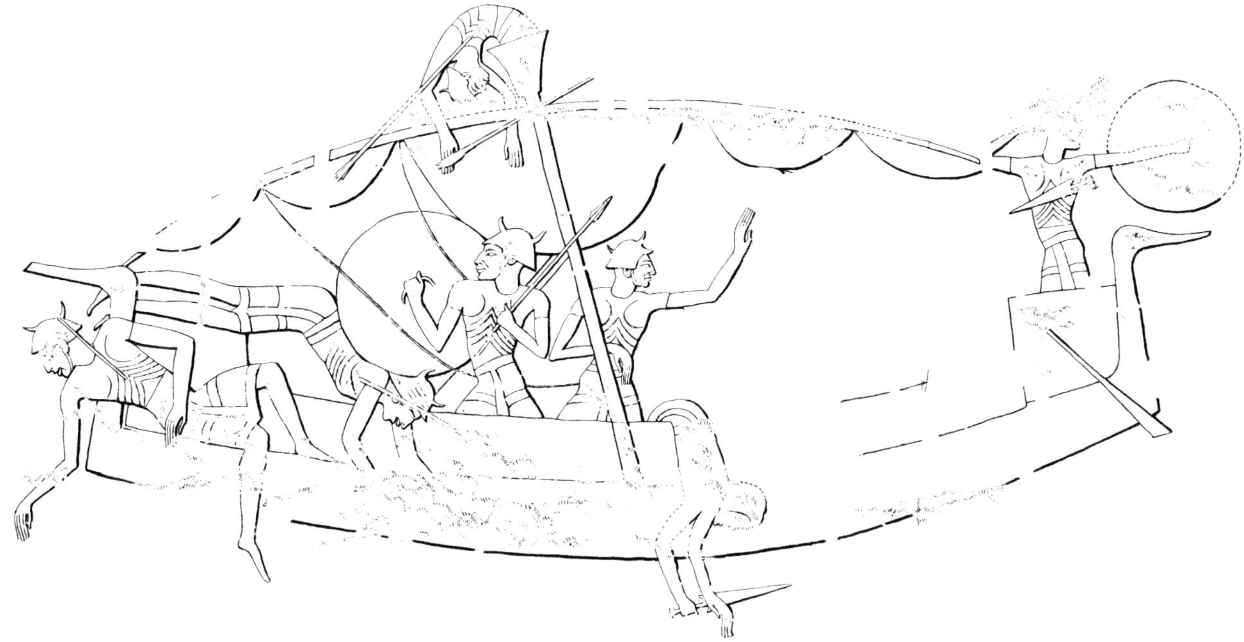

Fig. 2.3: Ship N.2 with extraneous bodies removed (NTS). After *MH* I: pl. 39.

Fig. 2.4: Ship N.4 with extraneous bodies removed (NTS). After *MH* I: pl. 39.

Fig. 2.5: Ship N.1 with extraneous bodies removed (NTS). After *MH* I: pl. 39.

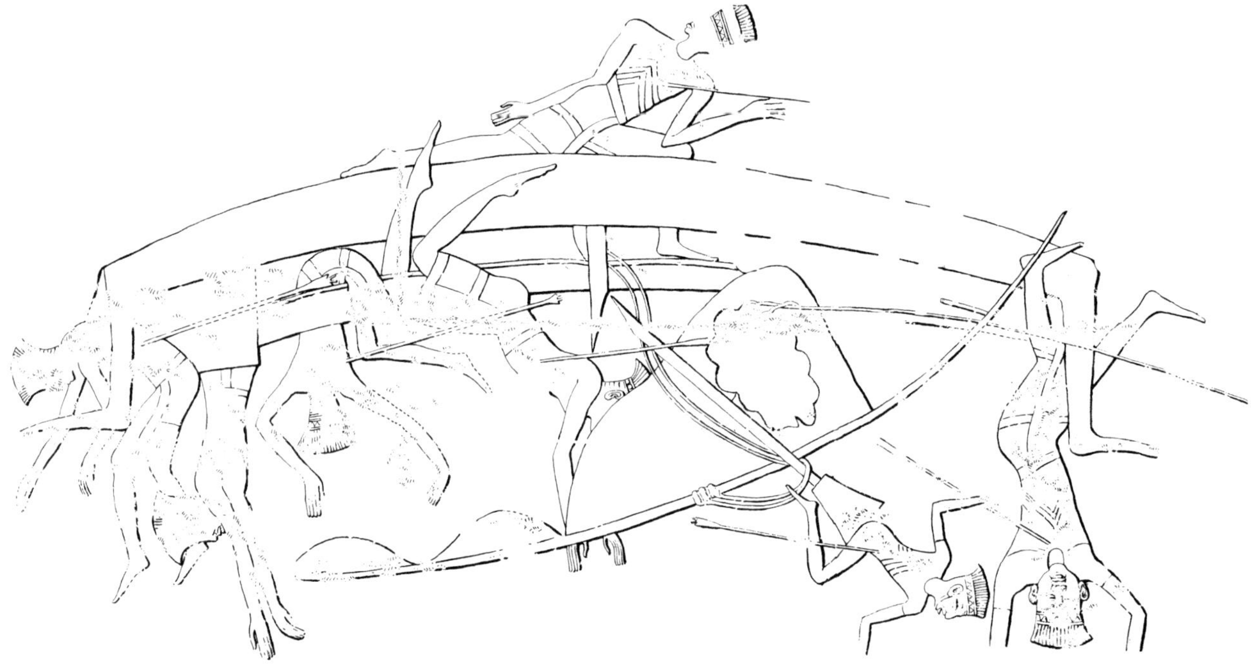

Fig. 2.6: Ship N.3 with extraneous bodies removed (NTS). After *MH* I: pl. 39.

Fig. 2.7: Ship N.5 with extraneous bodies removed (NTS). After *MH* I: pl. 39.

concepts.[17] Four specific moments of the battle are portrayed (Fig. 2.2). Ships E.1–N.1 are locked in mortal combat, and into his enemy's rigging an Egyptian marine has cast a grapnel, the only truly nautical weapon in evidence in the relief (Fig. 2.5).[18] Ships E.2–N.2 represent the midpoint of the battle, after the scales had tipped in favor of the Egyptians, whose ship is crammed with shackled prisoners while the opposing Sea Peoples' ship has begun to list.[19] Ships E.3–N.3 indicate the battle's conclusion, with the Sea Peoples' ship capsized. Ships N.4 and N.5 stand in place of the fallen and trampled enemy normally placed beneath the horses of the pharaoh's chariot in New Kingdom battle scenes, and E.4, jam-packed with captured enemy warriors, heads toward the victory celebration.[20]

Egyptian records reveal that the waves of coalition-based attacks inflicted on Egypt by the Sea Peoples varied in their ethnic makeup.[21] Two written sources list the groups that formed the coalition against Ramses III during his Year 8: the accompanying Year 8 inscription at Medinet Habu and Papyrus Harris. Although the latter is generally believed to be a copy of the former, the two sources are not entirely consistent. The Shekelesh of the temple text are replaced by Sherden in the papyrus (see Table 1).[22]

Papyrus Harris reveals that Sherden and Kehek fought on the Egyptian side.[26] Sherden are also listed among the foreign invaders during the reigns of Ramses II and Merneptah.[27]

The Medinet Habu reliefs and inscriptions are problematic from a historical interpretative point of view. Much of the material gives the impression of harking back to the XIXth Dynasty.[28] Some scenes have been copied outright from the Ramesseum with only minor changes.[29] This has led L. H. Lesko to conclude that Ramses III copied the naval battle scene from a now-missing mortuary temple of Merneptah.[30]

In presenting a cautious, minimalist, approach to the historicity of the Medinet Habu material, D. B. Redford notes

Medinet Habu[23]	Papyrus Harris[24]
1) Peleset	1) Denyen
2) Šikala[25]	2) Šikala
3) Sheklesh	3) Peleset
4) Denyen	4) Sherden
5) Weshesh	5) Weshesh

Table 1. Sea Peoples Listed at Medinet Habu and in Papyrus Harris

Fig. 2.8: Sekel leader among the enemies of Ramses III on the façade of the high gate at the entrance to Medinet Habu.

that "the Medinet Habu artists and sculptors witnessed, most likely directly, the prospects of strangely clad captives from the Aegean and Asia Minor paraded before Pharaoh. These they reproduced as faithfully as they could."[31]

One reaches an identical conclusion regarding the artists' multiple representations of the Helladic-style ship in the naval battle scene, a conclusion now confirmed by the Gurob model and further borne out by additional data, discussed later. Clearly, the source for these images derives from persons who actually saw and recorded at least one of the invaders' ships.[32]

A detailed sketch or drawing created by an artist either during or soon after the battle may have served as a basis for later depictions of the invading ship. Certainly the vivid narration of an enemy vessel in the process of being capsized by an Egyptian vessel by means of a grapnel lends a compelling element of reality (Fig. 2.5).[33] Artists accompanied Egyptian military campaigns and trade expeditions. This is evident in the attention to detail seen, for example, in Thutmose III's "Botanical Garden" in the Akhmenu at Karnak, in the misshapen Queen of Punt and the Red Sea marine life swarming in the waters beneath Hatshepsut's expedition to Punt, depicted on her mortuary temple at Deir el Bahri and in a relief of Ramses II at Luxor Temple.[34]

We cannot determine, from a distance of more than three millennia, what considerations led to the selection of this specific galley—with its high stem and stern topped with bird heads—as a prototype rather than other galleys that had taken part in the battle. Was the ship selected because it represented the predominant, generic, vessel type in the fleet, or did the artist(s) select this vessel because it was rare or unique and, therefore, stood out? The artist might have chosen to represent a specific vessel that the Egyptian forces had capsized during the encounter. Then again, we have no way of knowing, barring the discovery of an inscription, whether the intentional overturning of an enemy ship represents a unique event or a standard maneuver employed by the Egyptian fleet in the course of this battle.

Bird heads facing outboard affixed to the stem appear regularly on Late Helladic IIIB–C galleys.[35] The Medinet Habu Sea Peoples' vessels are *unique* within the corpus of Helladic galley representations, however, in showing the ship with outboard-facing bird heads capping both the stem *and* the sternposts.[36] As the bird heads would have had a numinous meaning for the vessel's crew, the crew must have had a *non Mycenaean* religious belief system.[37]

Who were these people? The motif of outboard-facing stem and stern bird-head ornaments points northward, for this ornamentation finds numerous and exact parallels with the roughly contemporaneous "bird-boat" (*Vogelbarke*) cult motif of the central European Urnfield culture.[38] The earliest bird-boats date to the European D period, ca. 1250–1200 (Fig. 2.9: A–B).[39] An ornament from Grünwald in Bavaria is only slightly later, dating to the Hallstatt A1 (twelfth century B.C.) (Fig. 2.9: D).[40] Bouzek identifies a *Vogelbarke* on a background of three sun disks on a Late Helladic IIIC sherd from Tiryns (Fig. 2.9: C).[41]

It appears, then, that the prototype Sea Peoples' ship on which the Medinet Habu artists patterned the ships in the naval battle relief had an Urnfelder crew. How big a contingent this Urnfield group constituted within the Sea Peoples' coalition is unknown and perhaps unknowable. Later on I examine the possibility of identifying the ethnic groups within the Year 8 coalition that can be associated with the Urnfield culture.[42]

Dakhla Oasis.—During a survey of rock art at Dakhla Oasis in 1936–1937, H. A. Winkler photographed a ship graffito at a location that he designated Site 69, situated in the vicinity of Teneida (Fig. 2.10).[43] In his notebook he describes the image thus: "Ship carrying men aboard et al. Grooved and incised engravings, patinated almost as darkly as the rock."[44] In his diary Winkler notes the following:[45]

> 300 m south/west of island—stone tablet—on it, a ship—on bow—soldier with rifle, or a sailor with periscope. Height of mast 300. M916
> surface: sherds, stone tools, some ostrich egg shells.

Winkler never published the photograph. This was left to L. Basch after he discovered it and identified the graffito as a Helladic ship while perusing the Winkler image collection, now housed at the Egyptian Exploration Society in London.[46]

Subsequent to its creation, the stone bearing the ship graffito broke, resulting in the loss of the upper parts of the image. The vessel faces right. A single line represents the straight hull. The vertical stempost ends just below the break. The bow continues past the stempost in a waterline projection. A horizontally bisected forecastle nestles behind the stempost. Its uppermost line continues beyond it aft. On this line once stood a figurehead, the only remains of which are traces of its feet and, astern of the forecastle, several lines that may represent the posterior of the image (Fig. 2.11).[47]

At left an oblique sternpost rises at a 129-degree angle from the hull. A single curving and nearly vertical quarter rudder descends from the sternpost. Two additional lines indicate the broadening of the blade, the bottom of which is left unfinished, giving it a tridentlike appearance. A diagonal line, which has been interpreted as rigging, leads from the stern to a vertical line that stands amidships, where one might expect a mast. This is undoubtedly the most interesting item in the entire scene. I return to it later on.

Nine complete stick figures inhabit the ship graffito. Horizontal lines that cross the torsos at right angles form their arms. Each figure has a "head appendage" (for want of a better term) that descends behind the head and ends in a protuberance signified by a drilled hole. The forwardmost figure, who presumably represents the group's leader, stands at the bow on the waterline projection. He alone sports three vertical lines, perhaps denoting feathers, rising from his head in addition to the appendage. Two of the figures—and what may be the head appendage and arms of an unfinished tenth figure—stand facing forward on the diagonal line, which also supports a chairlike object that has the appearance of a lowercase n.

The figures all have a wide stance, and the feet of all but two of the figures face the bow. The figure directly forward of amidships has his left foot (assuming he is facing the viewer) pointing toward the bow while his right foot faces the stern. The feet of the sternmost figure seem to be backward as they point toward the stern, although, like the others, he appears to be striding forward, with his long head appendage trailing left, toward the stern. Between the legs of each of these characters dangles a vertical line that reaches to their ankles. About midway down each line is a pair of holes, while a third hole indicates a widening at its lower tip.

Some of the men hold up ship models with stems and sterns that culminate in forward-facing bird-head devices (Fig. 2.12).[48] A vertical element stands slightly aft of amidships on each of the models. On the model held by the leader the line of the sternpost continues beneath the hull, perhaps indicating a quarter rudder. The figure standing immediately behind the leader and inside the lower part of the forecastle appears to be blowing on a trumpet.

Aft of the vertical element amidships is a line in the shape of an inverted capital L. This intersects a ship model, which suggests that the model had been exhibited on a standard. At the far right of the photo, in front of the ship, are some additional lines. These are best explained as part of the upraised arm of a figure similar to those in Winkler's Site 69, photos 883–884 (Figs. 2.13:B, 14). Beneath the stern are several additional unintelligible lines.

Of the other photos taken by Winkler at Site 69 only one, M890, has two figures made in a style somewhat similar to that of the figures standing in the galley (Fig. 2.15). A ship with a mast and stays, apparently made by a different hand, appears above the figures. Additional vessels appear in several of Winkler's other Site 69 photos (negatives M898, M903, and M907) (Fig. 2.16).[49] None of these other ship graffiti seem related in style or date to the graf-

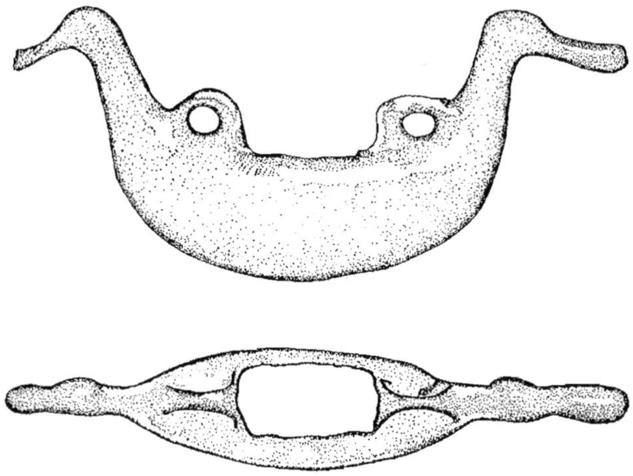

A

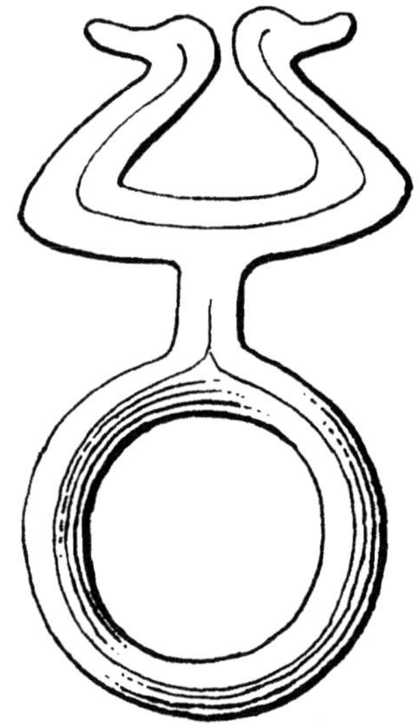

D

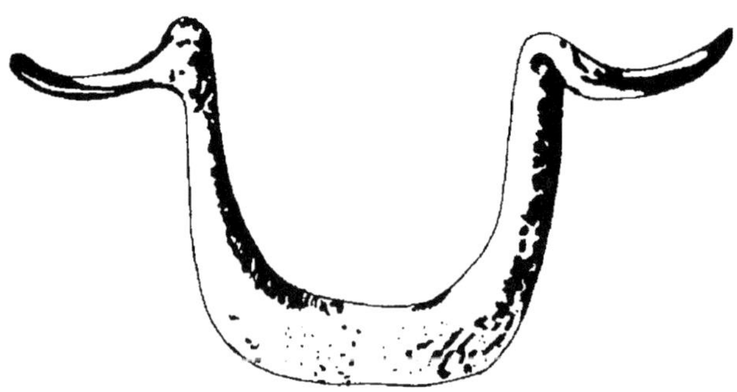

B

C

Fig. 2.9: Bird-boats *(Vogelbarke).*
(A) Ornament from the Somes River at Satu Mare in northern Rumania (European Bronze D?). (B) Ornament from Velem St. Vid in Hungary (European Bronze D?).
(C) Bird-boat painted on a krater sherd from Tiryns. Late Helladic IIIC. (D) Ornament from Grünwald, Bavaria (Hallstatt A1 (NTS)). A–B after Göttlicher 1978: Tafs. 33: 439, 34: 440. C after Bouzek 1985: 177 fig. 88: 6. D after Hencken 1968B: 516 fig. 478: f.

fito under consideration here. The ship in Winkler's negative M898 has the general shape of a Helladic galley but lacks a bird head at the stem: It is simply too schematic to allow for any firm conclusions (Fig. 2.16: A: arrow).

Let us now return to the lines identified as a mast stepped amidships, which is supported by a line of rigging running from the mast to the stern at a 45-degree angle (Fig. 2.10).[50] On closer examination this initial interpretation must be rejected. Two clues indicate that the diagonal line represents a long solid timber, which is supported on a crutch amidships. First, several figures are depicted standing on the diagonal line. While crewmembers in Egyptian ship depictions sometimes appear climbing up the rigging to the mast on Bronze Age ship depictions, they consistently hang upside down as they climb and never walk on the standing rigging, which is, of course, a physical impossibility.[51] One may contend that this anomaly is simply the result of the extremely schematic nature of the Dakhla graffito, but in this case we have another irrefutable piece of evidence that points to the same conclusion: The diagonal line crosses the "mast" and continues past it into the break at the top of the rock (Fig. 2.17). In other words, *the diagonal line is resting on the vertical line.* Figure 2.18 shows how this line might have continued originally, prior to the loss of the upper portion of the rock on which the graffito is carved.

Fig. 2.10: The Dakhla Oasis Helladic ship graffito carved on a rock at Teneida. Photo: H. A. Winkler. Egypt Exploration Society image of Winkler-Mond-Site 69 (negative M916). Courtesy Egypt Exploration Society.

Fig. 2.11: Detail of the upper-right portion of the Dakhla Oasis Helladic ship graffito (Fig. 2.10). Note what appear to be the legs and posterior of a figure (bird?) atop the forecastle deck. Photo: H. A. Winkler. Egypt Exploration Society image of Winkler-Mond-Site 69 (negative M916). Courtesy Egypt Exploration Society.

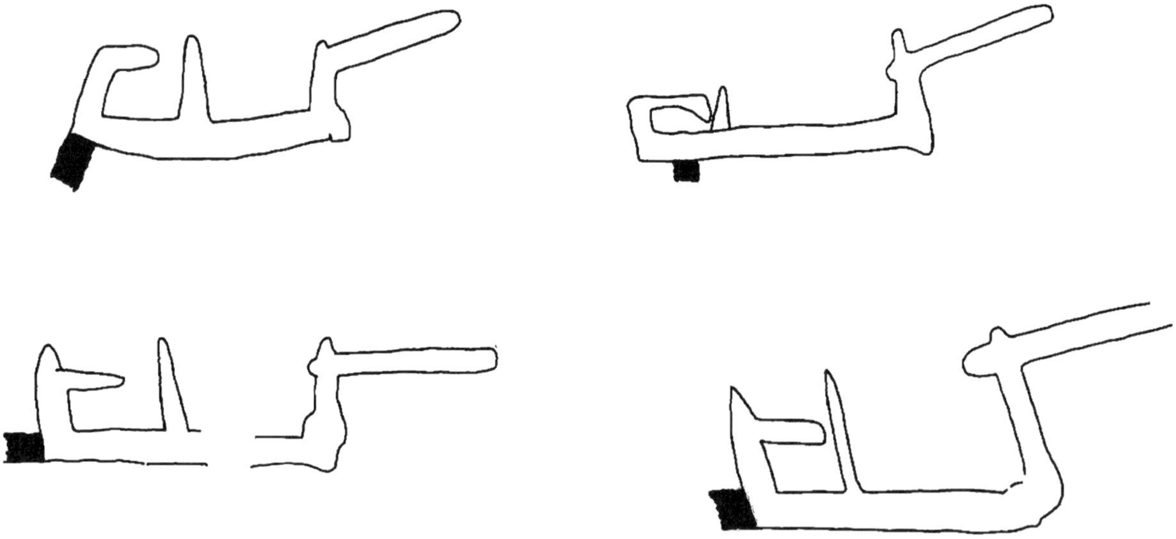

Fig. 2.12: Line drawing of the ship models held by the figures standing in the Helladic ship from Dakhla Oasis (NTS). After Basch 1994: 25 fig. 15.

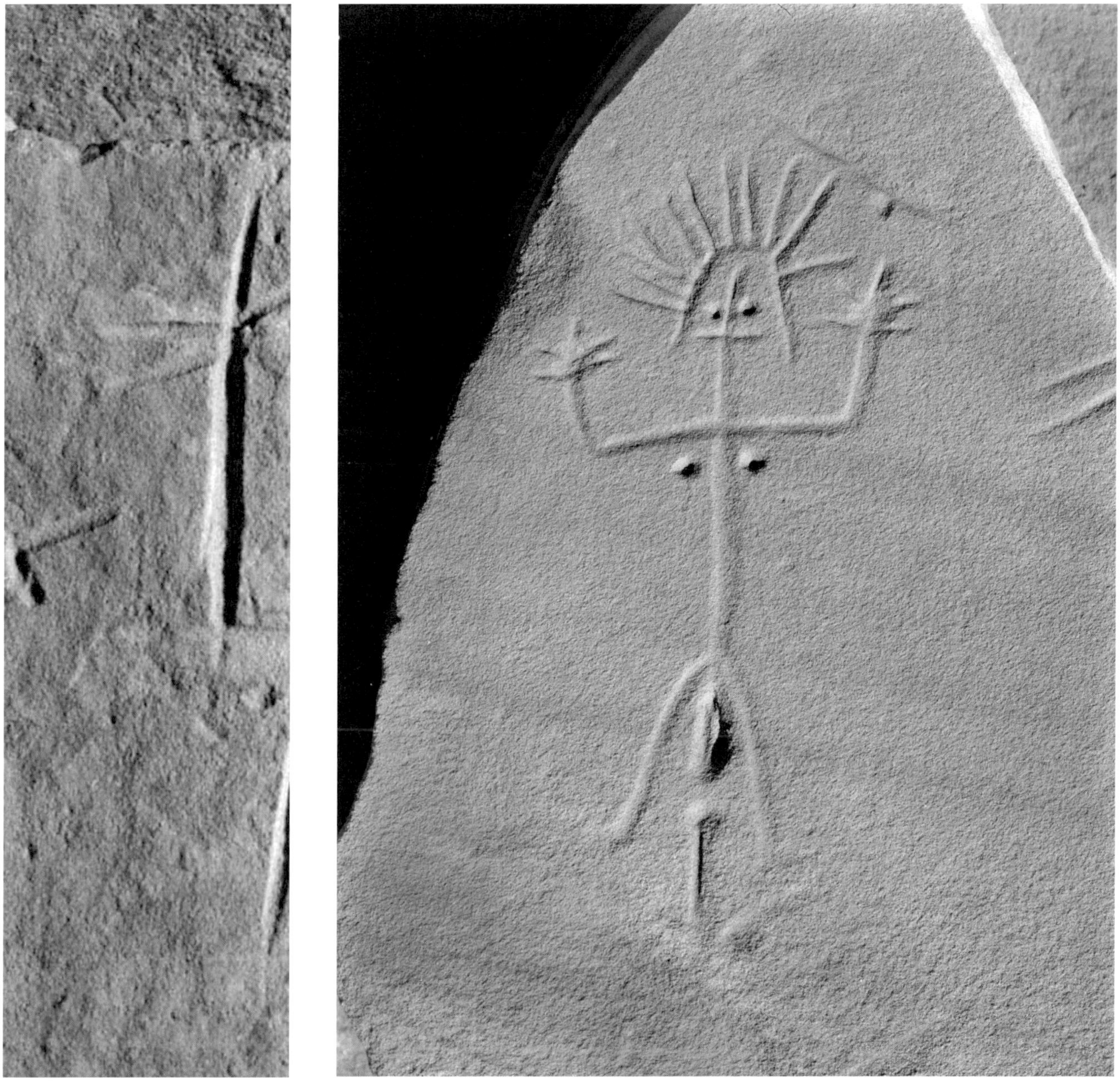

A B

Fig. 2.13: (A) The right edge of Winkler's photo M916. (See Fig. 2.10.) (B) Female figure from Winkler's Site 69 at Dakhla Oasis (M884) with raised hands. Photos: H. A. Winkler. Courtesy Egypt Exploration Society.

Fig. 2.14: Female figure from Site 69. This figure has two pairs of outstretched arms (M883) but is carved in the same style as Fig. 2.13. Photo: H. A. Winkler. Courtesy Egypt Exploration Society.

Fig. 2.15: Winkler's photo M890 from Site 69 at Dakhla Oasis includes a stick figure of a woman drawn in a style similar to those on the Helladic ship graffito. Near it are several other items carved perhaps by the same hand. Photo: H. A. Winkler. Courtesy Egypt Exploration Society.

What does this slanting timber represent? One possibility is that it signifies a lowered mast. In Egyptian art, ships often appear carrying their unstepped masts (Fig. 2.19).[52] The mast's heel is always positioned at the bow, however, with the mast lying either horizontally or at a diagonal angle rising toward the stern—*in other words, diametrically opposed to the manner in which the Dakhla Oasis artist portrayed his diagonal timber.* We may conclude, therefore, that it is not an unstepped mast.

A second possibility is that the diagonal line on the Dakhla Oasis graffito represents a large Nefertem symbol like that carried by priests as part of the panoply during the festival of Sokar, as portrayed in Ramses III's Second Court at Medinet Habu (Fig. 2.20: A).[53] One of the most colorful celebrations at Thebes was the festival in honor of

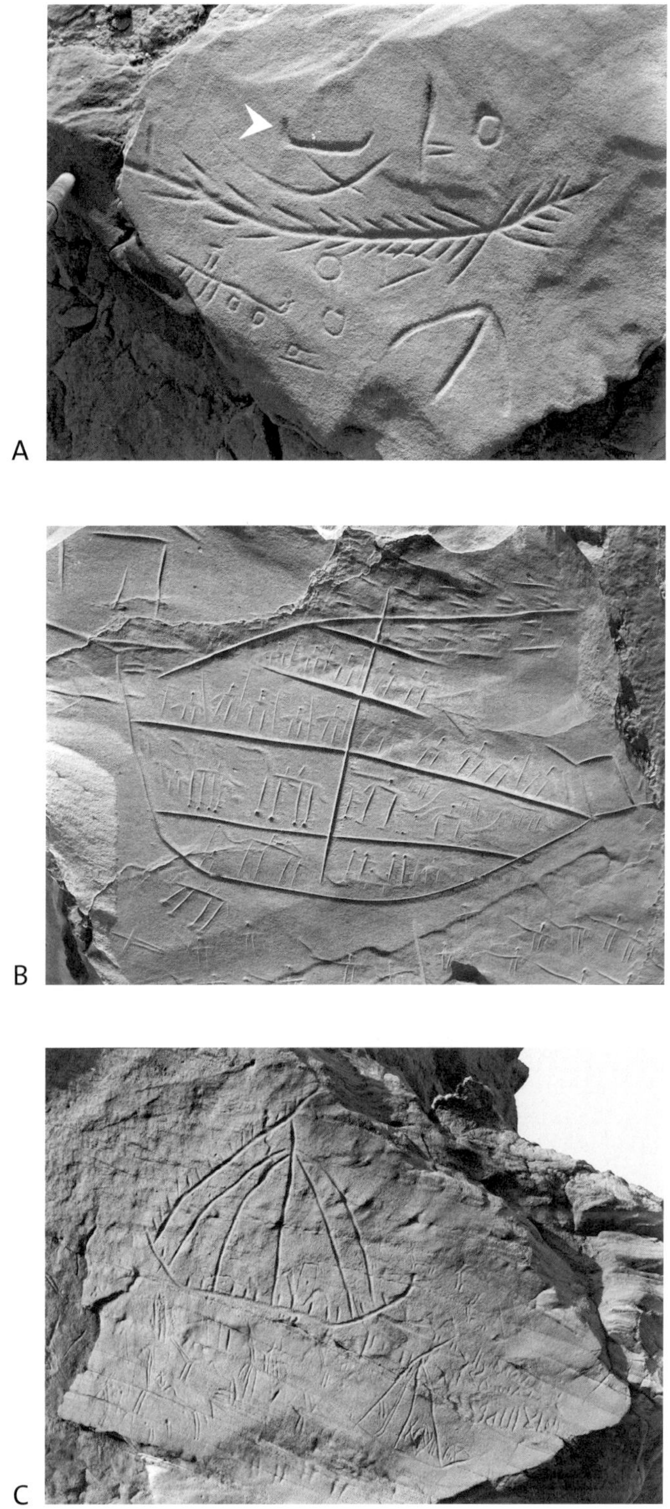

Fig. 2.16: Winkler's Site 69 at Dakhla Oasis contains additional graffiti of ships carved into the rocks. (A) M898, (B) M903, (C) M907. Photo: H. A. Winkler. Courtesy Egypt Exploration Society.

Fig. 2.17: Detail of the intersection of the vertical and oblique lines of the Dakhla Oasis Helladic ship graffito. Photo: H. A. Winkler. Egypt Exploration Society image of Winkler-Mond-Site 69 (negative M916). Courtesy Egypt Exploration Society.

Fig. 2.18: The Dakhla Oasis with the oblique line extended beyond the break. Photo: H. A. Winkler. Adapted from image of Winkler-Mond-Site 69 (negative M916). Courtesy Egypt Exploration Society.

Fig. 2.19: Ships carrying their unstepped masts and rigging. Note that the foot of the mast is in the bow and that the mast-head faces the stern. (A) Tomb of Amenemhet at Beni Hassan (T 2). XIIth Dynasty. (B) Tomb of Antefoker at Sheikh Abd el Qurna (TT 60). XIIth Dynasty. (C) Tomb of Ipy at Sheikh Abd el Qurna (TT 217). Ramses II (NTS). A after Newberry 1893: pl. XVI; B after N. d. G. Davies 1920: pl. XVIII; C after N. d. G. Davies, 1927: pl. XXX.

the god Sokar, during which the unusually shaped Henu barque, was transported overland by priest-porters (Fig. 2.20: B).[54] There are two additional parallels of possible interest between the Dakhla Oasis graffito and the Sokar festival as represented at Medinet Habu: A priest carries a standard topped by a falcon, which might be compared to the vertical standard(?) on the Dakhla Oasis graffito, and five boat models are carried in the Sokar procession (Figs. 2.20: A, 21). Contra this view one notes that the models carried in the Sokar festival are placed on mini-*pavois,* and four priests carry each one on poles.

The following example is offered here simply as a cautionary tale. A painting of a ship identified as a modern forgery by G. Brunton appears on a Predynastic jar now in the Otago Museum (Fig. 2.22: A).[55] It depicts a ship, which terminates in a strange "curlicue" at the hull's left extremity: Three men ascend what appears to be a yard attached to a mast stepped amidships, a peculiar location for a mast at this early date. The right tip of the yard touches, and continues past, the right extremity of the ship. The mast has a series of short slanting lines rising from its right side. Of the three lines visible, two run fore-and-aft from the masthead, while the third runs from two-fifths of the way between the left tip of the yard and the masthead to terminate at the curlicue. The latter line may be intended to represent a brace, while the pair of lines may indicate either fore and aft stays or possibly halyards. We have no internal information about the directionality of the Otago vessel. Six human figures, visible to about the height of their waists or upper thighs, stand in the hull. No sail is in evidence. A series of forty vertical lines beneath the hull presumably represent paddles.

Brunton's reasons for deeming this ship a forgery include the lack of a white undercoating, but more particularly the style of the decorations on the original Predynastic jars that he includes in this category. The paintings include numerous elements that are otherwise foreign to the Predynastic art repertoire although these elements derive their inspiration from it. The closest parallel to the Otago vessel is another forged ship on a Predynastic jar now in the Petrie Museum of Egyptian Archaeology (Fig. 2.22: B). Brunton concludes that the same hand, or at least the same studio, created both ships.[56]

Finally, a fourth possibility, and the most likely one in my view, is that the diagonal line represents a large phallus. Consider the telephone-pole-sized phalli depicted in two scenes on a later Attic cup now in Florence, which portrays naked porters supporting a horizontal base (Fig. 2.23).[57] The entire structure must have been quite heavy as the eight men—six along the "starboard" side and one at each extremity—suggest a total of fourteen porters, if we trust the depictions. The men seem to be bending under their burden.

The support structure has been identified as a "plough, a boat or a branch."[58] J. Boardman leans toward identifying it as a boat and compares it to the seventh-century B.C. wooden ship models from the Heraion on Samos.[59] Alternately, it might represent a simple, purpose-built framework, for which there are modern ethnographical parallels. Rising at diagonal angles from these objects' "sterns" are large festooned phalli, which are slightly longer and much thicker than their bases. As in the case of the Dakhla Oasis graffito, crutches positioned "amidships" support these items. On one representation a large statue of a fat nude man, perhaps a *komast* (reveler), leans against the phallus (Fig. 2.23: A–B).[60] A second scene depicts a statue of a satyr bent at the waist and hovering over the phallus (Fig. 2.23: C–D). A woman holding a horn in her left hand rides the satyr. In this comparison, the figures standing on the diagonal line of the Dakhla Oasis ship graffito are reminiscent of the concept of "riding the phallus" in the later Dionysian cult, as seen on the Florence cup.[61]

Phalli were paraded in Egypt under the Ptolemies. Athenaeus, quoting Callixeinus of Rhodes, describes a remarkable procession that took place in Alexandria under Ptolemy II Philadelphus (285–246 B.C.).[62] A number of display items were carried on wagons, including a colossal phallus:[63]

> In other carts, also, were carried a Bacchic wand of gold, one hundred and thirty-five feet long. . . in another was a gold phallus one hundred and thirty-five feet long, painted in various colours and bound with fillets of gold; it had at the extremity a gold star, the perimeter of which was nine feet.

Who might have created this remarkable image of a Helladic ship—depicted clearly taking part in a cult activity—at the seemingly unlikely location of Dakhla Oasis? Any

Fig. 2.20: (A) The Nefertem symbol carried by priests during the festival of Sokar. Medinet Habu. Ramses III.
(B) The Henu barque of Sokar carried by priests during the festival of Sokar. Medinet Habu. Ramses III (NTS). After Gaballa and Kitchen 1969: Tab. I fig. 2.

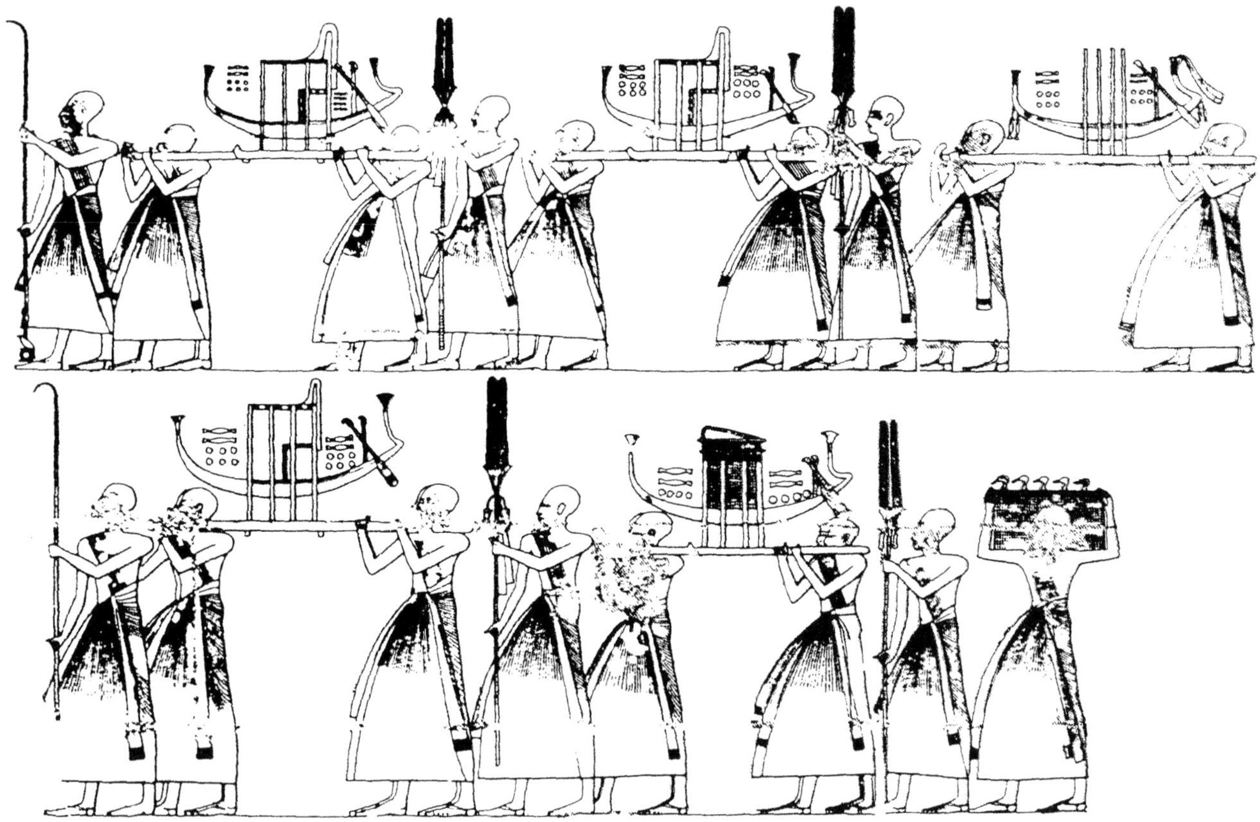

Fig. 2.21: Ship models carried by priests during the festival of Sokar. Medinet Habu. Ramses III (NTS). After Gaballa and Kitchen 1969: Tab. II fig. 3.

conclusions must be based on the evidence of the Helladic style galley, the identity of the head appendage, the leader's tripartite headgear, and the phallic objects between the figures' legs. The "usual suspects" in this case include the Mycenaeans, the Sea Peoples, and the Tjemhu (Libyans):[64]

Mycenaeans.—A papyrus from Amarna indicates that Mycenaean warriors, wearing traditional boar-tusk helmets, fought alongside Egyptian forces at some time(s?) during the XVIIIth Dynasty.[65] Presumably these forces could have arrived in Egypt in Helladic galleys. The head appendage of the Dakhla Oasis figures, in this alternative, might be compared to the plume that appears on several renditions of the Aegean boar-tusk helmet and on a bronze helmet of similar shape from Knossos.[66] In this scenario, the elongated head appendage of the Dakhla Oasis figures could represent a plumed helmet, which is probably what is represented on Late Helladic IIIC sherds from Kynos (Figs. 2.24, 26).[67] Although nothing in the Mycenaean milieu explains the phallic object, at Thera, male nudity is equated with cultic acts.[68]

Sea Peoples.—Michal Artzy suggests that the ship graffito may indicate that the Egyptian authorities settled some Sea Peoples inland in an attempt to keep them far away from the sea and out of trouble.[69]

The item dangling between the Dakhla Oasis figures' legs might be a phallus, and the pair of holes along the center of the line might represent the testicles/scrotum. If so, the hole drilled at the bottom of the line would seem to indicate that the figures are represented as circumcised. The Philistines' lack of circumcision is a defining and a particularly denigrated characteristic as seen through the biblical lens.[70] Not all the Sea Peoples were uncircumcised, however. Merneptah asserts that the Ekwesh—one of the Sea Peoples who joined forces with the Tjemhu tribes to invade Egypt during Year 5 of his reign—had been circumcised.[71] For this reason, in the count of the

A

B

Fig. 2.22: Modern painted forgeries on ancient jars. (A) Ship on a Predynastic jar now in the Otago Museum. (B) Ship on a Predynastic jar now in the Petrie Museum of Egyptian Archaeology. A courtesy of the Otago Museum. B after Petrie 1933B: 67 no.61.

enemy dead, a hand from each warrior was removed rather than the phallus.

Who were the Ekwesh? Along with the Sherden, Teresh, and Weshesh, the Ekwesh are defined as "of the sea," in opposition to the Philistines and the Sekels (Sikilu), who never receive this appellation.[72] D. B. Redford identifies the Ekwesh as coming from the island of Kos.[73] Most scholars, however, prefer to link this ethnic group with the Greek Achaeans and the Ahhiyawa of the Hittite texts.[74] Some, however, consider this identification difficult because of Merneptah's comments on Ekwesh circumcision. For example, W.-D. Niemeier writes that "there are problems with this identification, since the Ekwesh practiced circumcision, an Egyptian and Semitic custom which appears to have been alien to the Aegean."[75]

A

B

Fig. 2.23: (A–B) Diagonally positioned phallus supported by a crutch with a statue(?) of a fat man astride it. The base for this appears to be a framework that is carried by six naked porters on its right side.

C

D

Fig. 2.23: (C–D) A statue of a satyr rides the giant phallus, which is again supported on a crutch. The phallus and crutch sit on an elongated object ending in a phallus, which is carried by six naked, notably ithyphallic, porters on its right side and steadied by two additional figures at front and back. Scenes from Attic cup, Florence 3897 (NTS). Ca. 540 B.C. A and C from Carpenter 1986: pl. 22. B and D from Nilsson 1955: Taf. 35, 2–3.

Fig. 2.24: A warrior with shield appears on a sherd from Pyrgos Livanaton (Kynos). The appendage behind his head is reminiscent of the "head appendages" of the Dakhla Oasis figures in the Helladic galley graffito. Late Helladic IIIC. From Dakoronia 2002: 290 fig. 8. Courtesy Hellenic Institute for the Preservation of Nautical Tradition.

Fig. 2.25: Ship B from Pyrgos Livanaton (Kynos). Late Helladic IIIC (NTS). From Wachsmann 1998: 134 fig. 7.15.

Fig. 2.26: Ship C from Pyrgos Livanaton (Kynos). Late Helladic IIIC (NTS). From Wachsmann 1998: 135 fig. 7.16: A–B.

But might circumcision have been practiced in the Bronze Age Aegean? This question seems almost oxymoronic given the later Greek abhorrence of this custom.[76] However, the issue may not be as clear as it seems. Iconography helps little here, for the only useful images from this realm of male nudes, uncircumcised in this case, are from Akrotiri on Thera—a site inhabited by a non-Greek culture.[77] Notes Astour:[78]

> How can we be sure that the Achaeans of the Mycenaean period were not circumcised? No soft parts of their bodies have been preserved. We have seen how circumcision spread, with the expansion of Islam, to such non-Semitic peoples as Iranians, Indians, Malayans, black Africans and Southern Slavs: what is so inherently strange, for instance, in the transmission of this rite from Egypt to Minoan Crete (where Egyptian influence was very prominent), and from there to Mycenaean Greece—to mention only one possible route?

Tjemhu.—Basch proposes linking the figures with the Tjemhu.[79] While during the New Kingdom Tjemhu fought in coalitions with the Sea Peoples against the Egyptians, nothing in the textual or iconographic evidence suggests that the Tjemhu attacks were anything but terrestrial in nature.[80] Simply put, we have no other evidence linking Tjemhu to Helladic ships.

Basch argues that the line of defense built by Ramses II along the Mediterranean coast west of the Nile Delta must be related to concerns about seaborne attacks by Tjemhu.[81] These fortifications need not have been intended for this purpose, however. The line of Egyptian forts that stretched the length of the northern Sinai coast known to the Egyptians as the "Way of Horus"—the biblical "Way of the Philistines"—was built to enable the Egyptian armies to resupply themselves on their way to and from Asia.[82] A similar argument could be made for the western extension of this line. Alternately, the construction of the western arm of this fortification line may have been a response aimed at protecting Egypt's western Mediterranean coastline from northern ship-based marauders, like the Sherden, with whom Ramses II engaged at sea.[83]

Unfortunately, no representations survive of Merneptah's struggle with the coalition of the Tjemhu and the Sea Peoples. In two scenes at Medinet Habu in which Ramses III celebrates his victory over the Tjemhu in his Year 5, his retinue heap up before him mutilated body parts of the dead enemy warriors (Fig. 2.27).[84] Six of the piles consist solely of hands; two of the piles contain uncircumcised phalli only. One pile, however, contains a few hands and phalli with scrotums attached: some of these phalli are circumcised (Figs. 2.27: B bottom register, 28).[85]

To which ethnic group do the circumcised phalli belong? Papyrus Harris lists one northern group, the Karkisha, as having joined this Tjemhu attack.[86] One feather(?)-helmeted figure, lying upside down on the battlefield, is meant perhaps to represent a warrior of this ethnic group (Fig. 2.29).[87] If so, this does not help in clarifying the matter inasmuch as the warrior is uncircumcised, while his feathered helmet does not lend itself to identification with the long extension of the Dakhla Oasis figures' headgear.

We are left with the *possibility* that, following the incursions under Merneptah, one or more of the Tjemhu tribes that attacked Egypt had adopted circumcision.[88] This scenario seems reasonable, considering that the Tjemhu served in the Egyptian army as early as the reign of Akhenaten, that the Labu and the Meshwesh had settled in Upper Egypt in the interim between the reigns of Merneptah and Ramses III, and that the latter's Tjemhu wars subsequently resulted in the settlement of many of them within Egypt, apparently in the eastern Delta and the western oases.[89] Alternately, groups among the Tjemhu could have learned the custom from their allies, the Ekwesh. D. O'Conner offers another explanation:[90]

> From circumcised men, however, the Egyptians removed only a hand . . . , so one must assume that the Tjemhu shown in the Year 5 scenes must represent Sopedyu, apparently circumcised, since the Libyu and Meshweshyu explicitly were not. Osing also concludes that the Sopedyu were circumcised, but for a different reason, namely, that one of the Year 5 scenes . . . shows a pile of apparently circumcised phalli, which cannot be those of the Meshweshyu or Libyu. . . . The phalli in question are indeed circumcised, but the cutting off of such phalli would be against an Egyptian custom, referred to textually under Merenptah and apparently depicted in the Year 5 battle scenes. In fact, the pile of phalli in question are [*sic*] the only ones shown in the second narrative, and still have the scrotum attached, a most unusual circumstance. . . . I would suggest they are meant to represent the uncircumcised phalli, but the designer of the narrative has simply taken the Egyptian (and hence circumcised!) hieroglyph for phallus and turned it into a pictorial element.

Thus, in support of a Tjemhu identity for the Dakhla Oasis figures, a case can be made to interpret the head appendage as an exaggerated representation of the typical Tjemhu sidelock, which ends in a curl, and the phalli as either circumcised Tjemhu or, more likely, the typical phallic sheaths, or *penistasche,* with a knotlike protuberance at the lower end (Figs. 2.30–31).[91] The three feathers(?) on the head of the leader might be identified as ostrich

Fig. 2.27: (A–B) Nine piles of hands and phalli removed by Ramses III's troops from dead Tjemhu warriors (NTS). From *MH* I: pls. 22–23.

Fig. 2.28: One of the piles of mutilated limbs used as a body count following Ramses III's battle with the Tjemhu in Year 5 includes circumcised phalli. Medinet Habu (NTS). From *MH* I: pl. 23.

feathers like those worn by some Tjemhu depicted in Egyptian New Kingdom art, perhaps as indicators of high rank.[92] Normally only one or two feathers are depicted as worn by Tjemhu, but in at least one case a Tjemhu chief is shown with *four* feathers in his hair, so apparently the number could vary (Fig. 2.32).[93] A reasonable parallel of a Tjemhu combining the elements of the Dakhla Oasis figures—(in this case only two) ostrich feathers, the sidelock, and phallus sheath—appears on one of Tutankhamun's chariots.[94]

Helladic Ship Representations of the Syro-Canaanite Littoral

During the latter part of the Bronze Age and the beginning of the Iron Age the iconographic record is rich in depictions of Helladic ships complete with open rowers' galleries and/or bird-head stem devices.[95] Some of these representations, found in sites along the Syro-Canaanite littoral, indicate that the Sea Peoples introduced this ship type to this region.

Hama.—The Gurob ship model finds its closest parallel in a Helladic galley painted on a cremation urn from Hama, Syria (Fig. 2.33).[96] The ship has a rockered keel rising in a vertical stempost, which ends in a forward-facing bird head with an upturned beak. The bow continues in a waterline projection that follows the line of the keel, assuming one was present. The horizontal line that crosses the stempost may represent the continuation of a wale used to strengthen the hull longitudinally, which is also visible on two of the Pyrgos Livanaton (Kynos) galleys (Fig. 2.34).[97]

The ship's hull appears as a narrow, horizontal strip crossed by ten more or less vertical lines. Above it is an open rowers' gallery crossed by vertical stanchions. The vertical lines crossing the hull are puzzling. They do not continue past the bottom of the hull, suggesting that they do not represent oars. A somewhat similar motif appears on a galley painted on the side of a Late Minoan IIIB *larnax* from Gazi (Fig. 2.40: A).[98]

The urn on which the ship appears is one of some eleven hundred used for cremation burials uncovered in Level F, Early Period, and dated by the site's Danish excavators to ca. 1200–1075 B.C.[99] The material culture found in this level includes European-style fibulae and flange-hilted swords.[100]

In his review of Riis's excavation report W. F. Albright set the tone for interpreting the Hama cremations as a local phenomenon:[101]

> The reviewer is thus heartily in accord with the methods followed by the author in establishing the relative chronology of the four cemeteries with cremation burials. He is also in broad agreement with the absolute chronology proposed on 192ff. However, he considers these dates as tending to be too low, especially for the date at which Period I began. This is placed by the author about 1200 B.C., and attributed to the occupation of Hamath by the "Sea Peoples." It must be emphasized that Riis does not try to identify the latter specifically, but derives them in general from Anatolia. There can be no question that he is correct in recognizing a certain amount of Anatolian influence, but in the reviewer's opinion there is nothing to attach this phase to the "Sea Peoples." In the first place, it is improbable that the irruption of the Sea Peoples in the first decades of the twelfth century B.C., after the fall of the Hittite Empire, extended so far inland

A

Fig. 2.29: (A) A northern warrior, perhaps representing the Karkisha, lies fallen on the battlefield beneath the horses of Ramses III's chariot during his First Libyan War in Year 5. (B) Detail of the figure (NTS). After *MH* I: pl. 19.

B

Fig. 2.31: Tjemhu wearing phallic sheaths (NTS). From Bates 1914: 124 figs. 20–23.

Fig. 2.30: Tjemhu displaying their typical feathers, sidelock, and phallic sheath (NTS). From Bates 1914: pl. III.

> as the valley of the Middle Orontes—except perhaps for occasional forays from the coast. That Alalakh and especially Ugarit were destroyed at that time appears certain, but they were located on or near the coast. Moreover, northern Syria had been part of the Hittite Empire from its conquest by Suppiluliuma about 1370 B.C., and after the fall of the Empire the old traditions were maintained for centuries by local dynasties whose kings bore Hittite names and who wrote in Hittite characters and worshipped Hittite deities. Hamath was one of the most important of these successor states. Since the Hittites are known to have practiced cremation . . . it is hard to see any special reason why cremation should have been introduced after the fall of the Hittite Empire rather than before.

P. A. Bieńkowski, writing more than three decades after Albright, echoes his view:[102]

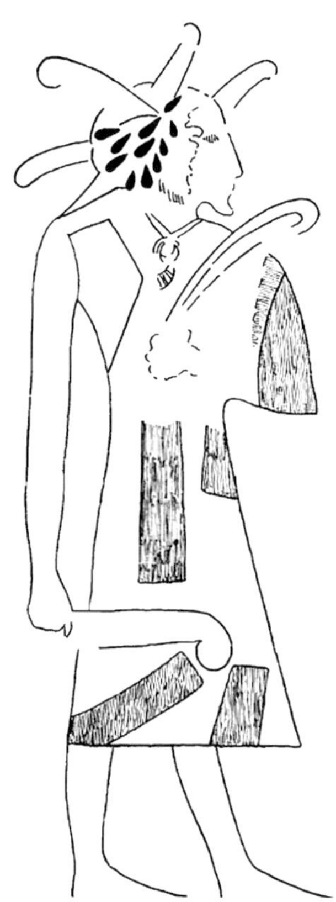

Fig. 2.32: Tjemhu chief wearing four ostrich plumes in his hair (NTS). From Bates 1914: pl. V: fig. 2.

Fig. 2.33: A Helladic galley painted on a funerary urn found at Hama, dating to ca. 1200–1075 B.C (NTS). After Ingholt 1940: pl. 22:2.

Hama is the only site where cremation definitely appeared at the beginning of the Iron Age. The city was not destroyed at the end of the Late Bronze Age, but according to the excavators, on the basis of the drastic changes in Level F (*c.* 1200–*c.* 925 B.C.) it was captured by the Sea Peoples. The new elements in Level F were iron, the Naue II cut-and-thrust sword, fibulae and cremation. There is absolutely no reason why the first three could not have been obtained through trade rather than capture. In fact, the main argument for postulating a conquest by the Sea Peoples was the appearance of cremation. Once it is admitted that the connection between the two is very tentative, there is no compelling reason to suppose that the Sea Peoples captured Hama (Albright also saw nothing to connect Level F to the Sea Peoples, as he did not think they extend[ed] so far inland). Cremation was being practiced in some neighbouring areas at about that time—the Urnfield culture in Europe, Troy VI (and possibly VIIa), and of course the LBA cremations at Alalakh—and although the Troy VI cremations are comparable with those at Hama, any definite link would be very hard to prove. There certainly seems to be no link in the pottery, which according to the excavators of Hama has sufficient new elements to suggest at least some immigration, but from where remains uncertain.

The Helladic-style galley on the Hama urn indicates that *the Hama urnfield must have been created by one of the groups of Sea Peoples who are known to have used this ship type.* This removes any possible arguments that cremation

Fig. 2.34: Sea battle scene on a krater from Pyrgos Livanaton (Kynos). Late Helladic IIIC. From Dakoronia 1999: 128 fig. 3. Courtesy Hellenic Institute for the Preservation of Nautical Tradition.

at Hama developed locally while also providing a logical explanation for the otherwise incongruous appearance of the European-style artifacts in Level F. The simplest explanation for this phenomenon (Occam's razor) is that a group of migrating central European Urnfelder folk or their descendants settled in Hama. There, as at Medinet Habu, we see a direct link between elements of the Urnfield culture and Helladic style galleys employed by the Sea Peoples.[103]

Nahal Meerot.—In 1967 E. Wreschner and M. W. Prausnitz identified a number of rock graffiti in Nahal Maarot on Israel's Carmel Coast. One of these depicts a ship.[104] Since then, Artzy reports finding many additional ship graffiti in this area.[105] She identifies three ship types. At present it is not clear whether any of these graffiti can be considered galleys, however. One hopes that this valuable group of ship representations will receive the attention they so richly deserve.

Beit Shemesh.—A limestone seal uncovered in the early excavations at Beit Shemesh bears an engraving of a Helladic galley (Fig. 2.35).[106] Despite its simplistic rendering, the representation contains sufficient detail to indicate clearly its engraver's intent.

The ship faces right. It has a vertical stem and a curving stern. The hull consists of a single line from which rise four vertical stanchions, indicating the ubiquitous open rowers' gallery. Below the hull are short lines indicating four corresponding oars. Standing amidships and in the stern two men hold to the viewer's left, that is, toward the stern, vertical curving objects (shields or, less likely, bows?). What might represent the open rowers' gallery of a second, incomplete vessel floats vertically over the ship's bow. At any rate, this element does not appear to be part of the first ship. A series of dots appears above and below the galley's stern.

Ekron.—Two locally made Late Helladic IIIC1 sherds found in recent excavations at the Philistine Pentapolis site of Ekron (Mikne) contain parts of a Helladic galley.[107] One sherd depicts the vessel's mast with a brailed sail furled to the yard (Fig. 2.36: A). Parallel lines on either side of the mast may represent halyards or perhaps, more likely, brailing lines tied to cleats at deck height adjacent to the mast. To the right of the mast and rigging, part of a freestanding wale survives. The second sherd bears a horizontal line with the remnants of three vertical stanchions attached to it.

Penelope A. Mountjoy and M. Wedde interpret a narrow triangle, placed on the diagonal in relation to the ship's structure, as the blade of the vessel's quarter rudder (Fig. 2.36: B). This reconstruction would place it at an incongruous angle to the hull, with the blade forward of the loom. They explain this anomaly as due to damage caused to the ship in battle, for which, however, the sherds themselves offer no evidence. Still, this identification finds support from ships represented on sherds from Phylakopi on Melos, which have quarter rudders appearing in this

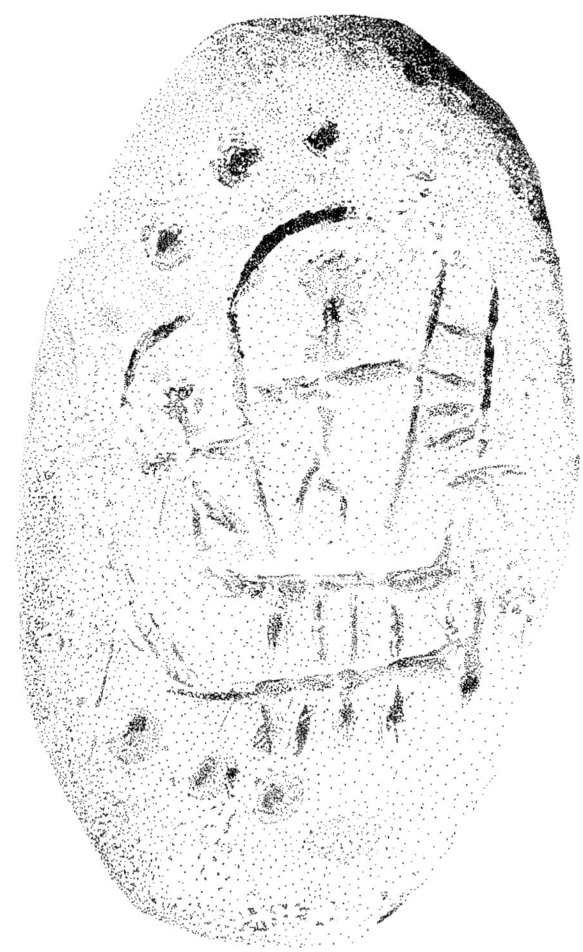

Fig. 2.35: Helladic galley engraved on a seal from Beit Shemesh, Israel (NTS). After Keel 1994: 34 fig. 20.

same unusual manner (Fig. 2.37: B).[108] I prefer to see this triangle as representing the galley's sternpost recurving over the hull in a manner for which good contemporaneous and later parallels exist.[109]

Ashkelon.—A post capped by a bird-head device appears on a sherd found in Ashkelon.[110] The sherd derives from fill. It was originally dated to the Late Helladic IIIC 1b period and considered to be of local manufacture, but more recently it has been assigned a Late Helladic IIIB date.[111] The stempost approaches the horizontal. In this it is reminiscent of the posts on the Late Helladic IIIB Levanto-Helladic-style krater from Enkomi, discussed later (Fig. 2.38). A figure, now preserved only up to his thighs, appears on the Ashkelon sherd stands on the post. Behind him is part of an arc, perhaps representing the profile of a shield or a bow.[112]

The Gurob Model: A Comparison of Characteristics

This section compares the Gurob model's characteristics to the corpus of Helladic/Sea Peoples' galley representations.

Rockered keel.—The Gurob model's long and narrow hull has an impossibly exaggerated rockered shape (Figs. 1.2, 9).[113] While the model probably reflects the prototype vessel to some degree, one should keep in mind that Egyptian representations of watercraft, whether drawn or in the round, tend to emphasize a severe—often overstated—sheer.[114] This tendency holds true for Egyptian depictions of seagoing ships, both indigenous and foreign.[115]

Helladic/Sea Peoples' galleys display two hull shapes. Many two-dimensional ship depictions show a clearly rockered hull (Fig. 2.39). Other images display galleys with straight hulls (Fig. 2.26, 40). Note, though, that the straight-hull shape tends to appear either on galley representations that are constrained in shape by the available space—as, for example, the Tragana galley, which appears on the side of a *pyxis,* and the Gazi ship, which appears on the narrow rectangular side of a *larnax*—or on ships represented in a severely linear manner, such as the Dakhla

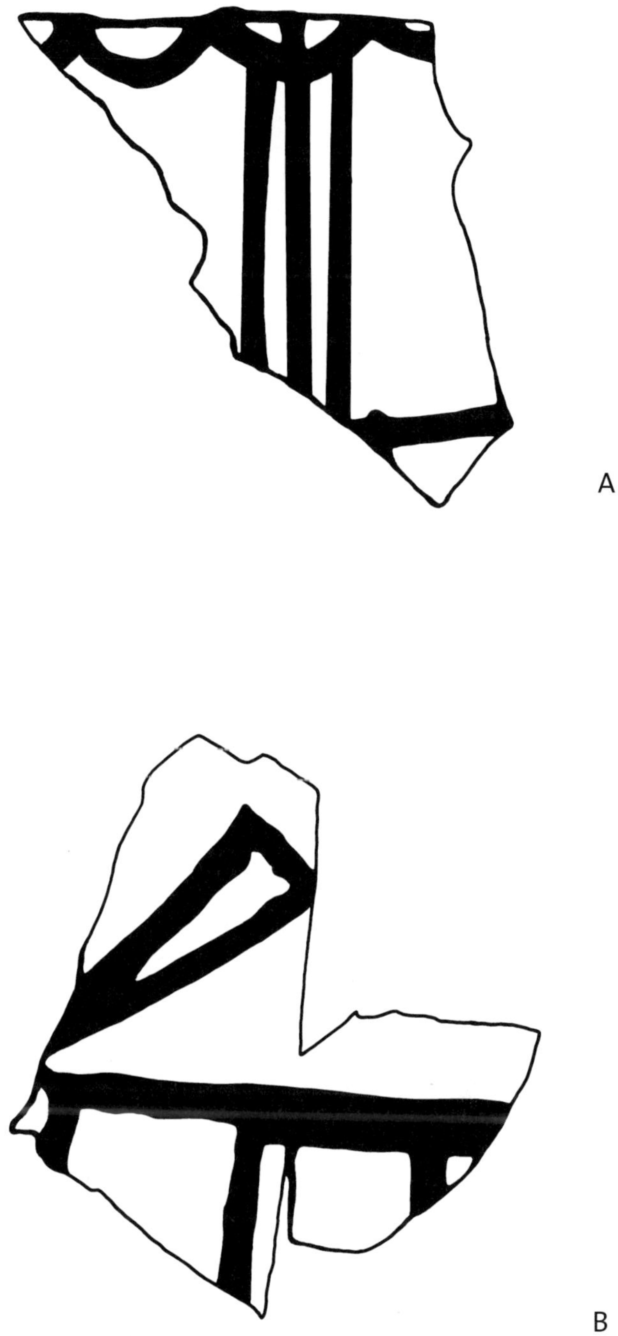

Fig. 2.36: Fragments of a galley on sherds from Ekron, Israel. Late Helladic IIIC1 (NTS). After Mountjoy 2006B: 85.

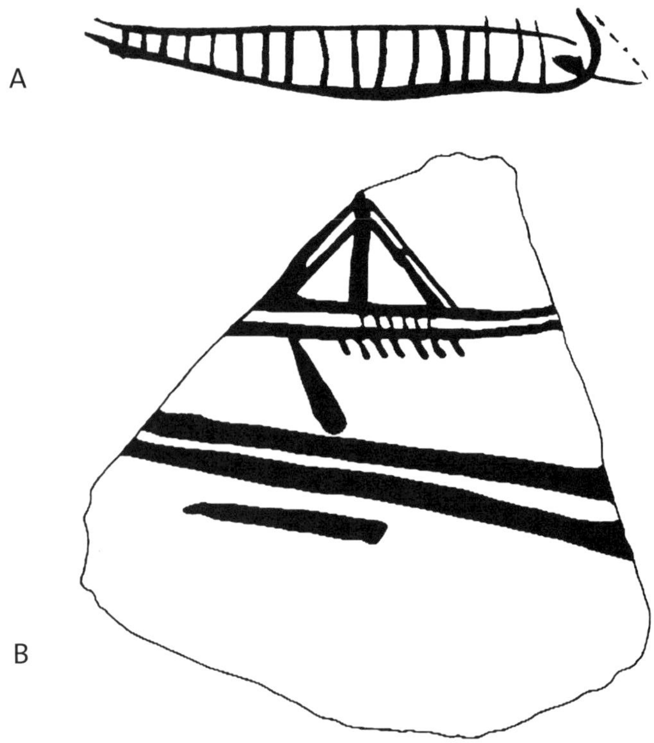

Fig. 2.37: Representations of Helladic craft sometimes expand the "horizontal-ladder" decoration (open rowers' gallery) out of all proportion to take up the length of the entire ship. (A) Galley on a Late Minoan *larnax*. (B) Galley on a sherd from Phylakopi (NTS). A after Gray 1974: G47, Abb 11. B after Marinatos 1933: pl. XIII: 13.

Oasis ship graffito (Figs. 2.10, 40: A, C). These considerations may have mitigated against portraying a rockered hull.

Models of the Helladic ship type usually show rockered hulls (Fig. 2.39: D).[116] Cypriot *askoi* in the shape of galleys have straight hulls with a curving stern (Figs. 2.40: E; 2.41: B–C). This may be a result of the generic *askos* shape. Given the evidence, the most we can say is that perhaps some of these ships had a lightly rockered hull when viewed in profile, whereas others did not.

Waterline stem projection.—In many ways Helladic-style galleys of the Late Bronze and Early Iron Ages are seen in a state of transformation, with certain components appearing on some ships but not on others. Such is the case of the waterline stem projection that appears prominently on the Gurob ship model (Figs. 1.6: A, 7, 9, 10: A, 11: B, 12: C–D).

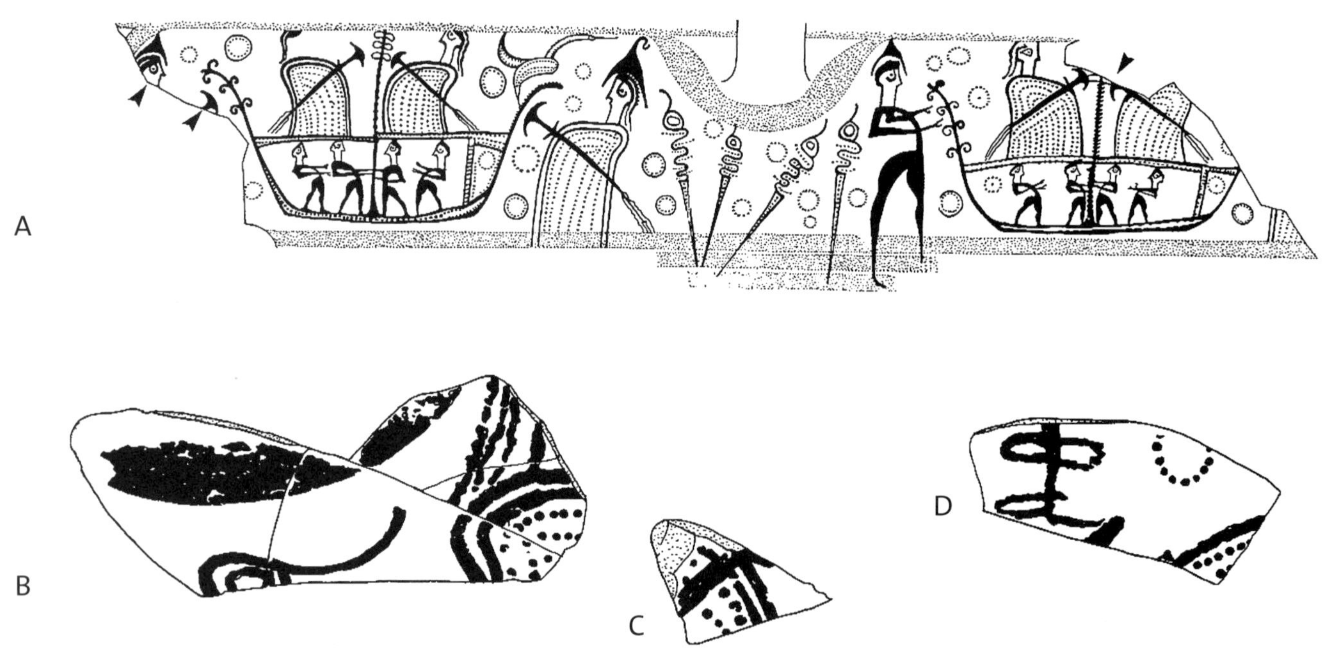

Fig. 2.38: (A) Two ships (galleys?) painted on an amphoroid krater from Enkomi, Tomb 3 no. 262. Late Helladic IIIB. (B) Additional sherds discovered after the original excavation (NTS). A from Sjöqvist 1940: fig. 20: 3. B–D after Karageorghis 1960: pl. X: VII.

A

B

D

C

E

Fig. 2.39: Examples of Helladic galleys represented with rocked hulls. (A) Sea Peoples' galley from Medinet Habu. Twelfth century B.C. (B) Ship painted on a stirrup jar from Skyros. Late Helladic IIIC. (C) Graffito of galley, facing right, carved on a stele from Enkomi. Late Cypriot III. (D) Terra-cotta ship model from Tiryns. Late Helladic IIIB. (E) Galley painted on a krater from Pyrgos Livanaton (Kynos). Late Helladic IIIC (NTS). A after *MH* I: pl. 39. B after Hencken 1968B: 537 fig. 486. C after Schaeffer 1952: 102 fig. 38: C. D after Kilian 1988: 140 fig. 37: 8. E after Wachsmann 1998: 131 fig. 7.8: A.

Fig. 2.40: Examples of Helladic galleys represented with straight hulls. (A) Galley painted on the side of a *larnax* from Gazi. Late Minoan IIIB. (B) Ship on a gold diadem found in Trench W-31 at Pylos. (C) Galley painted on a *pyxis* from a tomb at Tragana, Pylos. Late Helladic IIIC. (D) Galley painted on a sherd from Phaistos. Late Minoan IIIC. (E) *Askos* in the form of a Helladic galley from Lapithos. Late Cypriot III (NTS). Photo A: S. Wachsmann. B after Blegen et al. 1973: fig. 108: D. C after Korrés 1989: 200. D after Laviosa 1972: 9 fig. 1b. E after Basch 1987: 150 fig. 314.

Some ship depictions carry this element, whereas it is absent on others (Figs. [with] 2.10, 33, 39: C–D, 40: A, C–D, 41: B, 42: A, 43, 44: A–B; [without] 2.3–7, 34–35, 38, 39: B, 41: A, C, 45). The projection sometimes appears horizontal, while in other examples it curves upward. In all cases, it seems to be a continuation of the keel. Such horizontal bow projections are unknown on the iconography of Egyptian ships until the Late Period.[117]

The Late Helladic projection has led to considerable confusion: Some scholars claim that this represents the introduction of the waterline ram.[118] Two Late Helladic/Late Cypriot III ship representations have been identified as bearing waterline rams:

- A Late Helladic IIIC vase from Asine bears a ship with a long horizontal projection at its left, which has been identified as a waterline ram (Fig. 2.44: A).[119] Others believe the ship to be sailing to the right.[120] The direction of the billowing sail and other details indicate that the ship is indeed sailing to the right and that the horizontal extension is best understood as a quarter rudder strung out behind the ship (left), as on the Gazi ship (Fig. 2.40: A).
- Basch and Artzy identify a ship graffito incised at Kition in Temple 1 as depicting a ship with an immense ram (Fig. 2.44: B).[121] This ship, however, appears to be facing in the opposite direction (right), in which case the "ram" is best interpreted again as a quarter rudder.[122]

L. Casson notes that the fact that some of these ships have these bow extensions, while others do not, indicates that the naval weapon par excellence of the ancient Mediterranean world had not yet appeared, for just as the naval cannon changed the rules of warfare, so, too, did the waterline ram.[123]

Additional evidence also suggests that Bronze Age galleys were not outfitted with waterline rams. Late Bronze and Early Iron Age Helladic ships show little indication of the additional hull strengthening required to survive the enormous forces acting upon a ramming ship at the moment of impact. Albeit later in date, the surviving timbers of the third-century B.C. Athlit Ram

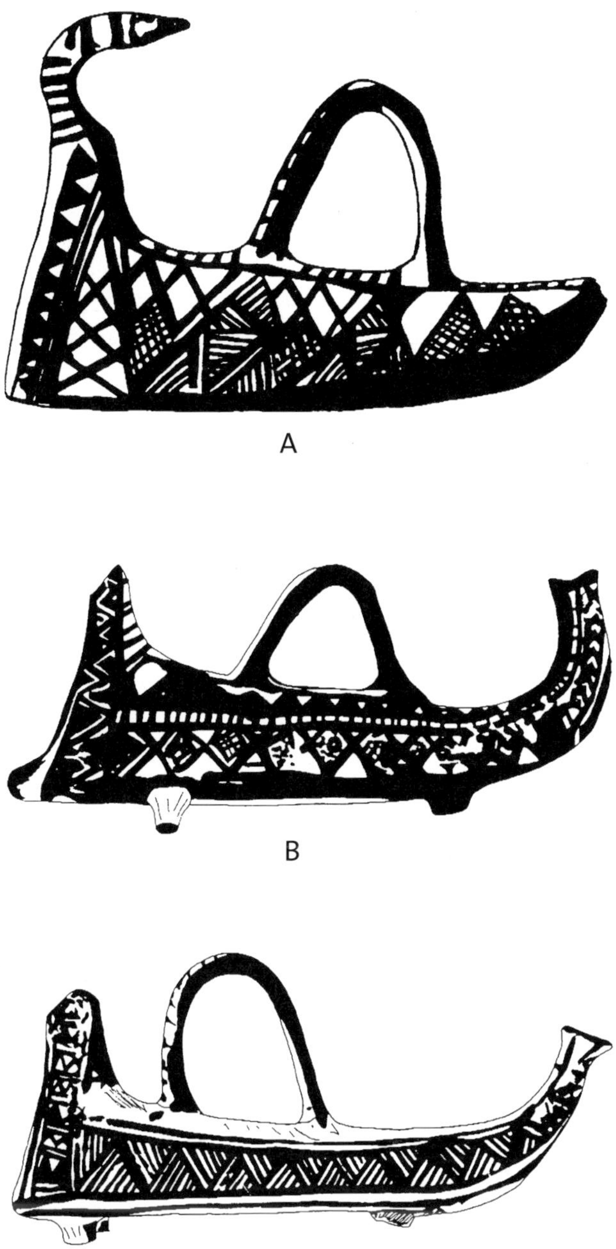

Fig. 2.41: Cypriot *askoi* in the form of Helladic galleys. Late Cypriot III (NTS). (A) From Lapithos. (B) Provenance unknown. (C) From Lapithos (NTS). A after Göttlicher 1978: Taf. 9: 149. B–C after Buchholz and Karageorghis 1973: 470 figs. 1720–1721.

demonstrate the immense effort made by its shipwrights to strengthen the ship's bow to allow it to withstand the impact of ramming.[124] A ramming timber forward of the keel helped dissipate the energy at impact by transferring it to two substantial wales that arched downward toward the stem/keel joint. The only hint of such construction in the period under discussion here is a bow fragment of a ship model found on the Acropolis with painted lines sweeping downward from the bow, *perhaps* indicating such wales.[125]

Furthermore, a number of the galleys represented with bow extensions have stem ornaments or, as in the Gurob model, a forecastle deck that protrudes out to, or even over, their horizontal stem projections (Figs. 1.7: A–B, 12: B–C; 2.33, 39: C–D, 40: A, D, 42: A). Such projecting features would have been damaged during a ramming event and, therefore, indicate that this was not a concern.

What purpose, then, did the bow extension serve, and what relationship does it bear to the eventual introduction of the ram? The appearance of the bow extension may be viewed in terms of the development of Helladic galley construction and the uses of these vessels. The textual evidence indicates that Mycenaeans and Sea Peoples used ships for acts of warfare and piratical attacks.[126] In these commando-style raids on coastal settlements, galleys may have been rowed bow first right up onto beaches. This type of landing is described in the *Odyssey* and appears as part of a cult celebration in the Miniature Frieze at Thera.[127]

G. S. Kirk believes that the bow extension was meant to protect the keel/stem scarf and to allow a vessel to slide up safely onto a sandy shore at speed.[128] J. Hornell, in his discussion of the bow and stern extensions of the fourth-century-B.C. Danish Hjortspring Boat, concludes that "these early boatbuilders had not yet learned to curve up the ends of the keel into a stem and stern post respectively. Instead they continued the keel into a long ram-like structure projecting several feet beyond the hull at each end."[129] In addition, J. R. Steffy proposes that extending the keel enhanced the galley's longitudinal control.[130]

The bow extension may simply have been a solution for handling the attachment of the stem to the keel of ships of all sizes, shapes, and purposes. In Classical antiquity, such bow extensions appear on every type of ship and boat from diminutive fishing boats to Brobdingnagian seagoing traders, none of which served a military purpose.[131]

Whatever the reason for its introduction, the bow extension on Helladic galleys may be considered a protoram, for this modest idiosyncratic constructional feature eventually evolved into the waterline ram. Scholars have debated when, where, and by whom the waterline ram developed as a weapon.[132]

We will probably never know exactly how the modest bow extension evolved into the greatest naval weapon of the Classical world. Once the concept of ramming took hold, however, it would have gradually led to the changes necessary to sustain the ship through a ramming event—changes that are evident in a mature form in the timbers of the Athlit Ram. In the early days of ramming, attacking galleys may have been in just as much danger of being damaged as were their victims.

Perhaps, as Casson suggests, Phoenicians invented the waterline ram as a weapon.[133] First-millennium-B.C. Phoenician oared ships evolved from the Helladic-style galley, originally brought to the shores of the Levant by the Sea Peoples. The Hama galley indicates that at least some of these introduced vessels bore waterline bow extensions (Fig. 2.33). Starting in the twelfth century B.C., galleys based on the common Helladic design developed independently in the Greek and the Phoenician spheres of influence.[134] If, indeed, the Phoenicians developed the waterline ram, this would fit a pattern of Levantine innovation in ship construction and seafaring.[135]

The waterline ram made its appearance no later than the end of the eighth century B.C. The earliest incontrovertible evidence of the use of a ram in battle, in my view, is a colorful, although poorly preserved, wall painting from Til Barsip dating to ca. 700 B.C. The painting portrays a Phoenician bireme with a copper or bronze (yellow) covered ram, apparently in the act of ramming an enemy ship, the oars of which appear at the bottom right of the scene.[136] The first reference to rams used in a sea battle is Herodotus's description of the battle of Alalia, which took place in 535 B.C.[137]

A

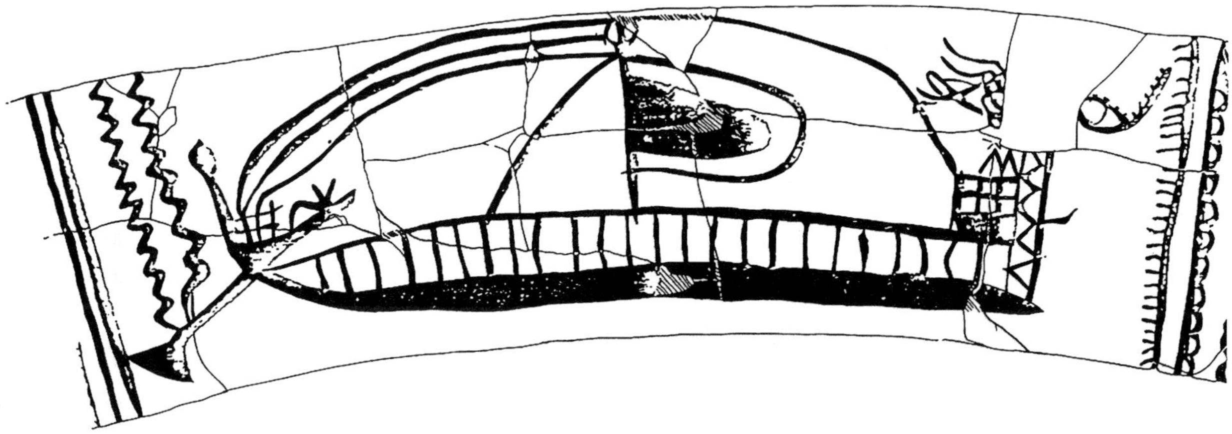

B

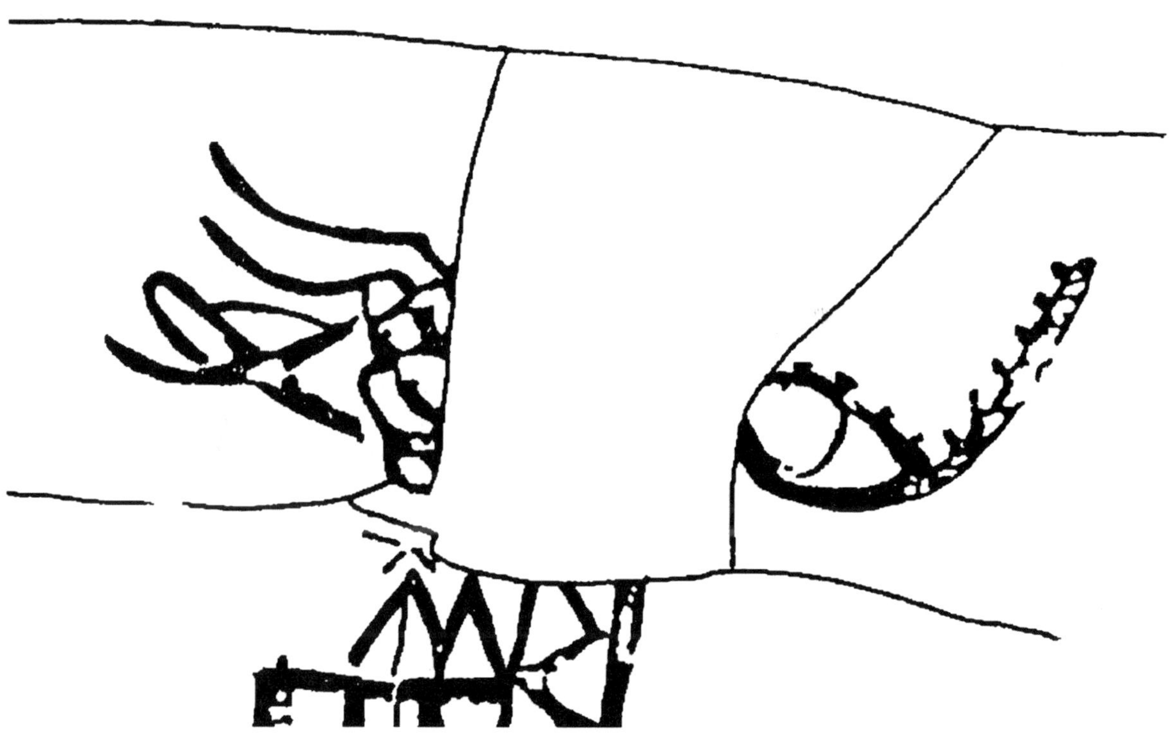

Fig. 2.42: (A) Helladic *pentekonter* painted on a pyxis from a tomb at Tragana, Pylos. Late Helladic IIIC (NTS). (B) Detail of stem bird ornamentation. After Korrés 1989: 200.

Fig. 2.43: The bow of a Helladic galley painted on a sherd from Pyrgos Livanaton (Kynos) has a series of downward-facing crescents (oarports?) above the oars. From Dakoronia 2002: 290 fig. 10. Courtesy Hellenic Institute for the Preservation of Nautical Tradition.

A

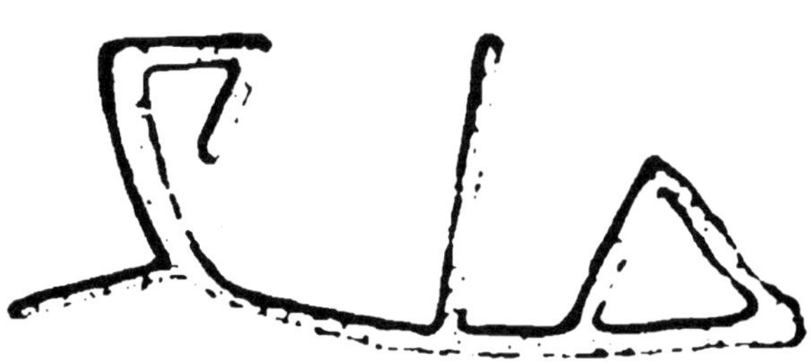

B

Fig. 2.44: Helladic galley representations suspected of bearing waterline rams. In both cases the suspected ram is best understood as a quarter rudder strung out behind the vessels to their left. (A) Galley painted on a vase from Asine. Late Helladic IIIC. (B) Ship graffito "P" from the southern wall of Temple 1 at Kition. Late Cypriot III (NTS). A after Casson 1995: fig. 29. B after Basch and Artzy 1985: 332 fig. 8B.

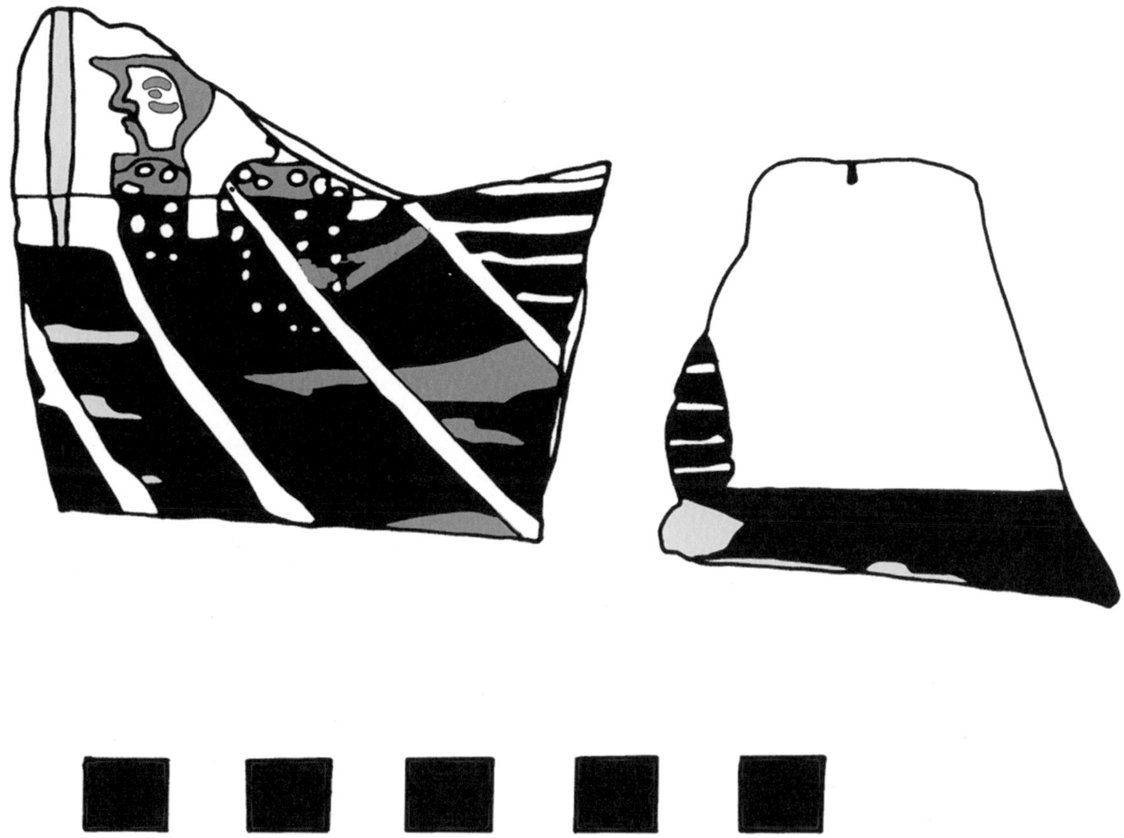

Fig. 2.45: Galley with rowers on sherds from Tiryns. Late Helladic IIIB2 End (NTS). After Güntner 2006: 179 fig. 5.

Forecastle.—The Gurob ship model has a predominant forecastle, or *ikria,* that once had a screen or fencing (Fig. 1.11, 12, 13: A).[138] Forecastles constitute a distinctive, as well as a common, element in many two-dimensional representations of Helladic craft (Figs. 2.3–7, 10, 34, 38: A[?], 39: C, 42: A, 44: B). Due to the basic nature of most Aegean-style galley models, castles rarely make an appearance. Several Cypriot galley-shaped *askoi,* however, have horizontal lines or a blotch of paint on the inboard side of their stem, perhaps indicating a forecastle (or the framework for one) nestled behind the stempost (Fig. 2.40: E, 41: B–C).

Often the fencing seems to be of light construction. On occasion a vertical structure in the bow ends at what appears to be deck level (Figs. 2.26, 38). The meaning that this is meant to convey is unclear, but this detail appears with sufficient frequency to indicate that it is not an artist's whim. Perhaps it indicates that the forecastle screen was removable.

The castle screens of the Medinet Habu Sea Peoples' ship representations rise to midthigh on the warriors standing behind them on the castle decks (Figs. 2.3–5, 7). The same holds true for the Egyptian ships in this scene.[139] This is probably a result of the figures' having been represented to a scale greater than that of the ships.

Geometric period representations of galleys, although presumably suffering from the same phenomenon of sizing, also suggest that the forecastle screen was a fairly low affair and offered scant protection (Fig. 2.46).[140] This forecastle screen may have been a light framework covered with fabric, as figures are hidden behind it.

Some New Kingdom ships carried fenced castles fore and aft. Hatshepsut's seagoing Punt ships, depicted on her mortuary temple at Deir el Bahri, have castles that come up to just below the waists of the figures standing in them.[141] The manner in which the forecastle deck overhangs the hull of the Gurob model seems to have been borrowed from New Kingdom Egyptian ship modeling.[142]

Stempost.—As already noted, the Gurob model's vertical stem is planted directly in the center of the forecastle deck (Figs. 1.2, 6: A, 7, 9, 11: A–C, 12: C, 13: A).[143] The stem's location in the middle of the forecastle deck is clearly the result of considerations related to the construction of the model—locking the deck—rather than to that of an actual ship. Most representations of Helladic galleys show stemposts that are vertical or canted slightly, usually with a forecastle nestled behind them (Figs. 2.3–7, 10, 12, 26, 34, 43, 45, 47).[144] The stem's bird-head finial and its perpendicular position relative to the hull fit well with what we know of Helladic galleys.

Open rowers' gallery/stanchions.—The most impressive element of Helladic galleys was a horizontal band bisected by a series of vertical lines located between the caprail and a (usually fenced) partial deck:

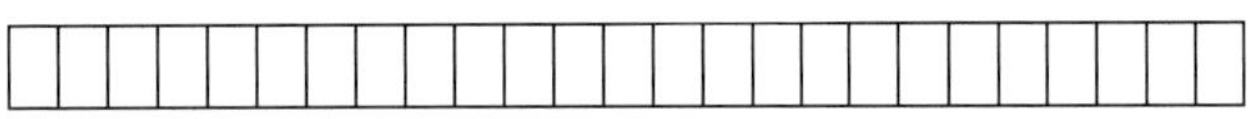

We know this because this motif, which looks rather like a ladder lying horizontally, appears on many, if not most, of the images of Helladic galleys that have come down to us. In some cases artists created images of galleys that consist of little more than this motif blown out of all proportion, with the hull itself drawn as little more than a narrow line (Figs. 2.33–35, 37, 40: A, 42: A, 47).[145] This same "horizontal ladder" motif also appears in paint on contemporaneous terracotta models of ships (Fig. 2.41: B).[146]

In 1981 I proposed that the horizontal ladder pattern represented an open rowers' gallery intersected with vertical stanchions between the rowers.[147] At the time, I based this hypothesis primarily on the evidence from Medinet Habu, where capsized ship N.3 has three crewmembers intertwined with elements of its architecture in ways that indicate that the vessel had an open rowers' gallery topped by a fence (Fig. 2.6).[148] Since that time, the discovery of more detailed representations of galleys, particularly from Bademgediği Tepe and Pyrgos Livanaton (Kynos), discussed later, have confirmed this hypothesis. The Gurob model, complete with stanchion holes in its caprails and some of the actual stanchions themselves, represents the only known three-dimensional evidence of this crucial feature of Helladic galley construction.

The stanchions are missing on Medinet Habu N.3, but the superstructure above the open rowers' gallery (fencing and a partial deck, along with the wales and beams that would have supported them) presupposes their existence, indicating that their absence is due to artistic, rather than structural, considerations. The stanchions may have been indicated in now-missing paint or, perhaps more likely, they may have been dropped entirely so as not to further confuse an already complicated presentation. When Geometric artists portrayed rowers at their benches, for example, they normally omitted stanchions but included them when the rowers are absent.[149] Oarsmen themselves appear in the following handful of Helladic galley representations:[150]

Pyrgos Livanaton (Kynos).—The site of Pyrgos Livanaton, identified as Homeric Kynos, in East Lokris, has revealed a remarkable collection of Late Helladic IIIC ship representations.[151] Two galleys on a krater from Kynos are among the most detailed representations of Helladic galleys (Fig. 2.34).[152]

The ships on the krater portray an unpainted register decorated with the typical horizontal ladder pattern but also with a series of "lunettes," each of which continues into an oar. The lunettes must represent the torsos of the rowers with their heads hidden behind the deck screen, a phenomenon for which good parallels exist in later iconography.[153] The lunettes in these galleys curve in opposite directions on the two ships, a detail that may indicate that one of the ships is backing water, that they are shown at different moments of their stroke, or that they are drawn antithetically. Another galley representation from Kynos shows oars attached at deck level (Fig. 2.26).[154]

Bademgediği Tepe.—A Late Helladic IIIC Pictorial-style krater from Bademgediği Tepe in western Turkey bears representations of two antithetical ships, of which the one on the left has been published (Figs. 2.47).[155] At present this is our clearest two-dimensional representation of rowers plying their oars below deck level from an open rowers' gallery intersected with vertical stanchions.

The galley faces right. Warriors, facing right, stand above

Fig. 2.46: Geometric period galley bow (NTS). Drawing: M. Kofahl. National Archaeological Museum, Athens.

the oarsmen, indicating the existence of at least a partial deck. The horizontal baseline on which the warriors stand must represent a midline deck on beams slotted into wales supported by the stanchions. A screen is not in evidence.

Horizontal bands cross the ship's stempost, which terminates in a forward-facing, bird-head device with a prominent eye, a recurving vertical beak, and numerous protuberances along the top of the head and beak.[156] The area where one might expect to find a forecastle is missing.

Two of the published sherds carry parts of what appear to be sternposts. One of these is reminiscent of the sternpost on one of the Khanniele Teke galleys.[157] The other sternpost stands upright and is bifurcated at the top (Fig. 2.47: B, leftmost sherd). This finds an almost exact parallel in the sternpost of one of the galley representations from Kynos (Fig. 2.26).

Both the rowers and the combatants wear what may be interpreted as feathered helmets similar to those worn by groups of Sea Peoples in the Egyptian monuments and by figures on sherds from the Seraglio at Kos (Fig. 2.40: D, 51).[158]

Tiryns.—The site of Tiryns adds two additional representations of rowers to the repertoire of Helladic galleys. Late Helladic IIIB2 End sherds bearing parts of a galley were found near the Stadt Nordwest (Fig. 2.45).[159] Remarkably, white paint is used for some details. The surviving part of the vessel faces right and shows a portion of the forward section of the hull. If it is indeed part of the same ship, a second fragment bears the bottom of the stem at the waterline and indicates that the prototype galley had a blunt bow lacking a horizontal bow extension. The bow section is crossed by a series of horizontal white lines, reminiscent of groups of lines visible behind the forecastles and stemposts on Late Helladic and Geometric galleys, which perhaps were intended to represent light railings (Figs. 2.26, 41: A–B, 44: A, 46).[160]

On the surviving part of the Tiryns ship, however, the lines do not continue to the aft side of the stem. The lines may simply represent a painted decoration, similar to that found on a Protogeometric model from Cyprus.[161] This raises the possibility that this detail is missing in other representations due to their monochromatic nature. Parts of three rowers and four oars survive. The oarsmen are represented at the beginning of their stroke. The two best-preserved rowers bear a white-dot decoration. Just aft of the best-preserved rower is a vertical line, presumably representing a stanchion or a mast.

A third sherd from Tiryns portrays the heads and upper torsos of four men who most likely represent rowers (Fig. 2.49).[162] Part of the hull, crossed perhaps by an oar, survives between the second and third figures. An oblique line may represent rigging.

Enkomi.—The groups of men standing below decks on two ships painted on a Late Helladic IIIB krater from Enkomi may represent rowers (Fig. 2.38).[163] The men in each of the ships stand in the same manner but face each other in pairs on either side of the vessels' masts. Due to their seemingly deep hulls, these ships have been repeatedly identified as round merchantmen.[164] As Kirk correctly notes, however, the hulls "clearly owe more to artistic freedom than to acute observation."[165]

Indeed, given the armed and helmeted personnel in, on, and around the ships, the lack of any evidence of cargo, and the similarities that these ships bear to other Helladic galleys (e.g., bow bird device and stern decoration), these vessels apparently represent galleys, albeit changed significantly by the artist. The men in these scenes are depicted heraldically. Thus, the facing pairs of men below decks are placed in this manner due to artistic considerations and may actually be facing in the same direction. We learn nothing from the figures' posture as this is conventionalized.[166] Given their serialized stance and their position below deck in what is probably a galley, it is likely, in my view, that these figures represent oarsmen.[167] If this interpretation is correct, then the vessels on the Enkomi krater are only two more in a series of examples of the aggrandizement of the vessels' open rowers' galleries at the expense of the overall integrity of Helladic galley images.

Mountjoy has identified ship fragments painted on sherds dating to the early Late Helladic IIIC Middle excavated by Dikaios at Enkomi.[168] These images may depict parts of two ships. One sherd bears a bird-head post decoration (Fig. 2.50: A). It has a large eye and appears to have two beaks, a situation otherwise undocumented on Helladic ships but one that should not surprise us, as in later times multiple beaks are a common occurrence.[169] The extremity of the "upper beak" ends in a splotch, which may be accidental, although at least two and possibly three additional examples exist of a stempost device bird's beak widening at its extremity (Fig. 2.25–26, 47?). The upper surface of the beak and head is covered with protuberances, a frequent occurrence on Helladic galley representations.[170]

Another sherd depicts most of one rower and an arm, oar, and part of the head of a second rower (Fig. 2.50: B). The sherds have been reconstructed as belonging to a single galley, but that may not be the case. Both oarsmen face right. *Assuming normal rowing circumstances,* that orientation indicates that their galley is advancing to the left. Thus, if the bird-head device on the first sherd faced outboard from the stem, it would have to belong to a second galley, although possibly painted on the same krater. On the other hand, we have evidence from Cyprus of the inversion of the stem bird-head device to face the stern around this time.[171]

Kos.—A galley portrayed on a Late Helladic IIIC sherd from the Seraglio at Kos bears rowers working their oars from what appears to be deck level (Fig. 2.51).[172] A second sherd perhaps represents parts of two rowers.[173] As noted earlier, one of the Kynos ships shows the oars attached at deck level, and the rowers on a sherd from Kos appear to be working their oars from deck level (Figs. 2.26, 51). This is a significant development as the ability to row these vessels either from the upper level or from the open row-

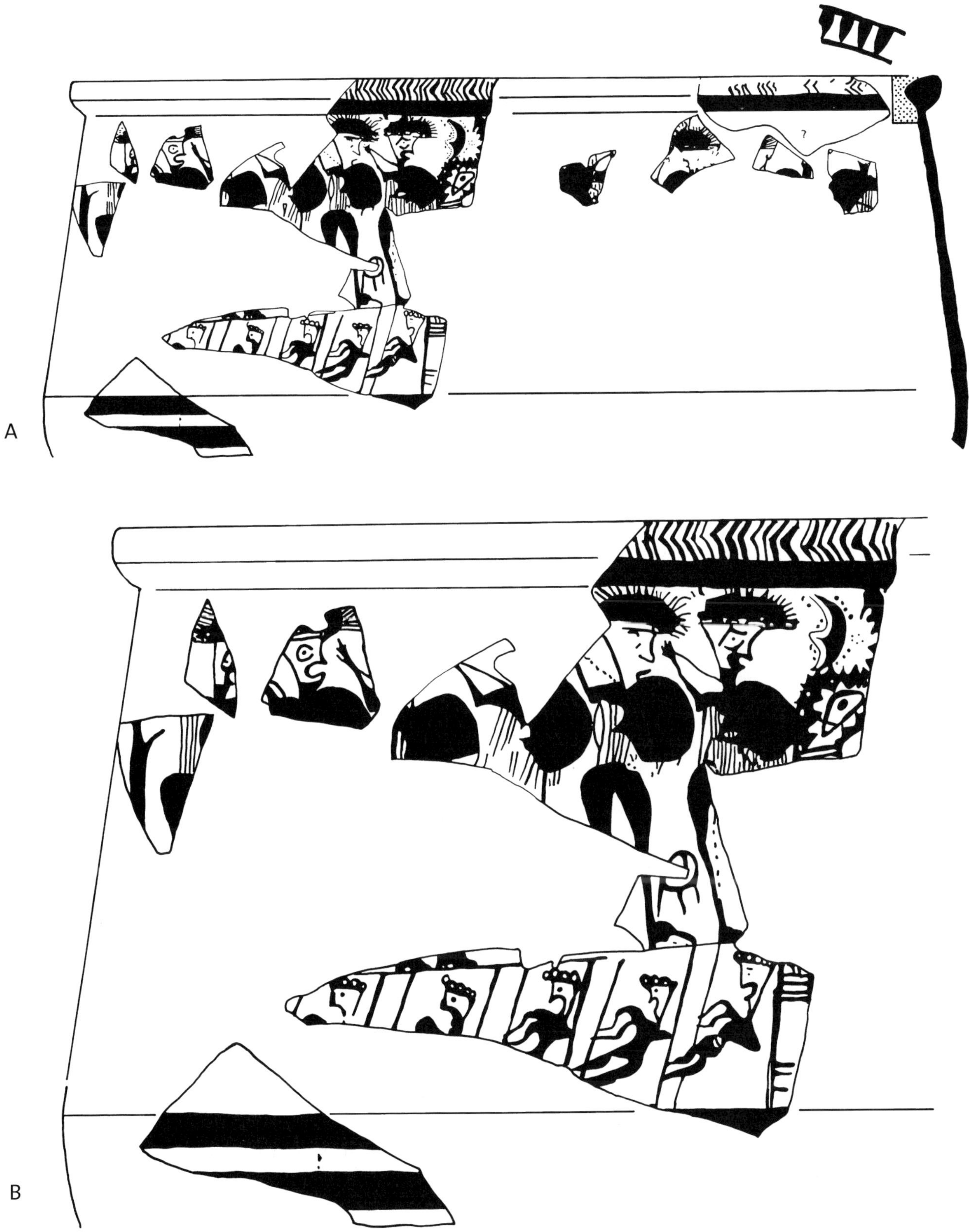

Fig. 2.47: (A) Helladic galley with oarsmen and warriors on Bademgediği krater FS 282. Late Helladic IIIC. (B) Detail (NTS). After Mountjoy 2005: pl. XCVI.

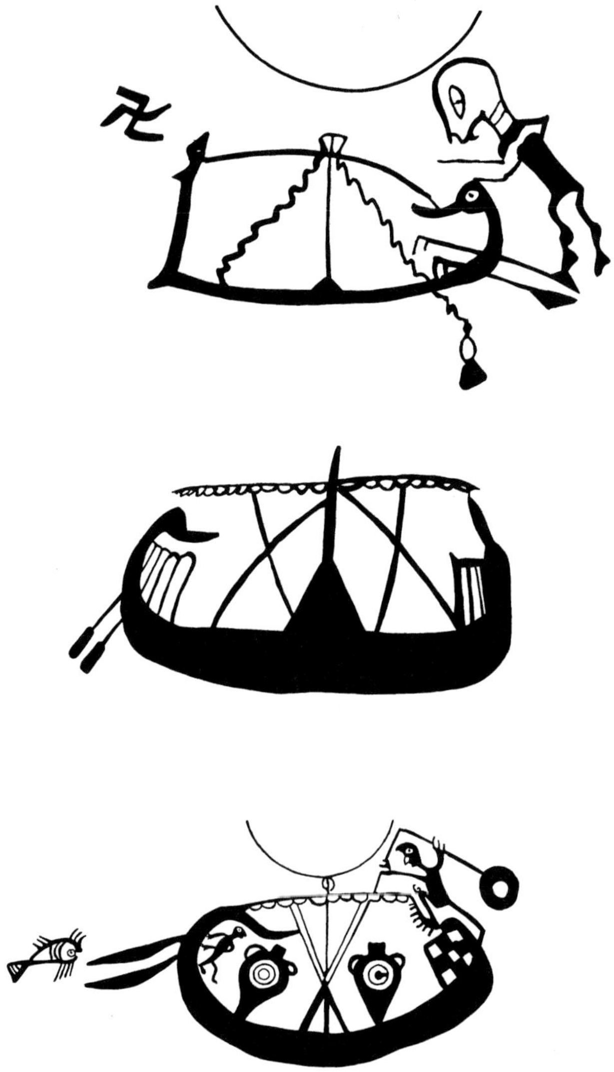

A

B

C

Fig. 2.48: Ships painted on three Cypriot jugs show a direct progression from naturalistic (A) to stylized (B) and then to abstract representations of the bird head. Seventh century B.C. (NTS). After Karageorghis and Des Gagniers 1974B: 122–123 no. 11: 1–3.

Fig. 2.49: Rowers(?) on a sherd from Tiryns. Late Helladic IIIB2 End (NTS). After Güntner 2006: 180 fig. 6.

ers' gallery would be the first step in the evolution from a single-banked to a double-banked galley. Casson notes the tendency to interchange the location of the rowers on Geometric-period galleys, but this flexibility in rower placement may have begun even earlier.[174]

Bird-head post finials.—Perhaps the single most telling detail in assigning the Gurob ship model to a Helladic tradition is its stempost, which culminates in a bird-head finial with a vertical beak (Figs. 1.4–6, 11: D). While the Gurob model's upturned beak diverges from any bird known in nature, it is in keeping with the manner in which the beaks or bird-head post finials are at times represented on Helladic galleys (Figs. 2.25–26, 39: B, 40: B, 42: B, 47: B).[175]

The stems of most Helladic ship representations terminate in outboard-facing bird heads. At times, however, a statue of a bird replaces the bird-head stem finial (Figs. 2.38: A, 42: B).[176] This may be the case with the Dakhla Oasis graffito, for, although most of the area above the forecastle is now missing, remains of feet survive, and several lines behind the forecastle could be the remains of a bird's posterior (Fig. 2.11).

There are parallels for the use of bird-shaped devices in the Bronze Age Aegean, as well as ethnographical materials from the recent past.[177] The manner in which the waterbird device sits on the stempost of the Enkomi krater, for example, is highly reminiscent of the swallow decoration attached to the ornamental bowsprit of one of the Theran ships taking part in the procession/race in the Miniature Frieze.[178] Some Theran ships carry angular decorations on their bowsprits that also appear on Minoan talismanic seals (sometimes in the form of a fleur-de-lis),

A

B

Fig. 2.50: Sherds from Dikaios's excavation at Enkomi may represent parts of two different ships. (A) Bird-head stempost device with protuberances along its upper edge facing right. (B) Remains of two rowers facing right, which means that they would normally be propelling the ship toward the viewer's left. A thick diagonal line above the oarsmen may represent a backstay or, less likely, a recurving stempost. After Mountjoy 2005: pl. XCVII: d.

which, as Basch demonstrates, represent swallow-shaped devices.[179] Their appearance on talismanic seals indicates that these swallow-shaped devices had numinous significance in Minoan religion as, we may assume, did the waterbird to the Mycenaeans and those Sea Peoples who placed it on their posts.[180]

The small protuberances on some of the Helladic bird-head devices perhaps represent diminutive bird heads as prophylactic "multipliers," which may have evolved eventually into the *aphlaston* of the Classical period.[181] If so, the bird's eye becomes the *aphlaston's* "shield," and the curving timbers represent a multiplication of beaks perhaps, again, as a strengthening of the device's prophylactic power.[182] Similarly, the volute that appears first on some sixth-century-B.C. ship representations evolved from a curved bird's head.[183]

The heads—and particularly the beaks of waterbirds—proved a remarkably resilient symbol on Mediterranean seacraft. These devices first appear on Helladic ships, whether used by the Mycenaeans/Achaeans and/or the Sea Peoples.[184] These same birds commonly appear on Mycenaean and Philistine pottery.[185] Over time we witness a remarkable, oft-repeated cycle in which the bird head first appears in its natural form, which then cycles through stylized and abstract renditions. This metamorphosis is perhaps best illustrated in a series of ship representations painted on a trio of sixth-century-B.C. Cypriot jugs, presumably all made by the same workshop (Fig. 2.48).[186] The abstract form seen here (C) is *identical* to the inward- and/or upward-curving stem and stern finials on Greek galleys in the Geometric period (Figs. 2.46, 52).[187] In other words, the Geometric galleys are capped with abstract bird heads, but now facing inboard. Little of the bird's head is left beyond the beak's compound curve. These abstract bird-head finials fit into a remarkably extensive and well-documented cultural continuum.[188] Furthermore, Homer uses the term κορωνίς to describe his warships.[189] A somewhat similar word, κορώνη, is a type of seabird. This might be an intentional play on words, suggesting that κορωνίς implies something like "having curved extremities that are bird shaped."[190]

The change in direction of the bird-head stem ornaments raises the question of the correct direction of the Gurob model's stem. Presumably the piece lay dislocated

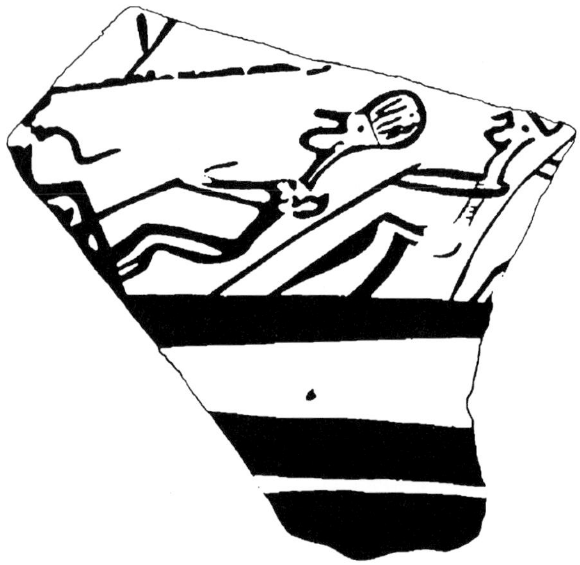

Fig. 2.51: Part of a galley with oarsmen wearing (feathered?) helmets on a krater sherd from the Seraglio, Kos. Late Helladic IIIC (NTS). After Morricone 1975: 360 fig. 358.

from the model when found, for, as already noted, Petrie reconstructed the model once with the bird head facing the bow and once with it facing the stern (Figs. 1.4: Λ, 5).[191] Which of these two reconstructions is correct?

This question is intertwined with the broader enigma of how the Helladic tradition of galley building survived the Dark Ages to reappear in a robust manner in the galleys of the Geometric period. Representations of ships are exceptionally rare during the Greek Dark Ages.[192] Yet, the warships in Greek art that reappear in the Geometric period (ca. 850 B.C.) show a clear line of descent—no doubt about it—from their Late Helladic IIIB–C ancestors.[193]

It requires a marked degree of social organization, resources, and leadership to build, maintain, and man a large galley. Yet, during the Iron Age, Greece underwent severe depopulation, a loss of its palace-based economy, and a movement to small villages.[194] The societal climate following the collapse makes it difficult to imagine a continuum of this shipbuilding, tradition for it seems inconceivable that any communities in Greece at that time could sustain the expense and effort required to build and maintain galleys. This hiatus presumably would have created a loss of the living knowledge base for constructing the Helladic style galley.

Wedde, however, proposes that the continuation of this quintessentially Greek galley tradition, along with that of chariots, indicates a limited but continued existence of the Bronze Age Mycenaean status quo during the intervening period:[195]

> The attested continuity in ship architecture, and the probable but inadequately documented use of chariots and other wheeled vehicles during the earlier Iron Age, indicate that the systemic collapse towards the end of the Bronze Age did not effect all forms of craftsmanship. Nor is it likely that the entire aristocracy abandoned Greece. . . .
>
> An alternative narrative stresses a basic continuity in occupation, at new sites, for short periods in conditions which left less durable traces of architecture, less splendid burials, but with a partially conserved, albeit comparatively impoverished and no doubt less heroic aristocracy, yet still capable of gathering the people to man the galleys, still motivated to engage in expenditures sufficient to cover the cost of a chariot or two, still keeping the flame of the epic past alive. To populate footnotes with concrete evidence supporting such a reconstruction of the "Dark Age" may prove difficult. Ship architecture provides eloquent data. The case for chariot use is at the moment tenuous. . . . The break in overall systemic continuity, the changes in settlement patterns, the decline in material culture are all evident. But the recovery is too rapid to support extensive discontinuity.

I propose a simpler explanation of how the Helladic galley survived the Greek Dark Ages: The tradition could have continued unabated in Cyprus, where a vibrant palace-based Achaean culture flourished throughout the Late Cypriot III and Cypro-Geometric periods.[196] There is abundant evidence for the survival and use of Helladic galleys on Cyprus in the form of representations and models, some in the form of *askoi* (Figs. 2.40:E, 41).[197] Cultural contacts between Cyprus and the Greek mainland were restored during the Submycenaean period, and it seems likely that knowledge of iron working arrived in Greece, focused in Athens, from Cyprus at this time.[198] In the same manner, galley construction could have been rein-

A

B

Fig. 2.52: (A) Geometric-period iron firedogs from Argos display inboard-facing abstract bird-head finials at bow and stern. (B) Detail of one of the bows, starboard view. A after Göttlicher 1978: Taf. 25: 338–339. B after Courbin 1957: 376 fig. 63.

troduced into Greece from Cyprus. This seems even more likely, considering that the introduction of iron into Greece in Submycenaean times coincided with a renewed interest in Greek seafaring in the Aegean. Writing of the impetus that propelled a devastated postpalatial Greece into a new and dynamic age, Desborough comments as follows:[199]

> [T]he concluding years of the early Dark Ages, down to about the middle of the eleventh century, mark a decisive turning point. The area mainly to be affected was central mainland Greece, and what took place was a liberation of energies, on the one hand in art and technology, on the other in an extension into and across the empty spaces of the Aegean. In consequence, the direction of things to come was set, and an Aegean outlook was established, on a firmer and more progressive basis than before, leaving the non-Aegean districts more in the background.
>
> As we have seen, the most compact and self-sufficient group that had emerged was that of the Sub-Mycenaeans, especially the Athenians. But some stimulus was needed. Where was the inspiration to come from? Not from the backward areas to the west and north, nor from anywhere among the 'old' Mycenaeans. The east Mediterranean was the most suitable source, but this was apparently not feasible while its first contacts were with the central Aegean group or with Crete, and thus the elimination of the former was extremely important.
>
> In the event, links with the east Mediterranean, specifically with Cyprus, are clear, visible in the acquaintance with Cypriot pottery, in the appearance of weapons of Cypriot type, and in the introduction of iron. But how were they effected? Was the initiative from the Sub-Mycenaean communities or from Cyprus? I am sure that it came from Cyprus, and also that it was not a matter of casual trading visits. Even if no potters came, there must have been technicians in the new metal, iron—otherwise how could the Athenians and other communities of the central mainland have learnt the processes? I think that here we have an example of the movement of a group from east to west rather than the other way around. What lay behind it we cannot tell, but it was something unforeseeable.
>
> As I have said, the two main developments related to pottery and iron-working. The influence of Cypriot pottery was primarily felt in Athens, and there is no doubt that it had a decisive stimulus on Athenian potters, and that the creation of the Protogeometric style, though essentially local, owed something to it. As to metal-working, the effects were wider, and probably included not only Attica but the Argolid and Euboea as well—it may not be pure chance that the process of silver extraction by cupellation first appears at this time in Argos. On the whole, however, it appears that the Athenians benefited most, and from the later situation that they were best fitted to reap the benefits. Of the other communities, Lefkandi was but newly reoccupied, and apart from this the only significant ones, so far as we know, were those of the Argolid, whose very proximity to one another could be a bar to the unimpeded progress of each. Athens stood by itself, and it is likely that it was at this time the most important town in the Greek world.
>
> The arrival of Cypriotes and what they brought with them will have encouraged the suitably placed communities of the central mainland to become more maritime in their outlook. This brings us to the other main feature of this short period—the exploitation of the Aegean, as seen in the diffusion of the Sub-Mycenaean culture. One example of it was at Lefkandi, but more important was a thrust eastwards, via Naxos, which terminated in the reoccupation of Miletus on the coast of Asia Minor, in the area which came to be known as Ionia. This seems to involve a further movement of people, in the usual direction, but it did not weaken thereby the district of origin.

There is an additional and quite compelling argument in support of a reintroduction of the Greek tradition of galley construction from Cyprus. The reversal in the direction of bird-head stem finials—from forward to sternward facing—must have taken place in the region where the galley-construction tradition carried over from the Late Bronze Age to the Geometric period: It is, therefore, significant that inboard-facing bird-head stem ornaments appear first in Cyprus in the form of galley-shaped Proto-White Painted (White Painted I) *askoi.* Although locally made on Cyprus, this ceramic type derives from Mycenaean and Cretan prototypes.[200] The inboard-facing stempost survives on only one model (Fig. 2.41: A). Other ship *askoi* now lack their stems; however, enough remains of the posts to strongly suggest that they, too, would have terminated in the same manner (Figs. 2.40: E, 41: B–C).[201] Due to the intrinsic nature of the *askois* ship models, they, unfortunately, cannot elucidate developments in sternpost finials. The earliest appearance of *inboard-facing bird-head stern ornaments* is, oddly enough, on the models held by the figures inhabiting the Dakhla Oasis galley (Fig. 2.10, 12): The galley's own sternpost, however, is undecorated.

In the Cypro-Archaic period we find firedogs with their stem and sternposts topped by inboard-facing abstract bird heads at Palaepahos and Salamis.[202] Similar firedogs are known from Argos in Greece (Fig. 2.52: A).[203]

Finally, returning to our question about the original direction of the Gurob model's bird-headed stem, given the earlier considerations and the lack of any Cypriot influence on the model, I believe that the evidence suggests that it faces outboard, as per Petrie's original reconstruction (Fig. 1.4).

Oarports vs. tholepins.—The rows of black dots along the outer sides of the hull are best explained as oarports (Fig. 1.7). Homer writes of tholepins, termed "keys"(κληῖδες).[204] These hook-shaped tholepins appear commonly on images of Geometric galleys.[205] The art of the Late Helladic period, however, has no representations of tholepins on galleys. This might be due either to the diminutive size of many of these ship depictions or to other artistic considerations. On the other hand, perhaps they are absent because they were not used on Helladic galleys.

Tholepin *straps,* have been identified either with the Linear B term *e-to-ro-qa-ta* followed by the *181 ideogram, or the term or ideogram alone, which appears on three tablets from Knossos (U 736, Oa 878 and M 757).[206] The juxtaposition of this item with 250 units of flax cloth listed directly above it on M 757 suggests that the *e-to-ro-qa-ta* were made of flax.

This interpretation stands alone in suggesting the use of tholepins on Mycenaean galleys and one may raise several objections to it. First, Homer refers to thole straps made of leather, not linen.[207] Second, in two of the tablets the term or the ideogram refers to only ten units each, which seems a peculiarly small number for an item that would have been needed by each rower on vessels with up to fifty rowers *(pentekonter)* and surely would have required frequent replacement due to wear, particularly if made of linen.[208] Third, and most importantly, tholepins are missing in the contemporaneous iconographic record, as already noted. Thus, their linguistic identification in the Linear B tablets, which can serve only as supportive evidence, is left with nothing to support.

Perhaps *e-to-ro-qa-ta* and the *181 ideogram refer to defined lengths of rope made from flax.[209] Recent shipwreck excavations now seem to confirm what has long been suspected: that the Aegean tradition of wood-plank ship construction used sewing rather than pegged mortise-and-tenon joinery.[210] Greek shipbuilders appear only to have adopted mortise-and-tenon joinery from the Phoenicians after an extended period of experimentation in the mid-first millennium B.C.[211] Thus, ropes would have been a vital component in the ship-construction process. This might explain why, at Mycenaean Pylos, shipwrights *(na-u-do-mo)* are exempted from supplying fifty measures of flax, the largest tax relief recorded in the entire series.[212] Additionally, the region of Pylos and Messene may have continued into Classical times as a center for rope making.[213]

Wedde argues that Mycenaean ships did not have oarports, an invention that, in his opinion, appears only ca. 700 B.C. on the Phoenician ship in the Til Barsip wall painting.[214] Exaggerated oarports may appear on a galley from Pyrgos Livonaton, however, in the form of a series of downward-facing crescents situated over the oars (Fig. 2.43).[215] As Phoenician galleys are descended from the Helladic galley-ship type, it seems that oarports continued in use on galleys in Phoenicia, but tholepins replaced them on later Greek galleys sometime between the Late Helladic IIIC and Geometric periods. Perhaps tholepins were a Cypriot innovation.[216]

Thwarts.—The Gurob model now has four thwarts (Figs. 1.13–14: A, 17, 19). The pattern of missing gesso amidships suggests that something had been placed there: an additional thwart, maststep block, or stanchion base.

Although no figures of rowers were found with the model, and there is no reason to assume that it originally had any, the artisan(s) who built the model skillfully used three different methods of indicating stations for the rowers: thwarts, rowers' stations between the stanchions, and the black dots representing oarports. By combining these different elements, the artisan fashioned a model that was not only pleasing and proportional to the eye but also hinted at the actual number of oarsmen in the ultimate prototype ship.

Quarter rudder.—The Gurob model has a single port quarter rudder that had been attached by a now-missing peg inserted through matching holes in the port stern quarter and the loom (Fig. 1.8: arrow; 18).[217] The absence of a hole on the starboard quarter indicates that the model originally had only one quarter rudder.

In his first reconstruction, Petrie drew the oar hanging over the stern notch in the manner of Nile traveling boats from the First Intermediate through the New Kingdom periods (Fig. 1.4).[218] He corrected this error, however, in his second illustration, placing it correctly over the port quarter (Fig. 1.5). It is not clear whether the quarter rudder had a stanchion. The piece that now serves as a quarter rudder stanchion (item no. 26) is affixed with modern glue: There is no evidence that it originally served this purpose (Fig. 1.19).

Late Helladic IIIB–C depictions of galleys consistently illustrate the use of a single quarter rudder placed well astern, sometimes shown as trailing behind the ship (Figs. 2.10, 37, 39: B, 40: A, D, 42: A, 44: A–B). The limitations of the iconography preclude determining on which side of the hull the quarter rudder was placed.

The five representations of a Sea Peoples' ship at Medinet Habu are portrayed carrying two, one, or no quarter rudders (Figs. 2.3–7). Presumably the artists intended to represent a prototype vessel equipped with a pair of quarter rudders, which in some cases were lost during the melee of battle. The Dakhla Oasis ship graffito carries a single quarter rudder (Fig. 2.10).

This use of a single quarter rudder is interesting, as Egyptian and Syro-Canaanite seagoing ships of the Late Bronze Age consistently carried two quarter rudders.[219] Homer mentions only a single quarter rudder.[220] Late Geometric galleys appear with either a single quarter rudder or one on each quarter.[221]

3 Wheels, Wagons, and the Transport of Ships Overland

The Gurob ship model was found together with four wheels (Figs. 1.21–22).[1] Wheels on a ship sound like the ultimate oxymoron. The model's wheels imply that the model—and hence the model's prototype vessel—was intended for travel on land. Watercraft, by definition, are meant to move over water, not land. So why did the model's maker feel compelled to include wheels?

In fact, there were a number of reasons and occasions when ships and boats were transported overland in antiquity and a number of ways in which to accomplish this. This chapter deals with the evidence from antiquity for ships, boats, and their replicas being transported overland and the different manners in which this was done, particularly, but not exclusively, in Egypt and Greece.

Egypt

The Egyptians depended on the Nile not just for agricultural sustenance but also as a transportation superhighway for watercraft. Given their central role in Egyptian culture, it is hardly surprising that various watercraft played a significant role in the Egyptians' psyche and religious life.[2] In some cases, this led to the perceived need to move vessels (or their replicas) overland.

Egyptians knew the wheel as early as the Old Kingdom: Siege ladders on wheels appear in battle scenes from the Vth and VIth Dynasties.[3] Despite this knowledge of the wheel, however, the Egyptians do not appear to have adopted it for general use until much later. Notes R. B. Partridge:[4]

> It is a curious fact that one of the greatest civilizations in the world flourished without the use of the wheel—an invention considered by many to be one of humanity's greatest inventions and essential to the development of any civilization. Why was the wheel not used in Egypt until relatively late in the country's long history? The answer is that there was simply no significant need for the use of a wheel.
>
> The importance of the river Nile as the main highway of Egypt has already been stressed. With the river running the length of the country and the numerous irrigation canals, there would have been few temples, palaces, or homes more than a hundred yards or so from a navigable stretch of water. All goods could be

A

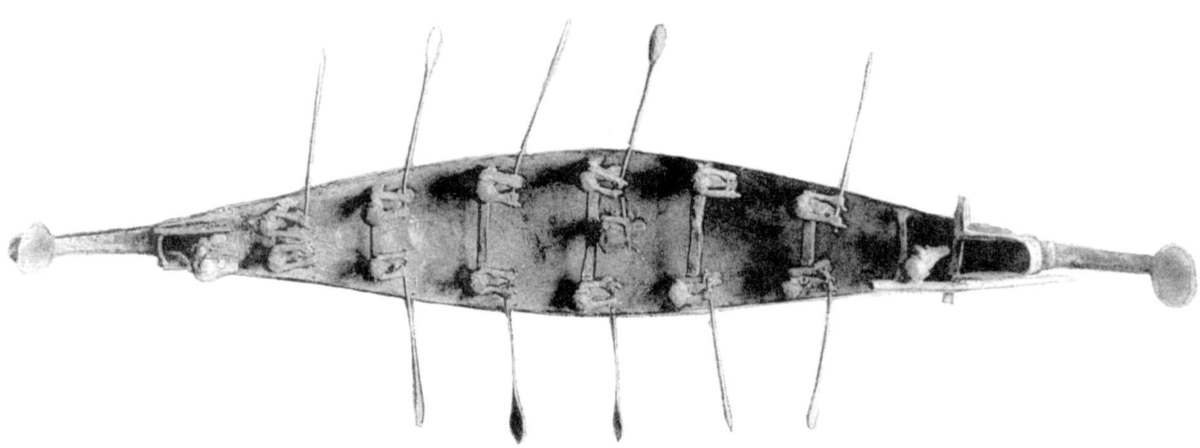

B

Fig. 3.1: The gold ship model from the tomb of Queen Ahhotep, mother of Ahmose, as originally published by von Bissing. From von Bissing 1900: Taf. X.

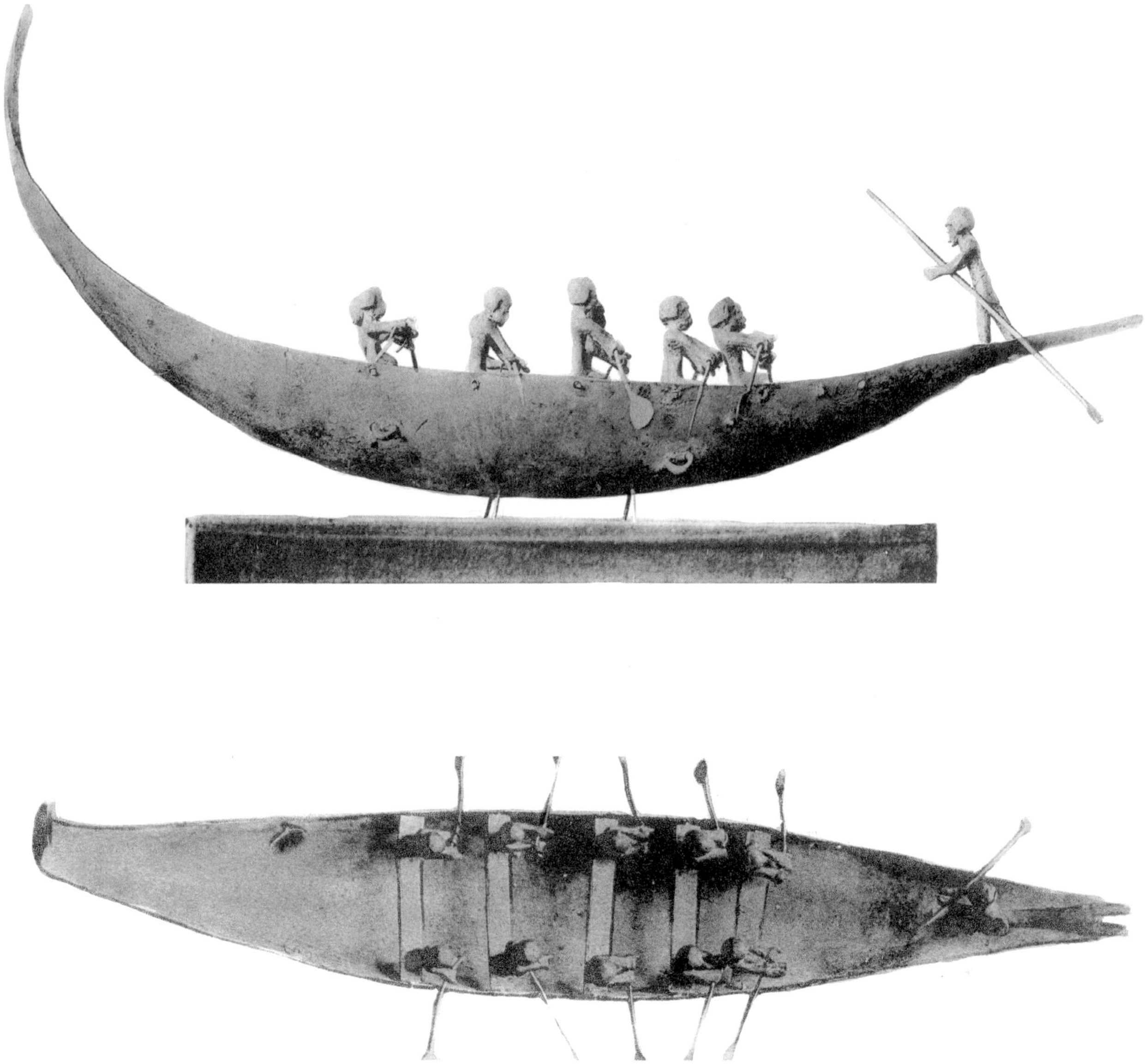

A

B

Fig. 3.2: The silver ship model found in the tomb of Queen Ahhotep as originally published by von Bissing. From von Bissing 1900: Taf. X.

transported easily by water and once on land could be carried by hand or on the backs of donkeys. There was no real need for carts or wagons.

Without using the wheel, the Egyptians still managed to build some of the largest and most spectacular buildings that the world has seen. In all probability wheels would have been of little practical use, for the building blocks utilized were far too large and too heavy to be carried on a wooden-wheeled cart. The relative scarcity of wood in Ancient Egypt would have made the building of such carts difficult, and overcoming the practical and technical difficulties of building carts to carry and move great weights would have proved impossible.

Wheels would have been, in any event, a far from practical method of transport on either the agricultural land or the desert, where they would have quickly become bogged down in either mud or sand. The additional labor required to lay a more solid road surface would have seemed unnecessary when alternative methods of transport were available.

The Hyksos introduced the general use of the wheel to Egypt, along with the horse, the chariot, and the composite bow.[5] Evidence is sparse for boats and ships or their facsimiles being transported on wheeled carriages in pharaonic Egypt, although the earliest examples appear there with the arrival of the wheel.

Ahhotep's ship models and carriage.—The tomb of Ahhotep (I), wife of the XVIIth Dynasty pharaoh Seqenre Tao II, contained two metal ship models—one made of gold, the other of silver (Figs. 3.1–3).[6] The tomb also held a four-wheeled carriage for use with the models (Fig. 3.4). The wagon, which is made of wood and bronze, has pairs of metal staples on either side for securing a model. Both of the ship models have complementary metal loops for this purpose. It remains unclear, therefore, whether the wagon was meant to serve double duty with either of the models or whether a second wagon either had not been interred in the tomb or had disappeared after the tomb's discovery. The regrettable circumstances of the tomb's discovery by A. Mariett's workers during his absence from the site in 1859 preclude any certainty regarding the exact details of the artifacts' in situ state.[7]

Originally F. W. Von Bissing published the three artifacts separately.[8] He speculates in the accompanying text as to which model might have been intended for the wagon. He notes that Mariette considered the wagon to have been an accoutrement of the silver model, but Von Bissing concludes that some of the structural details seem to favor its use with the gold one. G. Maspero and E. Vernier describe how the silver model had originally been placed on the wagon, but due to its poor state of preservation it was replaced by the gold model (Fig. 3.5).[9]

The models are unusual in three respects. First, Egypt did not have a tradition of metal ship models: Ahhotep's models are unique in the Egyptian pharaonic archaeological record. The only other Egyptian metal ship representations, which hardly qualify as models, are neckpieces in the form of Egyptianizing Phoenician galleys dating to the reign of Necho.[10] Second, Ahhotep's ship models are the only ones known from the entire Second Intermediate Period.[11] Third, both models are intended for display or use with a wheeled carriage: The only Egyptian parallel to this is the obviously foreign-inspired Gurob model. As discussed later, there is evidence of the overland transport of boats on wheeled conveyances beginning in the XVIIth Dynasty, but this subject was apparently not normally represented in Egyptian models.[12]

The gold model represents a typical wood-planked Egyptian papyriform boat with stem and stern papyriform finials.[13] Some scholars have identified the silver model as papyriform also.[14] Maspero suggests that it represents the type of vessel used by the deceased to visit Abydos by water during the afterlife.[15]

The silver model's hull is made of hammered sheet silver.[16] One extremity rises in an elegant arc that passes the vertical plane. The rounded shape of the hull continues up this high post. The other extremity ends horizontally, with a forked crutch attached to it. The model has an eleven-man crew: a standing figure, reconstructed as a helmsman, and ten figures who share five rectangular thwarts and face the horizontal end of the craft. The thwarts, made of beaten silver, are attached by means of silver wires that transfix the hull but lie flush against its outer sides. The

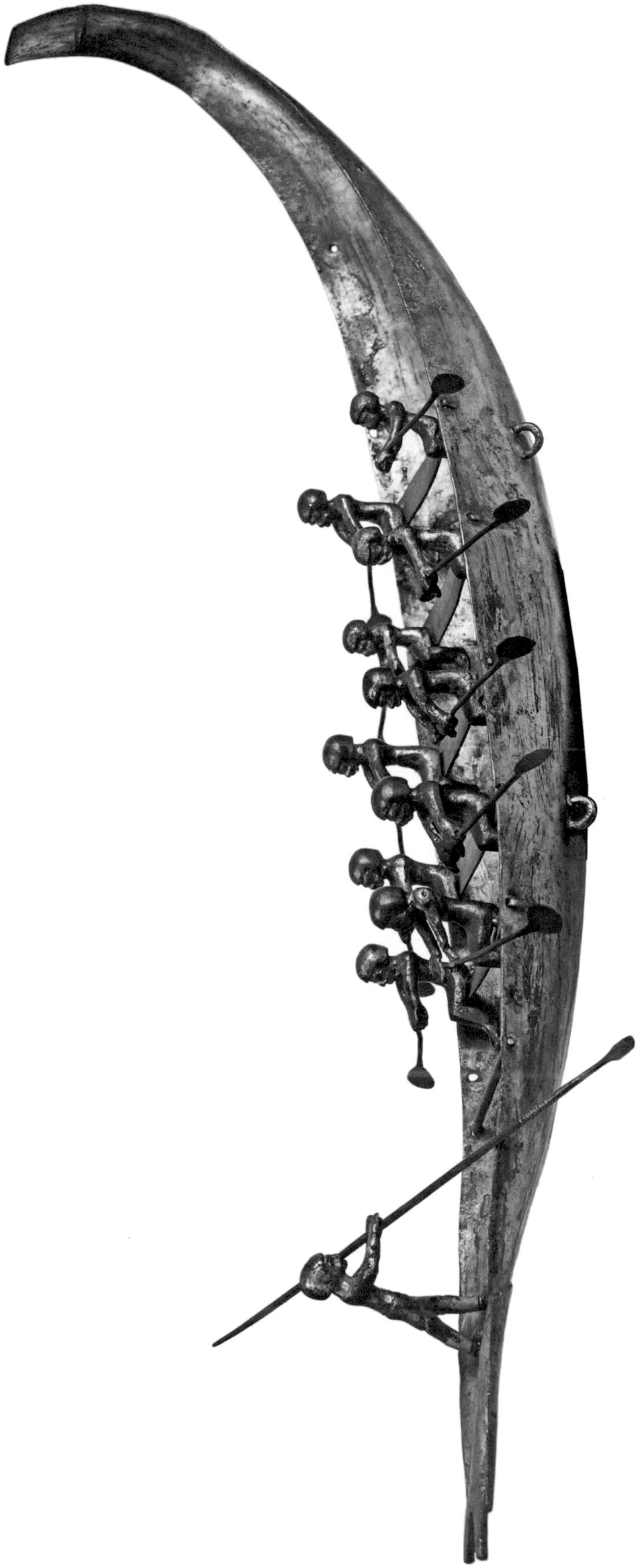

Fig. 3.3: Ahhotep's silver model. Courtesy of Egypt Memory.

Fig. 3.4: The carriage from the tomb of Queen Ahhotep. From von Bissing 1900: Taf. X.

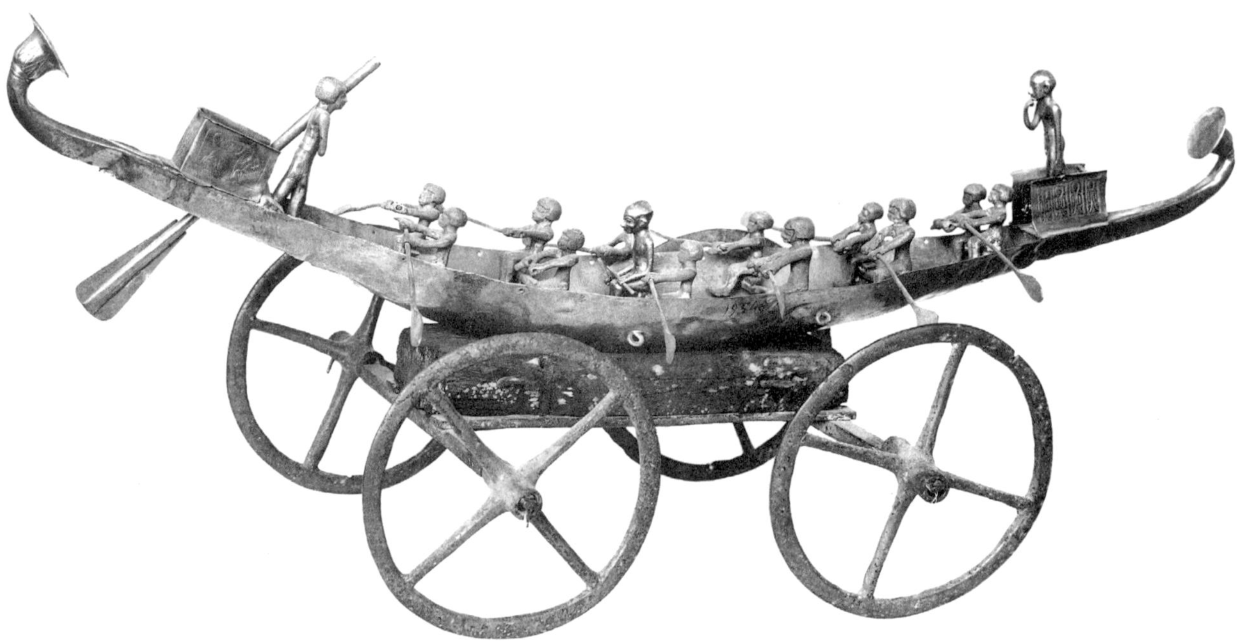

Fig. 3.5: Vernier's reconstruction of the golden ship model on the wagon. From Vernier 1927B: Pl. XLIX.

side rings for attaching the model to a carriage are made of twisted wires that penetrate the hull and are folded against its interior sides. The exterior diameter of the rings is 8 mm, although they are not regular.

The silver model now lacks some of its original parts. This may be surmised from additional pairs of holes—three at the high end and two at the horizontal end—that pierce the sides of the hull (Fig. 3.3). These sets of piercings might be interpreted as serving to attach additional thwarts for rowers, but the irregular spacing of the three sets near the high extremity argues strongly against such a reconstruction. One likely purpose may have been to attach decks to the model. Vernier notes a rod traversing the hull between the standing figure and the nearest crew pair. This rod is visible on a recent photograph but is missing in von Bissing's top view of the model (compare Figs. 3.2: B and 3). This is presumably a loose piece found with the model.

The silver model displays another anomaly in addition to those already noted. While the gold model's quarter rudder is of the type common on royal XVIIIth Dynasty ship models, the helmsman of the silver model holds a steering oar rather than the tiller of a quarter, or an axial, rudder.[17]

One is immediately struck by the frailty of the silver model's steering oar.[18] Such devices, which by definition lack tillers, appear on Predynastic and Old Kingdom vessels, but in Egypt these devices cease to be represented toward the end of the Vth Dynasty.[19] Also during the Vth Dynasty, tillers appear, and stanchions are portrayed supporting the looms of quarter rudders.[20] Subsequently, substantial quarter or axial rudders normally replace steering oars on Egyptian vessels. Following this, steering oars are relegated to occasional appearances on cultic vessels and reed rafts.[21]

The configuration of the silver model's helmsman's arms is unusual when compared with the different manners in which helmsmen are represented holding steering oars or quarter rudders in Egyptian iconography.[22] These considerations raise the possibility that the figure reconstructed as the helmsman originally did not serve that purpose on the model. Alternately, or in addition to this, the "quarter rudder" may actually be a cosmetic spatula added to the silver model after the tomb's discovery.[23]

Scholars disagree as to which of the silver model's ends represents its bow and which its stern. Von Bissing published the model with the "helmsman" at the horizontal end. If this is the correct positioning for the figure, then the vessel is being rowed, and the high end is the bow. Landström, on the other hand, omits the helmsman in his drawing of the model and considers the horizontal extremity to be the bow, thus turning the crew into paddlers.[24]

The seating of the figures remains one manner of determining the identification of a vessel's extremities. Paddlers face the bow, while rowers normally look toward the stern.[25] In his study of ship models in the Cairo Museum, G. A. Reisner notes that Middle Kingdom Nile boats (Type II) rowers are shown seated, while paddlers are depicted kneeling.[26] Following this rule, then, the *seated* crew of the silver model is rowing, and the model's high end is the bow (Fig. 3.6). Early Egyptologists, however, had the lamentable habit on occasion of arbitrarily rearranging the crews and the accoutrements of ship models—as W. M. F. Petrie had done in the case of the Gurob model—and Ahhotep's silver model may have suffered such attention.[27] Von Bissing notes that the crew appears to have been remounted in modern times, so the actual direction in which the rowers originally faced may now be lost.[28]

The forked device attached to the silver model's horizontal extremity bears comparison with the deep stern groove used to seat an axial rudder on some New Kingdom Nilotic craft.[29] A less likely parallel in my view, but one that should be considered nevertheless, is a device that has been termed a "bowstick" or a "bowsprit," which appears at the bows of Middle Kingdom ship models.[30] The purpose of these timbers remains enigmatic: Perhaps they served as a fairlead for a bow-anchor hawser.[31] They are dissimilar to the silver model's forked crutch in having a relatively small groove quite unlike the pronounced fork on the silver model.

When all is taken into consideration, it appears that rowers propel the silver model and that its high end represents the bow. Clearly, a characteristic Egyptian ship type did not serve as the source for Ahhotep's silver ship model even though it appears to have been made by Egyptian artisan(s). What foreign ship type served as a prototype

Fig. 3.6: Detail of an oarsman from Ahhotep's silver ship model. From von Bissing 1900: Taf. IX: 2A.

for the silver model? Two factors must guide our steps in addressing this question: (a) a comparison of contemporaneous water transport within the international cultural milieu of the Egyptians and (b) knowledge of which of these cultures had a tradition of metal ship models. Let us review the evidence:

Syro-Canaanite ships.—A wall painting from the tomb of Kenamun (TT 162), who lived under Amenhotep III, is the most detailed known representation of Syro-Canaanite merchant ships (Fig. 3.7).[32] Here the artists depicted a flotilla of seagoing merchantmen with identical stem and sternposts shown arriving at an Egyptian port. Other representations of Syro-Canaanite vessels, albeit less detailed, show the same high vertical stem of the Kenamun ships but have rounded sterns rising at various angles.[33]

A pertinent question with regard to the Kenamun ships is whether their extremities are represented in profile or in frontal view. The concavity that appears at the external edge of the stem may indicate that they are shown in frontal view, in which case they bear a similarity to the flattened shape of the silver model's vertical, recurving post. Despite the detail of the Kenamun wall painting, its creators clearly did not comprehend the ships' rigging, suggesting that they were working from copybooks and were at least once removed from the images of the ships that they created in the tomb.[34]

A tradition of metal ship models existed along the Syro-Canaanite coast, at least in Byblos, to judge from the bronze models found there at the Champ des offrandes.[35] Curiously, the best preserved of these Byblian models either represents an Egyptian Middle Kingdom ship or, more likely, duplicates a wooden model of an Egyptian ship.[36]

Cypriot ships.—A series of three terracotta models from Kazaphani *Ayios Andionikos* and Maroni *Zarukas* clearly represent deep-hulled cargo ships.[37] Nevertheless, they have several elements of interest for this study. The stems that survive on two of the models bear comparison to the high end of Ahhotep's silver model in that they have a similar flattened profile, which may be the result of the terracotta material of which the Cypriot models are made. Also, the models' sterns each end in a *vertical* bifurcation—best preserved on the model from Site A, Tomb 7, at Maroni *Zarukas*. Vertical orientation aside, this resembles the silver model's forked crutch.

I am unaware of any Cypriot metal ship models within the relevant time frame.[38]

Minoan ships.—The profile of Ahhotep's silver model bears a remarkable resemblance to that of the rowed ship accompanying the paddled flotilla in the Miniature Frieze from the West House at Akrotiri on Thera (Fig. 3.8: A).[39] Each of these two vessels also has five rows of seated oarsmen and uses a steering oar. Another representation of this ship type, identifiable by its shape and the five files of oarsmen, appears on a gold ring from Crete (Fig. 3.9).[40] The triangular object located at the bow of the ship on the ring should not be confused with a sixth file of rowers. It probably represents the splashguard that appears in greater detail on other representations of Cycladic/Minoan ships.[41] In general, the steering oars depicted on the ships in the Miniature Frieze seem inadequate for the vessels on which they are being used.[42] They are, however,

Fig. 3.7: Syro-Canaanite ships from the tomb of Kenamun (TT 162). Amenhotep III. From Daressy 1895: pl. XV.

an almost exact match for the steering oar of Ahhotep's silver model.

Furthermore, the Aegean has a long tradition of metal ship models, beginning with three Early Cycladic lead longship models from Naxos (now in the Ashmolean Museum).[43] A fragmentary Late Minoan IB/Late Helladic II bronze model from Keos also has a bow shape remarkably similar to that of the silver model (Fig. 3.8: B).[44]

When viewed from an Aegean perspective, the bifurcation at the horizontal extremity of the silver model may be compared to the curving forked stern on the Kolona ships from Aegina, which date to ca. 1700 B.C (Fig. 3.10).[45] Ships on minuscule Minoan seals often have what appears to be a stern bifurcation.[46] This detail, however, probably represents the post and the horizontal stern projection, which appear most clearly and in greater detail on the processional ships of the Theran Miniature Frieze.[47]

This short review suggests that Ahhotep's silver model may best be explained as replicating a relatively small and narrow ten-oared Minoan prototype vessel or, alternately, a model of such a craft. Although no ship models appear among the items brought by Minoans in the XVIIIth Dynasty tombs of the nobles at Thebes, high-status examples of such items clearly existed (e.g., the ship model from Keos and the one carried by a mourner on the Hagia

A

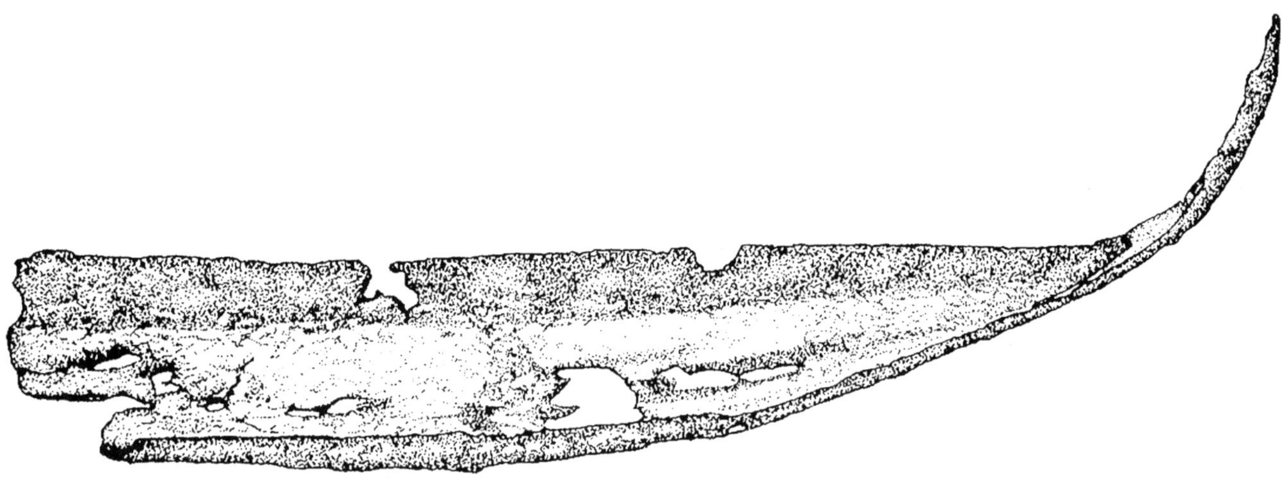

B

Fig. 3.8: (A) Rowed ship from the Miniature Frieze, West House, Akrotiri, on Thera. (B) Bronze model fragment from Keos. Late Minoan IB/Late Helladic II (NTS). A after Doumas 1983: 121 fig. 20. B after Göttlicher 1978: Taf. 25: 335.

Fig. 3.9: Gold ring from Crete depicting a vessel similar to the ship under oar in the Miniature Frieze in the West House at Thera (NTS). After *PM* IV: II: 953 fig. 923.

Fig. 3.10: L. Basch's reconstruction of a crescentic ship on a pithos from Kolona, Aegina, ca. 1700 B.C. Note the bifurcated stern at right (NTS). After Basch 1986: 427 fig. 10.

Triada sarcophagus) (Fig. 3.8: B, 11). Similar metal models could have been brought to Egypt by the Minoans or, as appears to be the case with Ahhotep's silver model, have been constructed by Egyptian artisans for Minoans residing in, or visiting, Egypt. Another type of Minoan model—bull statuettes, like those carried by two mourners on the Hagia Triada sarcophagus—was brought by Minoans in the Theban tombs of Useramun (TT 131) and Menhepperesonb (TT 86), both of which date to the reign of Thutmose III.[48]

A Minoan-inspired ship model would not be out of place in Ahhotep's tomb assemblage. The queen's dagger, as well as Ahmose's axe, shows clear Minoan influences. Warren summarizes these artistic relationships succinctly:[49]

> Although technically of New Kingdom date, two further pieces must be mentioned here because of their close Hyksos connection. These are the axe of Ahmose, conqueror of Avaris and the Hyksos, and the dagger of his mother Ahhotep, both found in her tomb . . . The griffin on the axe blade has wings decorated with the "notched plume" motif. The Minoan origin of this proposed by Evans and with details noted by Morgan . . . is now well confirmed by the notched plumed wings of the almost contemporary griffin guarding the seated goddess who presides over the crocus gatherers in the painting in Ashlar Building 3 at Akrotiri, Thera . . . The Aegean origin of a lion chasing a bull in a flying gallop position in a rocky setting on Ahhotep's dagger remains clear, again as proposed by Evans . . . the axe shows a powerful symbol of Minoan religion adopted and adapted as a symbol of political power in Egypt (even though the griffin as such was earlier established in Syria and Egypt . . .). The lion motif of the dagger, locally engraved, expresses the Aegean mode of symbolizing power and speed. Processes of iconographical transfer of ideology expressed in symbols are continuing between the two areas.

How might a model representing a Minoan craft find its way into the tomb of a XVIIth Dynasty royal consort? I believe that the simplest explanation is that it is plunder taken by the Egyptians during their attacks on, and conquest of, Avaris. Ahhotep was the mother of Kamose and Ahmose, the founders of the XVIIIth Dynasty. Kamose attacked the Hyksos capital of Avaris and claims to have taken much booty from its harbor, while Ahmose conquered the city and drove the Hyksos out of Egypt.[50] Even before the fall of the city itself to Ahmose, Kamose's Second Stele gives a vibrant view of the opulent spoils that he carried away from the harbor of Avaris:[51]

> I have cut down your trees, I have forced your women into ships' holds, I have seized ⌈your (?)⌉ horses; I haven't left a plank to the hundreds of ships of fresh cedar which were filled with gold, lapis, silver, turquoise, bronze axes without number, over and above the moringa oil, incense, fat, honey, willow, boxwood, sticks and all their fine woods—all the fine products of Retenu—I have confiscated all of it!

Fig. 3.11: Mourner carrying a high-status marble (?) ship model on the Hagia Triada sarcophagus. Late Minoan IIIA. Heraklion Museum.

The hypothesis that a model of a Minoan ship had been taken as booty from Avaris either by Kamose or Ahmose presupposes a Minoan presence at that site under Hyksos rule. M. Bietak's excavations at Tell el Da'ba have revealed remarkable fragments of Minoan frescoes, including bull jumpers, indicating the presence of Minoans at the site.[52] Bietak originally dated the Minoan fresco fragments found at Tell el Da'ba to the Hyksos period but has since revised his chronology and now places them in the early XVIIIth Dynasty.[53] E. Cline, Barbara Niemeier, and W.-D. Niemeier have disputed Bietak's redating, and he has responded vigorously.[54] The present discussion, unfortunately, does not contribute to this debate as Bietak's dating of the existent Minoan material at the site to the XVIIIth Dynasty does not preclude an earlier Minoan presence there that may yet remain undetected archaeologically. It certainly would not be surprising to find Minoans in Egypt at a time when the Delta was under Hyksos rule, as these Aegean seafarers make appearances in other eastern Mediterranean countries at this time or earlier, depending on the chronologies followed.[55] The alabaster jar lid engraved with the name of the XVth Dynasty Hyksos pharaoh Khyan found in a Middle Minoan IIIA level at Knossos also implies a royal connection between Hyksos Egypt and Minoan Crete.[56]

Based on this analysis then, the prototype vessel of Ahhotep's silver model would have been a relatively small craft, which would be better suited to coastal waters than to blue-water crossings. Its size and that of the rowed ship in the Theran Miniature Frieze may be estimated based on the approximately 1-m length *(interscalmium)* required by each of the rowers to work his oar. Assuming a relatively realistic scale of figures and vessels, these vessels probably would have been in the range of 12–14 m long.[57]

Also, if the silver model copies a Minoan ship type rather than a model of a Minoan ship, Minoans at Avaris may well have been building their own vessels on Egyptian soil. Egyptian texts refer to Keftiu ships in Syro-Canaanite ports and being built and/or repaired at the royal Egyptian shipyard of Prw-nfr, and it is tempting to identify any Minoan ships built in Egypt as the Keftiu ships referred to in these texts.[58] A study of the contexts in which these Keftiu ships appear, however, indicates that the term most likely refers to Syro-Canaanite ships of a type used on the trade route to the Aegean.[59]

The wheeled cart found with Ahhotep's models remains an enigma, however, even within a Minoan context, as there are no other known examples of Minoan ships depicted on wheeled carriages. Ahhotep's wagon may represent a Hyksos influence, as the wheels look like roughly contemporaneous chariot wheels.[60] After the fifteenth century B.C., Egyptian and Canaanite chariots became heavier and, after a short experiment with eight-spoked wheels during the reign of Thutmose IV, employed six-spoked wheels.[61]

Funerary boats.—Although by no means common, representations of ships on wheeled wagons do appear in pharaonic art depicting the transportation of funerary boats. Several representations of Egyptian tombs contain scenes of the deceased crossing to the west side of the Nile for interment, as well as transportation overland of the deceased to the place of burial, often in a boat-shaped catafalque.[62] Notes Jones:[63]

> Not only was a full-sized funerary boat used to transport the mummy of the deceased across the river but, on reaching the west bank, the coffin was transferred to a papyriform boat or boat-shaped bier, either actual or miniature, which was dragged on a sledge by men and oxen—in rare cases on a four-wheeled wagon—for its final journey across the desert to the tomb.

A remarkable expression of these multiple and entirely interchangeable methods of transporting the ship-shaped catafalque overland appears in the tomb of Nakhtamun (TT 341, Ramses II) at Kurnah, where porters carry the ship bier: However, just for good measure, the scene also shows the catafalque pulled by a group of men and a yoke of oxen (Fig. 3.12).[64] Similarly, a relief from the temple of a XXVth Dynasty pyramid at Begerauieh, near Meroë, portrays a uraeus-topped shrine transported on a four-wheeled chariot-like structure, which also has the typical bars porters used to carry cult boats (Fig. 3.13).[65]

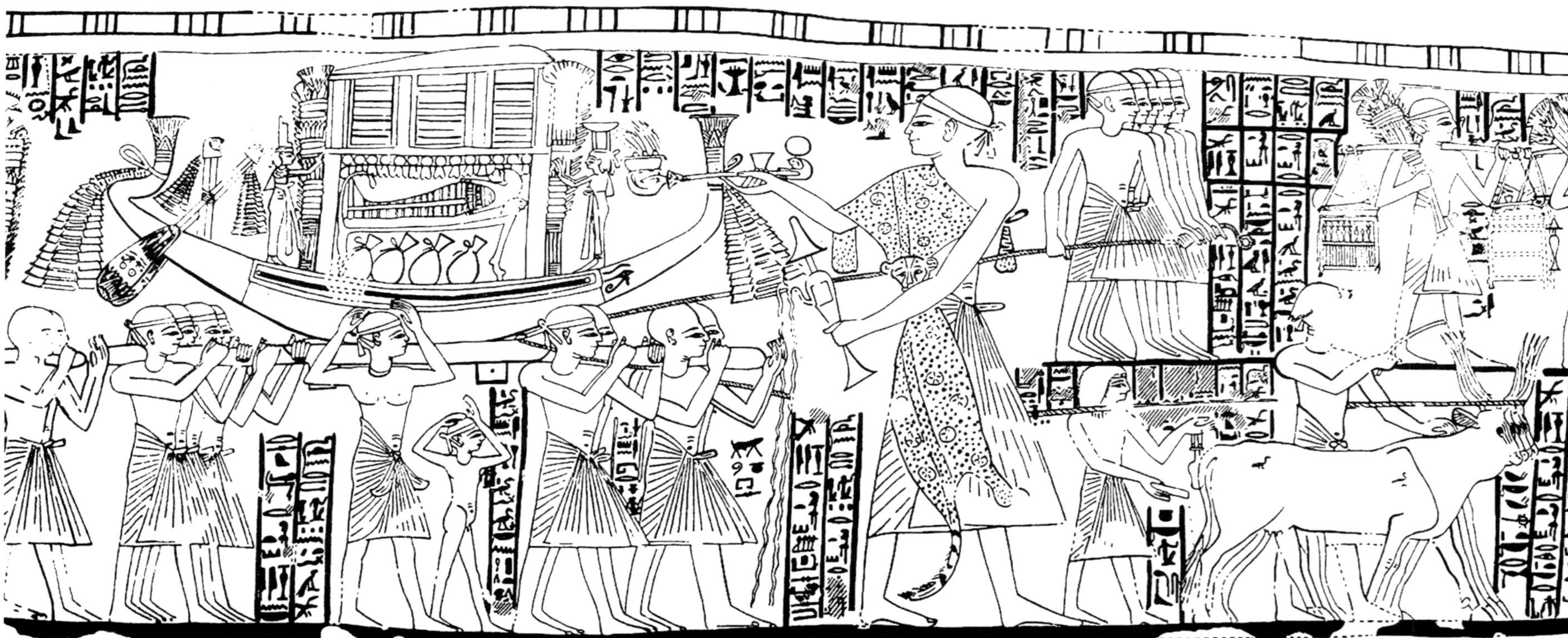

Fig. 3.12: Boat-shaped catafalque being carried by porters while also being pulled, as if on a sled, by men in the upper register and by oxen in the lower register. Tomb of Nakhtamun (TT 341). Ramses II (NTS). After N. d. G. Davies 1948: pl. XXV.

The Bible also contains an interesting example of the various ways of transporting a portable cultic object. The Ark of the Covenant had been constructed with poles, which enabled Levite porters to carry it.[66] When the Philistines returned the Ark to the Israelites after the battle of Ebenezer, however, as a result of a plague of *'opalim*—normally translated as tumors or hemorrhoids—they placed it on a wagon yoked to cows.[67]

Fig. 3.13: Four-wheeled chariot-like wagon, with support bars for porters attached, used to transport a shrine topped by a *ureaus*. From a XXVth-Dynasty pyramid temple in Nubia (NTS). After Dittman 1941: 67 Abb. 6.

Tomb of Sobeknakht.—A wall painting from the tomb of Sobeknakht (T. 10) at El Kab contains the earliest-known Egyptian representation of a boat transported on a wheeled carriage (Fig. 3.14).[68] This is also one of the earliest representations of a wheeled conveyance known from Egyptian art. Sobeknakht served as Overseer of Prophets and may have been a contemporary of Sobekemsaf II (Sekhemre Shedtawy), the first pharaoh of the XVIIth Dynasty.[69]

In the painting, Sobeknakht's sarcophagus, with his mummy visible inside, lies in a papyriform boat-shaped catafalque carried on a flat carriage that stretches the length of the vessel. A yoke of oxen pulls the wagon, which has an upturned front end, identical to that of a typical Egyptian sledge. In fact, the "wagon" appears to be a sledge outfitted with wheels. The diameter of the wheels—situated at the wagon's extremities—seems disproportionately small. The tops of the wheels do not rise above the wagon bed. The lack of spokes and the exaggerated axle holes give the wheels a decidedly doughnutlike appearance.[70]

A man walking ahead of the wagon pours liquid from an elongated jar onto the roadway in front of the carriage. In similar scenes, this action is carried out before sledges to ease their movement over the ground.

A row of parallel slanting lines appears beneath the wagon and continues rising in front of it. Joseph J. Tylor identifies these as log rollers, while noting the incongruity of using them to facilitate the movement of a wheeled vehicle. He considers the manner in which these parallel lines are portrayed as a rare example of perspective in Egyptian art. Tylor suggests, therefore, that this might be due to the artist's unfamiliarity with the use of the wheel and proposes that wheeled vehicles may have been so novel at the time that the artist might never actually have seen one.

Norman de Garis Davies notes, however, that the "logs" are painted blue, a color used to represent water but never wood, and compares this scene to one from the early XVIIIth Dynasty tomb of Tetaky (TT 15) at Thebes, in which a man pours liquid in front of the deceased's funerary sledge.[71] The accompanying inscription there reads "casting water under the [sled?] and milk (?) for the *amakhy,* the royal son, Tetaky."[72] Based on this evidence, Davies similarly interprets the blue stripes in Sobeknakht's scene to represent splashes of libational liquid tossed by the man with the jar.

Tomb of Amenmose.—A wheeled carriage transports a ship-shaped catafalque in the tomb of Amenmose, a priest of Amun who lived during the early XXIInd Dynasty.[73] Amenmose rests in state in the *naos.* The wagon has an upturned sledge-shaped front end, and its painted decoration is reminiscent of the Gurob ship model's *pavois* plaque, while the daubed decoration on Amenmose's wheels resembles those of the Gurob model. A row of five men tows the boat: Their rope is attached to the bow of the funerary boat.

Wooden coffin of Tetmaat.—The wooden coffin of the XXIInd Dynasty priestess Tetmaat, now in the Vatican, depicts her husband's funerary boat moving on a wagon.[74] This also has an upturned sledge-shaped wagon

A

B

Fig. 3.14: (A) Scene of funerary boat on a wagon from the tomb of Sobeknakht at El Kab. XIth Dynasty. (B) Detail (NTS). From Tylor 1896: pl. II.

A

B

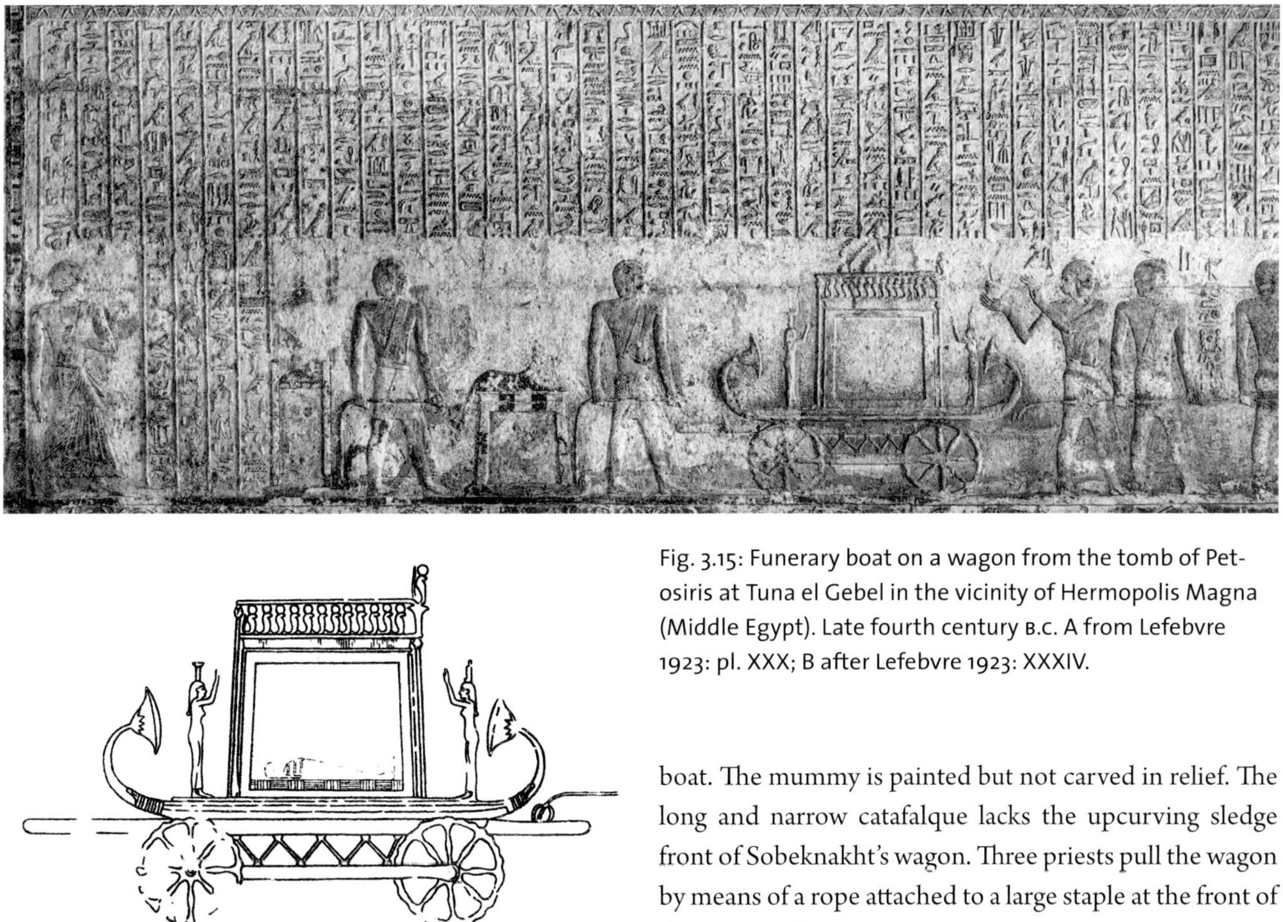

Fig. 3.15: Funerary boat on a wagon from the tomb of Petosiris at Tuna el Gebel in the vicinity of Hermopolis Magna (Middle Egypt). Late fourth century B.C. A from Lefebvre 1923: pl. XXX; B after Lefebvre 1923: XXXIV.

carriage and moves on two pairs of wheels. The forward right wheel has six spokes, while the rear right wheel has only five spokes. Two rectangles visible between the wheels attached to the bottom surface of the chassis may indicate strengthening beams. P. Creasman and Noreen Doyle reasonably suggest that the "ship" may have existed only in bow and stern look-alikes attached to the burial shrine.[75]

Tomb of Petosiris.—The tomb of Petosiris at Tuna el-Gebel, the necropolis of Hermopolis, contains a bas-relief of a funerary scene in which the deceased appears in a papyriform boat transported on a four-wheeled wagon (Fig. 3.15).[76] Petosiris was the High Priest of Thoth. The tomb dates to the last quarter of the third century B.C., soon after Alexander's conquest of Egypt, and displays an extraordinary combination of Egyptian and Greek artistic styles and detail.[77]

Petosiris's mummy lies in the *naos* at the center of the boat. The mummy is painted but not carved in relief. The long and narrow catafalque lacks the upcurving sledge front of Sobeknakht's wagon. Three priests pull the wagon by means of a rope attached to a large staple at the front of the carriage.[78]

A trellis that fills the space beneath the wagon bed between the two wheels suggests some form of chassis. A member extends from outside the axle hole of the front wheel to the forward edge of the rear wheel: The meaning of this item is unclear and may be due to an artist's error. The large eight-spoked wheels have studded felloes.[79]

Tomb of Siamun.—The mummy of Siamun is transported on a wheeled cart in a scene from his tomb at Siwa Oasis (Fig. 3.16).[80] The tomb dates to the fourth or third centuries B.C.[81] Siamun was apparently of mixed parentage: His father, Peritu, was probably a Greek merchant, while his mother was Egyptian. Greek influence is evident in the tomb's scenes and in the physical appearance of Siamun and his son.[82]

Fakhry notes that the boat's prow ends in a lotus flower, but this does not appear in the accompanying figure. The wagon wheels have eight spokes, and their felloes have studs, similar to the wheels on Petosiris's wagon. The outer termination of each spoke ends in a papyrus

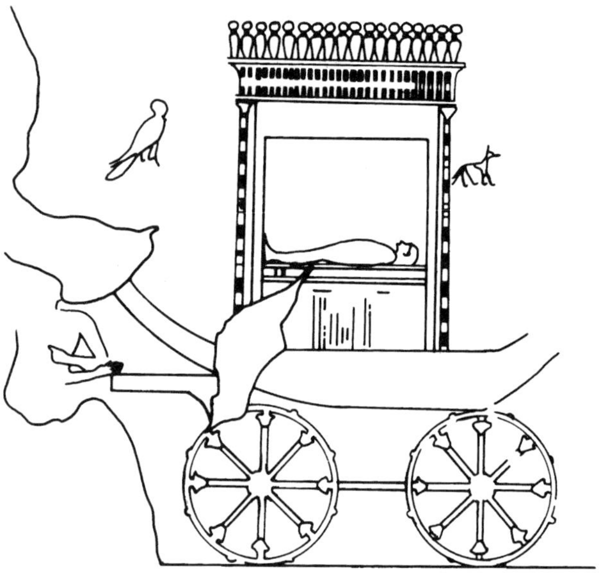

Fig. 3.16: Funerary boat on a wagon from the tomb of Siamun, Siwa Oasis. Fourth–third centuries B.C. (NTS) After Fakhri 1944: 144 fig. 24.

umbel.[83] Two men on the adjacent western wall pull the wagon by means of a rope.[84]

MUMMY WRAPPING.—Wilkinson published a drawing on a Ptolemaic mummy wrapping from the British Museum in which a funerary ship has been placed on a wagon with eight-spoked wheels (Fig. 3.17).[85] A priest pulls the wagon with a rope while turning to offer incense.

VEHICLES FOR DEITIES.—Shiplike palanquins served as the preferred conveyances for Egyptian deities when they voyaged overland.[86] During sacred celebrations, statues of deities traveled to the lands owned by their temples or to the temples of other gods (Fig. 3.18). A billowing linen cloth hid the god's image, stowed in a shrine amidships, from the public. The portable boat-shaped shrines were secured to purpose-built bases equipped with poles or bars that were shouldered by priestly porters.[87] G. Legrain terms such a base a *pavois* (French: "shield" or "bulwark") (Figs. 3.19–21).[88] These devices often have a recurving forward edge reminiscent of a sledge, from which they perhaps evolved (Fig. 3.22).[89] As noted, the Gurob model was found with a *pavois,* and similar *pavois* appear on other Egyptian wooden ship models (Figs. 1.20, 3.24).[90]

Ship palanquins were heavily and lavishly decorated.[91] Nakht-Thuty, an artisan active during the reign of Ramses II, specialized in their construction. The Temple of Amun at Thebes loaned him out to build more than two dozen portable boats for other gods.[92] Nakht-Thuty notes that he adorned them "with gold, silver, real lapis-lazuli, turquoise. . . ."[93]

Perhaps the best known of these remarkable vessels are the portable ship shrines of the Theban Triad at Karnak: Amun, Mut, and their son, Khonsu (Fig. 3.18). During the Opet Festival, the apex of the Egyptian religious cycle during the New Kingdom, statues of these deities rode in their portable boat shrines to their heavily decorated sacred Nile ships, which were then towed upriver to Luxor Temple.[94] At the end of the Opet Festival the statues of the deities returned in their portable ship shrines in the same manner to Karnak Temple.

Processions in which the deity took part in his boat shrine were an essential element of religious worship in Egyptian temples and notably at Thebes: Processional ways created expressly for this purpose were an integral aspect of temple architecture.[95] Special shrines in the temples housed these cult boat shrines.[96] In fact, the existent structure at Karnak temple, constructed by Philip Arrhidaeus, Alexander's brother, evolved around the storage shrine for the portable boat of Amun (Fig. 3.18: B).[97]

Upon its death, the Apis bull was mummified in an

Fig. 3.17: A funerary boat on a wagon with eight-spoked wheels painted on a mummy cloth (NTS). From Wilkinson 1837: 341 no. 243.

A

B

Fig. 3.18: The boat of Amun carried by priest-porters. (A) Karnak/Thutmose III; (B) Karnak/Arrhidaeus.

Fig. 3.19: (A) The barque of Amun under Pinudjem II. (B) Detail of the *pavois* (NTS). From Krutchten 1986: foldout.

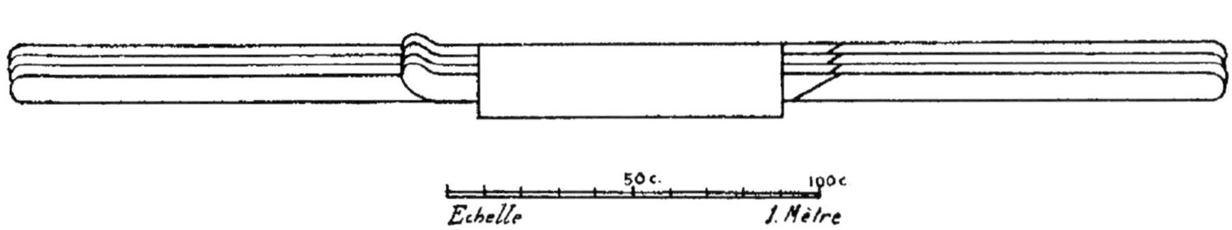

Fig. 3.20: The *pavois* and four bars of the ship shrine of Mut at Luxor. From Legrain 1917: 5 fig. 1.

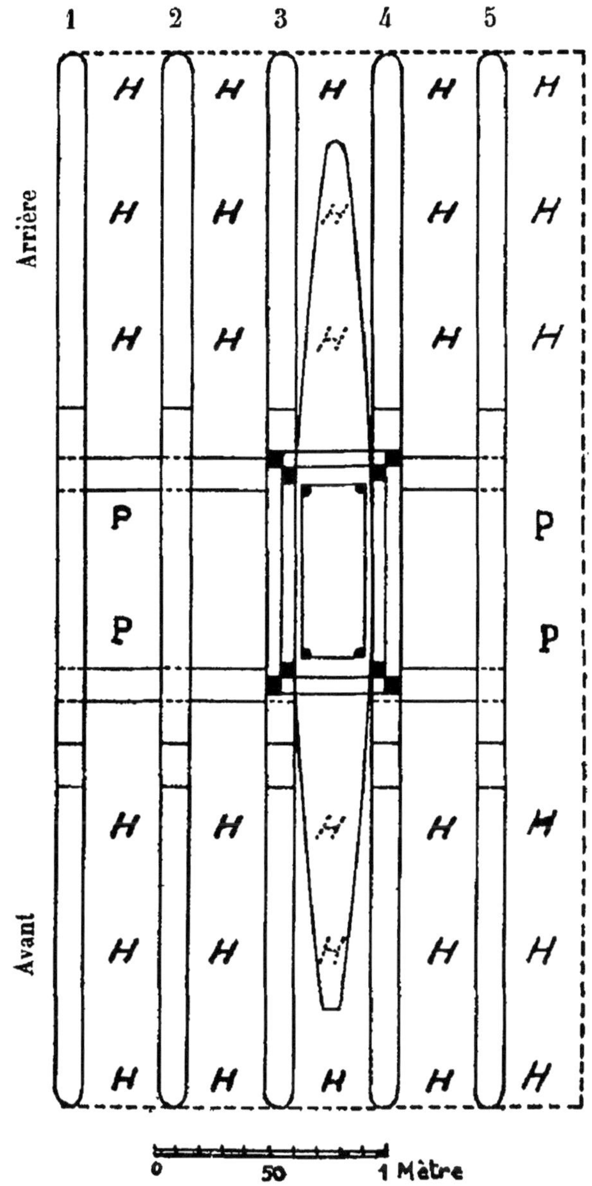

Fig. 3.21: Legrain's reconstruction of the ship shrine of Amun, with its *pavois* and bars. (H = man; P = prophet) (NTS). From Legrain 1917: 12 fig. 2.

elaborate ritual, described in detail in Papyrus Vindob. 3872, which dates to the latter half of the second century B.C.[98] Following the mummification process, the Apis was removed from the House of Embalming and transferred across Lake Abusir to the tent of purification.[99] During this transfer the Apis was placed on a vessel described in the papyrus as "the Bark of Papyrus with the shrines," which included statues of the two goddesses Isis and Nephthys, depicted as mourning for the deceased Apis.[100] From the house of embalming to the lake, the funerary ship was transported apparently on a wheeled conveyance.[101]

Limestone carved reliefs depict such an Apis funerary boat on a wheeled carriage with various degrees of detail (Fig. 3.25).[102] The vessel is papyriform. The façade of the portable shrine has been depicted facing the viewer to reveal the head of the embalmed Apis bull. Located at the bow and the stern and facing the double shrine are kneeling figures of Isis and Nephthys. The ship rests on a *pavois,* with support poles extending fore and aft of the boat and wagon. The front end of the wagon curves upward, replicating the shape of a sledge.[103] The wagon moves on *four* pairs of wheels. There is no evidence of spokes, suggesting that the wheels are solid wooden disks, which are locked on to the axles by pairs of linchpins.

We have no way of knowing how many Egyptian Graeco-Roman cults utilized boats in their processions. Herodotus, for example, describes a cult of Ares at Papremis, a site perhaps located near Damiette, in which the deity was towed on a wagon, although Herodotus does not specifically mention a boat in this case:[104]

> The Egyptians do not hold a single solemn assembly, but several in the course of the year. Of these the chief, which is better attended than any other, is held at the city of Bubastis in honour of Artemis. The next in importance is that which takes place at Busiris, a city situated in the very middle of the Delta; it is in honour of Isis, who is called in the Greek tongue Demeter (Ceres). There is a third great festival in Saïs to Athena, a fourth in Helioplois to the Sun, a fifth in Buto to Leto, and a sixth in Paprêmis to Ares . . .
>
> At Heliopolis and Buto the assemblies are merely for the purpose of sacrifice; but at Paprêmis, besides the sacrifices and other rites which are performed there as elsewhere, the following custom is observed:—When the sun is getting low, a few only of the priests continue occupied about the image of the god, while the greater number, armed with wooden clubs, take their station at the portal of the temple. Opposite to them is drawn up a body of men, in number above a thousand, armed, like the others, with clubs, consisting of persons engaged in the per-

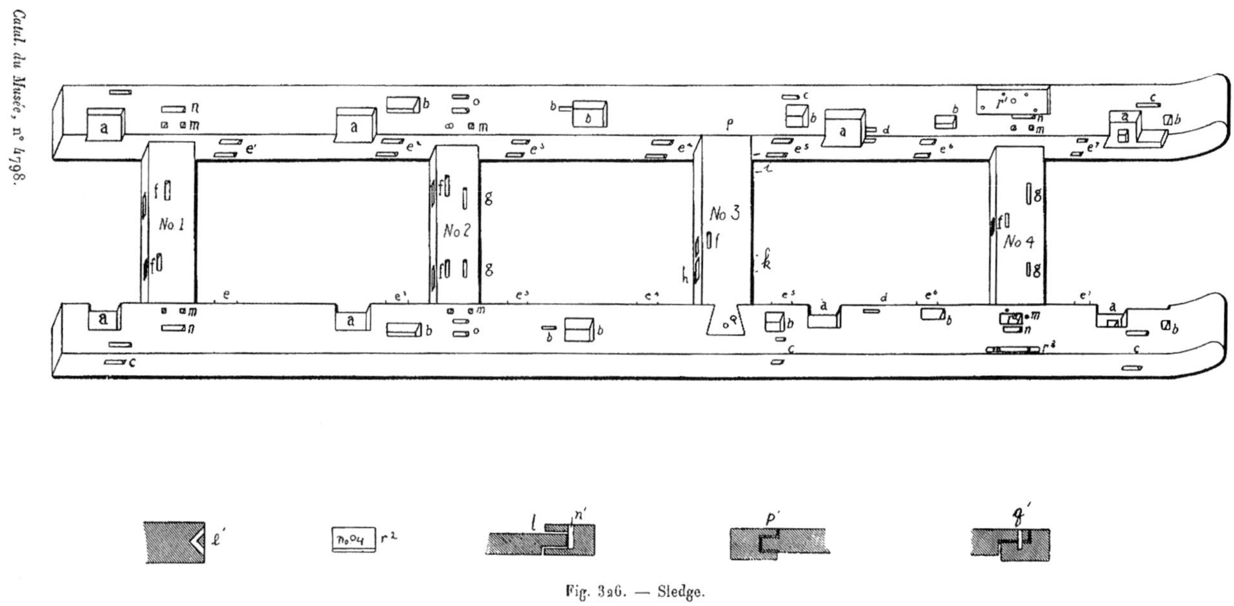

Fig. 3.22: Sledge found by J. de Morgan at Dahshur (NTS). From Reisner 1913: 89 fig. 326.

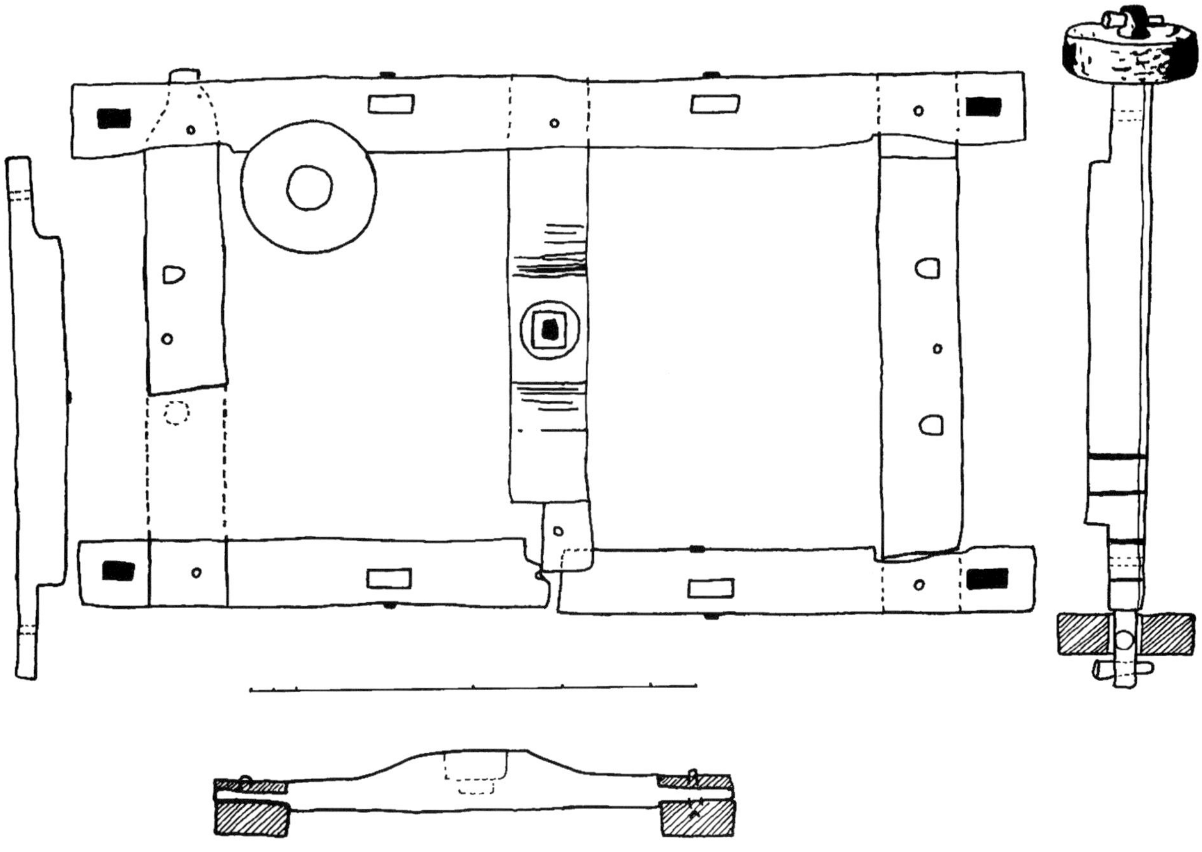

Fig. 3.23: The wheeled trolley from Medinet Madi. Ptolemaic period (NTS). From Dittmann 1941: 63 Abb. 1.

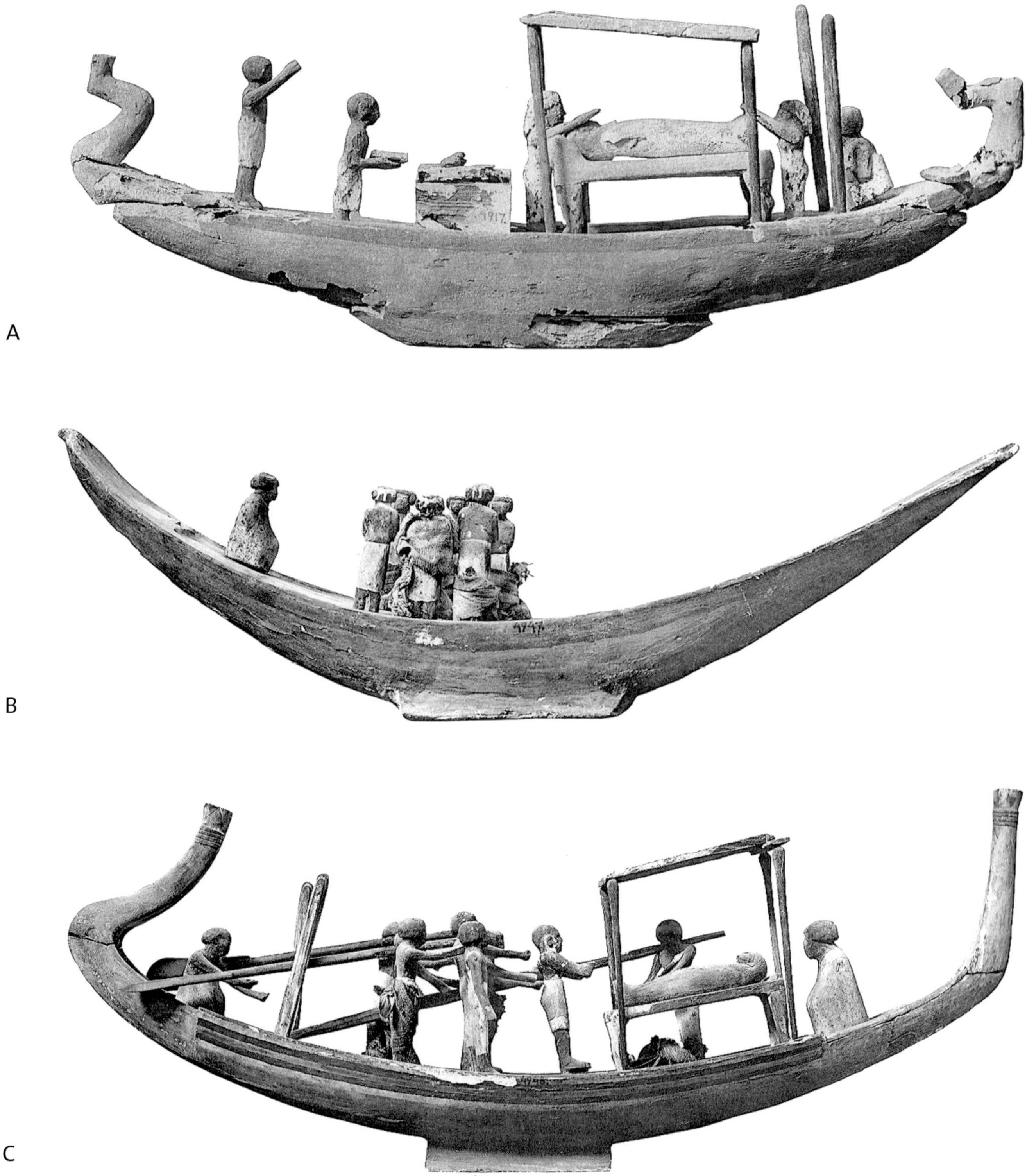

Fig. 3.24: Model ships of the dead with a *pavois* base. From Reisner 1913: pl. XVIII no. 4917, pl. XXI no. 4947, pl. XXII no. 4948.

A

B

Fig. 3.25: Apis bull funerary boat transported on an eight-wheeled wagon. (A) From Kom el Fahri. XXVIth Dynasty. (B) Similar relief now in the Louvre. Ptolemaic period. A from Vigneau and Drioton 1949: pl. 192.

formance of their vows. The image of the god, which is kept in a small wooden shrine covered with plates of gold, is conveyed from the temple into a second sacred building the day before the festival begins. The few priests still in attendance upon the image place it, together with the shrine containing it, on a four-wheeled cart, and begin to drag it along; the others, stationed at the gateway of the temple, oppose its admission. Then the votaries come forward to espouse the quarrel of the god, and set upon the opponents, who are sure to offer resistance. A sharp fight with clubs ensues, in which heads are commonly broken on both sides. Many, I am convinced, die of the wounds that they receive, though the Egyptians insist that no one is ever killed.

The natives give the subjoined account of this festival. They say that the mother of the god Ares once dwelt in the temple. Brought up at a distance from his parent, when he grew to man's estate he conceived a wish to visit her. Accordingly he came, but the attendants, who had never seen him before, refused him entrance, and succeeded in keeping him out. So he went to another city and collected a body of men, with whose aid he handled the attendants very roughly, and forced his way in to his mother. Hence they say arose the custom of a fight with sticks in honour of Ares at this festival.

How many cults with similar practices were not recorded? In truth, we do not know. Thus, when evidence linking a representation to a specific cult is lacking, it is best to leave their identity open. Several Egyptian representations of ships on wheels fall into this category:

- Two coins from Alexandria, an epicenter of Isiac worship, depict a boat on wheels carrying a *naos* (Fig. 3.26).[105] It is tempting to try to link these images with the ship of Isis and the festival of the *Ploiaphesia,* but we have no other evidence of the appearance of a *naos* on the ship used in these celebrations.[106]
- A Ptolemaic-period, molded terracotta ship model with wheels is reminiscent of Isiac boat-shaped lamps, but here, again, no other features definitely identify it as belonging to that cult, so caution is advised (Fig. 3.27).[107]
- Two sherds that appear to be mold siblings (now in the Benaki Museum) depict a cult ship on wheels bearing a *naos* (Fig. 3.28).[108] The ship has a bull-head stem ornament and a gazelle(?)-head stern ornament with forward-curving horns, both facing forward.[109] The ship carries two quarter rudders. These sherds, which were purchased in Egypt, are attributed to the Athribis School and dated to the first half of the second century B.C.

I am aware of only one other representation of an Egyptian ship that combines gazelle(?) and bull iconography (Fig. 3.29).[110] I. F. Moll describes the image as coming from "Palast Ramses IV, zu Theben XX. Dynastie (Champollion III, 257)."[111] L. Bell comments on this representation:[112]

> This image—said by Moll to come from the "palace" of Ramses IV—actually represents one of seven boats found in the wall paintings of the Tomb of Ramses III (No. 11) in the Valley of the Kings (*Monuments de l'Egypte et de la Nubie* III, pls. 255–257; utilizing an old numbering system, Champollion identifies the king as Ramses IV). Unfortunately, this tomb has not yet been published in modern times. But according to Porter and Moss (*PM*), *Topographical Bibliography of Ancient Egyptian Hieroglyphic Texts, Reliefs, and Paintings,* vol. I.2 (1964), pp. 519 (9: Side-room D), 510 (plan of the tomb), these ships occur in the unparalleled decoration of one of the ten side chambers carved off the first two entrance corridors of the tomb. *PM* states that the Oriental Institute has photos of the ships (probably taken in the 1930s)—as they survive today. They are listed as negatives 3392, 3405–6, and 6500. (Neg. 6500 is misidentified in *PM:* that negative actually documents a scene from the tomb of Queen Twosret.)
>
> Added under Ramses III, the vessels apparently have nothing to do with the text of the Litany of Re (one of the Netherworld Books) carved in the corri-

A

B

Fig. 3.26: Alexandrian coins depicting a cult ship on a four-wheeled wagon, perhaps the ship of Isis used in the *Navigium Isidis*. (A) A.D. 160. Obv. Laurate bust; inscription: (in Greek): ΜΑΥΡΗΛΙΟC ΑΝΤΩΝΙΝΟCCΕ (Antonius Pius). (B) A.D. 107. Obv.: Laurate bust with aegis; inscription: (in Greek): ΑΥΤΤΡΑΙΑΝC ΕΒΓΕΕΡΜΔΑΚΙΚ (Hadrian). From Dattari 1901: pls. XXVII no. 3557, XXX no. 1158.

Fig. 3.27: Terracotta molded ship model with wheels from Egypt. Ptolemaic. From Weber 1914: 256 Abb. 128.

A

B

Fig. 3.28: Ships transported overland on wheeled carriages. First century B.C. Photos courtesy of the Benaki Museum, Athens.

Fig. 3.29: Ship with stern-facing gazelle(?)-head stempost, and a bull standard planted amidships. Painted in a side chamber in the tomb of Ramses III (NTS). From Moll 1929: pl. A IV h: no. 484.

> dors under his father Sethnakht. Reeves and Wilkinson, *The Complete Valley of the Kings* (1996), p. 160, describe these chambers: "On the walls of the side chambers added by Rameses III, unique secular scenes were painted, including the royal armoury, representations of boats, the famous blind harpists of James Bruce . . . and, in the scenes of the king's treasury, many luxury items, some of which were clearly imports from the Aegean." This last reference, of course, is to the well known Mycenaean pottery, including IIIC(?) stirrup jars (whether real or imitation) depicted here. In Kent Weeks, ed., *Valley of the Kings* (2001), p. 239, Ted Brock writes: "Decorative innovations in the tomb are primarily in the subject matter of the side rooms off the first and second corridors. These include food preparation, offerings from different parts of the realm, and funerary equipment.

Unfortunately, this brings us no closer to understanding the cultic identity of the ship represented on the Benaki sherds.

Divination.—Beginning in the New Kingdom, portable boat shrines also served as a means by which the gods could deliver oracles to their supplicants.[113]

Some pharaohs—including Hatshepsut, Thutmose III, Thutmose IV, and Ramses III—report receiving guidance from Amun in this manner.[114] Psusennes II (Pasebakhaenniut II), the last pharaoh of the XXIst Dynasty, requested an oracle from Amun to determine whether Shishak I, the future founder of the XXIInd Dynasty, could establish a funerary cult for his father, Nemrat, at Abydos.[115] Pinudjem II, a high priest of Amun, records an oracle regarding Amun's displeasure with a deceitful priest named Djuhtemose.[116] The Banishment Stele records another high priest of Amun, Menkhperre, receiving an oracle from Amun, one that permitted the return of persons banished from Thebes to Kharga Oasis, and the deity's agreement to desist from any future deportations.[117] These consultations were not limited to the concerns of living pharaohs. When Siamun moved the mummies of Seti I, Ramses I, and Ramses II to the cache in which they were eventually discovered in the nineteenth century, he did so only after Mut had approved this arrangement.[118]

At Thebes, under the theocratic rule of the high priests of Amun, who dominated Upper Egypt, concurrent with the rule of the weakened XXIst Dynasty in Lower Egypt, almost all state decisions appear to have been made by inquiring of Amun, Mut, and Khonsu.[119] Černý notes that during this period, "Thebes and Upper Egypt were virtually ruled by the gods."[120] Amun also seems to have intervened in the legal transactions of his high priests.

A certain degree of cynicism seems prudent regarding the results of such oracles, at least at the state-sanctioned level, and particularly under the sacerdotal rule of the high priests of Amun.[121] As J. H. Taylor notes wryly:[122]

> It is difficult to avoid the supposition that kings and high officials manipulated this system for their own benefit. The ruling High Priests of the Twenty-First Dynasty were not slow to see the advantage the process could hold for them . . . they made use of oracular decrees to sanction their policies; they developed what had begun as a popular practice, carried out in impromptu fashion during a procession, by instituting a regular 'Beautiful Festival of the Divine Audience', lasting for several days, in which the statue of Amun of Karnak, the most prestigious of the divine images of Thebes, was brought forth from his sanctuary specifically to give oracular pronouncements. The solemnity of the occasion was increased by the additional presence of the images of Mut and Khons, and by the staging of the event on the 'Silver Floor of the

House of Amun', an especially sacred spot probably located within the court of the Tenth Pylon at Karnak. Records of important oracular judgments were carved on the walls of courts and on the surfaces of pylons along the processional route leading southwards from the sanctuary to the Tenth Pylon.

How was the will of the gods made known to mortals in this manner? Regarding the mechanics of oracles in ancient Egypt, Černý concludes that the movement of a deity's boat palanquin indicated the correct response. Movement forward equaled a positive answer, whereas backward movement indicated a negative response.[123]

Due to Amun's preeminence during the New Kingdom, his worship eventually spread in the south as far afield as Napata in Nubia and in the northwest to Siwa Oasis, on the Libyan border.[124] The oracular aspect of the Amun cult was present at both locations. Undoubtedly, the most famous of these oracles was the one Alexander the Great received at Siwa in 331 B.C.[125]

Such numinous decisions were not limited to the Theban Triad, however. Oracles appear to have been a phenomenon prevalent among other Egyptian deities as well. Records exist of oracles given by Montu, Isis, and Ptah.[126] During the XXIInd Dynasty the chief god of Dakhla Oasis, Setekh, gave an oracle that determined a priest's fate.[127]

By requesting oracles, commoners occasionally had an opportunity to interact with gods traveling in their portable barques. The people submitted to the transported deity questions written on sherds or stone flakes.[128] A document from Deir el Medineh, P. BM 10335, is most instructive regarding oracles given to the general public.[129] It records garments stolen while under the supervision of a guard named Amenumia, who asked the *local* Amun, named "Amun of Pakhenty," to identify the culprit. After the god agreed to do so, the names of the villagers were called out, and the god "moved" at the reference to one of the names, a certain Petauemdiamun. This led to the suspect's denying his culpability, which reportedly upset the god. In the ancient equivalent of challenging a court decision, Petauemdiamun proceeded to get second opinions from two other gods, both of whom denounced him and sent him back to Amun of Pakhenty. In the end Petauemdiamun confessed to the crime.

Amenhotep I, who was immensely popular at Deir el Medinah, was deified there after his death. He was called on, as a god, to determine lawsuits. Unlike other deities, when his statue was transported overland, however, Amenhotep was not hidden and did not ride in a boat.[130] Ahmose and Ramses II, following their deaths and deifications, also delivered oracles.

The Moulid of Abu el Haggag.—No discussion of Egyptian cult vessels transported overland would be complete without reference to the boats drawn in procession to this day during the *moulid* (birthday celebration) of Sheikh (Sidi) Abu el Haggag el Uqsuri.[131] This occurs each year in Luxor a fortnight before the beginning of Ramadan, on the fourteenth day of the Moslem month of Shaaban.[132] As part of this celebration display, boats secured on wagons are towed through the streets of Luxor. Of the many other *moulids* celebrated today in Egypt, that of Abu el Haggag is unique in this regard.[133]

The festival honors Luxor's medieval Islamic patron saint. Abu el Haggag descended from a noble Meccan family related to the Prophet Mohammed. Born in Mahdia, Tunisia, in the twelfth century (sixth century H), he became an Islamic scholar and immigrated to Luxor, where he founded a school in the ruins of Luxor temple within the court of Ramses II. This developed into the present-day mosque of Abu el Haggag, which is the focal point of the *moulid*. The mosque was built at a time when Luxor Temple was buried to a considerable depth under debris. The temple's subsequent excavation resulted in the mosque's present and rather bizarre location atop a high podium (Fig. 3.30).[134]

The mosque, thus, is located in the former epicenter of the pharaonic Opet Festival, in which the transport of deities in their boat palanquins played a major role.[135] Egyptologists generally agree that the transport of boats in the Abu Haggag procession represents a faint memory of the Opet Festival.[136] Today the celebration is intensely Islamic in nature, however. Perhaps understandably, given the present-day Islamic tenor of the *moulid*, modern celebrants are adamant in disassociating the modern practice of transporting boats from pharaonic traditions. Members of the Haggag family (the *Hajjajjia*) whom I interviewed in 1998 gave me numerous explanations as to why the

Fig. 3.30: The mosque of Abu el Haggag. View from west southwest. Luxor, 1998.

Fig. 3.31: Sufi dancing in the back streets of Luxor as a prelude to the *dura*. Luxor, 1998.

Fig. 3.32: A horseman displays his equestrian skills in the *murmah.* Luxor, 1998.

Fig. 3.33: Placing the Abdul Sharif boat on its wagon prior to the *dura.* Luxor, 1998.

Fig. 3.34: The El Husein boat on its wagon. Luxor, 1998.

Fig. 3.35: Boat of the felucca men taking part in the Abu el Haggag pageant *(dura)*. Note the men holding model feluccas (arrows). Luxor, 1998.

Fig. 3.36: Detail of one of the members of the felucca men's processional boat holding a model of a felucca. Luxor, 1998.

boats are pulled in procession, while denying that the custom had any ties at all to pharaonic times (Figs. 3.35–39).[137]

The highlight of the festival is the *dura* (circuit), which refers to the course of the carnival-like parade.[138] In anticipation of the *dura* the boats are repainted and decorated. With one exception, these are wood-planked vessels that differ markedly from modern river craft. The processional craft vary in length from 4 to 6 meters and have an unusual beak at the bow. J. Hornell speculates that the boats of his day replicated Turkish galleys.[139]

Various celebrations take place during the week preceding the *dura*. The festival begins with a convocation led by the present head of the *Hajjajjia*, which can trace its lineage to the saint and funds both the mosque and the *moulid*. Celebrants from near and far descend on Luxor during the week of festivities. *Zakr* (Sufi dancing) is carried out to the mesmerizing beat of Arabic music in large tents constructed to the east of Luxor Temple: Dances also take place spontaneously in the streets (Fig. 3.31). Men spend the evenings studying the Koran and Haditha in the mosque.

The entire area surrounding the mosque takes on a carnival atmosphere. In the *murmah* riders display horsemanship by galloping while dragging a long stick *(zana)* along the ground, raising clouds of dust (Fig. 3.32). The mosque itself is decorated with lights, and vendors sell pastries and cheap festival souvenirs. On the evening before the *dura* the *Hajjajjia* hosts an all-male tent meeting featuring Egypt's leading Islamic scholars and muezzins.

On the morning of the *dura,* in anticipation of the procession, the boats are placed on wagons, but their use represents a relatively recent innovation (Figs. 3.33–34). Lady Duff Gordon, writing in 1864, records a boat *carried* in the procession of the *moulid* of Abu el Haggag:[140]

> *Friday, January 29.* The moolid (festival) of the Shaikh terminated last Saturday with a procession, in which the new cover of his tomb, and the ancient sacred boat, were carried on men's shoulders. It all seemed to have walked out of the royal tombs, only dusty and shabby instead of gorgeous. . .

The number of vessels transported in the *dura* has grown in recent times. Duff Gordon describes a single boat in use. In 1925, when Harry Burton filmed the *dura,* he documented two boats, for which Hornell also supplies photos.[141] By 1983 three boats were in use.[142] Six vessels, including one made of iron belonging to the felucca men, took part in the *moulid* in 1998 (Fig. 3.35). These boats are built only for the *moulid* and never sail the Nile. Men standing on the iron boat held scale models of feluccas (Figs. 3.35: arrows; 36).

On the morning of the *dura* the boats receive final preparations, and their owners mount them on their wagons, after which the boats are wheeled to a side street next to the mosque. Children, an integral part of the festivity, crowd into the boats. A mass of humanity descends on the mosque and the grounds opposite its entrance to the east of Luxor Temple.

Fig. 3.37: The *dura* on the Corniche at Luxor during the 1998 *moulid*.

Fig. 3.38: The El Husein boat during the *dura*. Luxor, 1998.

Fig. 3.39: The boats are towed in the *dura* by means of two long ropes in a manner reminiscent of the towing of the *Amunuserhet* during the Opet Festival. Luxor, 1998.

A

B

Fig. 3.40: (A) Towing cult ships by means of cables during the Opet Festival depicted in Tutankhamun's tableau in the Colonnade Hall, Luxor Temple. (B) Detail. From Epigraphic Survey 1994: pls. 18, 23.

The heads of the *Hajjajji* family lead the parade, followed by camels carrying wood-and-cloth mockups of sheikhs' tombs (Fig. 3.37). The flotilla of boats follows (Figs. 3.35, 38). Celebrants tow the vessels with thick ropes, thus evoking the manner in which the *Amunuserhet* was towed from shore in pharaonic times (Figs. 3.39–40).[143]

Greece

The ship-cart (carrus navalis) of Dionysos.—Ships on wheels or carried by porters do not normally appear in the Minoan or the Mycenaean record. The one exception—if it can be counted as such—is the ship model carried by a mourner/porter on the Late Minoan IIIA Hagia Triada sarcophagus (Fig. 3.11).[144]

Within the Helladic world, cultic artifacts on wheels make their appearance in the Iron Age. One of the earliest and most curious examples is a Late Helladic IIIC three-wheeled, bull-head chariot figurine from Karphi.[145] For the earliest reported representation of a ship on wheels in Greece, we must return to the site of Pyrgos Livanaton (Kynos), which revealed fragments of terracotta ship models in addition to the remarkable painted depictions of Late Helladic IIIC ships already discussed.[146] One ship-model fragment has a piercing in the lower part of the hull aligned perpendicular to the keel (Fig. 3.41).[147] As noted by F. Dakoronia, this piercing serves for an axle to attach a pair of wheels to the model, and, indeed, a terracotta wheel was uncovered nearby (Fig. 3.42).[148] Dakoronia identifies the ship model as a child's toy, while calling attention to the fact that such toys on wheels are otherwise unknown from this period.[149] Given that the Kynos ship model on wheels dates roughly contemporaneously with the Gurob ship-cart model and the Karphi chariot, I believe we have good reasons to identify it not as a toy but rather as a cultic object.

If the Kynos wheeled boat model is indeed of cultic nature, there is no indication of which cult it belonged to. Following its appearance, the curtain descends for a number of centuries during which, to my knowledge, no representations of ship carts have appeared in the Greek archaeological or the iconographical record.

When ships on wheels first reappear in Greece in the mid-first millennium B.C., they are associated specifically with the god Dionysos.[150] Galleys fitted with waterline rams or replicates of these vessels played a role in the cult of Dionysos from Archaic to Roman times. When it appears, the Dionysian ship-cart copies an Archaic galley prototype. In some cases, these cultic galleys were transported overland, either on a wheeled ship-cart or carried by porters. Galley races have also been linked to the Dionysian cult.

Three Attic *skyphoi* dating to ca. 500–480 B.C. bear scenes of Dionysian ship-carts in cult procession (Figs. 3.43–45).[151] All three representations show the ship-cart moving to the right. The "galleys" carry the typical galley's boar-head waterline rams of Archaic warships and screened

A B

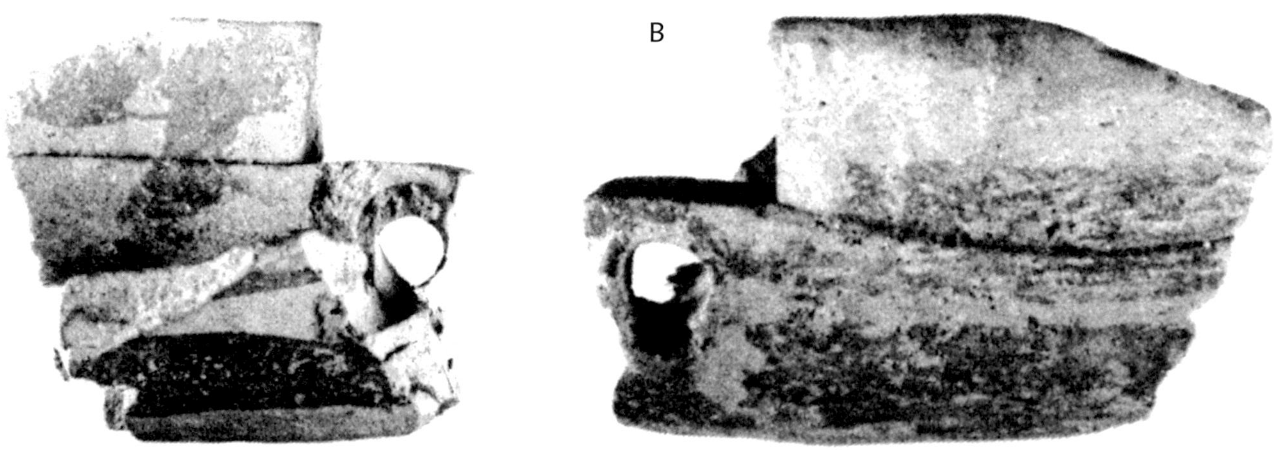

Fig. 3.41: Terracotta fragment of a ship model from Pyrgos Livanaton (Kynos) with a hole for an axle. Late Helladic IIIC1b (NTS). From Dakoronia 2002: 289 figs. 1–2. Courtesy Hellenic Institute for the Preservation of Nautical Tradition.

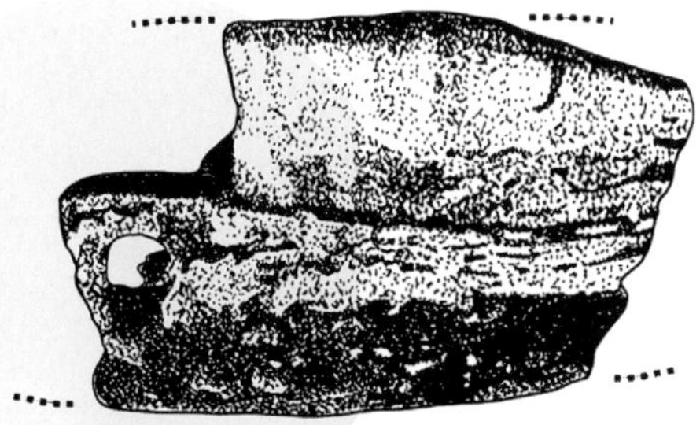

Fig. 3.42: The Pyrgos Livanaton ship model fragment with a wheel reconstructed (NTS). After Dakoronia 2002: 289 figs. 2 and 4.

forecastles familiar from representations of actual contemporaneous ships.[152] Beginning under the center of the forecastle, a row of closely fitted vertical lines descends from below the sheer strake, representing the prototype galley's bank of oars. I interpret the rigid manner in which these banks of "oars" are portrayed as indicative of a wooden panel with the oars painted and/or carved on it. The panel, if such it was, would have angled out from the "hull," probably supported by struts placed judiciously between the hull or cart and the inner side of the panel.

The panel extends over the wheels, hiding them to below the axles so that only about the lower third of the wheels is visible. In two of the ships—those of the British Museum and the Athens Museum—the wheels have two parallel spokes (Figs. 3.43, 45). Depictions of similar wheels on two-wheeled wagons used in bridal processions clarify the actual structure of the wheels on the ship-carts.[153] This type of wheel had a single long spoke crossing its diameter and two subsidiary spokes bisecting it at right angles midway between the axle and the wheel. The Athens Museum ship-cart is unique among this group in displaying an *ophthalmos* at its bow (Fig. 3.45).

Stern endings vary on the ship-carts. The stern section of the Athens Museum ship-cart is missing. The British Museum ship-cart terminates in a recurving inboard-facing bird head (Fig. 3.43: B).[154] The Bologna ship-cart's stern ends in a rising bifurcation, which is not common in contemporary galleys (Fig. 3.44: B). It is not possible to determine from the image whether the two rising stern elements are stacked above the other or lie horizontally next to each other.[155]

The sternward extremity of the Bologna ship-cart's oar panel is unique among the ship-cart depictions in showing the stern end of the oar panel. The oars at the stern curve up, and their lower ends are level with the bottom of the "hull." This ship-cart has four-spoked wheels as opposed to the type of wheels on the other two ship-carts.

A fourth scene of a Dionysian ship-cart procession appears on a sheet of lead thought to have been stolen from Montagna di Marzo in central Sicily (Fig. 3.46).[156] This ship-cart is similar to those on the *skyphoi.* The ship-cart faces right. The wheels have the same double bar visible on the ship-carts in both the British Museum and the Athens Museum. Unique to the Montagna di Marzo scene, however, are two rings, one at the bow and the other at the stern, to which are attached ropes by means of which celebrants pull and maneuver the cart. The stern ring includes the bar that secures it to the ship-cart, a detail lacking at the bow. The ship lacks a sternpost.

As the ship-carts are two-dimensional representations, it is not possible to determine whether the actual contraption was simply a cart with ship elements built on to it or an actual purpose-built galley replicate transported on a cart, similar in concept to the boats in the Abu el Haggag festival discussed earlier. The latter possibility is the more likely in my view.

All four scenes of the Dionysian ship-cart display an object hanging over the sterncastle fencing.[157] The items bear a crosshatched decoration. A similar object covers the stern of a galley at sea on a black-figure vase (now in the Museum of Tarquinia) in which a sitting Dionysos appears in the vessel surrounded by his retinue (Fig. 3.47: A).[158] Black-figure vase artists used similar crosshatching, sometimes decorated with dots, stars, or circles, to represent the forecastle screens of galleys, including on the Tarquinia vase.[159]

The object transported by the Bologna ship-cart differs from the other representation in being spread abaft the stern on a horizontal spar affixed to the stern (Fig. 3.44). On the Montagna di Marzo ship-cart the object appears to be wrapped around the vessel's stern.

A single sherd from a fourth *skyphos,* now in Tübingen, is anomalous in showing the forepart of the vessel,

A

B

Fig. 3.43 (A): Dionysos in a ship-cart painted on a *skyphos* now in the British Museum. (B) detail (NTS). From Kerényi 1976: ils. 59: A–B.

A

B

Fig. 3.44 (A): Scene of Dionysos in a ship-cart painted on a *skyphos* now in the Bologna Museum.(B) detail (NTS). From *CVA* Bologna 2 (Italy 7): Tav. 43: 88.4.

Fig. 3.45: Dionysos in a ship-cart on a black-figure *skyphos* now in the Athens Museum (NTS). After Kerényi 1976: il. 57.

with this item reaching all the way forward to the boar-head ram (Fig. 3.47: B).[160] In this case the object bears what appears to be a design taken from the world of fabric decorations.

On several ancient terracotta models of galleys that have been raised from the sea, this object is shown in its three-dimensional form draped over the sterncastle (Fig. 3.48).[161] The fact that these ship models come consistently from the sea indicates that they did not reach the seabed accidentally: We may reasonably presume that they were jettisoned into the sea as part of a cultic ceremony. The only *Greek* representations of which I am aware in which these items appear unconnected to Dionysos are two scenes showing Odysseus lashed to the mast.[162] O. Höckmann notes that some representations of later Phoenician galleys appear to carry a similar item hanging off their sterns.[163]

Remnants of a scene depicting a Dionysian cult ship—in this case carried by porters rather than a ship-cart—also appear on sherds from a black-figure neck amphora reported to have come from Karnak.[164] The amphora dates to 550–540 B.C. One panel portrays a boy and two dogs in a vineyard.[165]

A second panel survives only in a handful of sherds. One sherd depicts a man holding a large phallus in his right hand and wearing a kilt with a horsetail attached, role-playing a satyr (Fig. 3.49: A).[166] The kilt signifies that the scene portrays an actual ceremony rather than a mythological one, as the kilt allows for the attachment of the satyr's tail.[167] Behind this first figure J. Boardman identifies the tips of a double flute played by a now-missing second figure. A contorted third figure holds a phallus in each hand while balancing precariously on the ship's boar-head waterline ram (Fig. 3.49: B).[168]

Additional sherds bear remnants of four kilted men whom Boardman reconstructs as porters carrying the vessel (Fig. 3.49: C–D). These men vary in height. Boardman explains this variation in their height by placing the taller pair at the missing stern, where the hull would have curved upward.

These meager remains allow Boardman to propose a convincing reconstruction of the entire scene (Fig. 3.50). He notes that the satyr role-players, as well as the phalli they hold, indicate that the scene is connected to Dionysos. The vineyard scene on the vase's opposite side further solidifies this identification with the deity. Presumably a statue of Dionysos or someone playing the part of the god would have been ensconced in the now-missing central part of the vessel.

The possibility that the sherds of this vase came from Karnak is of particular significance given that site's important role in the Opet Festival, with its emphasis on boat-shaped palanquins carried by priests.[169] The fact that the ship is carried rather than moving on wheels represents just one more example of the interchangeability of methods of transporting cultic ships and their replicates overland.

A particular type of one-handled vessel, known as a *chous* is linked to the Attic Dionysian Anthesteria.[170] R. Hamilton cautiously proposes that the Anavysos Chous displays a scene of a Dionysian ship-cart (Fig. 3.51):[171]

> A quite different possibility is that the parallel curved lines represent the curved stern of a boat. The picture then could represent the Dionysian *carrus navalis* found on a number of vases. The bearded spectator "wrapped in his mantle, wreathed" bears a striking resemble to the Dionysus seated in one of the "floats" [Fig. 3.44]. The grape vine has vanished and flute players have been replaced by our dancing Perseus, but Dionysus's ivy crown remains. The shape of the pot suggests that it is appropriate for the Anthesteria and perhaps our dancing Perseus is to be explained as one of the comic actors who were said to compete for a place in the City Dionysia during the Anthesteria. The second spectator, holding a staff, remains a mystery.
>
> It must be admitted that there are difficulties with such an interpretation. First Dionysus is facing not the prow but the stern, which itself lacks the usual, double, swan-headed *aphlaston*. Secondly, he is holding none of the usual implements: cantharus, thyrsus or grape vine. Finally, there is no firm evidence connecting the *carrus navalis* with the Anthesteria.

The curving lines at the bottom right of the scene could reasonably be interpreted as either the rising lines of a vessel's port stern quarter or the starboard bow of a ship. The narrower groups of curving lines in that case could represent strengthening wales and a caprail. The diagonal double line with rungs rising to the platform, in this line of thought, may represent a boarding ladder like those commonly shown hanging off the sterns of ships or in use.[172] As the Dionysian ship-carts lack the platform supported by column(s) on which the Perseus-actor is performing, it seems more prudent to identify this as a shiplike stage structure purpose-built for a performance.[173]

The Anthesteria celebrated the return of Dionysos in triumph from overseas in the spring, at the time of the first wine in the Attic month of Anthesterion.[174] Apparently, during this festival a celebrant role-playing the god entered the city in a mock galley on wheels, decked out with Dionysian paraphernalia and administered to by flute-playing actors dressed as satyrs, for a *hieros gamos* with the *basilinna*.[175]

J. Bremmer notes that the oldest indications of the activities of a god are his or her festivals.[176] Dionysos, in this regard, seems to be involved in the breakup of social order: His festivals contain elements that divide men from women, give equality to slaves, and emphasize phalli during the Dionysia. The rural Attic Dionysia *en agrois* religious festivals, for which Aristophanes gives the most detailed description, appear to have emphasized phalli.[177]

In the Late Roman period Dionysian ship-carts participated in religious processions known as *katagogia* in the cities of Ionia. At Smyrna, the local equivalent of the Dionysian ship-cart took part in the Anthesteria. Describing the honors that the Sophist Polemo received from Smyrna, Philostratus (ca. A.D. 230–238) notes the following:[178]

> [F]or they bestowed on him and his descendants the right to preside over the Olympic games founded by Hadrian, and to go on board the sacred trireme. For in the month Anthesterion a trireme[179] in full sail is brought in procession to the agora, and the priest of Dionysus, like a pilot, steers it as it comes from the sea, loosing its cables.

A

B

Fig. 3.46: (A) Dionysian ship-cart depicted on a lead sheet. Date: sixth-fifth centuries B.C.(?) Looted and now in a private collection. Said to have come from Montagna di Marzo, central Sicily. (B) Detail of ship-cart (NTS). From Wilson 1987–1988: 135 fig. 36.

A

B

Fig. 3.47: (A) Ship of Dionysos on an amphora. Note the fabriclike object hanging over the stern, ca. 510 B.C. Museum of Tarquinia, Amphora no. 678. 3.47B. The starboard side of the forward section of a Dionysian ship-cart appears on this sherd now in the Tübingen Institute for Classical Archaeology. This representation is unique in depicting the typical fabriclike object along the entire surviving starboard side of the hull, all the way forward to the boar-head waterline ram (B) at the far right. A from Basch 1987: 226 fig. 470: A. Courtesy Hellenic Institute for the Preservation of Nautical Tradition. B courtesy Tübingen, Institut für Klassische Archäologie, Inv. S./10 1497.

The Ionian painter of the Karnak sherds presumably would have known of similar boat processions from Dionysian festivals that incorporated a war galley in his homeland, as documented later at Smyrna. Thus, one can trace this interconnection of Dionysos and ship-carts in iconography and texts stretches from the sixth century B.C. to the third century A.D. The Homeric Hymn to Dionysos gives some insight into the connection between this god, galleys, and the sea:[180]

> I will tell of Dionysus, the son of glorious Semele, how he appeared on a jutting headland by the shore of the fruitless sea, seeming like a stripling in the first flush of manhood: his rich, dark hair was waving about him, and on his strong shoulders he wore a purple robe. Presently there came swiftly over the sparkling sea Tyrsenian pirates on a well-decked ship—a miserable doom led them on. When they saw him they made signs to one another and sprang out quickly, and seizing him straightaway, put him onboard their ship exultingly; for they thought him the son of heaven-nurtured kings. They sought to bind him with rude bonds, but the bonds would not hold him, and the withes fell far away from his hands and feet: and he sat with a smile in his dark eyes. Then the helmsman understood all and cried out at once to his fellows and said:
>
> "Madmen! what god is this whom you have taken and bind, strong that he is? Not even the well-built ship can carry him. Surely this is either Zeus or Apollo who has the silver bow, or Poseidon, for he looks not like mortal men but like the gods who dwell on Olympus. Come, then, let us set him free upon the dark shore at once: do not lay hands on him, lest he

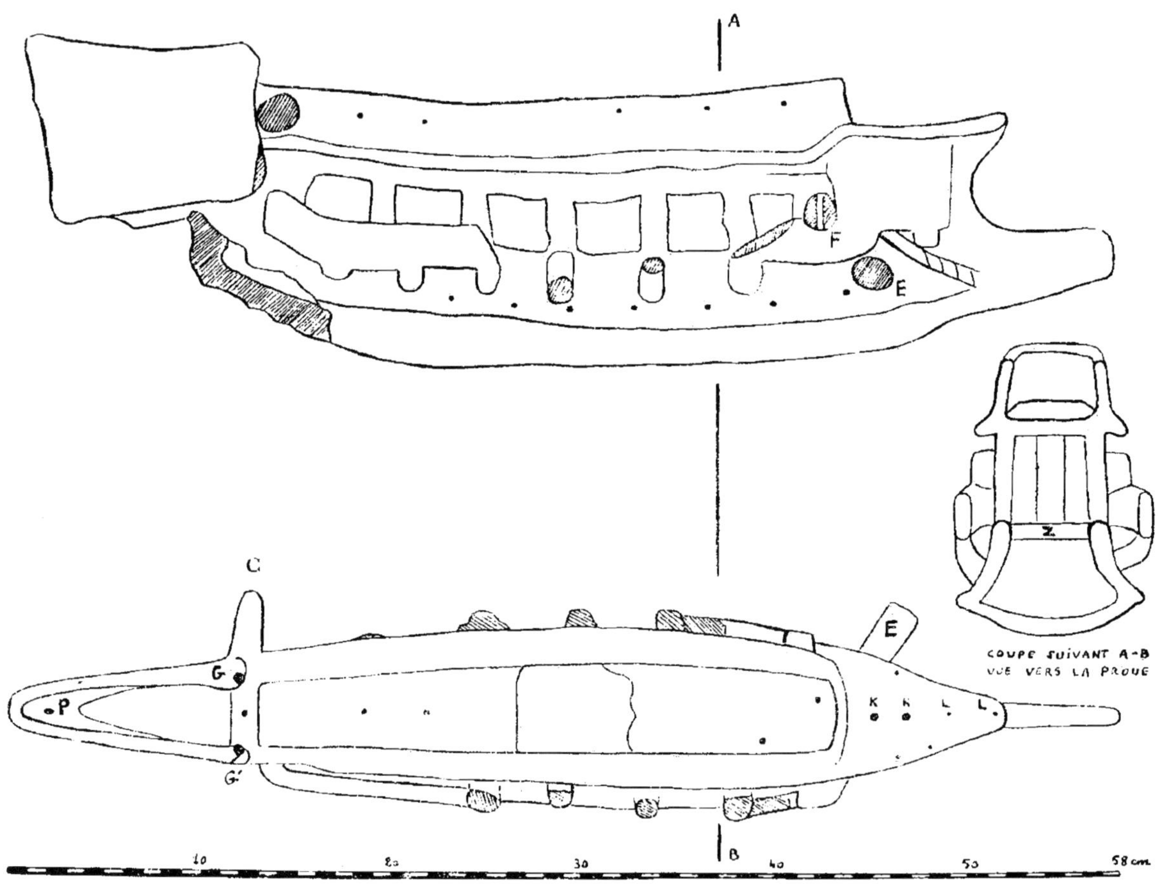

Fig. 3.48: Terracotta model of a galley with an object covering the stern castle. Museum of Sparta. From Basch 1987: 433 fig. 943. Courtesy Hellenic Institute for the Preservation of Nautical Tradition.

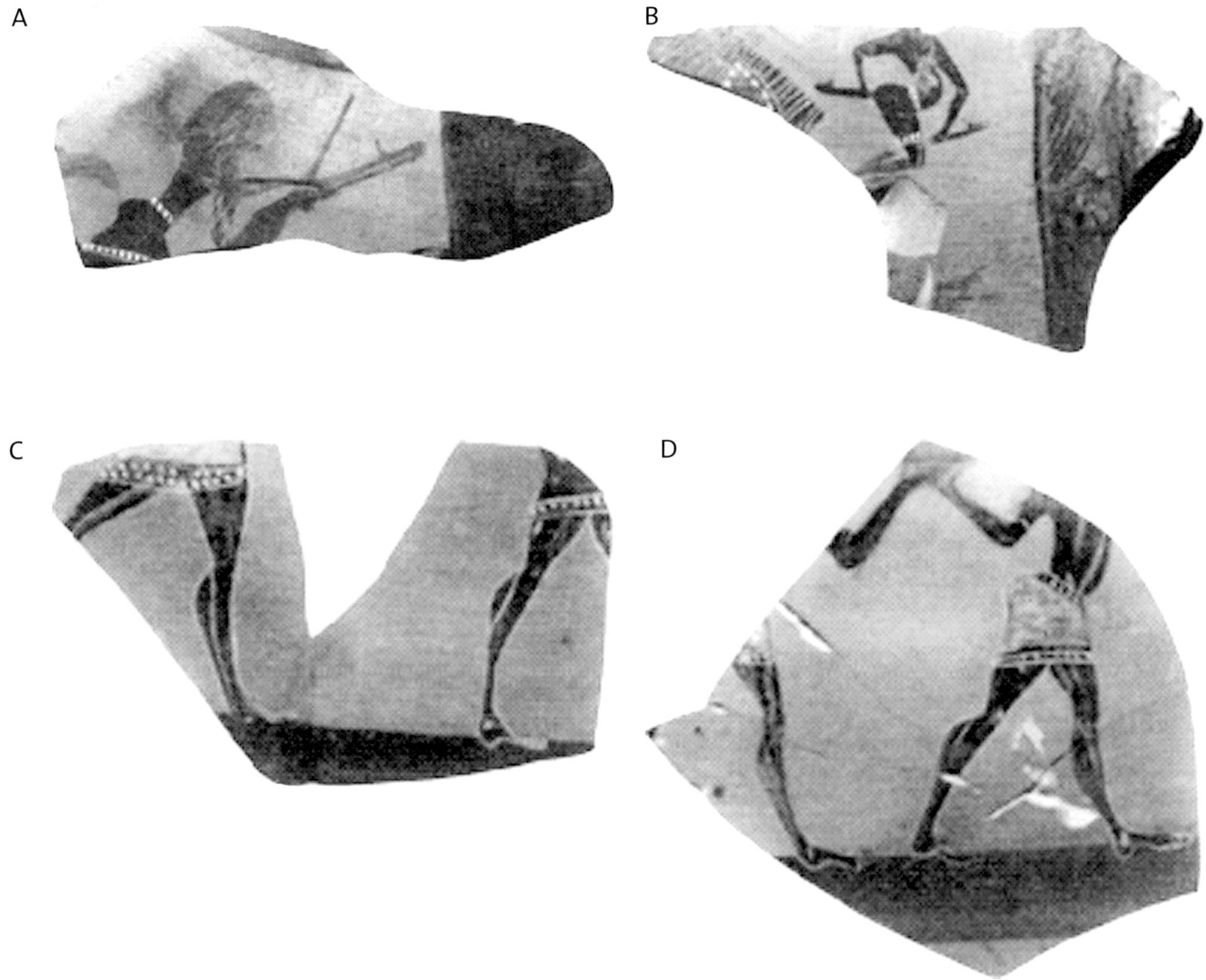

Fig. 3.49: Ashmolean Museum, Oxford, sherds 1924. 264. From Boardman 1958: pl. I: b–f.

grow angry and stir up dangerous winds and heavy squalls."

So said he: but the master chid him with taunting words: "Madman, mark the wind and help hoist sail on the ship: catch all the sheets. As for this fellow we men will see to him: I reckon he is bound for Egypt or for Cyprus or to the Hyperboreans or further still. But in the end he will speak out and tell us his friends and all his wealth and his brothers, now that providence has thrown him our way."

When he had said this, he had mast and sail hoisted on the ship, and the wind filled the sail and the crew hauled taut the sheets on either side. But soon strange things were seen among them. First of all sweet, fragrant wine ran streaming throughout all the black ship and a heavenly smell arose, so that all the seamen were seized with amazement when they saw it. And all at once a vine spread out both ways along the top of the sail with many clusters hanging down from it, and a dark ivy-plant twined about the mast, blossoming with flowers, and with rich berries growing on it; and all the thole-pins were covered

Fig. 3.50: Boardman's reconstruction of a Dionysian cultic ship with boar-head ram and satyr-actors carried by men, based on East Greek black-figure sherds from Luxor (Ashmolean Museum, Oxford, 1924. 264). After Boardman 1999: 138 fig. 162; 1958: 6 fig. 2, pl. I: c–f.

with garlands. When the pirates saw all this, then at last they bade the helmsman to put the ship to land. But the god changed into a dreadful lion there on the ship, in the bows, and roared loudly: amidships also he showed his wonders and created a shaggy bear which stood up ravening, while on the forepeak was the lion glaring fiercely with scowling brows. And so the sailors fled into the stern and crowded bemused about the right-minded helmsman, until suddenly the lion sprang upon the master and seized him; and when the sailors saw it they leapt out overboard one and all into the bright sea, escaping from a miserable fate, and were changed into dolphins. But on the helmsman Dionysus had mercy and held him back and made him altogether happy, saying to him:

"Take courage, good . . . ; you have found favour with my heart. I am loud-crying Dionysus whom Cadmus' daughter Semele bare of union with Zeus."

Hail, child of fair-faced Semele! He who forgets you can in no wise order sweet song.

This hymn appears as a motif in pagan art from the sixth century B.C. Exekias Cup to a third-century-A.D. mosaic from Dougga, Tunisia.[181]

The connection of Dionysos to the sea may also have been expressed in ship races. J. E. Harrison proposes that black-figure drinking vessels that display groups of galleys juxtaposed with Dionysian cult symbols may be connected to galley races carried out in his honor.[182] Toward the end of the production of black-figure vases, a motif of several ships appears around the vertical edge of the inner lip of vessels.[183] When full, the ships would have appeared to sail on a wine-dark sea. Furthermore, if we consider that the Greeks regularly mixed seawater with their wine, *symposia* members would have been correct in considering the ships on the rims of their mixing cups to be "at sea."[184]

Fig. 3.51: The Anavysos chous. From Hamilton 1978: 385 fig. 1.

Harrison further argues that the large eyes *(ophthalmoi)*, which often appear on black-figure vessels represent galleys participating in Dionysian races: The *kylikes* (cups) on which these decorations appear would have made them particularly appropriate for this purpose.[185]

Pausanias, writing in the second century A.D., supplies literary evidence of ship races in honor of Dionysos at the settlement of Hermione:[186] "Near by is the temple of Dionysus Melanaegis. In his honour is yearly held a musical festival, with swimming-races and boat-races."

With the advent of the red-figure style, nautical scenes all but disappear from the vase painters' repertoire. J. Hale notes that the link between the Athenian maritime world and democracy in the fifth century B.C. perhaps made ship scenes unappealing to the wealthy Athenian patrons who commissioned the works of art.[187]

The Panathanaic ship on wheels.—The Athenian Panathenaia was the most important festival in the Athenian religious calendar: It celebrated the traditional birthday of Athena, patron goddess of Athens, and was also associated in ancient sources with the defeat of the giants by the Olympian gods.[188] This festival occurred at the end of the month of Hekatombaion, the first month of the Athenian calendar, which fell during the summer.[189] On the final day of celebrations (28 Hekatombaion) the goddess received a special garment *(peplos)*, brought to her in a suitably impressive procession, parts of which appear on the Parthenon frieze.[190] For much of antiquity, a ship on wheels brought the *peplos* to the Acropolis. Indeed, the custom of using a ship in the celebration continued down to the last recorded celebration of the Panathenaia in A.D. 410, although the custom of weaving the goddess a *peplos* probably ceased earlier, perhaps during the reigns of Constantine (A.D. 306–387) or Constans (A.D. 337–350).[191]

Unfortunately, although Athena's *peplos* appears on the Acropolis in the Parthenon Frieze, the Panathenaic ship is missing, perhaps because the ship was not part of the procession that took place on the Acropolis or was so large, in relation to the human figures in the frieze, that it would have caused problems in proportionality.[192]

The Panathenaia is thought to have been established no later than the seventh century B.C., although legend relates that Erichthonios/Erechtheus or Theseus created it.[193] In the mid-sixth century B.C. the Peisistratids initiated athletic games, which included ship races, every four years.[194] This special celebration became known as the "Great Panathenaia" to differentiate it from the "Lesser Panathenaia," which took place during the three off years.[195]

The preparation of the garment was an intensive group effort carried out by young women who came from the aristocratic families of Athens.[196] Up to a hundred women toiled on the *peplos*, prompting some scholars to believe that the *peplos* was enormous; alternately, the *peplos* may have been of normal—or nearly normal—size, but the honor of weaving the garment induced many aristocratic women to participate in the process.[197]

N. Robertson proposes that the preparation of the *peplos* for the goddess was introduced into Athens in the sixth century B.C., together with the Panathenaic ship, as part of an adoption of the customary New Year celebration of Lemnos and that the garment would have been consigned to Athena's treasury.[198] J. Hurwit suggests that after the celebration the new *peplos* may have been hung as a tapestry on a wall, perhaps the otherwise bare southern wall of the Erechtheion.[199]

The ship was towed along a route from the Kerameikos to the base of the Acropolis.[200] Philostatus describes the route of the ship that Herod Atticus built for the Panathania of A.D. 138–139.[201] This vessel may have been unique in being moved by an underground mechanism: "[T]he ship, as it took its course, was not hauled by animals, but slid forward by means of underground machinery."[202]

Jennifer Tobin posits that Herodes Atticus constructed a special structure to house his remarkable Panathenaic ship.[203] She identifies this as the long and narrow (42 × 9.5 m) monument commonly known as the "Tomb of Herodes Atticus" next to the Panathenaic stadium. Convincing arguments against this interpretation have been put forward, however, by Katherine Welch and J. L. Rife.[204]

Views vary regarding the Panathenaic ship's introduction into the festival.[205] Some consider the vessel to have been an integral part of the Panathanaia from its inception.[206] Mansfield believes that the Panathenaic ship memorialized the Athenian contribution to the Battle of Salamis and that one of the Greek or enemy triremes that fought there had been absorbed into the festival.[207]

Naomi J. Norman prefers to link the introduction of the ship to the Panathanaic procession to the phenomenon of rulers' dedicating ships during the Hellenistic period.[208] She emphasizes that literary references and iconographic representations of the vessel are few in number and late in date.[209] Norman identifies Demetrios Pollorketes as a likely candidate for this contribution, which he would have instituted for the Great Panathenaia of 302 B.C.[210]

A

Scholars have proposed the following four ship representations as signifying the Panathenaic ship:

THE CALENDAR FRIEZE SHIP.—Iconographic evidence for the appearance of the vessel remains scanty.[211] The Calendar Frieze, which contains the best-known representation, now graces the Small Metropolitan Church (Church of St. Eleutherios) in central Athens (Fig. 3.52).[212] The date of the frieze is debated: Scholars place it anywhere from the Late Hellenistic period to the second century A.D.[213]

The frieze contains a visual display of the twelve Athenian months, with representative symbols of the main festivals that took place each month.[214] The relief is composed of two long and narrow blocks that had been removed from an older building. Unfortunately, their total combined length proved to be greater than the church's façade, so to fit the available space each block had a section removed.[215] The left block lost the month of Anthsterion, which may have contained a representation of the Dio-

B

Fig. 3.52: (A) The Small Metropolitan Church (of St. Eleutherios) in central Athens. The depiction of the Panathenaic ship is located to the left of the entrance (arrow). (B) Detail of the month of Hektombaion.

nysian ship-cart. The right block had its right side shortened, with the loss of the month of Gamelion. This is all the more lamentable as the persons who removed these decorations could have instead shortened the inner sides of the two blocks, which had been left undecorated.[216]

The Panathenaic ship image has suffered severe degradation (Fig. 3.53). Over the left side of the frieze Byzantine artisans carved three roundel-enclosed crosses, one of which lies directly over the month of Hekatombaion, represented by the Panathenaic ship. The cross virtually obliterated the ship. The spaces inside the roundel between the arms of the cross were chiseled down, and here all detail is lost (A). The cross and most of the surface of the roundel appear to have been smoothed, removing most—*but not all*—of the plastic relief while leaving visible some deeply carved details of the ship on these raised parts (B). On the lower left side of the raised part of the roundel—from 5 to 9 o'clock—plastic relief survives. Additionally, a horizontal crack begins to the right of the cross above the ship's bow and continues across the entire area that would have contained the ship (C). This damage resulted in the loss of part of the cross's right arm, further degrading the original surface and destroying part of the upper superstructure of the ship's forecastle screen. A stempost was probably also lost in the break. A white material, which may be lichen or plaster, covers parts of the cross and the ship (Fig. 3.52: B).

Although little of the ship on the frieze survives, enough remains to identify it as Archaic in style and form.[217] The ship faces right, and the bow is visible to the right of the roundel (Fig. 3.54). The ship has a chisel-shaped waterline ram (A) but displays no evidence of the boar-head design that commonly appears on Archaic galleys.[218] A barely visible, downward-curving line ends at the tip of the ram (B): This presumably represents a massive starboard wale, which in an actual war galley engaged the ram to further strengthen the hull from harm when crashing into an enemy ship.[219] Above the ram is a thick line that rises at a slight angle (C). This represents the *proembolion,* or "fore ram," a subsidiary projection capping the massive longitudinal wales that met here.[220] The ship's forecastle sits above the *proembolion* and is surrounded by a screen. A line bisects the forecastle horizontally (D): This is a common feature in Archaic galley depictions.[221] The screen's upper rail is high at the stem but then arcs downward aft before rising slightly as it approaches the roundel (E). The line of the stempost above the *proembolion* angles slightly forward before disappearing into the crack. It is important to remember that the Calendar Frieze would originally have been painted and that these details may have been more strongly emphasized in color.[222] There is no evidence of *ophthalmoi* (ships' eyes).

The keel line continues across the bottom of the roundel to the left of the ship's bow (Fig. 3.55: A–B). Several additional marks allow, in broad terms, for the reconstruction of the line of the keel. These include a line cut into a raised element extending below the bottom of the lower arm of the cross (C) and another slanting line that passes over the lower left corner of the cross's lower arm (D).[223] This line appears to continue in a series of pockmarks in the space to the left of the cross's lower arm, although I suspect that this is fortuitous (E). A horizontal line on the upper part of the lower arm of the cross may represent the caprail, or upper boundary of the hull (F).

At the stern, a deeply cut sinuous vertical line cuts through the lower left edge of the cross's left arm (Fig. 3.56: A). Below it, remnants of the starboard quarter rudder survive outside the roundel and on the surface of the roundel itself (B1–2). This also permits the reconstruction of the angle of the quarter rudder's loom, a detail that will be of significance later in this discussion. On the left side of the roundel above the quarter rudder appears the hand of a female figure: Some scholars tentatively identify this personage as Pompe, the anthropomorphic incarnation of the Panathenaic procession (Figs. 3.56: C, 52: B).[224]

The *peplos* served as the vessel's sail.[225] Only remnants of the vaunted rig and *peplos*/sail survive, however.[226] These include a short molded protrusion at the top of the roundel, just to the right of center (Figs. 3.52: B, 57: circle, 58: A). When the line of this protrusion is extended downward, it slants toward the bow on the right. This is apparently all that remains of the ship's forward-raking mast, which must have been stepped aft of amidships. Additionally, a slanting line surviving on the upper arm of the cross may also be part of this system (Fig. 58: B).

Along the outside of the upper right quadrant of the circle and rising above the ship's bow are three protuberances (Fig. 3:58: C): These are all that remain of the ship's *peplos*/sail. The projections replicate the billowing folds of a brailed sail along its leading edge, a detail often present in Archaic representations of ships depicted under sail (Fig. 3.58: insert).[227] Apparently Philostratus had this phenomenon in mind when he wrote the following: "Moreover, I have been told the following facts concerning this Panathenaic festival. The robe of Athene that was hung on the ship was more charming than any painting, *with folds that swelled before the breeze . . .* "[228]

The ship moves on four pairs of wheels (Fig. 3.59).[229] Two wheels appear below the bow, to the right of the roundel: Remnants of two other wheels survive on the lower edge of the roundel. The tops of the two wheels are in line with, or just slightly above, the level of the keel. There is no indication of a wagon or chassis, so it is not clear what type of system supported the vessel. The four wheels lack spokes. Unless they had been added in paint, now lost, this indicates that these are constructed of solid wood.

The Panathanaic vessel's four sets of wheels indicate that the vessel's weight must have been substantial and that it was dramatically larger and more massive than the Dionysian ship-cart, which appears to have been a relatively flimsy affair that required only two sets of spoked wheels or could even be carried by porters (Figs. 3.43–46, 47: B, 49–50).[230] I am aware of only one other case from Mediterranean antiquity for the four-axle configuration for transporting a ship on wheels represented on the Calendar Frieze: the Ptolemaic-period funerary ship of the Apis bull (Fig. 3.25).[231]

Combining these clues allows us to define—in broad terms—the shape of the hull and rigging (Fig. 3.60). The galley that appears on the Calendar Frieze remains the only one for which no doubt exists regarding its Panathenaic identification. Thus, despite its poor preservation and obviously skewed proportions, this vessel becomes the "gold standard" against which all other potential Panathenaic ship representations must be measured.

The Erechtheion ship model/lamp.—A fourth-century-B.C. model of the Panathanaic ship may have been hiding in plain view for a century and a half, now in the form of the famous bronze ship model/lamp that was discovered during the nineteenth-century excavation of the Erechtheion (Figs. 3.61: A–B, 62: A). This identification was first proposed by K. S. Pittakis in 1862 and more recently has been endorsed by H.-G. Martin.[232]

The model's excavator, C. Boetticher, originally identified the model as an *ex voto* to Poseidon and, on the basis of that perceived link, proposed that the Erechtheion housed a cult for that deity.[233] Unfortunately for Boetticher's theory, subsequent conservation of the artifact revealed a dedicatory inscription to Athena in stippling on the hull's port side, which reads ΙΕΡΟΝ ΤΗΣ ΑΘΗΝΑS, meaning "Sacred to Athena": Based on paleography, the inscription dates to the fourth century B.C. (Fig. 3.61: A, 62: A).[234] D. Blackman believes that it is not possible to date the inscription so precisely and prefers a date from the late fifth to the early third centuries B.C.[235]

The circumstances of the model's discovery indicate that it represents an *ex voto* contributed to the Erechtheion and not an accidental artifact from an earlier era at the site.[236] Thus, the model's interment must postdate the completion of the Erechtheion in 406 B.C. This is curious, for the Erechtheion model replicates a much earlier Archaic galley.[237] J. S. Morrison and R. T. Williams note that the model, of course, could have been in someone's possession for an extended period before having been donated to the temple, at which time it could have received its dedicatory inscription.[238] A simpler explanation to this conundrum may apply, however. The model could represent a contemporaneous fourth-century-B.C. version of the Panathenaic ship: In that case the model could have been manufactured contemporaneously with its dedication.[239]

Might the Erechtheion ship model replicate a Panathenaic ship? The answer to this question must be based, first and foremost, on a comparison between the model and the Calendar Frieze's Panathenaic ship.

The Erechtheion model has a somewhat rockered keel as it rises gently in the stern. The bow has a remarkable waterline ram, broad when viewed in profile but chisel shaped in the vertical plane (Figs. 3.61: A–B, 62: A). The lamp sits above and behind the ram. In profile this could be mistaken, perhaps intentionally, for a *proembolion.*

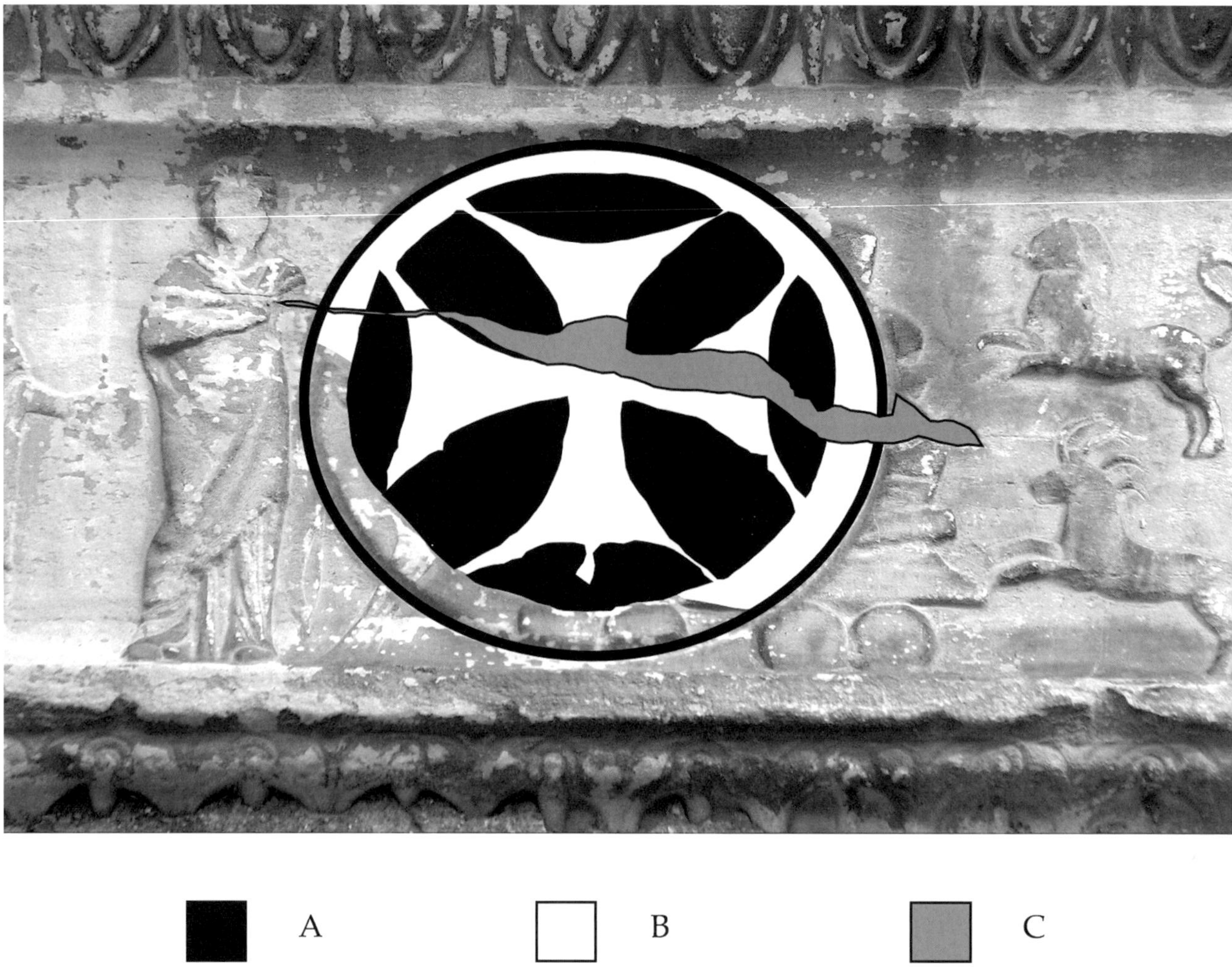

Fig. 3.53: Much of the Calendar Frieze Panathenaic ship and its rig has been lost. Some areas (A), which include a ring around the roundel and the spaces between the arms of the cross and the inner edge of the roundel, were entirely carved away, and here nothing at all remains of the vessel. Additionally, the entire area of the cross and part of the roundel itself (B) has been smoothed, destroying any molded features that existed: Here deeply carved elements still survive. A deep horizontal crack (C) crosses the upper parts of the vessel and its rig. Finally, parts of the figure have a white covering, perhaps lichen or remains of plaster. Photo and design: S. Wachsmann.

A fenced forecastle, bearing a concave screen stretched between the stempost and a pair of raised posts astern, nestles behind the lamp. Of course, in an actual ship the forecastle would have been situated forward, directly in the bow, but due to the location of the lamp, the model maker was forced to move it astern. The shape of the forecastle, however, appears regularly in the bows of representations of Archaic galleys.[240] No *ophthalmoi* are in evidence.

Parallel engraved lines run the length of either side of the hull. Morrison is probably correct in interpreting these as wales (Figs. 3.61: A, 62: A). A series of circles positioned between these lines and the caprail presumably indicates oar ports.

A rope motif tops the vessel's caprail (Figs 3.61: A–B, 62: A). Morrison and Williams identify this as a *hypozoma,* which served to control the torsion in the hulls of Classical galleys.[241] L. Basch, however, notes that ropes placed in this manner could serve no functional role and reasonably proposes that these ropes were solely decorative.[242] Alternately, this motif may represent a wooden caprail decorated in a rope motif.

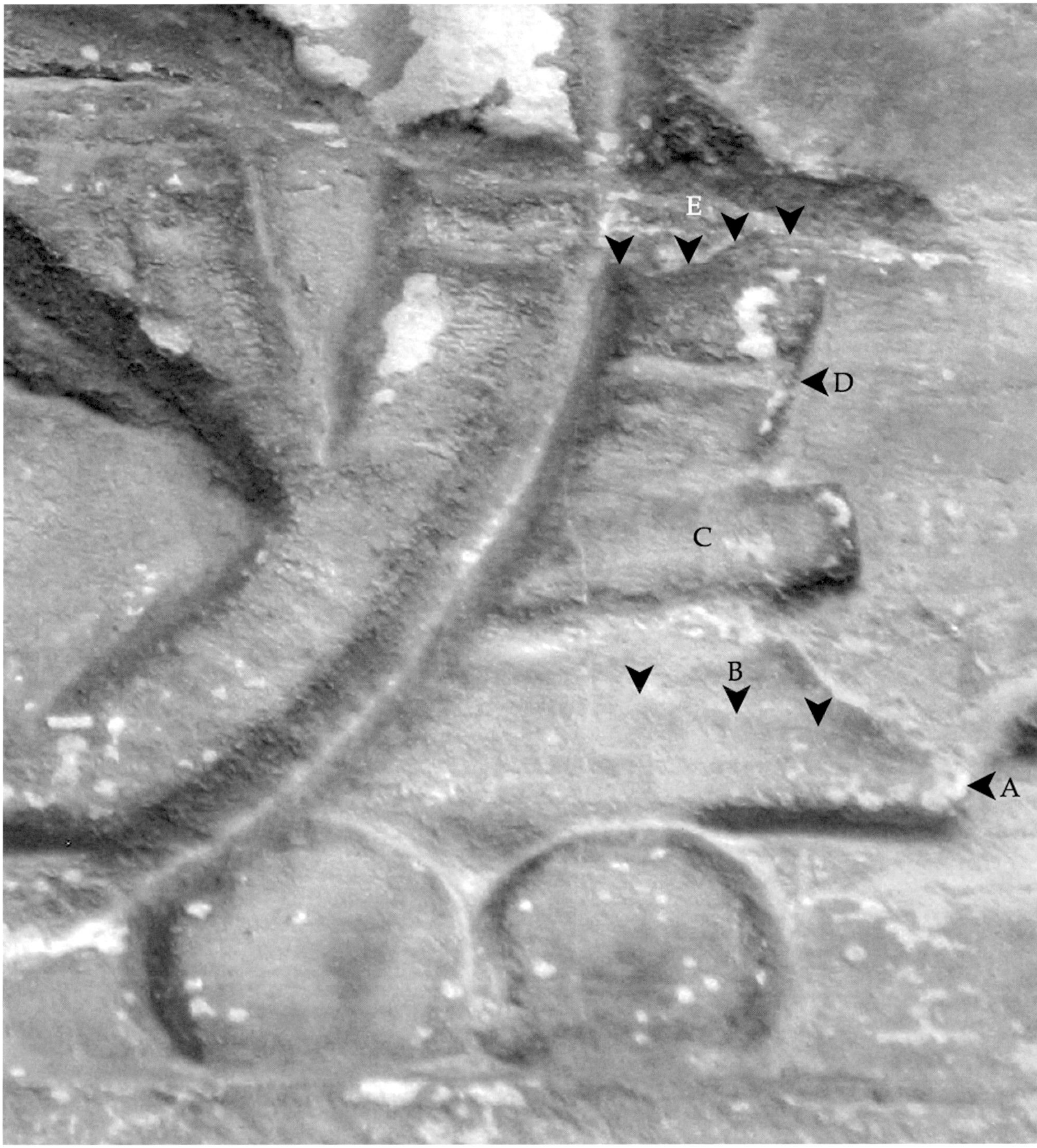

Fig. 3.54: Detail of the bow section of the Calendar Frieze Panathenaic ship.

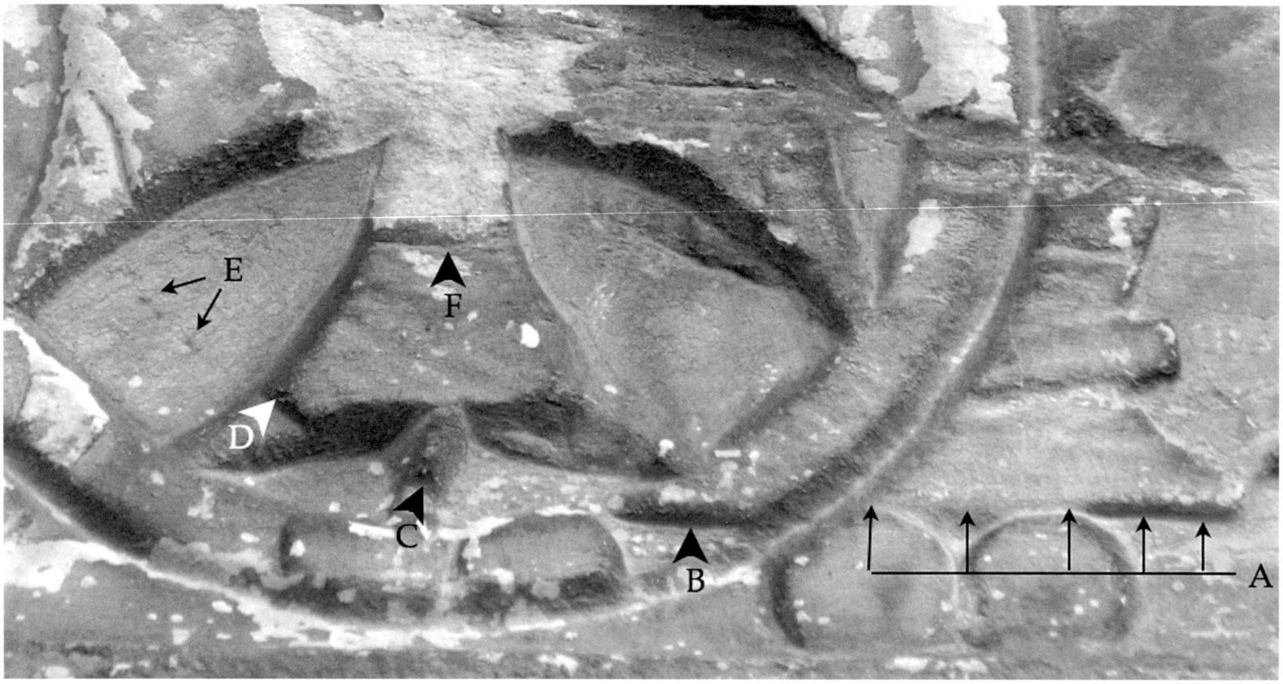

Fig. 3.55: The shape of the hull may be reconstructed based on a few surviving marks.

A horizontal guardrail, supported forward on two stanchions, loops *behind* the sternpost railing in the poop. This feature also appears frequently on depictions of Archaic galleys (Fig. 3.63).[243]

The ship has three beams and a single throughbeam at the stern for attaching the quarter rudders (Fig. 3.62: A).[244] Only the protruding ends of the throughbeam jutting out from the hull are depicted: The beam is missing inside the hull but must have existed in the prototype vessel. The sternpost recurves, ending over the stern in a pair of abstract bird heads, one above the other, with a single pair of eyes between them (Figs. 3.61: A, 62: A). Representations of Archaic galleys contain numerous parallels to similar arrangements of bird heads in renditions varying from naturalistic to completely abstract.[245]

Finally, perhaps the most striking element of the Erechtheion model is one that is missing: the outrigger. Any ship worthy of dedication to Athena in the fourth century B.C. might reasonably be expected to have been a trireme or larger. Beginning with the trireme, a massive outrigger *(parexeiresia)* allowed for the seating of three banks of rowers in Greek Classical galleys: Today this is best represented by the oarboxes on the contemporaneous galley of the Nike of Samothrace, now in the Louvre (Fig. 3.64: arrows).[246]

In comparing the Erechtheion model to the Calendar Frieze ship, we must be aware that the artisans who created the Calendar Frieze had limited horizontal space in which to portray each individual Athenian month. As a result of this, the Panathenaic ship is "horizontally challenged."[247] The ship's length has been curtailed disproportionately, and the depiction is much squatter than the original vessel it represents, almost as if the ship's midsection had been removed and then stem and stern were attached to each other. This becomes immediately clear when the Calendar Frieze Panathenaic ship is compared to the majority of other known representations of Archaic galleys.[248]

It follows, then, that in order to correctly overlay the profile of the Erechtheion model on the Calendar Frieze ship, the former must be similarly shrunk longitudinally (Fig. 3.65: A–B). Furthermore, consideration must be given to the displacement of the forecastle (Fig. 3.65: A: 3A–3B) as a result of the addition of the oil lamp (2) placed at the model's bow. When we include these two

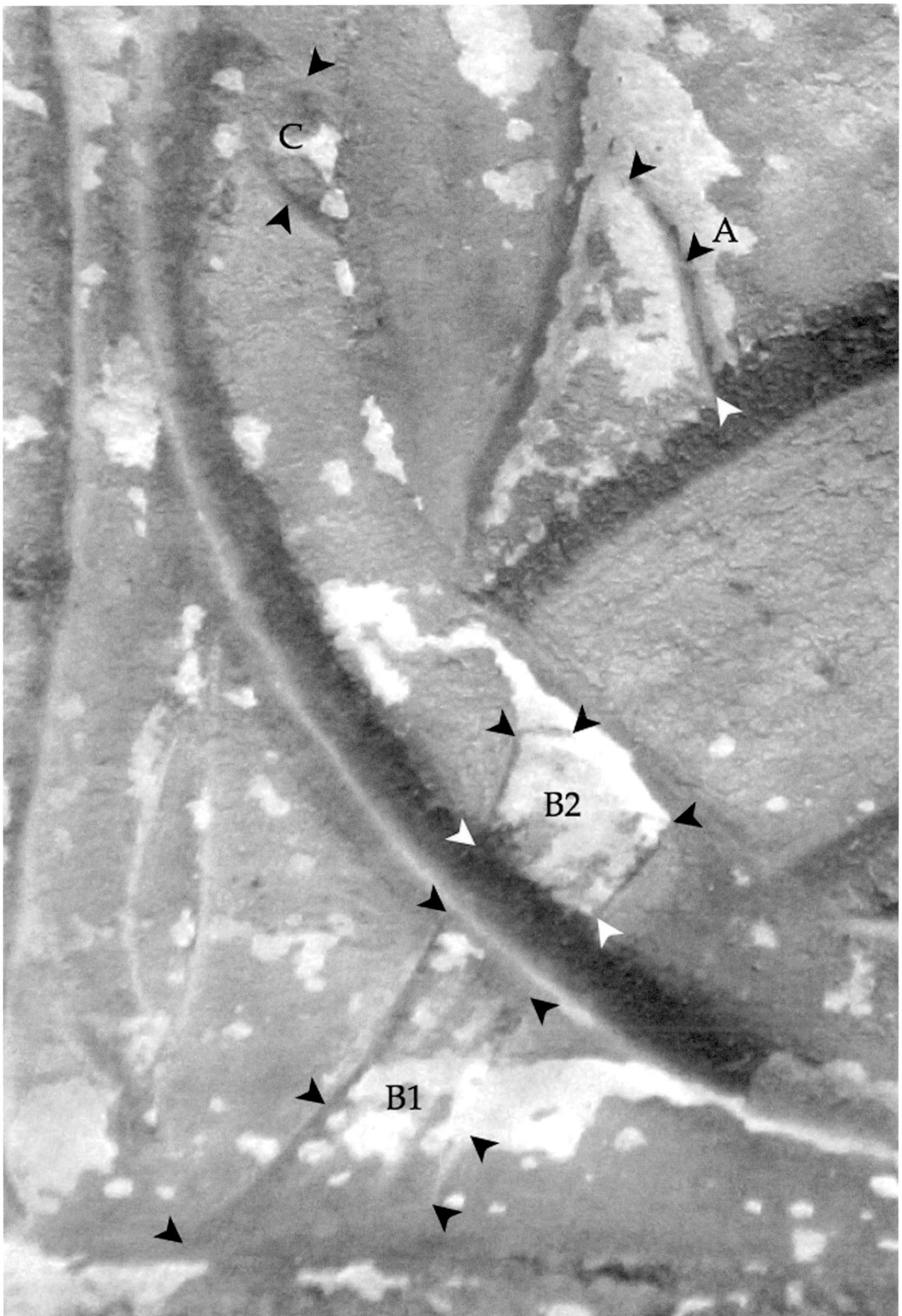

Fig. 3.56: A fragmentary sinuous, roughly vertical, line is visible (A) above the quarter rudder, which survives to the lower left of the roundel (B1) and on the roundel itself (B2). Above it, the hand of a figure, which some scholars identify as Pompe (C) is visible on the roundel.

Fig. 3.57: Slant view of the Calendar Frieze of the Panathenaic ship indicating the location of the surviving masthead fragment (circle).

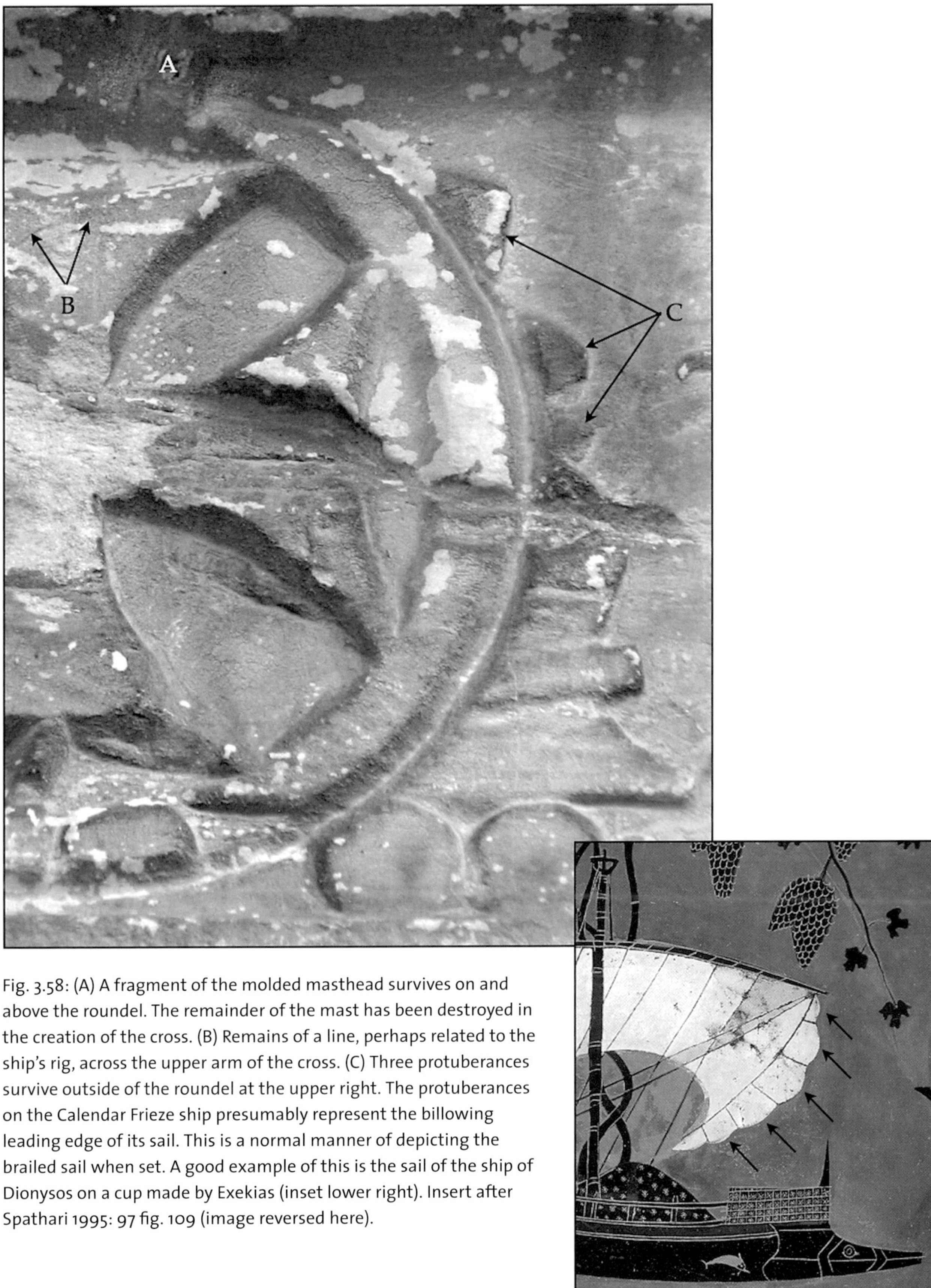

Fig. 3.58: (A) A fragment of the molded masthead survives on and above the roundel. The remainder of the mast has been destroyed in the creation of the cross. (B) Remains of a line, perhaps related to the ship's rig, across the upper arm of the cross. (C) Three protuberances survive outside of the roundel at the upper right. The protuberances on the Calendar Frieze ship presumably represent the billowing leading edge of its sail. This is a normal manner of depicting the brailed sail when set. A good example of this is the sail of the ship of Dionysos on a cup made by Exekias (inset lower right). Insert after Spathari 1995: 97 fig. 109 (image reversed here).

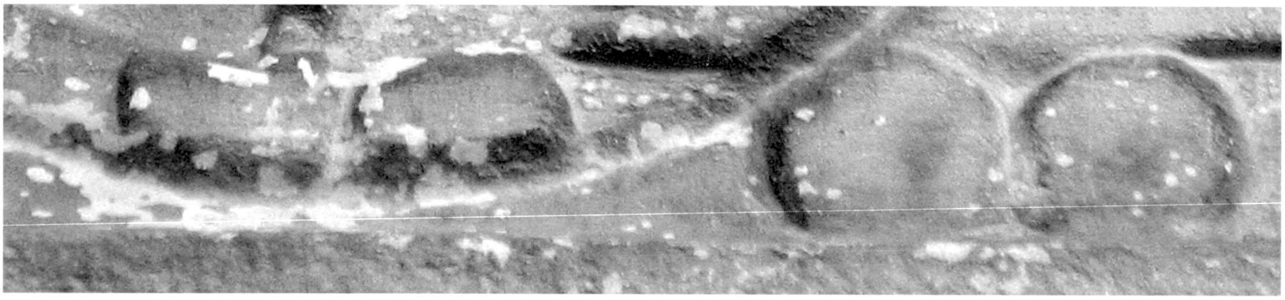

Fig. 3.59: The four wheels visible on the ship's starboard side.

considerations in the evaluation and overlay the resultant outline of the model over the relief, we find nearly a perfect fit. The results are most dramatic at the stern. We can size the stern section and position it over the frieze by placing the quarter rudder's reconstructed loom against the throughbeam (5) and by aligning the model's keel line with a surviving mark of it at the bottom left of the cross's lower arm (4). When we so align the two images, the birds' beaks surmounting the stern of the Erechtheion model align with the sinuous line carved into the lower left extremity of the cross's left arm on the Calendar Frieze (Figs. 3.56: A, 65: A: 6). Archaic period bird-head sternpost decorations regularly ended in a sinuous curve that represented, at times if only in the abstract, the head, eye, and beak of a waterbird (Fig. 3.63).[249] This is perhaps the most remarkable of the correspondences between the two representations. Given the chronological period separating the two ship figures and the different mediums in which they appear, the similarity between them is truly remarkable.

As we have seen, both vessels carry a blunt, chisel-shaped ram (Fig. 3.65: A: 1). On neither ship do we see either the boar-head-shaped ram typical of the Archaic period or the quintessential three-level ram of war galleys of the fifth century and later.[250]

Two important differences exist between the images, however, and these require explanation:

- The *proembolion,* so prominent on the Calendar Frieze, is entirely missing on the Erechtheion model. One can suggest two possible explanations for this seeming disparity. Perhaps the model's lamp was intended to do double duty by also standing in for the *proembolion,* for, as already noted, when seen in profile, the lamp has a somewhat similar shape (Fig. 3.65: A: 2). A second possibility, and the more likely one in my opinion, is that the Panathenaic ship would have been repeatedly repaired and/or replaced through the ages as, for example, in the case of the ship built by Herodes Atticus and the refurbishing of the timbers of the Athenian sacred ship of Theseus:[251]

  The ship on which Theseus sailed with the youths and returned in safety, the thirty-oared galley, was preserved by the Athenians down to the time of Demetrius Phalereus. They took away the old timbers from time to time, and put new and sound ones in their places, so that the vessel became a standing illustration for the philosophers in the mooted question of growth, some declaring that it remained the same, others that it was not the same vessel.

  This phenomenon of the replacement, in part or in whole, of a venerated vessel finds an interesting parallel in pharaonic Egypt, where the *Amunuserhet,* the cult ship of the god Amun, which transported the deity's cult statue on the Nile, was built anew at great expense whenever deemed necessary.[252] The most famous of these was the iteration commissioned by the high priest of Amun Herihor, who sent Wenaumun to Byblos for the cedar timbers with which to build it.[253]

A

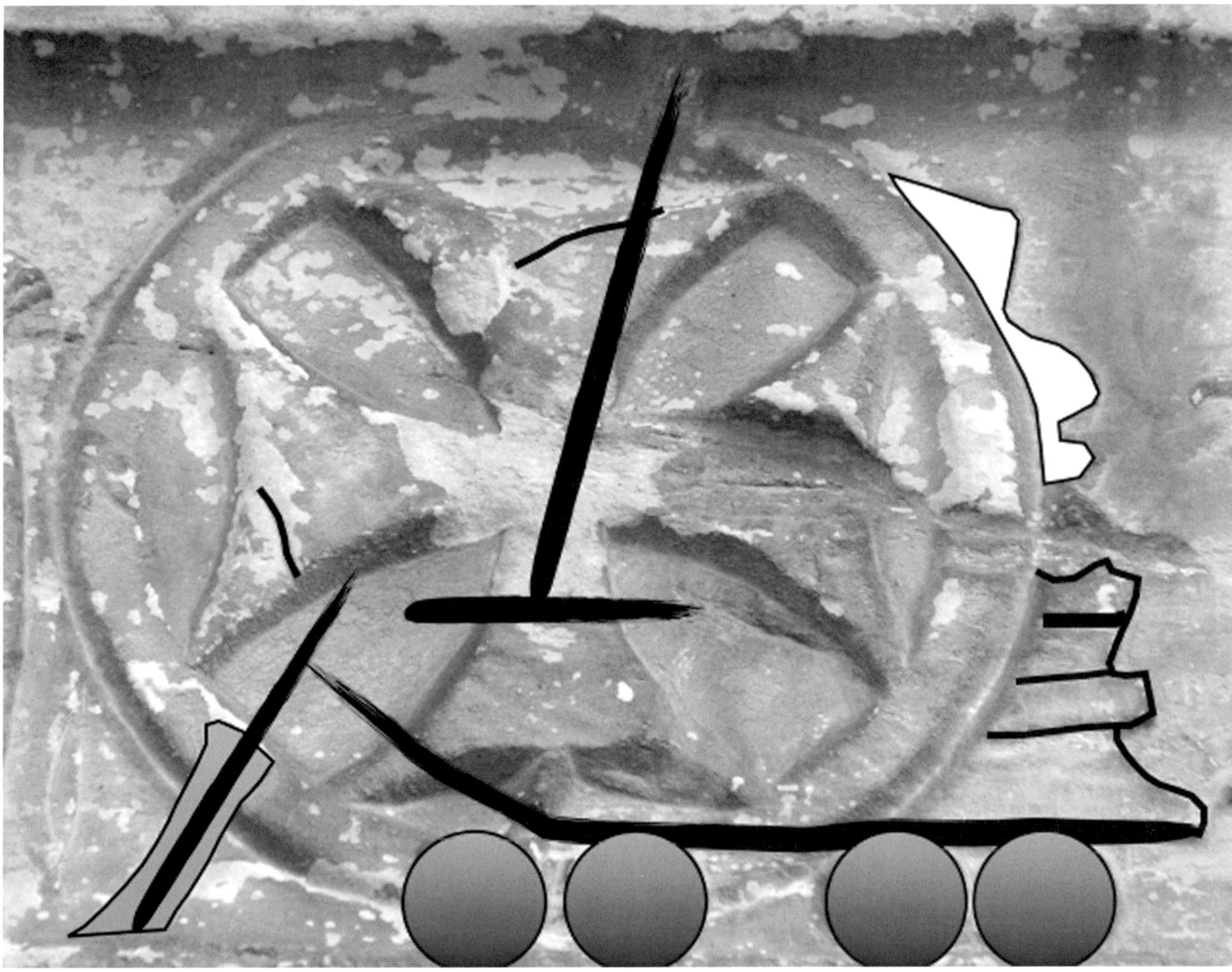

B

Fig. 3.60: Surviving fragments of lines permit, in broad lines, the reconstruction of the Calendar Frieze Panathenaic ship's hull. Due to a lack of horizontal space, its creator(s) eliminated the amidships section of the hull and basically attached the bow to the stern. Photo and reconstruction: S. Wachsmann.

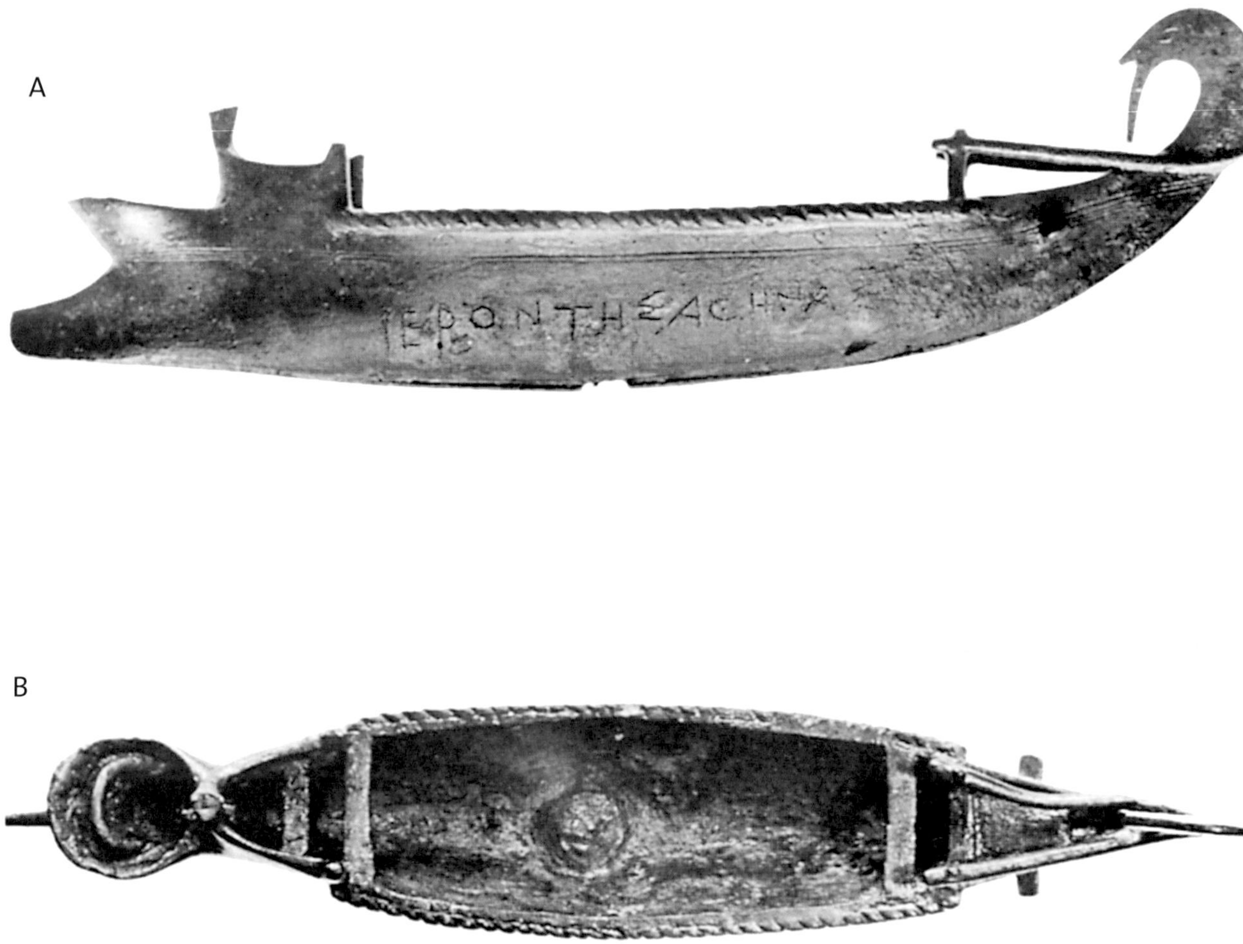

Fig. 3.61: (A) The Erechtheion model, starboard side. (B) Top view. From Basch 1987: 229 fig. 477. Courtesy Hellenic Institute for the Preservation of Nautical Tradition.

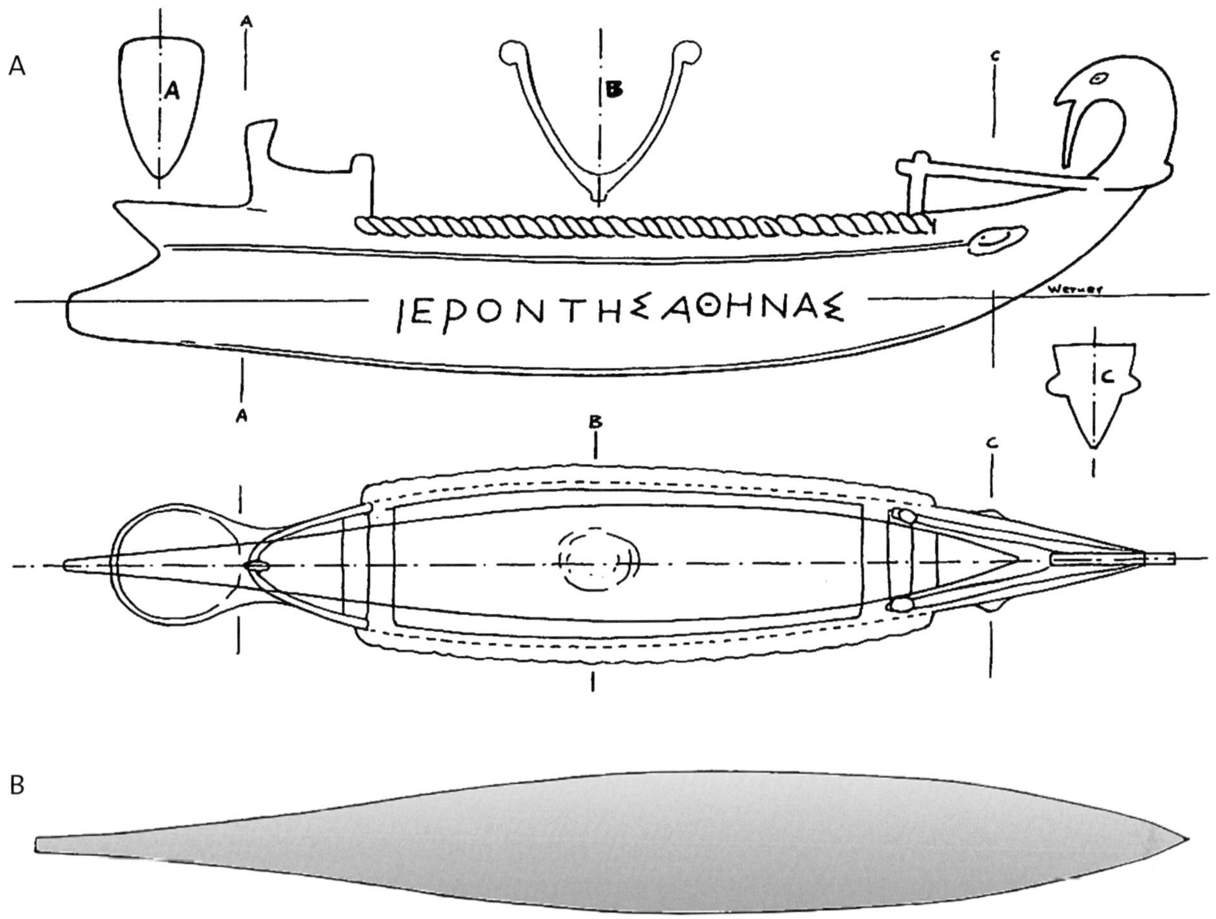

Fig. 3.62: (A) The Erechtheion model, line drawing, port side and plan view. (B) Waterline plan view. Note that the model is beamiest aft of amidships (NTS). From and after Göttlicher 1978: Taf. 27 no. 362.

Thus, the Erechtheion model and the Calendar Frieze ship may—*and probably do*—represent different iterations of the Panathenaic ship, separated by centuries, with the Erechtheion model representing the older of the two. The *proembolion* was not a common element on Archaic ships. It appears to have been introduced in the fifth century B.C. as a strengthening measure on galleys when ramming warfare became the norm; therefore, a *proembolion* may conceivably have been incorporated into a later iteration of the Panathenaic ship.

- The second apparent difference is that early depictions of the Erechtheion model show it with a stump amidships that survived to about the height of the caprail (Fig. 3.66).[254] Subsequently, this element has been lost, and only a slight indication of it survives (Fig. 3.62: A). Morrison and Williams identify this feature as a tabernacle *(histopede)* for a mast.[255] A mast stepped amidships would be at variance with the manner in which it is depicted on the Calendar Frieze Panathenaic ship. I suspect, however, that the stalk served to hang the lamp at its center of gravity and has nothing at all to do with a mast.[256] If correct, we witness here again, as in the case of the Erechtheion model's lamp, a situation in which the needs of the model superseded accuracy with regard to the prototype ship.

Fig. 3.63: Horizontal guardrail and bird-head decoration at the stern of an Archaic galley, ca. 530–510 B.C. After Spathari 1995: 87 fig. 95. Drawing: M. Kofahl.

Thus, the evidence supporting the hypothesis that the Erechtheion model copies a contemporaneous Panathanaic ship may be summarized as follows:

- The model comes from the Erechtheion and is dedicated to Athena. Simply put, the likeness of which vessel would be more appropriate to dedicate to Athena than that of her Panathenaic ship?
- While the model replicates an Archaic galley, the dedicatory inscription dates to the fourth century B.C. The simplest explanation of this paradox is that the donor commissioned a lamp in the form of the contemporaneous Panathanaic ship, had the model engraved with a dedicatory inscription, and then gave it to the Erechtheion, all during the fourth century B.C.
- The Erechtheion model agrees remarkably well with the Panathenaic ship portrayed on the Calendar Frieze.

THE KERAMEIKOS SHIP MODEL.—H. G. Martin proposes that a ship model made primarily of Pentelic marble found at the Kerameikos may represent a Panathanaic ship.[257] The find location is of interest for, as noted earlier, the Panathenaic procession began at the Kerameikos.

Unfortunately, the two most diagnostic areas of any ship model—the stem and the stern—have been lost. For convenience, then, in the following discussion I call the model's two ends "Extremity X" and "Extremity Y" (Fig. 3.67).

The model is now in three parts, and, although the bow and stern are missing, it can be identified as a galley based both on the oarports along the sides of the hull and its long and narrow proportions. The vessel has two rows of oarports in which every third hole is larger than the neighboring holes and has bronze fragments in it, presumably all that remain of the original metal oars.[258] The lower row exists only on Extremity X (Fig. 3.67: A–B). The upper row appears at both extremities but not in the center

Fig. 3.64: The fourth-century B.C. galley on which the Nike of Samothrace stands demonstrates the appearance of the outrigger *(parexeiresia)* (arrows), a detail missing on all the potential representations of Panathenaic ships discussed here. Louvre.

of the hull, a consideration that leads Martin to conclude that this area had originally been covered by part of a now-lost chassis support system that extended up the sides of the model.[259]

Martin identifies Extremity X as the ship's bow based on the one-word inscription MINOKIA, which appears on either side of it: He considers this to be the prototype ship's name (Fig. 3.67: A–B).[260] The inscription's paleography dates it to the fourth century B.C. or later.[261] The name MINOKIA seems a strange one for a Panathenaic ship, but none of the names of these vessels have come down to us, and replacement ships may have received individual names to differentiate them from their predecessors.[262] J. M. Hurwit and H. Williams have independently suggested to me that the vessel's name may hint that the Kerameikos ship model replicates the sacred ship of Theseus.[263]

While admitting that the Kerameikos model might have been a grave gift or a dedication, Martin concludes that it probably represents a Panathanaic ship primarily on the basis of two considerations: (a) the evidence of its having been mounted on a carriage and (b) a hole in the upper surface at Extremity X (Fig. 3.67: C arrows).

Martin identifies the hole as the vessel's maststep.[264] However, the direction of the bore of the hole would cause any mast placed in it to slant toward Extremity Y.[265] In other words, according to Martin's reconstruction, the mast was stepped in the bow and raked toward the stern—the exact opposite of the manner in which it appears to be depicted in the Calendar Frieze (Fig. 3.60).[266] There are at least four possible explanations for this seeming discrepancy:

- My proposed reconstruction of the mast on the Calendar Frieze Panathenaic ship is incorrect.
- The Kerameikos model does not represent a Panathanaic ship.
- The various iterations of the Panathanaic ship handled the mast in divergent fashions, and the Calendar Frieze and the Kerameikos models represent two different Panathanaic ships.
- Martin has reversed the model's bow and the stern, and Extremity Y is actually the bow. This scenario is supported by several considerations: The Kerameikos model would parallel the evidence from the Calendar Frieze; The broadest part of the Kerameikos model is near Extremity X (Fig. 3.67: C–D): on the Erechtheion model, as Basch notes, the widest part of the hull is at the stern (Fig. 3.62: B).[267] The appearance of the inscriptions on Extremity X might seem at first to argue against this interpretation, but naming devices can be situated at both the bow and the stern: The latter was the location of the *stylis,* a ship identification device that appears in the Classical period.[268]

  I prefer this solution for the reasons stated. While the Kerameikos ship model may represent the Panathenaic ship, the evidence is somewhat more equivocal than in the case of the Erechtheion ship/lamp model.[269]

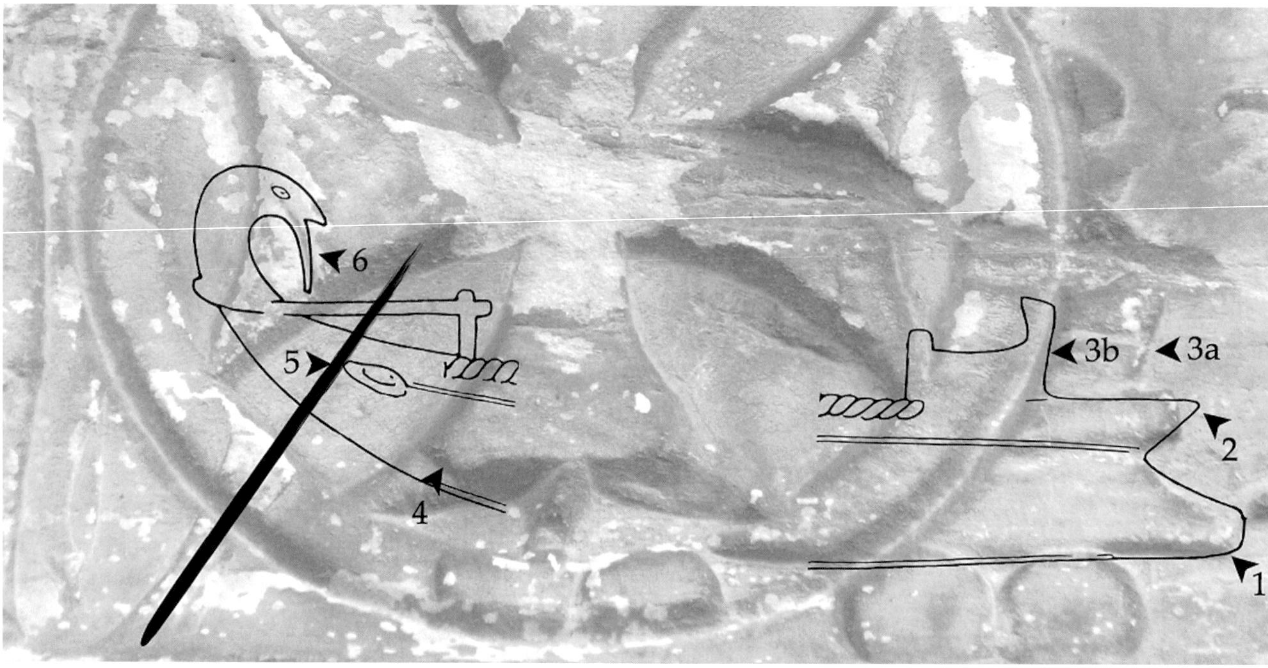

A

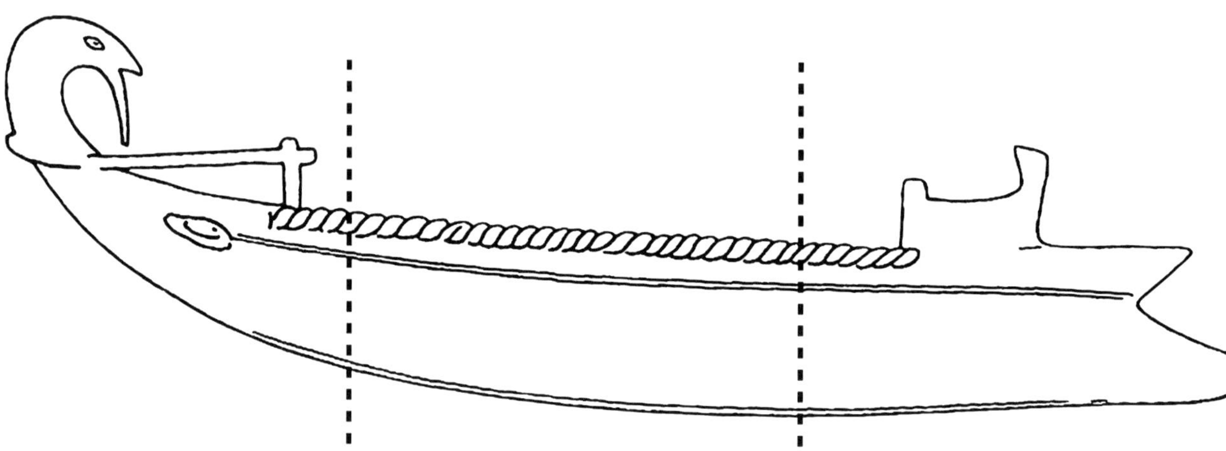

B

Fig. 3.65: When profiles of the stem and stern of the Erechtheion model are overlaid on the Calendar Frieze Panathenaic ship, the similarities are striking. Photo and design A: S. Wachsmann. Line drawings A and B after Göttlicher 1978: Taf. 27 no. 362.

A

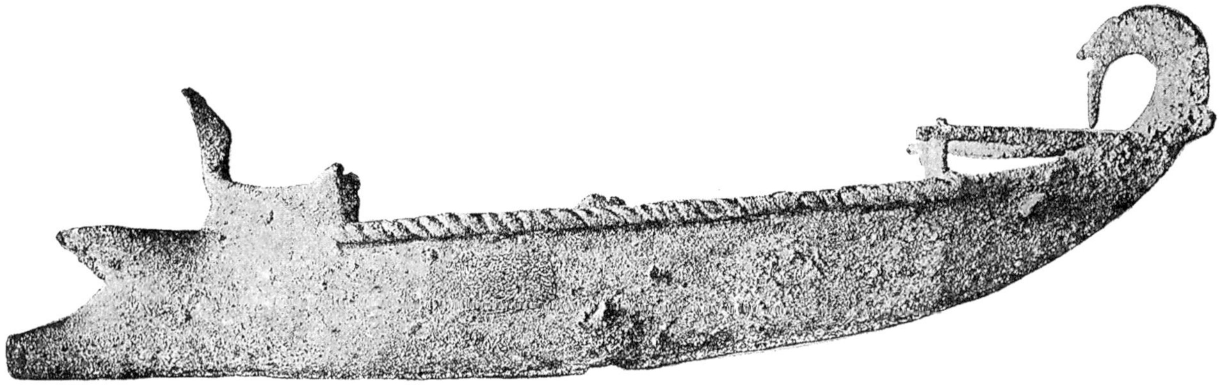

B

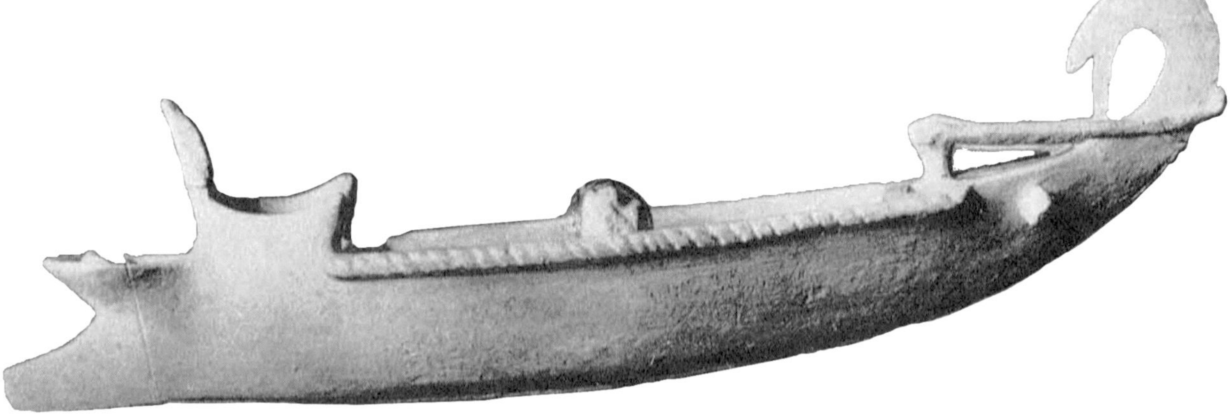

C

Fig. 3.66: Early line drawing and photos of the port side of the Erechtheion model, showing the stump amidships. From Pittakis 1862: 91–92; de Ridder 1896: 140 fig. 95 (425); Moll 1929: no. B VII: 1.

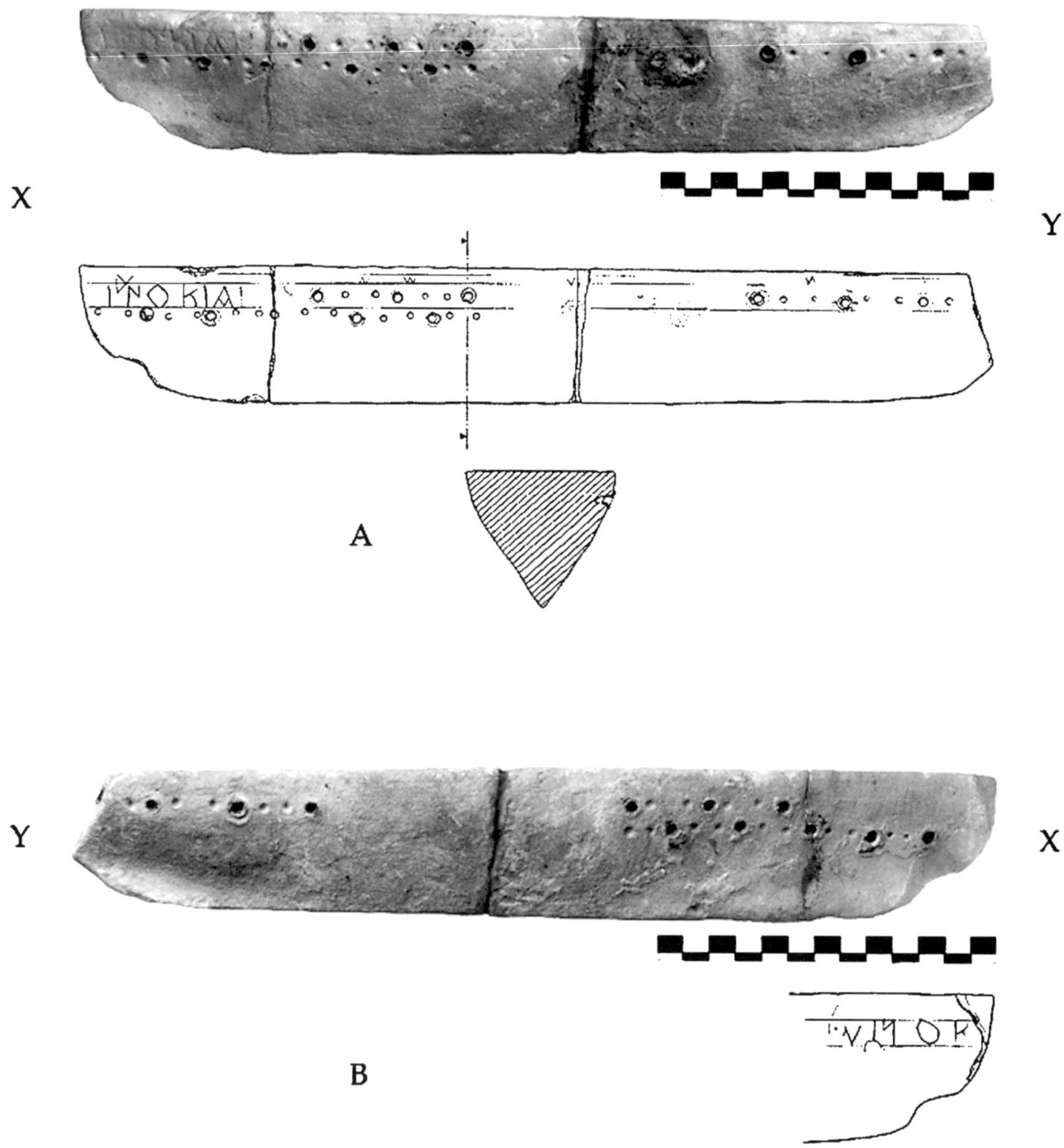
X
Y
A
Y
X
B

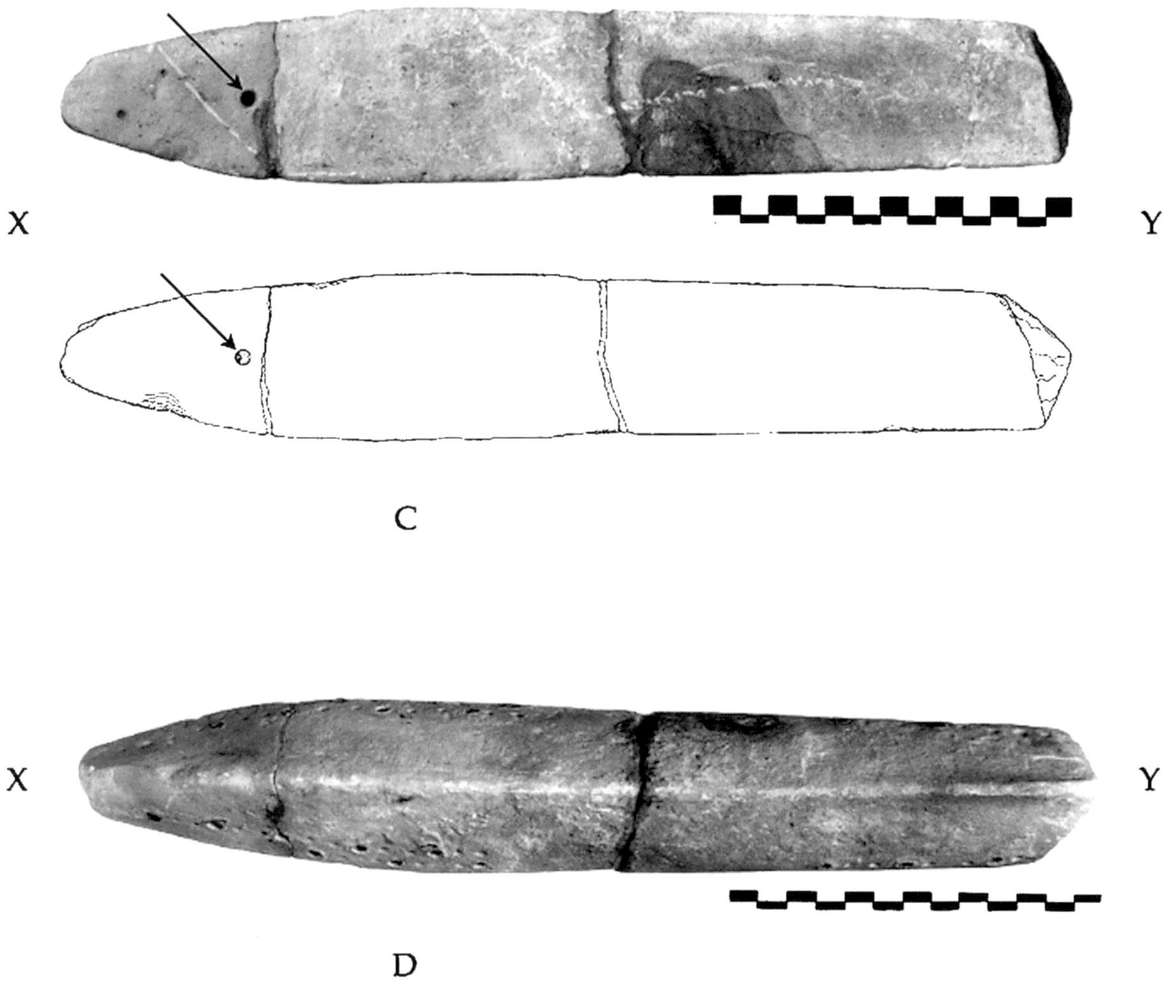

Fig. 3.67: Marble ship model from the Kerameikos, Athens (A–B) side views. (C) top view. (D) bottom view. Date: fourth-century B.C. or later. After Martin 1999: 138 Abbs. 1–3, 139 Abbs. 4–7, 141 Abb. 8.

Ship on a marble relief fragment from the City Eleusinion.—Alkistis Spetsieri-Choremi identifies part of a ship's stern that appears on a Pentelic marble relief fragment from the region of the City Eleusinion as a representation of a Panathenaic ship (Fig. 3.68).[270] Two registers are visible on the fragment (Fig. 3.69). The upper register contains part of the midship section and stern of a galley. The lower register contains the effaced heads of several figures carrying objects in a procession.

The whole scene is badly abraded, making any interpretation difficult. The most prominent detail on the ship is a broad and flat horizontal feature running its entire surviving length. I take this to represent a wale. As with the representations discussed earlier, no outrigger is visible.

The upper-right part of the hull has been chipped off. Beneath the ship is a narrow, rectangular carved depression, which may represent the space beneath the hull between two badly worn items, which Spetsieri-Choremi interprets conditionally as the wheels (Fig. 3.68: arrows).[271] These objects are, however, too small—when compared to the wheels on the Calendar Frieze ship—to represent wheels. I take them for chocks or, more probably, the ends of logs on which the hull rests.[272] To the left of these are two long, narrow objects that slant upward from the register's baseline at a 60-degree angle from the horizontal and culminate at the ship's keel line. These items probably represent struts supporting the ship. Thus, the limited evidence available suggests that, if this vessel represents the Panathanaic ship, which seems likely, then it appears here in a state of storage, dismounted from its wheeled transport system. The City Eleusinion seems to have been the customary way station during the Panathanaia for the ship while the pageant continued up to the Acropolis.[273] Thus, the Eleusinion ship relief raises the question of whether the actual ship might have been quartered there on occasion. Against this view, one notes the extremely fragmentary nature of this relief and the consideration that Athens had many sacred ships and that at least some of these may have had Archaic attributes.[274]

The person on the right of the lower register holds a long, narrow object, the left end of which rises at a slight angle (Fig. 3.70). Its right extremity is missing in the break. Spetsieri-Choremi interprets this as a basket held in a procession, and this is certainly possible, but H. Tzalas proposes that this may represent a model ship.[275] As we have already seen, model ships often play a role in cults that use ships in their festivities (Figs. 2.10, 12, 3.35–36).[276]

Thus, one reaches the following conclusions about the iconographic evidence for the Panathenaic ship:

- Beyond a reasonable doubt the Erechtheion ship/lamp model joins the Calendar Frieze ship relief as a bona fide representation of a Panathenaic ship.
- The attributes of the Kerameikos ship model are consistent with information available on the Panathenaic ship from the Calendar Frieze. Given the name MINOKIA engraved on the Kerameikos ship model, however, one cannot rule out Hurwit and Williams's suggestion that the name may hint that the model instead represents the sacred ship of Theseus.
- The ship on a marble relief fragment from the City Eleusinion may be interpreted as representing a Panathenaic ship on stocks, but again one should keep in mind the incomplete nature of this representation, as well as the possibility that it may represent another of the many sacred ships of ancient Athens.
- The fact that the Panathenaic ship appears as a clearly Archaic galley seems to support a sixth-century-B.C. date for the introduction of the ship into the Panathenaic festival. However, as opposed to the ship-cart of Dionysos, which appears in artwork from that century, we have no specific references in art or in text that early for the Panathenaic ship. In theory, the ship could have been introduced later but have been built to represent an older form of the ship to add gravitas at its introduction. Consider, for example, the ship sent to Delos by Athens each year, said to be the original one used by Theseus.[277] Note, however, that this ship remained a *triakonter,* a thirty-oared ship, which, in the Classical period, was apparently the Athenians' closest approximation to an "ancient" ship from the time of Theseus.[278]

Fig. 3.68: Fragment of a relief portraying a ship's stern. Region of the City Eleusinion. From Spetsieri-Choremi 2000: 2 fig. 1.

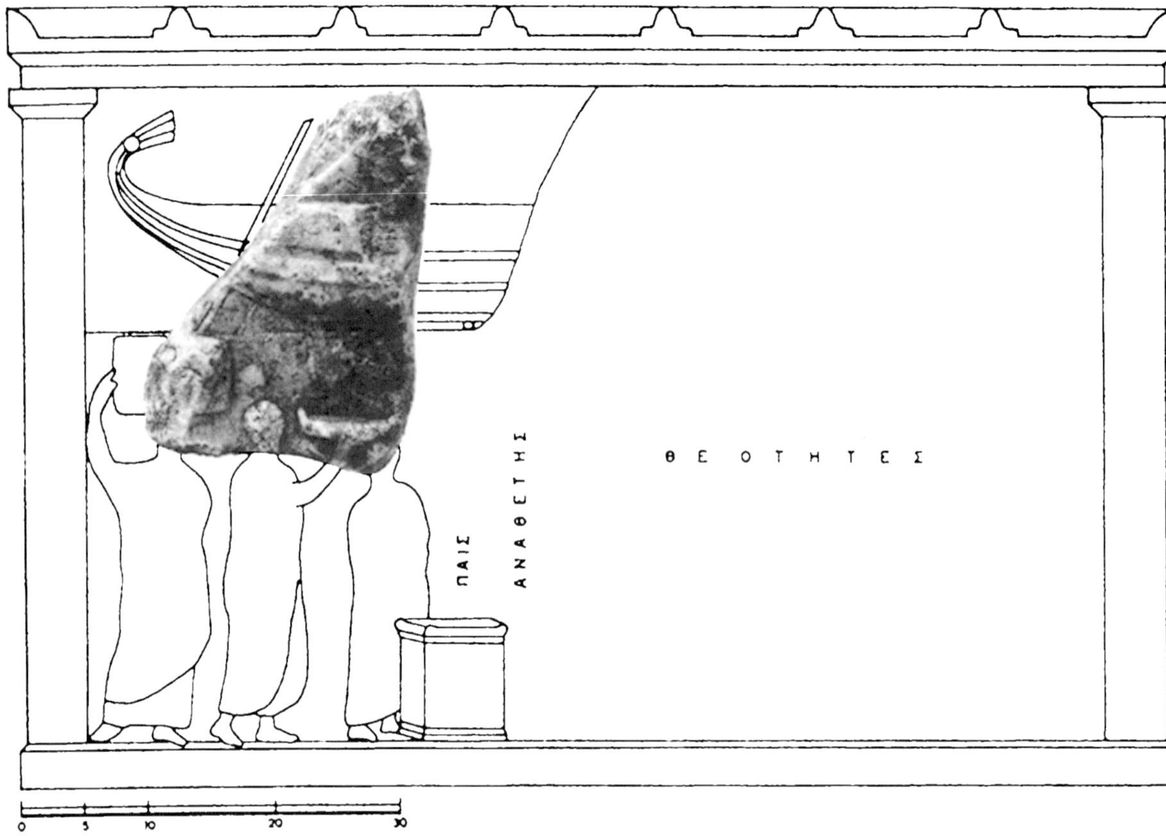

Fig. 3.69: Reconstructions of the Athenian Eleusinion scene in Fig. 3.68. From Spetsieri-Choremi 2000: 17 fig. 5.

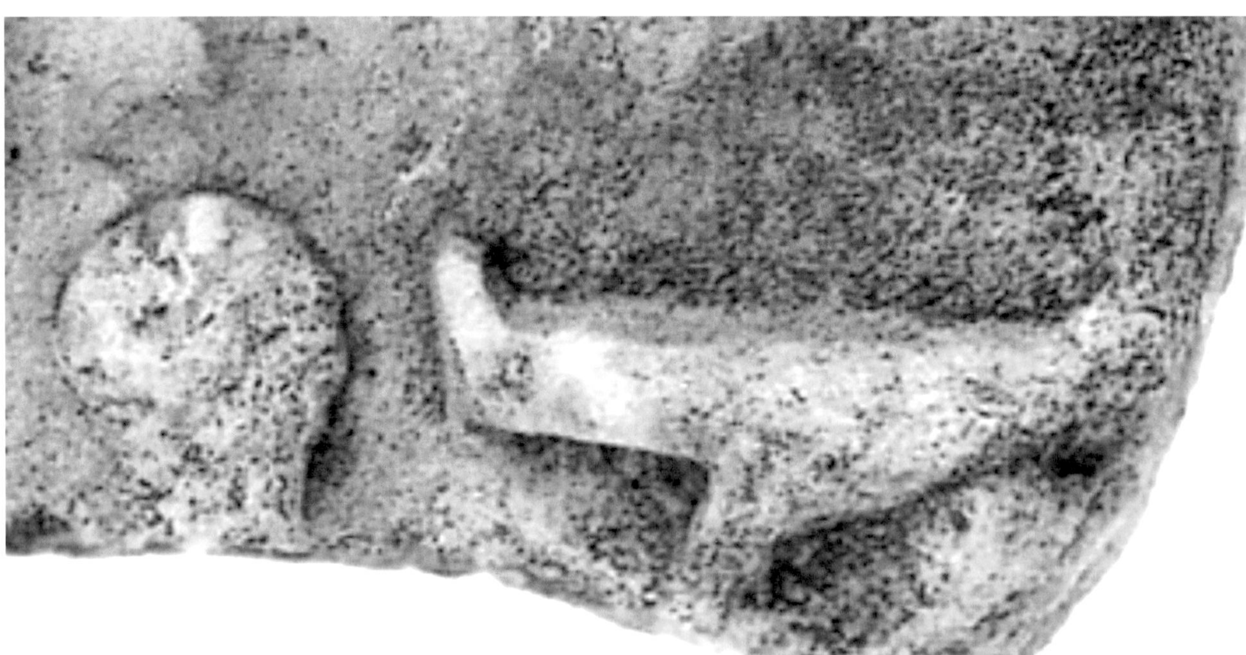

Fig. 3.70: Detail of a celebrant holding an object in the Athenian Eleusinion relief fragment. From Spetsieri-Choremi 2000: 2 fig. 1.

- Deborah N. Carlson, in a detailed study of marble *ophthalmoi,* has offered several explanations as to why three fragmentary ship's eyes were found at the Athenian Agora.[279] One possibility, which Carlson considers remote, is that these objects may have been attached to Panathenaic ships. *Ophthalmoi* do not appear either on the Calendar Frieze ship or on the Erechtheion model.[280] Moreover, while an *ophthalmos* may have been painted on the former, the latter argues against this conclusion, suggesting that the Panathenaic ship did not bear *ophthalmoi;* this makes it decidedly unlikely that decorating this vessel was the original purpose of any of the Agora *ophthalmos* fragments. Another possible explanation for their find location is that they derive from Dionysian ship-carts, as an *ophthalmos* appears on one black-figure Attic representation of this type of cult vessel (Fig. 3.45). This is also unlikely in my view, however, due to the inherent flimsiness of the Dionysian wheeled vessels. I suspect that such *ophthalmoi* would have been applied to them in paint only.

The ship of Isis (Ploiaphesia, Navigium Isidis).—Throughout antiquity, sailing on the Mediterranean remained primarily a seasonal activity limited to the summer months between March and November, when northwesterlies predominate.[281] Although ships sailed in the off-season, such activity remained at a minimum.[282]

Recognition of the sailing season and its initiation in the spring appear to have existed by the Bronze Age, if not earlier.[283] Linear B tablet Tn 316 from Pylos begins with the month name *po-ro-wi-to-jo,* which, as L. R. Palmer notes, indicates a cultic awareness of the start of the sailing season: He proposes that the month was named Plowistos—meaning the "sailing month."[284]

Also known by its Latin name, Navigium Isidis, the *Ploiaphesia* festival celebrated this opening of the sailing season. The festival took place in many Mediterranean locations on March 5.[285] The earliest reference to the *Ploiaphesia,* from Eretria in Euboea, dates to the first century B.C., and it continued as late as the mid-sixth century A.D.[286]

It is hardly surprising that Isis would be the goddess chosen in antiquity to secure the sailing season as she receives credit for the creation of navigation and the invention of the sail.[287] The goddess in her nautical aspect is known as Isis *Euploia* or Isis *Pharia* but most often as Isis *Pelagia.*[288] She is represented in various nautical themes but mainly as a goddess standing on a ship's bow while using her hands to spread her veil like a sail.[289]

The Ploiaphesia/Navigium Isidis ceremony included a cultic parade that culminated in the launching of a fully provisioned ship, usually considered a specially built ship, or a large model of a ship—without a crew. Apuleius, writing ca. A.D. 170, gives a remarkably vivid account of this festival as it was practiced at Cenchreae, near Corinth.[290] He is probably writing about a festival in which he himself had participated.[291] Apuleius describes the vessel of Isis and the ceremonies related to it in detail:[292]

> In the meantime, amid the clamour of vows made in festive spirit, we gradually passed onwards and now, approaching the sea-shore, we reached the very spot where my double the ass had lain the day before. When the images of the gods had been duly set down there, from reverent lips the chief priest first uttered the most exalted prayers over a ship that had been built with exquisite skill and decorated round about with wonderful Egyptian paintings. With a bright torch, with an egg and with sulphur he purified the ship so well that it was purity itself. Then he publicly named it and dedicated it to the goddess. The gleaming sail of this auspicious barque bore golden letters woven into its texture; these signified the inaugural prayer for fortunate sailing in the new year's commerce. A rounded pine rose as its mast, lofty in its radiance and with a finely resplendent top. The stern had a curving beak and shone with a covering of gold leaf. In short the whole ship was aglow with the polish of smooth citrus-wood. Thereupon all the people, both the devotees and the unattached alike, vied in loading the ship with baskets heaped with spices and similar offerings and they poured on the waves libations of meal mixed with milk, until the ship, laden full with generous gifts and votive tokens of good omen, was freed from its anchor cables and launched

to sea with a favourable breeze that blew especially for it. When by reason of its movement it had faded from our sight, the bearers of the images took up again their respective loads and with a will made their way back to the temple, observing the same dignified order of procession.

Apuleius makes no reference to the ship of Isis being transported overland. Apparently, at least at Cenchreae, the parade met the ship at the harbor. The Navigium Isidis appears on a fragmentary Roman fresco found near Ostia that dates to the reign of Septimius Severus (A.D. 193–211) and is thought to have derived from a calendar; it is now in the Vatican Museum (Fig. 3.71).[293] The fragment shows the ship of Isis launched in the water while two porters return with the cart used to carry the ship (or ship model) down to the water.

A delightful, albeit broken, relief depicting the Navigium Isidis is now exhibited in the Palazzo Altemps in Rome (Fig. 3.72: A).[294] It is said to have come from a tomb near the Via Appia at Ariccia in the vicinity of Rome and to date to ca. A.D. 100 In the relief a procession, including exuberant female dancers, passes before a reviewing stand (Fig. 3.72: B). In the uppermost register a seated figure presumably representing Isis sits next to a series of baboons, the god Bes, and the Apis bull. Beneath the procession appear a row of ibises and a crab, the latter perhaps representing Neptune. Unfortunately, no ship is visible in the scene although one may have existed originally on the missing part at left. The excitement visible in the scene supplies a visual aid to better understanding Apuleius's description of the festival.

D. Levi identifies a scene depicting this festival on a mosaic from the House of the Mysteries of Isis at Antioch.[295] The mosaic is largely destroyed, but it includes a scene of two ships, positioned stern to stern, in a bay.[296] The surviving scene presents no evidence of a wheeled conveyance.[297] Some *vota publica* coins from Rome show the image of Isis in her carriage *(thensa),* but without a ship.[298]

This festival played an important part in the pagans' efforts to continue their way of life despite the vigorous opposition of the ascendant Christian church, particularly in Rome at the end of the fourth century.[299] The pagans may not have been *entirely* unsuccessful, however. Although the English word "carnival" is commonly believed to derive from the Latin for "flesh" *(carnem)* and "to put away" *(levare),* referring to the advent of Lent, its

Fig. 3.71: The ship of Isis used in the *Navigium Isidis.* From a wall painting now in the Vatican Museum. Courtesy Vatican Museum.

A

B

Fig. 3.72: (A) *Navigium Isidis* scene. From a relief in the Palazzo Altemps, Rome. (B) Detail of the female dancers.

meaning has also been derived from the Latin terms *carrus* and *navalis,* meaning "ship-cart," referring to the ship of Isis used in the Navigium Isidis.[300] A number of Isaic elements were absorbed into the church, and A. Evans notes that modern coastal Greek sites named after the Christian saint Hagia Pelagia only vaguely hide the name of Isis *Pelagia*.[301]

Cyprus

The bow section of an Archaic-period terracotta ship model from Cypriot Salamis has a surviving hole intended for an axle (Fig. 3.73).[302] The bow of the ship is stylized with a ram positioned well above the keel/stempost juncture. A molded horizontal strip at the height of the ram presumably represents wales used to strengthen the prototype galley when ramming.[303] Remains of a round shield attached to the starboard side of the forecastle indicate that the prototype ship was Phoenician.[304] The excavations at Salamis also revealed numerous terracotta model wheels, but none can be linked to the ship model.[305]

Mesopotamia

A terracotta ship model on four wheels dating to the Neo-Babylonian (612–539 B.C.) period was found at Uruk.[306] It contained a model quadruped and a boxlike object identified as a model censor.

Miscellaneous

Military and commercial vessels carried overland.—Ships were also moved overland for noncultic purposes such as trade and war.[307] In some cases the ships were stripped down and carried in pieces, while in other cases entire ships were loaded on wagons or moved across land on rollers. At Mirgissa in Sudan, J. Vercoutter uncovered remains of a slipway used by the Egyptians to transport ships around the Second Cataract.[308] To facilitate this, the Egyptians poured Nile mud over the sand and added wooden crossties.

The Egyptian Red Sea coast is a dry treeless area lacking any evidence of sustained settlement. P. E. Newberry first suggested that the ships used on Hatshepsut's expedition to Punt were built on the Nile, disassembled to be carried across the Eastern Desert in pieces, and then reassembled on the shores of the Red Sea.[309] The XIth Dynasty official Henu records the building of such a ship for the Punt run:[310]

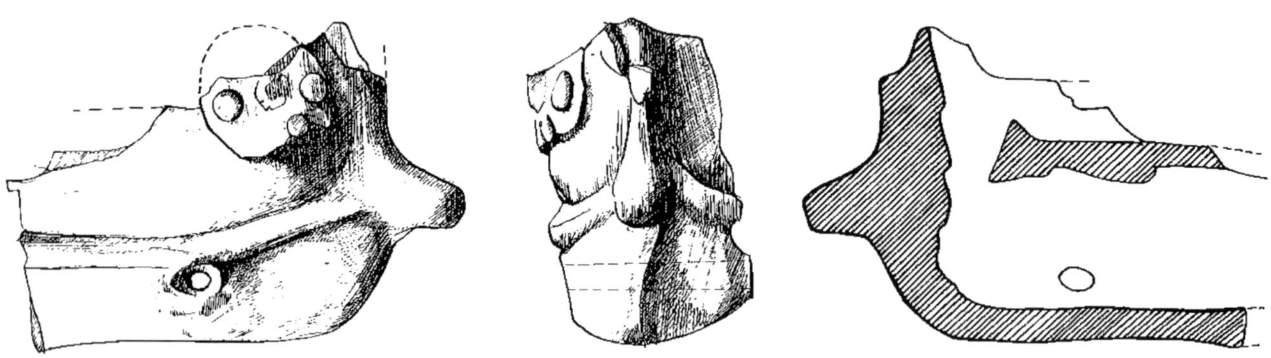

Fig. 3.73: Bow section of an Archaic ship model from Salamis with holes for an axle. From Monloup 1984: 159 no. 601.

[My lord, life, prosperity], health! sent me to dispatch a ship to Punt to bring for him fresh myrrh from the sheiks over the Red Land, by reason of the fear of him in the highlands. Then I went forth from Koptos upon the road, which his majesty commanded me . . . I went forth with an army of 3,000 men . . . Then I reached the (Red) Sea; then I made this ship, and I dispatched it with everything, which I had made for it a great oblation of cattle, bulls, and ibexes.

Another stele raised at Wadi Gawasis on the Red Sea by Antefoker, who served under Sesostris I, records that ships departing from there had been built at Koptos on the Nile.[311] Recent excavations at Mersa Gawasis and at Ayn Soukhna have revealed caves containing ship timbers, ropes, and anchors of such seagoing vessels.[312]

Egyptian transport of ships was not limited to the Red Sea, however. During his eighth campaign, Thutmose III had entire boats carried overland on wagons in order to permit his army to cross the Euphrates:[313]

When my majesty crossed over to the marshes of Asia, I had many ships of cedar built on the mountains of God's Land near the Lady of Byblos. They were placed on chariots with cattle drawing (them). They journeyed in [front of] my majesty, in order to cross that great river which lies between this foreign country and Naharin.

Fig. 3.74: One of two functional ships placed on carts and used to carry military equipment. The four-wheeled carts are pulled by yokes of oxen. Column of Marcus Aurelius, Piazza Colona, Rome. From Caprino et al. 1955: Tav R.

A

B

Fig. 3.75: (A) The Diolkos at the Corinthian end of the track. View to west. (B) Portions of the Diolkos in the process of collapse as a result of damage caused primarily by the bow waves of passing ships. View to east.

The transport of ships overland, under remarkable circumstances, has continued throughout history.[314] Semiramus brought together shipwrights from a variety of lands, including Phoenicia, Syria, and Cyprus, and ordered them to construct vessels that could be taken apart so that they could be transported anywhere she wished. To cross the Euphrates, Alexander the Great used vessels carried in sections overland from the Mediterranean.

Of course, one way of moving ships over land is to remove the land, that is, dig a transportation canal. This, too, was done in antiquity. The best known of these canals are the "Canal of the Pharaohs," which connected the Nile via Wadi Tumilat to the Red Sea, and Xerxes' canal, which cut across the isthmus of the Athos Peninsula in the years prior to his failed 480 B.C. invasion of Greece.[315]

While literary references to the transport of ships overland in times of war abound, such motifs are rare in the iconographic record. Two boats appear on the column of Marcus Aurelius in the Piazza Colona in Rome, depicted on four-wheeled carts pulled by yokes of oxen (Fig. 3.74).[316] Boats of this type, identified by P. Romaneli as *scaphae,* had two main purposes.[317] On the column they are shown either used as river transports or as pontoons to support floating bridges assembled over rivers.[318] The boats were placed parallel to each other and covered with a superstructure of beams to construct the bridges.[319] Armor and weapons are shown above the boats, indicating that, once loaded on the carts, they did double duty as storage facilities for transporting equipment. These boats and the equipment they held required yokes of oxen to pull them and had wheels constructed of solid disks of wood. Armor and weapons were also transported on two- or four-wheeled wagons with six-spoked wheels pulled by donkeys and in two-wheeled carts with solid-wood wheels pulled by oxen.[320] Supply wagons that carried other materials are shown with eight-spoked wheels and drawn by mules.[321]

The Romans also transported watercraft as part of the pageantry of imperial triumphs. In *The Jewish War,* Josephus Flavius records the triumphal parade for Vespasian and his son Titus, with which Rome honored them following their destruction of Judea and the fall of Jerusalem.[322] Numerous ships took part in the parade: Presumably these were transported on wagons, although Josephus does not supply details.[323]

Josephus does not explain the significance of the ships vis-à-vis the triumph, but they presumably commemorated the battle of Migdal (A.D. 67).[324] Titus minted a *sesterius* as part of the Judea Capta series, in which he stands with his right foot on the bow of a ship, perhaps commemorating the same event.[325] Josephus mentions only two other minor episodes related to ships in his description of the First Revolt. At Jaffa, a Jewish fleet harbored there had been molesting Roman shipping: It was destroyed in a storm when it put out to sea to avoid the Romans. During clean-up operations near the Dead Sea the Romans captured a number of Jews who had taken refuge on the water.[326]

Fifteen hundred years after Alexander, the disreputable Crusader Reynald of Châtillion launched vessels on the Red Sea that he had moved across the Arava after testing them on the Dead Sea.[327] The Venetians transported a flotilla of war galleys overland to Lake Garda in 1439, crossing mountains nearly four hundred meters high in order to relieve the blockade on Brescia.[328]

The Diolkos.—The narrow Isthmus of Corinth separates the Gulf of Corinth from the Saronic Gulf. Today ships pass from one body of water to the other through the Corinthian Canal. Although ancient in conception—construction was first started under Nero—the canal was not completed until 1893.[329]

A stone-paved cart road known as the Diolkos existed in antiquity to transport cargoes, and at times ships as well, between the Corinthian and Saronic Gulfs (Fig. 3.75: A).[330] This route wound its way for some eight kilometers across the isthmus, taking advantage of the topography. The inception of the Diolkos is unknown: Thucydides ascribes to it a venerable age.[331] B. R. MacDonald proposes a sixth century B.C. inauguration date and links its creation to the beginning of the use of marble for statuary and temples.[332] The road appears to have gone out of use in the twelfth century A.D.[333] W. Werner notes that ancient authors repeatedly describe the use of the Diolkos, or at least its general vicinity, for the transport of warships, but merchantmen, other than those of super size, would probably have been transported over the Diolkos, also.[334] MacDonald, however, considers the track's primary purpose to be the transporting of cargoes rather than merchant ships.[335] It seems likely that any ships carried along this route used some form of wheeled conveyance.[336]

Much of the Diolkos disappeared as a result of the construction of the Corinthian Canal in its various iterations. The continued destruction of this unique ancient, historically significant monument as a result of insufficient conservation efforts is to be sincerely lamented (Fig. 3.75: B).

4 Foreigners at Gurob

The plaque, or *pavois,* found with the Gurob ship model indicates that it had cultic significance.[1] Thus, while we know virtually nothing about the model's owner, we may assume beyond a reasonable doubt that the model reflects aspects of that person's religious beliefs. Clearly, these beliefs were foreign to Egypt, although the *pavois* does indicate a degree of syncretism with Egyptian norms.

To what ethnic/cultural group did the model's owner belong? This chapter examines the evidence for aliens resident at Gurob with the aim of identifying the *most likely* foreign candidate for the model's owner. *From the outset it should be clear that the best answer that we can expect to give to this question can be little more than an educated guess.* In attempting to determine the identity of foreigners in Egypt we have three categories of information—archaeological, iconographical, and textual:

Archaeological evidence.—This includes foreign artifacts, or evidence of foreign customs, uncovered at the site. Archaeology, in general, is a poor tool for determining the presence of individuals or enclaves in a foreign setting. The classical problems with the archaeological approach are admirably expressed in the twentieth-century B.C. Assyrian trading station uncovered at Karum Kanesh, in the lower city at Kültepe, where, but for the cuneiform archives found in them, the layout and material culture of the Assyrian merchants' residences were indistinguishable from those of the surrounding indigenous houses.[2]

With archaeological data, the question arises as to which artifacts indicate the presence of foreigners at any site as opposed to simply representing imports that arrived at the site through various forms of trade and exchange. The presence of foreign pottery by itself at a site is insufficient to posit foreigners there. On the other hand, nonindigenous phenomena and cultural artifacts for specific purposes lacking local equivalents can reasonably be assumed to indicate a foreign presence.[3]

Foreigners arriving in Egypt tended to acculturation; thus, over time their material culture melded with the local milieu. Rachael T. Sparks's comments on the problem of identifying Syro-Canaanites in Egypt should hold true for other nonindigenous cultures, also:[4]

> A gap may also exist between the Egyptian ideal of how a Canaanite should appear, and the reality of how individuals presented themselves. Ethnicity should

be viewed as a flexible concept, very much in the eye of the beholder, rather than a static fixed principle of identity. It is something that probably changed over time. As resident Canaanites began to assimilate with Egyptian culture, it seems likely that many of the foreign aspects of dress and hairstyle would have disappeared, and it may become increasingly difficult to isolate their descendents in Egypt through their material culture. Thus while their "foreign" identity might be preserved in aspects of language, such as people's names, or the epithets applied to them by Egyptians, the visual correlate of this and our ability to detect it archaeologically may become lost.

Iconographic evidence.—Egypt, more than any other Bronze Age land, abounds in iconographic evidence of foreigners visiting, living peacefully, or invading. One can only wish for "ideal" iconographic evidential situations indicative of foreigners in Gurob, as in the example of a stele from Tell el Amarna, belonging to a Semite named Trr, who appears in Syro-Canaanite dress (Fig. 4.1).[5] His wife also has a Semitic name but was apparently an ancient *fashionista,* as she appears in Egyptian garb.

During the XVIIIth Dynasty in particular, scenes of foreigners became a popular motif in the tombs of nobles, perhaps encouraged by the emphasis Hatshepsut placed on her international pursuits.[6] The Aegeans depicted in some of these tombs and defined as coming from "Keftiu" or the "Isles in the Midst of the Sea" represent Minoan visitors who arrived during the reigns of Hatshepsut and Thutmose III.[7] The latest Theban Tombs scene in which actual Minoans appear is that of Rechmire (TT 100), which includes a scene of Amenhotep II's inauguration.[8] This is significant as it is synchronous with the end of the Late Minoan IB period.[9] In later scenes the terms "Keftiu" and the "Isles in the Midst of the Sea" are used loosely to define patently hybrid characters or non-Aegeans: Characteristics inherent in Egyptian art can confuse or hide the reality behind these images.[10]

In recent years, excavations at Tell el Da'ba have revealed fragments of frescoes from early XVIIIth Dynasty deposits depicting Minoan motifs, including bull-jumping scenes, which indicate that Minoans resided at that site.[11] Following this and until the advent of the Sea Peoples in Ramesside art, the appearance of "northerners" in Egyptian art is limited to the Mycenaean warriors on a papyrus from Amarna.[12]

Textual evidence.—This includes any written documentation that defines the foreign ethnic, cultural, or geographical origins of individuals or groups at Gurob and its surrounding region. Documents referring to persons or groups with non-Egyptian names are particularly significant, although one indication of the rapid acculturation of foreigners into local society is evident in the trend for foreigners to adopt Egyptian names and burial customs.[13]

The term "ethnicity" in itself requires definition in this context.[14] Perhaps the best determiner of ethnicity is a combination of cultural kinship, language, and religious beliefs.[15] For this discussion, the single most important indicator of an ethnic group is the common name by which they are known to all, their *ethnonym.*[16]

The various Sea People groups clearly were not ethnically monolithic by the time they reached the eastern Mediterranean. During the process of the Sea Peoples' migrations, the groups composing the various coalitions were permeable, readily absorbing nonmembers into their ranks. A number of texts refer to Sea Peoples' raids for the specific purpose of acquiring captives. In RS 34.129, the Hittite king requires that Ibnadušu, a man from Ugarit who had escaped from a Sekel raiding party, be sent to him for debriefing regarding the foreign invaders "who live on their ships."[17] The Ahhiyawa of the Hittite archives, generally identified as Achaeans, also appear as slavers.[18]

Some of these indigenous folk captured by the marauding Sea Peoples defected to them en masse. For example, in EA 38 the king of Alashia writes the following, presumably in response to an accusation by the pharaoh that his countrymen had attacked Egypt:[19]

> Why, my brother, do you say such a thing to me, "Does my brother not know this?" As far as I am concerned, I have done nothing of the sort. Indeed, men of Lukki, year by year, seize villages in my own country.
>
> My brother, you say to me, "Men from your country were with them." My brother, I myself do not know that they were with them. If men from my country were (with them), send (them back) and I will act as I see fit.

Fig. 4.1: Scene on the Stele of Trr, a Semitic soldier, and his wife from Amarna. Akhenaten. From Spiegelberg and Erman 1898: fig. XVII.

You yourself do not know men from my country. *They* would not do such a thing. But if men from my country did do this, then you yourself do as you see fit.

A message from the final days of Ugarit, addressed to the king of that city from the chief prefect of Alashia, suggests a similar situation: "As for the matter concerning those enemies: (it was) the people from your country (and) your own ships (who) did this! And (it was) the people from your country (who) committed these transgression(s)."[20]

The Sea Peoples would have certainly mixed to some extent—both formally through marriages and informally through general liaisons—with the populations they encountered during their migrations and following their settlement. An iconographic study of the representations of the women of the invading Sea Peoples in Ramses III's land battle scene at Medinet Habu seems to support this conclusion. Some of the migrating women in the scene have Syro-Canaanite features, while others resemble female figures portrayed in Aegean art.[21] Thus, whatever the ultimate origins of the various Sea People groups, it would be a mistake to view any of them as homogeneous in ethnic background. By the time they become visible in the historical record, they would have been in a state of flux, and the original groups would have formed the nuclei of each of these seminebulous entities.

In analyzing the ability to isolate foreigners at the text-rich site of Deir el Medineh, W. A. Ward raises a number of relevant concerns in determining the meaning of foreign names in Egypt.[22] He notes that care must be taken with words that appear foreign but are in fact Egyptian. As an example, he points to the "throwstick" determinative, which defines a foreigner or a foreign location. He emphasizes that the opposite, however, is not necessarily true, as this determinative is at times missing from names that are clearly non-Egyptian. As a foreign name does not necessarily indicate that someone has been born abroad, Ward warns that in dealing with nonindigenous names we should think in terms of "names" rather than "foreigners."[23]

Deir el Medinah is particularly rich in textual sources. A number of foreign names, mainly West Semitic, appear there.[24] Some foreigners in Egypt are defined by their home country, such as "the Nubian stonemason Trkl" known from Deir el Bahri and "the Syrian Basiya" mentioned at Deir el Medineh.[25] An ostracon with a West-Semitic account written in Hieratic script from Deir el Medineh is additional evidence of a Syro-Canaanite who actually spoke the "mother tongue" at Deir el Medineh.[26] The ostracon mentions "the gods of Sidon Maritima."[27]

Foreigners could be given Egyptian names. The daughter of the Hittite king Hattusilis III, who married Ramses II in his Year 34 (1246 B.C.) was given an Egyptian name, Maat-Hor-Neferure, which means "She who beholds the Falcon (King) that is the visible splendour of Re."[28]

Some foreigners had both a foreign and an Egyptian name, such as the vizier Bay, who served Siptah (1194–1188 B.C.) during the XIXth Dynasty.[29] In Egypt he used his local name, Ramses-Khamenteru, but employed his Semitic name elsewhere. Perhaps most important for the present discussion, when a person is given an identifiable foreign designation, beyond reasonable doubt, such is the case. Even here it is important to differentiate between a foreign label (e.g., "PN, the Sherden") and a foreign (non-Egyptian) name.[30] Ward considers it unlikely that Egyptians adopted alien names and thus concludes that persons with such names in Egypt must be foreigners.[31] G. A. Gaballa, however, notes that Egyptians indeed did adopt foreign names and supplies the example of Ramses II's eldest daughter and consort, Bint-Anath, whose name is decidedly Semitic.[32] Other considerations include the use of nicknames, as well as names written in "group writing."[33] It is also important to remember that Egyptian scribes transcribed foreign names as they heard them.[34] There can be a significant difference between how the names actually sounded and what has come down to us.[35]

Egyptologists generally agree that, by the New Kingdom, Egypt had become home to numerous foreigners of diverse backgrounds, who permeated all levels of society.[36] Foreigners were present in Egypt for numerous reasons: They were imported slaves, attackers made into slaves, nomads arriving to pasture their animals there (Beni Hassan), and mercenaries absorbed into the army, to name but a few.[37]

Some foreigners arrived by sea. The Tomb of Kenamun at Thebes depicts a vibrant scene of independent merchants trading with Syro-Canaanite sea merchants who are seen arriving by ship (Fig. 3.7).[38] In addition to the women and youth in the top register identified as slaves, a group of men is being recorded by a port official in the register above the men offloading Canaanite amphoras. N. de G. Davies and R. O. Faulkner take these to be mariners brought before the port authorities by an official, but in my view they are better interpreted as male slaves that are being offloaded as part of the ship's cargo.[39]

Identifying ethnic elements at Gurob, however, is difficult, as is determining the relative percentage of foreigners available within the overall population at the settlement. This is due partly to insufficient available information for determining the size of the population through time at Gurob.[40]

W. M. F. Petrie concludes that numerous foreigners lived in Gurob.[41] Clearly, however, the majority of Gurob's inhabitants were Egyptian.[42] Martha Bell describes the complexity of the problem:[43]

> [I]n spite of the great varieties of nationalities that seem to have been present in the vicinity at various times, the lack of non-Egyptian remains on the site (except for the Burnt Groups) indicates that all these people must have been Egyptianized. Their stelae, burials, and other monuments appear perfectly normal.

Can we identify the foreign background of immigrants at Gurob against the milieu of indigenous Egyptians? Any discussion of foreigners at Gurob must begin with a short overview of the site itself.

The Site of Gurob

Gurob lies at the juncture of the Nile Valley with the Fayum. Thus, in a real sense, Gurob served as a gateway to the Fayum.[44] The site nestles next to the Bahr Yusuf, which is the channel of the Nile that, together with natural springs, supplies water to the Fayum (Figs. 4.2–3).[45] In antiquity these sources fed the large Lake Moeris, now shrunken to the Birket Qarun located on the Fayum's northwestern side.[46]

The story of excavations at Gurob is not a happy one. Between 1888 and 1920 Gurob saw a number of excavations undertaken by archaeologists and antiquity dealers.[47] Then, for a time, the site was out of bounds to archaeologists due to its location in a military-restricted zone.[48] Archaeological exploration led by I. Shaw has resumed in recent years.[49] This research aims to clarify the ambiguities of work at Gurob in the late nineteenth and early twentieth centuries by creating a comprehensive map of the site, developing accurate plans of the surviving architecture and tombs, reviewing the site's ceramics, and using modern methods to better understand the site.

Petrie concludes that the site of Gurob lived its life from the reign of Thutmose III through that of Merneptah, or, at the latest, of Ramses III, and that after this period, the area remained uninhabited until Ptolemaic times.[50] The Wilbour Papyrus, discussed later, indicates, however, that Gurob existed at least into the reign of Ramses V.[51] While scarabs postdating Ramses II are rare at the site, examples bearing the names of Mernteptah, Seti II, and Ramses III–V have been recovered there.[52] Gurob's abandonment probably came during or soon after the reign of Ramses V, as Egypt continued its decline into a period of deepening socioeconomic and political woes.[53]

Tomb 611, in which Burton and Englebach discovered the ship model, was one of ten tombs that they opened during their 1920 excavation in a cemetery located on a rise southwest of Gurob, which is sandwiched between their Group D on the northwest and Group G in the east (Figs. 4.4–5).[54] By that time much of the cemetery had already been looted. Aligning the excavators' plans on a Quickbird Satellite photo of Gurob makes it possible to identify the main structures and the general location of Tomb 611 (Figs. 4.6–9).[55]

The caliber of the grave goods in this burial ground suggests that, overall, their owners held high status.[56] The excavators dated these tombs without exception to the XVIIIth or XIXth Dynasties.[57] Bell assigns the majority of the tombs in this group to the XIXth Dynasty.[58] Also in this cemetery, Tomb 605 has received special attention as it was found undisturbed and contained both Late Helladic IIIB ceramics and a scarab of Ramses II, thus making it an important synchronism for dating this phase of Mycenaean ceramics.[59]

Gurob is identified with the pharaonic harem settlement of Mi-Wer.[60] Numerous texts found at the site

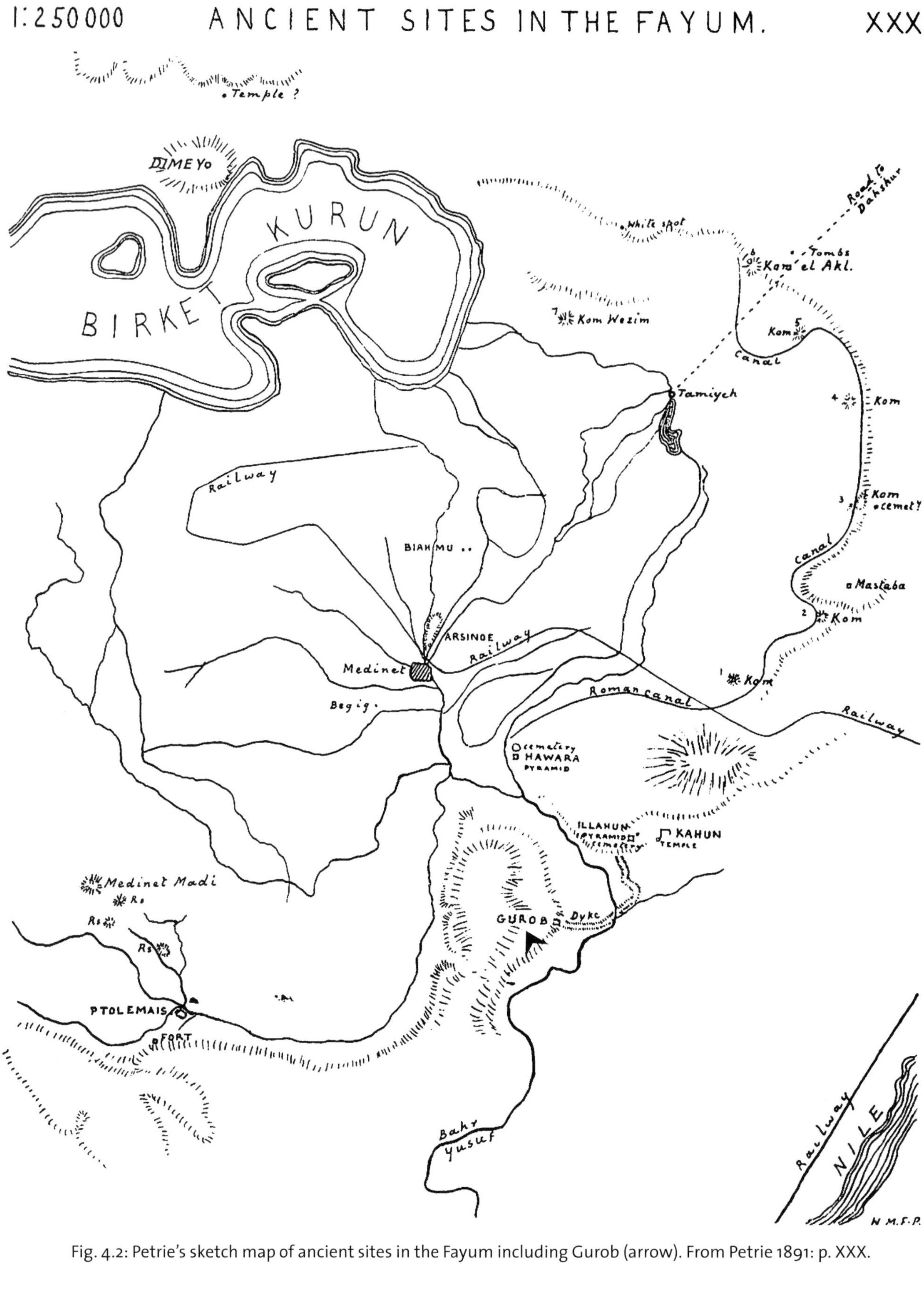

Fig. 4.2: Petrie's sketch map of ancient sites in the Fayum including Gurob (arrow). From Petrie 1891: p. XXX.

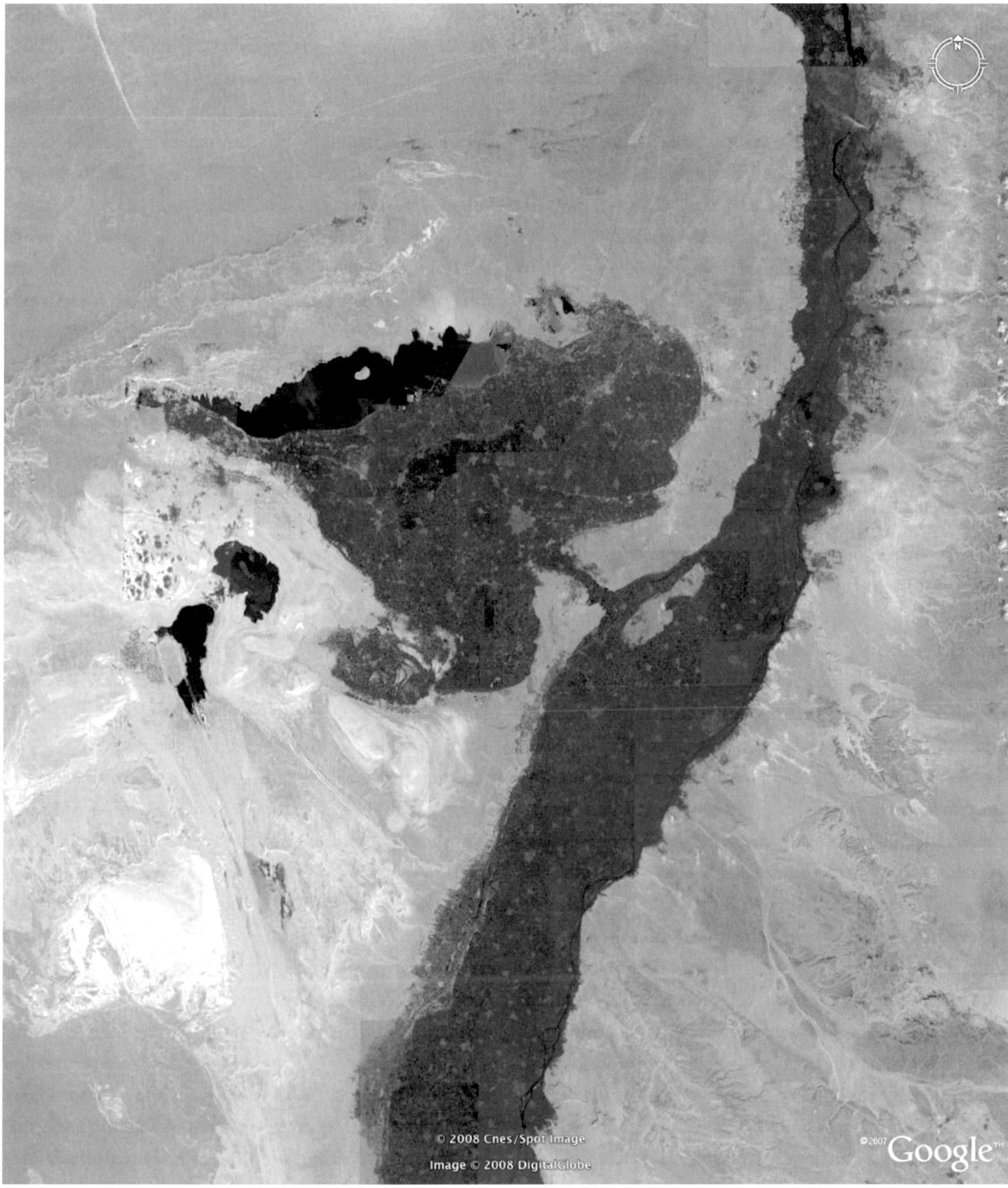

Fig. 4.3: Satellite view of the Fayum. Courtesy of Google Earth.

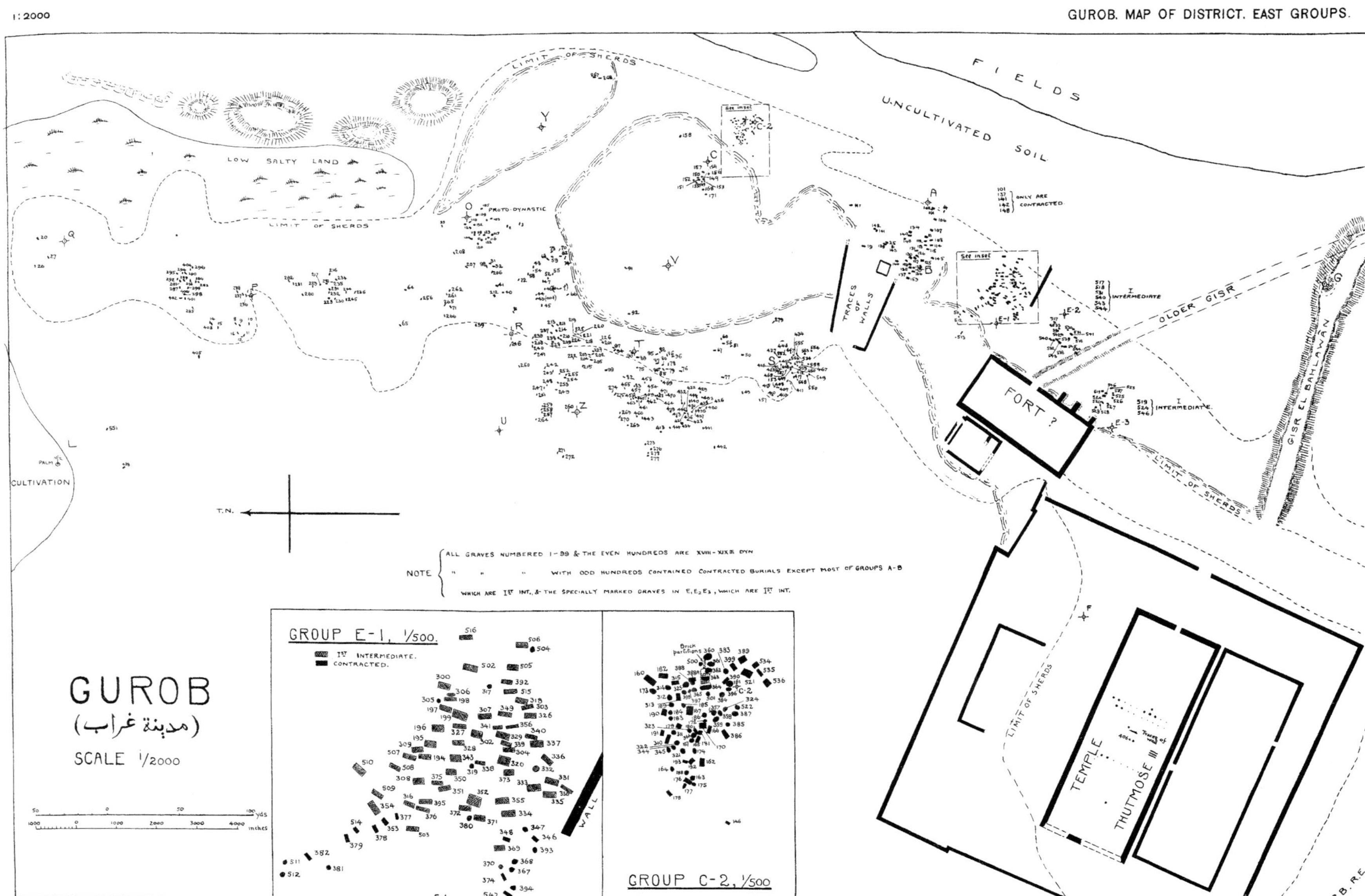

Fig. 4.4: Brunton and Engelbach's map of the area north of the harem of Mi-Wer. From Brunton and Engelbach 1927: pl. I.

mention either Mi-Wer or officials involved in its administration, thus corroborating the site's identification with that harem city.[61] The location of Gurob, on the border of the Fayum marshland, was well situated for the establishment of a harem for New Kingdom pharaohs wishing to take advantage of the fishing and birding that the region provided and to enjoy other pleasures as well to allay their fatigue after the hunt.[62]

The name Mi-Wer means "Great Channel," which refers to the nearby Bahr Yusuf, flowing from the Nile into the Fayum.[63] Lake Moeris appears to have derived its name from an earlier term that defined the lake by its association with Mi-Wer, that is, the Lake of Mi-Wer.[64] It is not clear, however, whether the lake received its name from the site or from the Bahr Yusuf itself.[65]

In the Golenischeff Onomasticon, which lists settlements, Mi-Wer and the settlement of She appear separately, presumably on opposite sides of the Bahr Yusuf (Figs. 4.10–11).[66] Land assessments for both Mi-Wer and a second harem settlement at Memphis occur in the Wilbour Papyrus, where Mi-Wer appears as a consequential institution with administrative domains in various areas under its control.[67] From this Gardiner concludes that both of these names refer to separate towns and that Mi-Wer was probably located somewhere to the north of She.

Among the institutions that owned land assessed in the Wilbour Papyrus and that would have been located on the Bahr Yusuf was the "Landing Place of the Pharaoh at Mi-Wer."[68] This may be an early form of the name of the later Ptolemaic site of Ptolemais Harmos (Ptolemais Harbor),

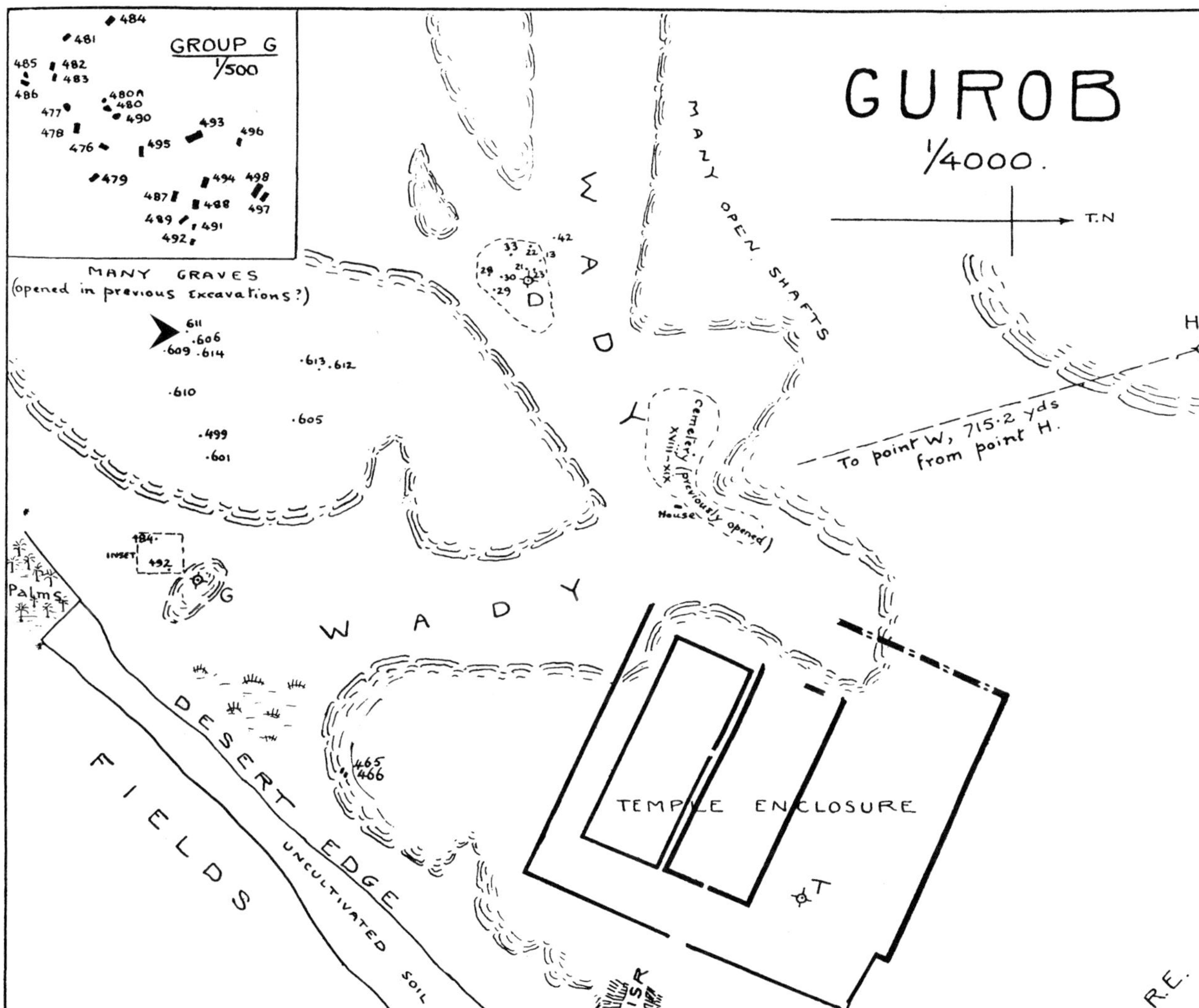

Fig. 4.5: Brunton and Engelbach's map of the area south of the harem of Mi-Wer, which includes the region of Tomb 611 (arrow). From Brunton and Engelbach 1927: pl. II: Lower.

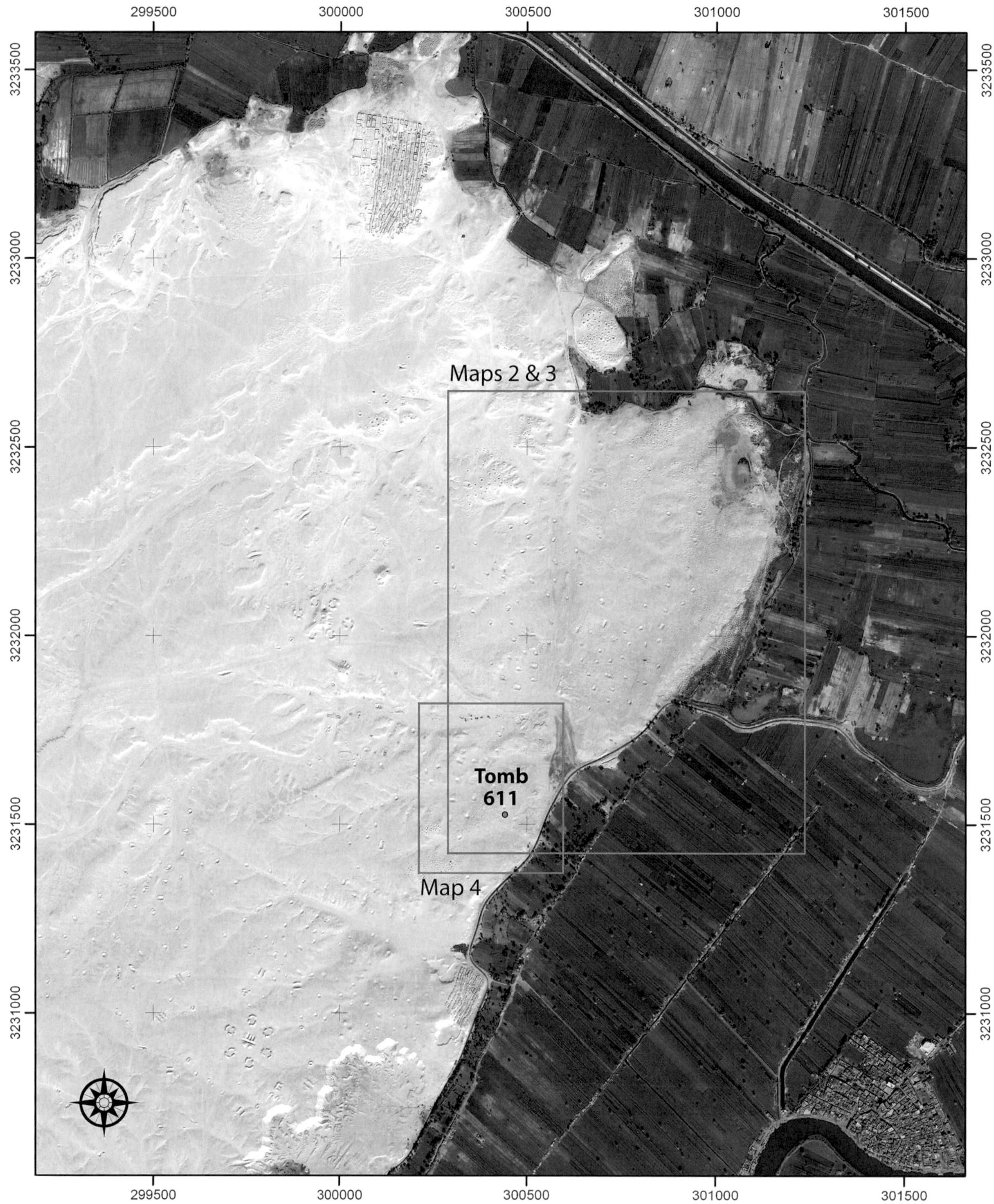

Map 1

Fig. 4.6: QuickBird satellite image of Gurob and its region. Boxes represent areas covered in Figs. 4.7–9. Prepared by D. Davis. Includes copyrighted material of DigitalGlobe, Inc.

Map 2

Fig. 4.7: QuickBird satellite image of the Gurob area. DigitalGlobe, Inc.

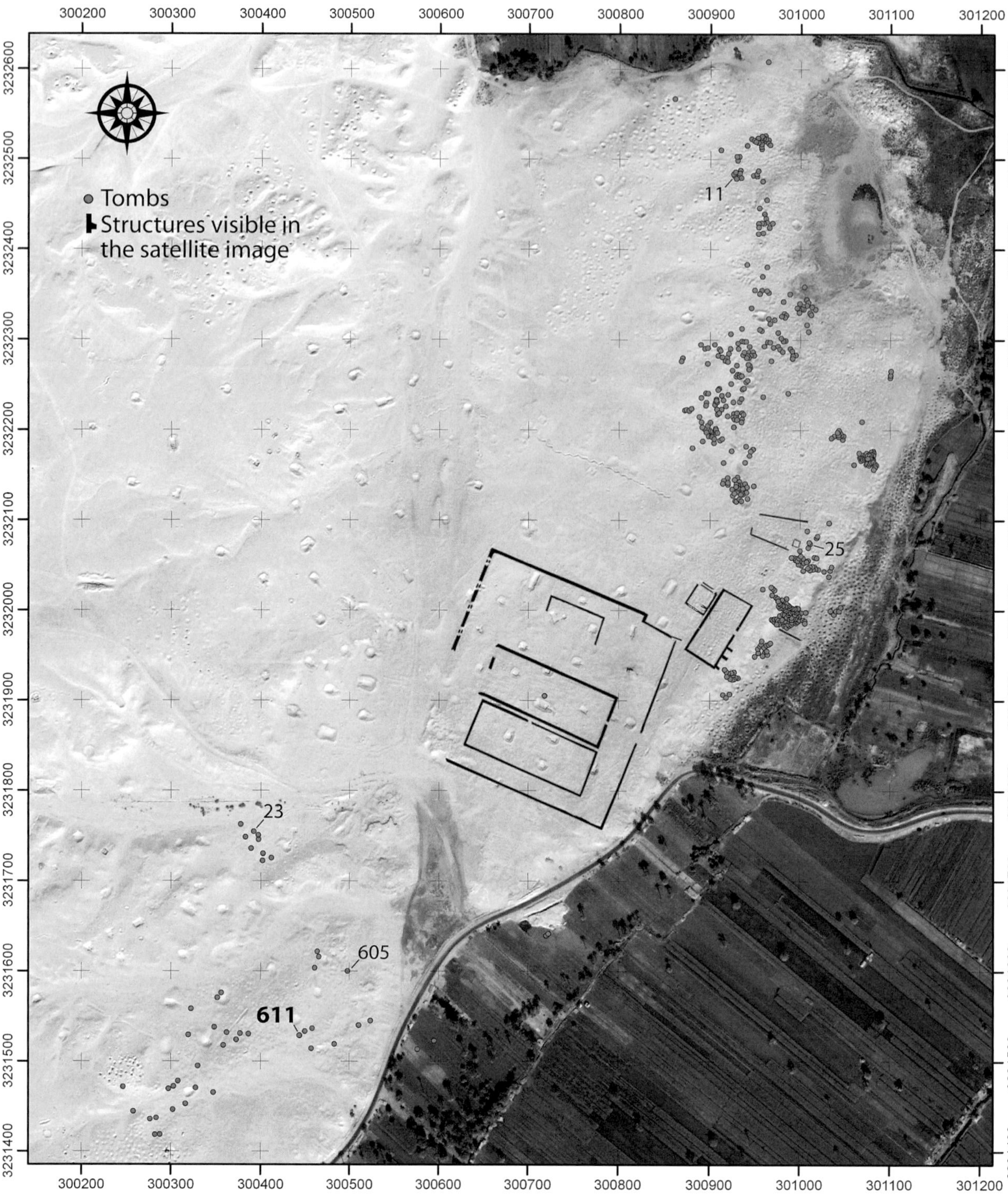

Map 3

Fig. 4.8: QuickBird satellite image of the Gurob area, with structures (in black outline) and tomb sites (represented by red dots) in the digital component superimposed on a UTM grid with 100-m spacing (WGS 1984). The structure lines are traced over features visible in the satellite image and conform to the excavation plans (Figs. 4.4–5). As the tombs were not visible in the satellite image, they were georeferenced from the plans using ESRI's ArcGIS 9.2. The structures and scale in Petrie's plans were used as baselines. Note Tomb 611 at the lower left of the photo. Prepared by D. Davis. Includes copyrighted material of DigitalGlobe, Inc.

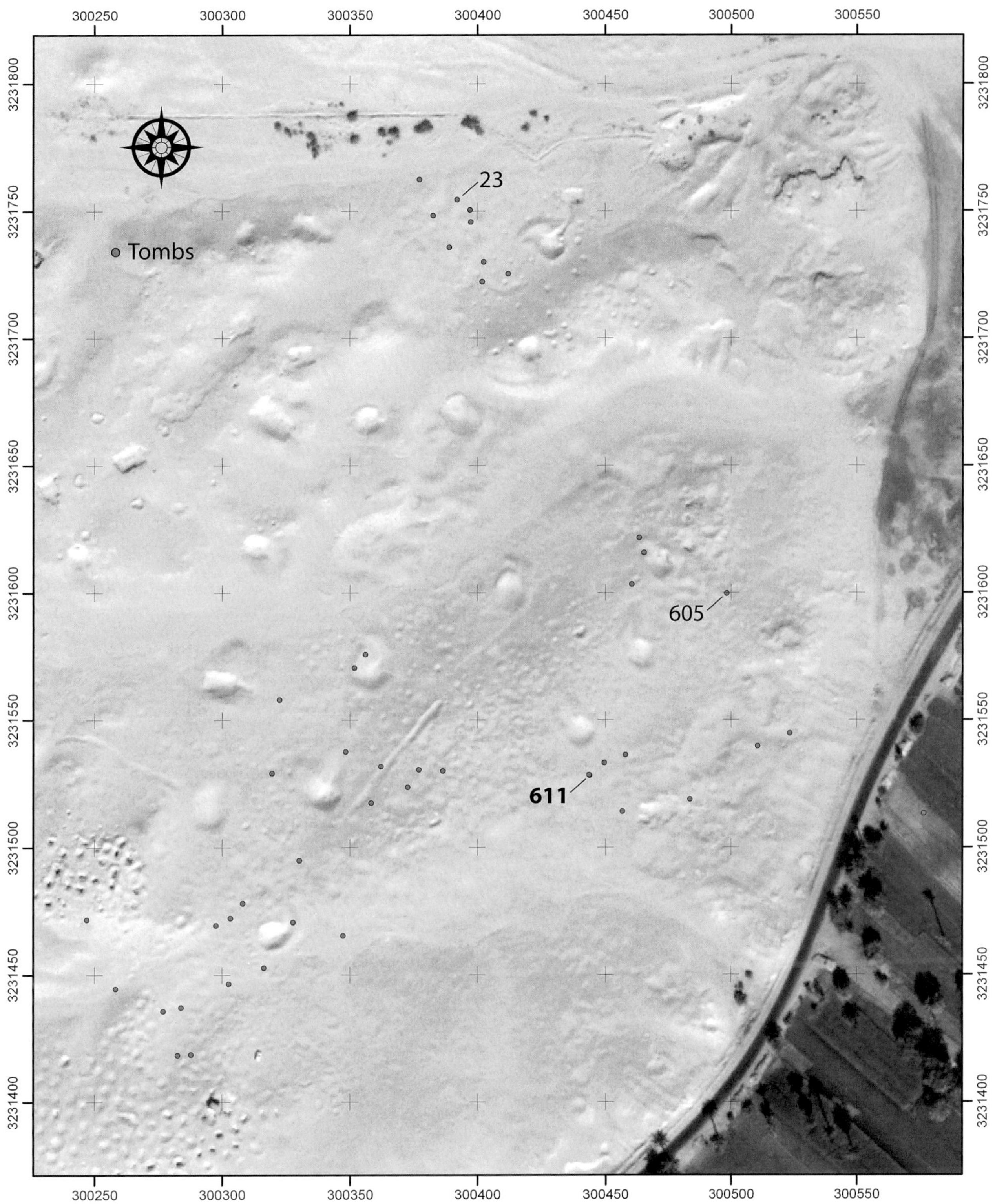

Map 4

Fig. 4.9: Satellite image of the area of Tomb 611. Prepared by D. Davis. Includes copyrighted material of DigitalGlobe, Inc.

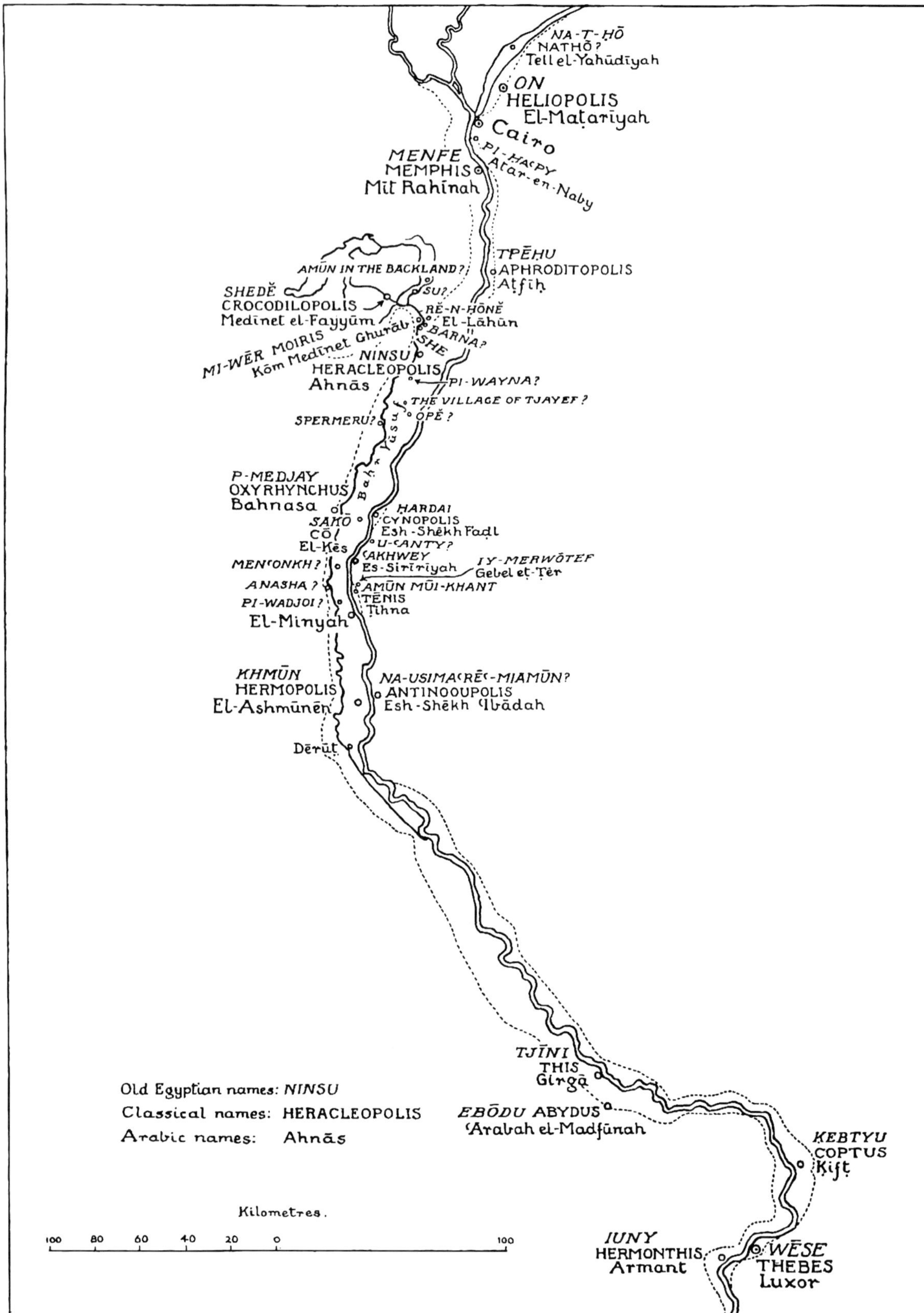

Fig. 4.10: Gardiner's map of settlements with temples holding land recorded in the Wilbour Papyrus, Text A. After *WP* II: Map 1 (opposite p. 54).

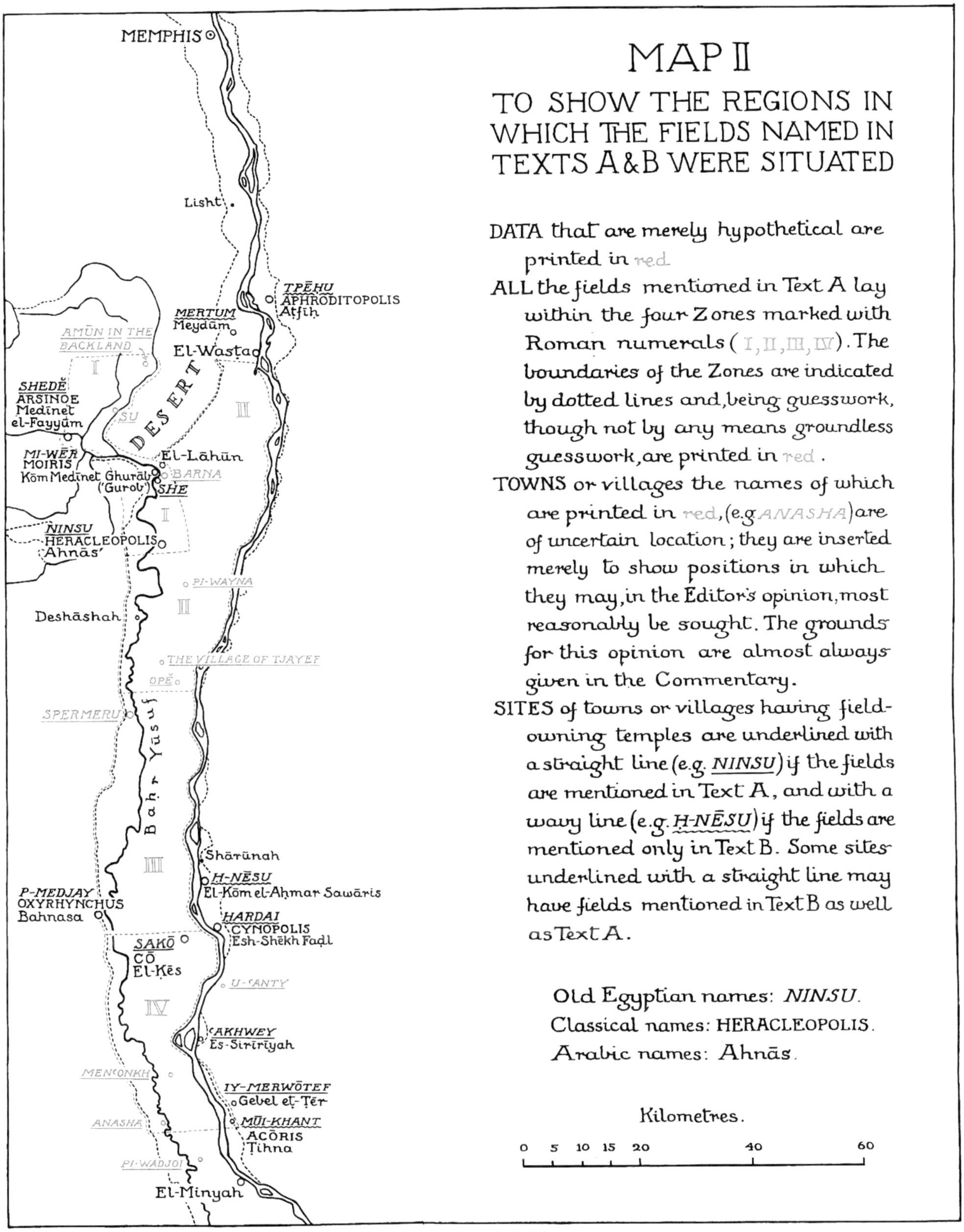

Fig. 4.11: Gardiner's map of regions with lands recorded in the Wilbour Papyrus, Texts A and B. After *WP* II: Map 2 (opposite p. 55).

perhaps located on the bend of the Bahr Yusuf between Gurob and Lahun.[69]

The harem was not a women's area in an occasional palace but rather a palace in its own right.[70] It consists of two parallel rectangular structures located on the southern side of a surrounding rectangular enclosure wall (Fig. 4.7–8, 12). P. Lacovara compares the plan of the harem at Gurob to that of the North Palace at Malkata, which may have served as a harem complex adjacent to Amenhotep III's royal palace.[71] We gain an impression of its appearance from a scene in the tomb of Neferhotep (TT 49) at Thebes depicting the harem of Ay (Fig. 4.13).[72]

In its heyday Mi-Wer, with its royal harem located at the entrance to the Fayum, clearly was an important settlement. References to various commercial activities appear in fragmentary papyri from Gurob.[73] Artifacts found by Petrie argue for workshops there that manufactured faience, glass, metal items, and wooden statues.[74] Wooden artifacts from the site include high-quality articles.[75] The most outstanding example of these is an exquisite wooden head thought to represent the older Queen Tiy.[76]

Most important, a textile industry flourished here.[77] Weaving is normally an occupation undertaken by women and likely practiced or supervised by the women of the harem.[78] At Mi-Wer textile workers were trained in weaving, and papyri fragments from the site refer to harem textiles sent to various locations.[79]

Thus, the surrounding settlement existed to support the harem of Mi-Wer, which would have supplied the main focus of the local work force, served as a significant manufacturing center, and acted as an important economic focus for the surrounding area.[80] Angela Thomas notes:[81]

> Such an institution would require a whole network of officials, administrators, artisans, workmen, and servants and these people and their families would need to be accommodated in the area. Labourers would have been employed to till the lands of the harem, the temple, and the landing place of Pharaoh. Produce from the Fayum was probably brought here and also the catches from the fishing grounds. Over a period of time the town must have become an important centre for the area. The town would not have been as tightly State-controlled as the necropolis workers' towns and may have possessed some degree of autonomy. Nevertheless, a State institution provided much of the work and must have exercised a considerable influence.

Foreigners at Gurob

Syro-Canaanites.—Petrie published three typical Middle-Late Bronze-Age Syro-Canaanite toggle pins from Gurob, one made of gold and two of bronze (Fig. 4.14).[82] While such pins are common along the Syro-Canaanite littoral, they are rare in Egypt.[83] As these are personal items, which were unlikely to have arrived as articles of trade, they presumably represent a foreign presence at Gurob. However, toggle pins could also arrive in Egypt as royal gifts, as in the case of a pair sent by the Mitannian king Tushratta to his sister Kelu-Kheba, who was a wife of Amenhotep III.[84] Even in this case, however, the toggle pins still represent the foreign presence of Kelu-Kheba in Egypt. The gold toggle pin from Gurob presumably belonged to someone of high status connected to the harem (Fig. 4.14: right).[85] At least one Syro-Canaanite of high rank is known to have officiated in the harem of Mi-Wer: the royal butler Ramessesemperrē̓, who served in the harem's administration during the reign of Merneptah.[86] Two Syro-Canaanite amulets found at Gurob also suggest a foreign presence.[87]

Petrie discovered a bronze mirror of Syro-Canaanite or Cypriot origin at Gurob, which led him to conclude that the artisan who made it lived there.[88] The mirror's presence does not necessarily equate with its having been made at the site, and there is no strong argument for seeing in this artifact anything other than trade.

Tjemhu (Libyans).—Libyans are hinted at in the general region of Gurob in the Wilbour Papyrus by the Tjuk, a group identified with Libyan soldiers serving the Egyptians, probably from the region of the Dakhla and Kharga Oases.[89] The Tjuk People have been identified with the Sukiim (סוכיים), who, together with Libyans and Nubians, composed part of the force that Shishak I brought against Rehoboam following the death of Solomon in 926 or 925 B.C.[90]

The Kehek are identified as Lybian, also.[91] They are

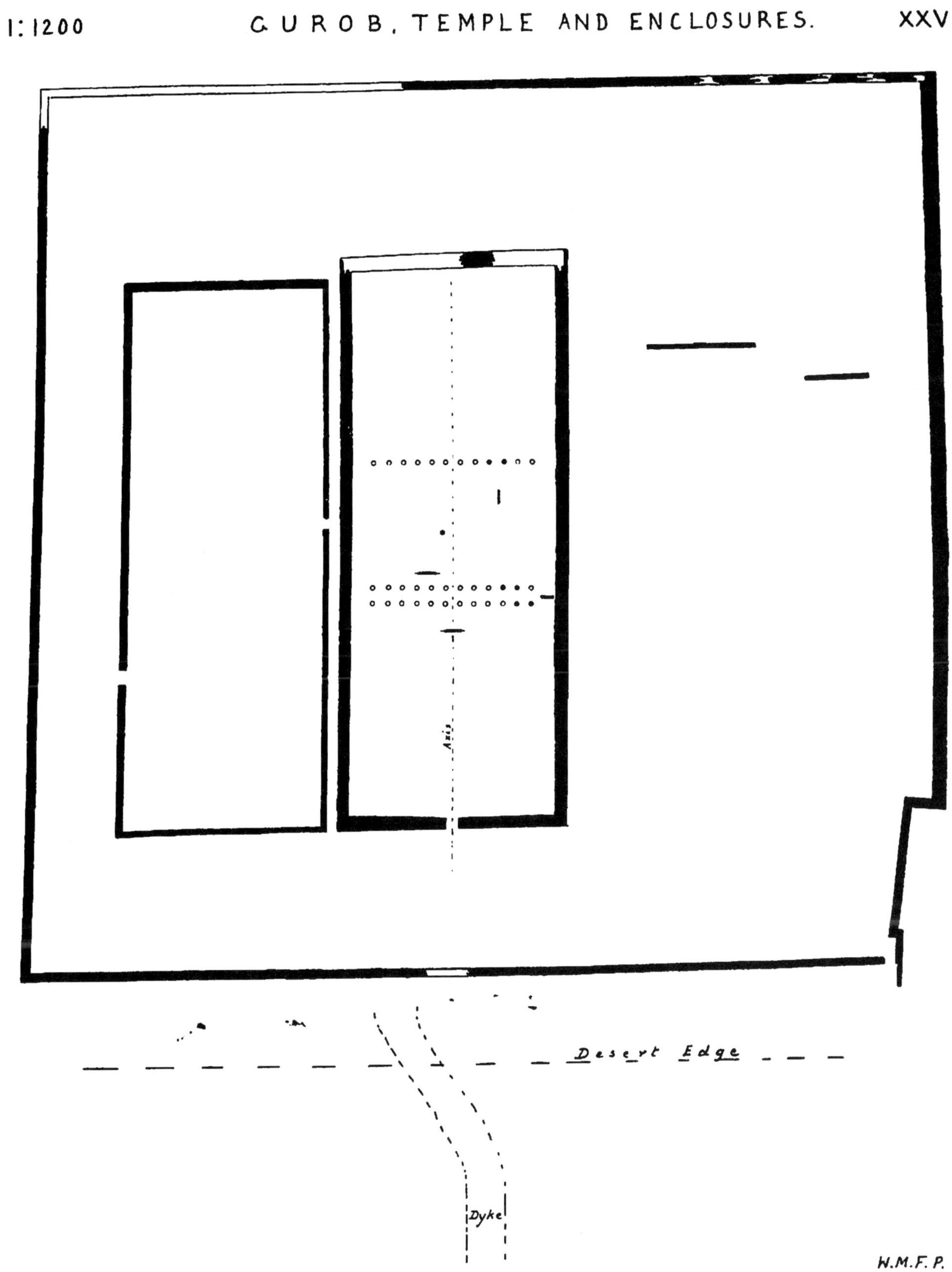

Fig 4.12: Petrie's plan of the structures that he identified as a temple and an enclosure (NTS). From Petrie 1891: p. XXV.

Fig. 4.13: The harem of Ay in the tomb of Neferhotep (TT 49). From Davies and Davies 1973: pl. I.

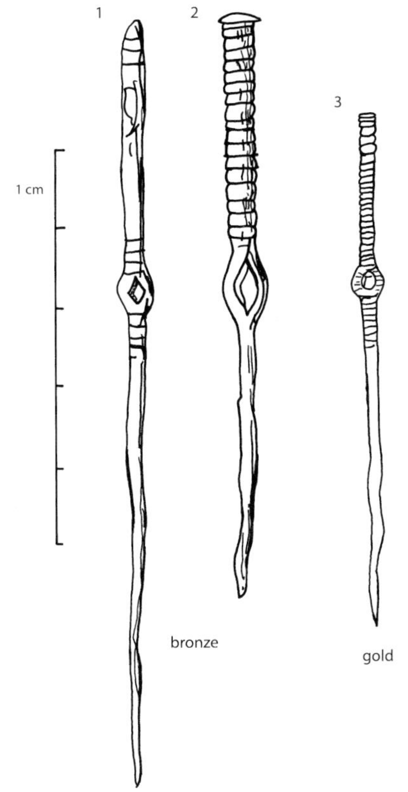

Fig. 4.14: Three typically Syro-Canaanite toggle pins found at Gurob. After Petrie 1891: 19 no. 39, pl. XXII: 1–3.

the largest foreign military contingent in the Egyptian army (1,600) mentioned in Papyrus Anastasi I.[92] Merneptah lists 218 "Kehek and Libyans, carried off as living prisoners."[93] Ramses III refers to Kehek in the Egyptian army.[94] Unlike the Tjuk, however, one finds no trace of the Kehek in the Wilbour Papyrus.

Ramses III appears to have been instrumental in resettling Tjemhu in Egypt. His Year 5 inscription at Medinet Habu notes that "Their leaders were made into gangs through victories, and branded with the great name of his majesty."[95] Papyrus Harris reports that Tjemhu groups had been living in Egypt, primarily in the west Delta but that Ramses III conquered them and then resettled them in the land:[96]

> Behold, I will inform you of other things, done in Egypt since my reign. The Libyans and the Meshwesh (*M-š’-w’-š’*) were dwelling in Egypt, having plundered the cities of the western shore, from Memphis to Kerben (*Ḳ’-r’-b’-n’*). They had reached the great river on both its banks. They it was who plundered the cities of Egwowe (*G-wt-wt*) during very many years, while they were in Egypt. Behold, I destroyed them, slain at one time. I laid low the Meshwesh, the Libyans, the Esbet (*’-s’-b’-t’*), the Keykesh (*Ḳ’-y’-ḳ’-š’*), the Shai (*Š’-y*), the Hes (*H’-s*) and the Beken (*B’-ḳ’-n’*); they were overthrown in their blood and made heaps. I turned them back from trampling the border of Egypt. I carried away those whom my sword spared, as numerous captives, pinioned like birds before my horses, their wives and their children by the ten-thousand, their cattle in number like hundred-thousands. I settled their leaders in strongholds in my name. I gave to them captains (*ḥryw*) of archers, and chief men of the tribes, branded and made into slaves, impressed with my name; their wives and their children were made likewise. I led their cattle into the house (*pr*) of Amon; they were made for him into herds forever.

From a historical perspective then, we have no reason to doubt that large numbers of Libyans lived in Lower Egypt and that at least one group, if not more, dwelled in the Gurob region.[97] Identifying them remains problematic, however.

Ward emphasizes that, although we know a great deal about the languages of the literate cultures with which the Egyptians interacted, virtually nothing is known of the names of the illiterate cultures with which they came in contact.[98] Thus, Nubian, Libyan, and European names remain largely beyond our grasp.[99] Ward's comments on the invisibility of Libyans at Deir el Medineh also holds true for Gurob:[100]

> In other words, though we can be reasonably sure that Libyans and Nubians lived at Deir el Medina and that some Libyan and Nubian names appear in the village archives . . . it is only rarely that we can define a specific personal name as African—and even then, we are unable to produce a cognate in an African language. This means, of course, that we are unable to assess accurately the extent of African penetration into the native population of Egypt. Unless contemporary inscriptions in the African languages are discovered,

the African elements in Egyptian society—other than individuals specifically stated to be Libyan or Nubian—will remain largely anonymous.

Mycenaeans and Cypriots.—Excavations at Gurob revealed Mycenaean and Cypriot ceramics and figures, which Petrie identified as Mycenaean.[101] This led him to posit "a first settlement of Aegean races, and a later influx of Cypriotes."[102] The Mycenaean pottery at Gurob must be considered within the context of the other New Kingdom sites where it has surfaced.

Some scholars believe that Mycenaean ceramics, in isolation and lacking additional cultural indicators, can signify a foreign presence at their discovery sites.[103] By themselves, however, trade goods such as ceramics found on foreign soil cannot be taken as indicators of persons from the country of origin transporting the items or living with the artifacts at the discovery sites.[104]

This situation is analogous to the problem of identifying the origins of shipwrecks when documentary evidence is lacking.[105] As G. F. Bass notes, a vessel's identity cannot be determined by its cargo, which could have been taken on board anywhere along its route: Rather, the crew's *personal items* are one of the most diagnostic data sets for determining a ship's cultural identity and geographic homeport.[106] The meticulous excavation of the Uluburun shipwreck has led to new insights into Bronze Age trade and demonstrates that this could be far more complex than has been previously contemplated. A comprehensive study of these types of artifacts enabled the excavators to identify the ship as Syro-Canaanite and also to determine that two high-status Mycenaeans and possibly a Balkan mercenary were passengers on its last voyage.[107]

Papyrus fragments from Amarna depict Mycenaeans supporting the Egyptian army in fighting Libyans in a manner reminiscent of the later use of Sherden by the Ramessides.[108] This does not, however, necessarily place such Mycenaean troops at Amarna, and, in fact, nowhere do Mycenaeans (or Sherden for that matter) appear as personal guards of the pharaoh, as is the case with the Sherden in Ramesside times.[109]

Ugarit is also worth considering in any discussion regarding the linkage, or, more correctly, the lack thereof, of Mycenaean artifacts with Mycenaean traders. Although the site has copious amounts of Mycenaean pottery and a rich onomasticon, no names from Ugarit can be identified with those appearing in the Mycenaean Linear B onomasticon.[110] Indeed, the only evidence of *direct* trade between Ugarit and the Aegean comes from a text dealing with tax relief for Sinaranu, *an Ugaritian* sea merchant, when his ship returns from Caphtor (Crete or the Aegean).[111]

A stronger argument can be made for the presence of Cypriot (Alashian) traders in Egypt.[112] The Amarna tablets document a robust trade between Egypt and Cyprus. Seven surviving Amarna tablets are addressed from the king of Alashia to the pharaoh, and one is from the "Minister of Alashia" to his Egyptian equal.[113] These documents show a complex trade and diplomatic connection between the two countries.[114] Some of the tablets refer specifically to Alashians living in Egypt. For instance, EA 35 mentions an Alashian who had died while in Egypt.[115] The Alashian ruler requests that the pharaoh return his countryman's possessions. The retention of Alashian traders in Egypt is a bone of contention: The

Fig. 4.15: The coffin face of Anen-Tursha. After Petrie 1890: pl. XIX.

king of Alashia asks repeatedly that his people be returned to him.[116] Other complaints deal with the taxation of Alashian merchants in Egypt.[117]

Closer in time to the period we are considering, Ward identifies a woman named "the Cypriot" at Deir el Medineh.[118] When Wenamun shipwrecked on Alashia during his escape from the Sekels, he found a man there who understood Egyptian, which suggests that the fellow had spent time in Egypt.[119]

Finally, at Karnak, L. Basch identified a stone anchor of typical Cypriot shape but made of local stone.[120] The simplest explanation for this anomaly is that a Cypriot who was familiar with the island's anchor styles carved one while in Egypt, presumably for dedication at the temple.

The Teresh.—Petrie revealed the coffin of Anen-Tursha, a palace official, which he found together with Sadi-amia's ushabti at Gurob.[121] Describing the face portrayed on Anen-Tursha's coffin Petrie writes the following (Fig. 4.15):[122]

> On turning to personal characteristics, as well as the names, we see the evidence of foreigners. The face of the coffin of Anen-Tursha . . . is far from Egyptian in the type of it; the long nose, and the close, slightly sloping eyes come from abroad; further, the pierced ears do not belong to Egyptian men, nor is the piercing of the lower lobe an Egyptian custom as their ornaments belonged to the outer or upper edge.[123]

Petrie identified Anen-Tursha as a Teresh:[124]

> In the opposite chamber of the same tomb was the coffin of Anen-Tursha . . . , apparently one of the foreign Tursha race, formerly identified with the Etruscans, but perhaps rather to be connected with the Turseni or Tyrrhenians of Lemnos and the Dardanian coast. The face is certainly non-Egyptian.

Bell notes that Anen-Tursha's name may be Hurrian and that the most that can be determined is that he appears to be a foreigner.[125] Anen-Tursha was a deputy overseer of the harem during the XIXth Dynasty.[126] His burial at Gurob was entirely Egyptian in nature and showed that this foreign-named individual had risen high in government administration while also becoming entirely Egyptianized.[127]

The Teresh took part in the Sea Peoples' attack on Egypt during the reign of Merneptah.[128] At Karnak he lists 742 men and 790 hands of Teresh in his body count after the battle.[129] The Teresh do not appear at Medinet Habu or in Papyrus Harris in connection with the invasion during the reign of Ramses III, but a chief of the Teresh nevertheless appears in the façade of the Migdol at the entrance to the temple (Figs. 4.16–17).[130] There he receives the designation "of the sea," an epithet not given to the Teresh by Merneptah.[131] As this group of worthies also includes leaders of other "enemies" of Egypt with whom Ramses III did not contend, the historical value of this figure is questionable. The Teresh have been linked with the Lydian Tyrsenoi/Tyrrhenians, whom Herodotus identifies as the ancestors of the Etruscans, as well as Tarsus and Tarwisa in Anatolia.[132]

The Sherden.—Biblical studies of the Sea Peoples naturally focus on the Peleset/Philistines, who appear prominently in the Bible as archrivals of the Israelites: Of the various documented Sea Peoples, the material culture of the Philistines after their settlement in Canaan has been the most studied.[133] Yet, of all the Sea Peoples that strode across the Egyptian stage, appearing in their monuments and mentioned in their texts, during the New Kingdom, from an internal Egyptian perspective, the Sherden were undoubtedly the most significant by far.[134] The Egyptians record the Sherden, of all the Sea People groups, playing the dual roles of foreign aggressors and warriors for Egypt, both at home and in foreign conflicts. At times these skilled combatants fought against other groups of Sea People invaders—including their own kin.

Once settled in Egypt, the Sherden adopted local customs and names. Over time they successfully integrated and were eventually absorbed into Egyptian society, although perhaps a certain cohesiveness surrounded their continued service in the military.[135] We can identify Sherden only when they are given their ethnonym, as their personal names are typically Egyptian.[136] It is possible to compare the Sherden settlement in Egypt with other historical migrations in which a culturally inferior

Fig. 4.16: Kneeling and bound enemy leaders on the eastern High Gate of Medinet Habu. Left to right: Hittite, Amorite, Sekel, Sherden, Shasu, and Teresh. The title of a Philistine chief appears at the far right, but the figure is now hidden behind a Ptolemaic wall. From Epigraphic Survey 1970: pl. 600. Courtesy of the Oriental Institute of the University of Chicago.

group enters a region claimed by a higher culture.[137] The following is a short chronological tour of references to the Sherden with regard to their interaction for and against Egypt:

Amenhotep III/Akhenaten.—The Sherden first appear in Egyptian awareness in the Amarna tablets. In one case a Sherden participated in an assassination attempt against Rib Addi, the king of Byblos, but managed to escape to Abdi Ashirta when the plot went awry.[138]

Two additional texts pertain to a second incident. Rib Addi writes from Byblos to the pharaoh, complaining that Sutû mercenaries, under the orders of an Egyptian official named P'-Ḫurú, working out of Damascus, killed a single Sherden and carried off three men from Byblos to Egypt.[139] Rib Addi pleads with the pharaoh to send back the kidnapped men to Byblos as their capture is causing an uproar in his city, and he fears open rebellion if the men are not returned. It is not clear why P'-Ḫurú has taken this action, why the Sherden was murdered, and why the Byblian men were kidnapped, but it was clearly in the pharaoh's interest, as the men were sent to Egypt.[140] The fact that Rib Addi assumed that the pharaoh would know who or what a Sherden was indicates that they were a known ethnic group by this time and may indicate that the earliest textual evidence considerably postdates first contact. J. Yoyotte notes that Sherden do not appear in Akhenaten's personal guard, which was composed solely of Syro-Canaanites and Nubians, to judge from the monuments.[141] This suggests that while the Sherden are found along the Syro-Canaanite coast at Byblos and perhaps Ugarit, they may not yet have been absorbed into the Egyptian army at home.[142]

Ramses II.—Following the Amarna tablets, the next reference to them appears in Ramses II's "Sherden Stele" from Tanis, which describes how he repulsed their seaborne invasion early in his reign: "Sherden rebellious of heart. . . . battle-ships in the midst of the [sea]."[143] Ramses II's Year 2 Aswan Stele may also allude to this clash.[144] Regrettably, Ramses does not supply us with representations of this engagement, which is the earliest recorded sea battle in history.[145]

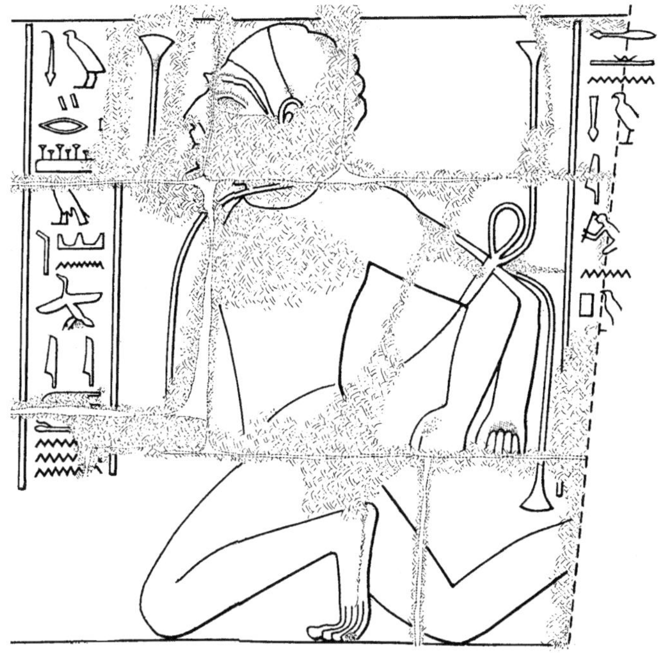

Fig. 4.17: The Teresh Leader on the entrance to the eastern High Gate of Medinet Habu. After Epigraphic Survey 1970: pl 600.

In a phenomenon that was to be replayed centuries later by Psammetichos I, Amasis, and Psammetichos II, Ramses II incorporated these "brazen" foreign men into the standing Egyptian army.[146] By the Battle of Kadesh Sherden served alongside the Egyptians.[147] They also appear in scenes of battles against the cities of Tunip and Dapur.[148]

In the "Poem of Pentaur," Ramses II describes how he prepared for battle with the Hittites: "Behold, his majesty prepared his infantry and chariotry, the Sherden (*Š'-r'-dy-n'*) of the captivity of his majesty from the victories of his sword."[149] The singling out of these foreign troops indicates the high value the Egyptians placed upon their military prowess.

In Papyrus Anastasi I, which probably dates to the reign of Ramses II, 520 Sherden are listed—together with 1,600 Kehek, 620 Meshwesh, and 880 Nubians—among the forces of foreign ethnic groups for an Egyptian campaign in Asia.[150]

Fig. 4.18: The Sherden chief on the entrance to the eastern High Gate of Medinet Habu. After Epigraphic Survey 1970: pl. 600.

Merneptah.—Following the death of Ramses II, Merneptah received the brunt of a major incursion in which groups of Sea Peoples joined forces with the Tjemhu.[151] Sherden were among the invading groups.[152] Unfortunately, no representational art of this conflict survives.

Papyrus Anastasi II mentions Sherden twice. A homage to Merneptah refers to his victory over the Sherden: "The Sherden thou didst carry off through thy strong arm have plundered the tribes of foreign lands."[153] A second reference, which echoes Merneptah's Great Karnak Inscription, mentions the capture of Sherden, their absorption into the Egyptian military, and their paying of taxes:[154]

> Sherden of the Great Green[155] that are captives of His Majesty (l.p.h.), they are equipped with all their weapons in the court, and bring a tribute of gallons of barley and provender for their chariotry, as well as chopped straw.

Ramses III.—The Medinet Habu nautical battle relief depicts Sherden fighting on both sides. Oddly enough, though, they are not mentioned in the accompanying inscription listing hostile Sea Peoples (Figs. 2.3–4).[156] This is all the more surprising as eight Sherden appear in the top register of bound prisoners directly beneath the naval battle scene (Fig. 2.1).[157]

A Sherden prince "of the sea" appears among the vanquished enemy leaders kneeling on the façade of the Migdol entrance to Medinet Habu (Figs. 4.16, 18).[158] While little can be made of this character as the list includes leaders of peoples who cannot be seriously considered as enemies of Egypt at that time, he does link the horn-helmeted warriors of the monuments with the Sherden of the texts.[159]

Sherden warriors man two of the five northern ships in the naval battle relief. Warriors of a Sea People group wearing feather helmets man other ships: These have been identified uncritically as Philistines due to their feather helmets.[160] But a number of Sea Peoples wore feather helmets.[161] In the parallel land-battle scene Sherden warriors appear fighting solely on the Egyptian side against the migrating invaders.[162]

Ramses IV.—Found in a cliff tomb near Deir el Medineh, Papyrus Harris is primarily a record of Ramses III's donations to temples, but it ends with a short history of that pharaoh's accomplishments.[163] His son and successor, Ramses IV, prepared the report at the time of Ramses III's death. Sherden appear prominently in the historical account.

Ramses III addresses this section of the papyrus "to these princes, and the leaders of the land, the infantry and chariotry, the Sherden (*Š'-r'-d'-n'*), the numerous archers, and all the citizens of Egypt."[164] Then, in the assessment of his organization of the country he notes:[165]

> I made Egypt into many classes, consisting of: butlers of the palace, great princes, numerous infantry, and chariotry, by the hundred-thousands; Sherden (*Š'-r'-d'-n'*) and Kehek (*Ḳhḳ*), without number; attendants by the ten-thousands; and serf-laborers of Egypt.

Papyrus Harris lists Sherden among the Sea Peoples who attacked Egypt in Ramses III's Year 8.[166] These references to the Sherden, taken together with their numerous appearances on the walls of Medinet Habu as both protectors and invaders, speak to the important role played by the Sherden in Ramesside Egypt.

In describing life under his rule, Ramses III paints a rather idyllic picture, which includes a snapshot of the Sherden:[167]

> I made the infantry and the chariotry to dwell (at home) in my time; the Sherden (*Š'-r'-d'-n'*) and Kehek (*Ḳhḳ*) were in their towns, lying the ⸢length⸣ of their backs; they had no fear, (for) there was no enemy from Kush, (nor) foe from Syria. Their bows and their weapons reposed in their magazines, while they were satisfied and drunk with joy. Their wives were with them, their children at their side; they looked not behind them, (but) their hearts were confident (for) I was with them as the defense and protection of their limbs.

Ramses V.—The Wilbour Papyrus is the single most important source of information for Sherden settled in Egypt. This document dates to Year 4 of Ramses V (ca. 1143 B.C.) and is the third largest of all known ancient papyri, measuring over ten meters.[168] It contains an exceptional list of land ownership, which A. H. Gardiner compares in some aspects to the English Domesday Book. The papyrus deals with a main basis for Egyptian taxation: the calculation and the evaluation of lands.[169] Documents like this must have been common in antiquity, but to date this is the only known specimen of such length. The entries follow a formula in which the names of the responsible persons are followed by the locations of the areas measured, their sizes, and the taxes levied on them. The lands are organized in the papyrus under the principle holders of property, mainly temples and royal lands.

The papyrus covers an area of more than 140 kilometers, from north of Medinet el Fayum (Crocodilopolis) to somewhere north of El Minya (Figs. 4.10–11).[170] Gardiner situates at Gurob the "Mansion of Ramʿesse-miamūn, Beloved like Rēʿ."[171] This name appears on a fragmentary papyrus from Gurob and on a stele from Abydos belonging to Ramessesemperrēʿ, which describes him as living at the Mansion of Ramʿesse-miamūn.[172] A second stele of Ramessesemperrēʿ was found at Gurob.[173]

Most important to the present discussion, the papyrus indicates that numerous Sherden were living throughout the region surrounding Gurob more or less contemporaneously with the date of the Gurob ship model.[174]

In the Wilbour Papyrus the Sherden appear settled as mercenaries responsible for lands. They probably held land as a reward for serving in the Egyptian military, or perhaps they received land as pensioned veterans.[175] They also appear here in a variety of vocations, not only as workers or possibly managers of fields owned by others but also as landholders in their own right.[176] Some Sherden had both farmers and slaves.[177] Others are further identified as "retainers" or "standard bearers."[178] The papyrus lists Piuiu, a Sherden herdsman, among the "herdsman's territory" belonging to Ramses IV's Theban temple.[179] References simply to "standard bearers" and to "retainers" may refer to Sherden, although Egyptians also held these titles.[180] All of the Sherden listed in the Wilbour Papyrus appear to have normal Egyptian names.[181] One man is given the title of "herdsman of the Sherden."[182] This does not necessarily identify the man as a Sherden but, again, does indicate their presence in the area. Undoubtedly, the most unusual vocation ascribed to a Sherden is "approacher of crocodiles," which sounds very much like a crocodile wrangler.[183] Another Sherden is responsible for lands belonging to the House of the Queen.[184] None of the Sherden in the Wilbour Papyrus have nautical connections: One soldier is listed as belonging to "the warships," but he is not identified as a Sherden.[185]

Ramses V or VII.—Papyrus Amiens, which dates to the reign of either Ramses V or Ramses VII,[186] deals with the collection of grain by ship from various locations.[187] In two instances grain is contributed from a domain founded by Ramses III specifically for the "people of the Sherden."[188]

Ramses XI.—The "Adoption Papyrus" comes from the Middle Egyptian site of Speremu, probably located south of Herakleopolis Magna (Fig. 4.10).[189] In a section of the papyrus dating to Year 1 of Ramses XI (ca. 1099 B.C.), two Sherden named Pakamen and Satameniu, as well as the latter's wife, Adjedaa, appear as witnesses.[190]

A number of Late Ramesside letters have survived of the correspondence between the necropolis scribe Djut-

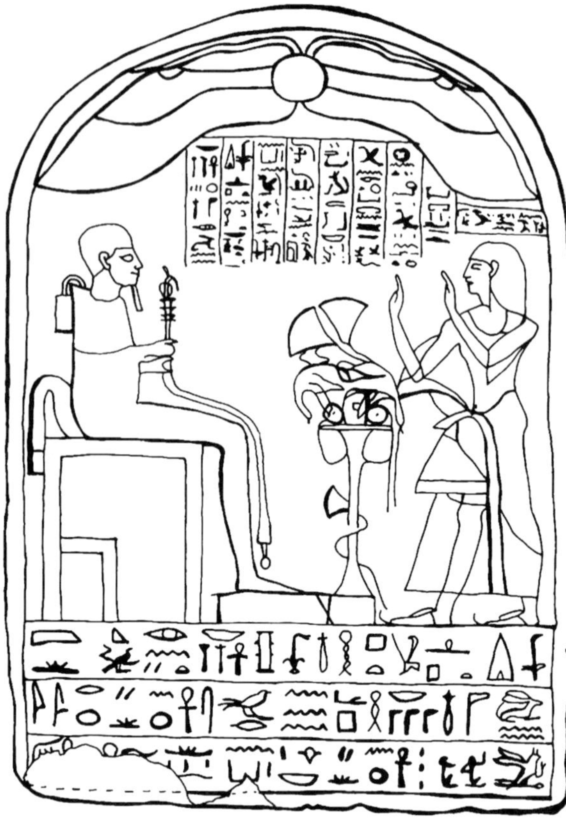

Fig. 4.19: Stele of Setemhebu discovered by Petrie at Ehnasya, which mentions a "Fortress of the Sherden." After Petrie 1905: pl. XXVII: 1.

mose and his son Butehamon.[191] Several of the letters, dating to Year 10 of the "Renaissance" (ca. 1070 B.C.), mention a trusted Sherden messenger named Hori.[192]

Late XXth Dynasty.—At Ehnasya (Heracleopolis), located eleven kilometers south-southwest of Gurob, Petrie uncovered a stele belonging to Setemhubu that mentions a Sherden fortress (Fig. 4.19).[193]

XXIInd Dynasty.—The term "Sherden Fields" appears in the XXIInd Dynasty Helwan Donation Stele from Tell el Minieh.[194] A second stele found by Petrie at Ehnasya belongs to Pazesef and identifies him as a Sherden soldier at the fortress of Usermarres (Fig. 4.20).[195] Pazesef appears in local clothing and worships local gods, demonstrating complete assimilation into the Egyptian mainstream culture. Petrie suggests that the fortification's name may refer to either Osorkon II (874–850 B.C.) or Shishak III (825–773 B.C.). If so, then this would be the latest known reference to the Sherden.[196] When Shishak I invaded Judah and Israel in 926 or 925 B.C. after Solomon's death, Sherden do not appear in his list of troops, which suggests a *terminus ad quem* date for the dissolution of the Sherden as a significant fighting force within the Egyptian army.[197]

Thus, although at times arriving in Egypt as invaders, the Sherden became productive members of Egyptian society, serving in the military and in numerous other vocations for about half a millennium. Over time they became integrated into the cultural fabric of Egypt, adopting its customs, clothing, names, rituals, and burial customs until they were entirely absorbed into Egyptian society and disappeared from view.[198]

The Weshesh.—The Weshesh are the most enigmatic of all the Sea Peoples. Egyptian documents mention them only twice.[199] They first appear in the list of invading Sea Peoples on the Year 8 inscription at Medinet Habu:[200]

> The countries— —, the ⌜Northerners⌝ in their isles were disturbed, taken away in the ⌜fray⌝ —at one time. Not one stood before their hands, from Kheta (*Ḫt'*), Kode (*Ḳdy*), Charchemish (*Ḳ-r'-ḳ'-m-š'*), Arvad (*'-r'-ṯ w*),[201] Alasa (*'-r '-s'*), they were wasted. [The]y [⌜set up⌝] a camp in one place in Amor (*'-m-r'*). They desolated his people and his land like that which is not. They came with fire prepared before them, forward to Egypt. Their main support was Peleset (*Pw-r'-s'-ṯ*), Thekel (*Ṯ-'k-k'-r'*), Sheklesh (*Š'-k-rw-š'*), Denyen (*D'-y-n-yw* [*sic*]), and Weshesh (*W'-š'-š'*). (These) lands were united, and they laid their hands upon the land as far as the Circle of the Earth. Their hearts were confident, full of their plans.

They appear again in the historical section of Papyrus Harris:[202]

> I extended all the boundaries of Egypt; I overthrew those who invaded them from their lands. I slew the Denyen *(D'-yn-yw-n')* in their isles, the Thekel *(Ṯ'k-r')* and the Peleset *(Pw-r'-s'-ty)* were made

ashes. The Sherden and the Weshesh *(W'-š-š)* of the sea, they were made as those that exist not, taken captive at one time, brought as captives to Egypt, like the sand of the shore. I settled them in strongholds, bound in my name. Numerous were their classes like hundred-thousands. I taxed them all, in clothing and grain from the storehouses and granaries each year.

This latter text has caused much discussion regarding whether it indicates that Ramses III actively settled the Philistines in Canaan as mercenaries after the conflict.[203] One way of approaching this question is to establish what within this text can actually be confirmed from other sources. While the documentary trail of the Weshesh tapers out, the Sherden can be used to determine how these comments in Papyrus Harris compare to actual reality. The Sherden are a particularly useful resource for this evaluation, for only they wore horned helmets, which facilitates identifying them on the monuments.

The Harris Papyrus singles out the Sherden and Weshesh for two reasons: Only they are considered "of the sea," and only they are listed as having been brought as captives to Egypt and taxed. Earlier we saw the ample documentary evidence available that demonstrates that the Sherden indeed were settled in Egypt and that they were taxed.[204] True, not all of them had been settled by Ramses III, but the statement reflects a documented reality.

Why does the Harris Papyrus assign the term "of the sea" only to the Sherden and the Weshesh? A similar situation exists with regard to Merneptah's Libyan war, in which various Sea Peoples participated, but, both in the Great Karnak Inscription and on the Athribis Stele, *only the Ekwesh* are defined as "of the sea."[205] Clearly, all of the Sea Peoples participating in that coalition had to reach Libya by ship, so it makes no sense to assign the designation "of the sea" to one group and not to the others based on that consideration alone.

I propose that the term "of the sea" in the inscriptions of both Merneptah and Ramses III refers to those groups who invaded Egypt on their ships—that is, "by sea"—as opposed to those groups of Sea Peoples who arrived by land. This is, of course, the *modus vivendi* of the Sea Peoples as seen at Medinet Habu during their invasion of Egypt during the reign of Ramses III. I suspect a similar situation existed under Merneptah.

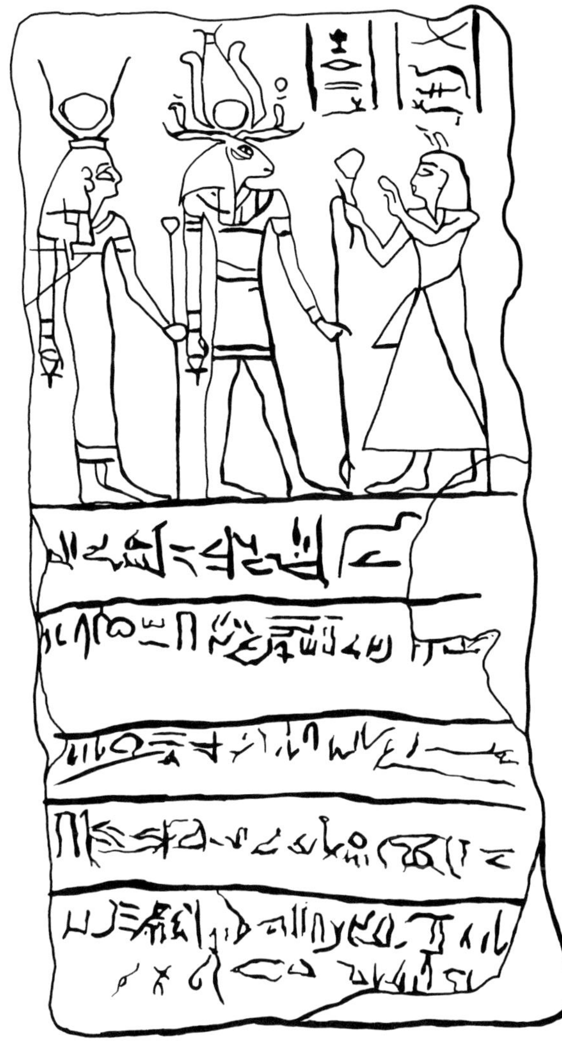

Fig. 4.20: Stele of Pazesef, who is identified as a Sherden warrior of the great fortress of Ussermarres. Possibly XXnd Dynasty. After Petrie 1905: pl. XXVII: 2.

Is it possible to test this hypothesis? I believe so. If this paradigm is correct, one would expect to see the Sherden arriving by sea in the Medinet Habu reliefs, and, indeed, horn-helmeted Sherden warriors man two of the five Sea People ships in the naval battle scene (Figs. 2.3–4). On the other hand, we *should not see* Sherden in the invading land force, and, in fact, we do not: In the land-battle scene involving the invading Sea Peoples, the Sherden appear fighting only on the Egyptian side.[206]

Following this logic, then, the feather-helmeted warriors in the other three galleys are not Philistines as has so often been assumed but rather they *must be Weshesh* (Figs. 2.5–7).[207] A corollary of this conclusion is that the Weshesh wore feather helmets—just as did the Peleset/Philistines, Sikila (Sekels/Tjeker), and the Denyen.[208]

In this scenario Papyrus Harris is to be understood in the following manner: A coalition of Sea Peoples attacked Egypt in Ramses III's Year 8. After the battles Ramses III settled in Egypt those groups who came by ship and had taken part in the nautical invasion (i.e., the Sherden and the Weshesh). These became part of his military and were taxed. The other groups, who had invaded (migrated) by land—the Peleset/Philistines, Sikila (Sekels/Tjeker), and the Denyen—were driven outside of Egypt and, at least in the Egyptian mind, "were made ashes." As it turned out, history proved this to be an overly optimistic assessment of the situation.

Furthermore, the northern ship depicted repeatedly with the *Vogelbarke*-like shape created by its outboard-facing bird-head bow and stern ornaments at Medinet Habu indicates that at least one of the two groups of seaborne invaders—the Sherden and/or the Weshesh—held to the belief system of the central European Urnfield Culture.[209] Also, if the Sherden and the Weshesh were the only two groups who participated in the naval battle, then it follows that the Sea Peoples' ships must be patterned after either a Sherden or a Weshesh ship. The following considerations make it likely that it was a Weshesh vessel:[210]

- The Sherden had been active in the Eastern Mediterranean since the fourteenth century.[211]
- Weshesh appear only twice in the Egyptian record, both times in relation to the last attack of the Sea Peoples, in Ramses III's Year 8.
- Had the Sherden employed a ship with the outboard-facing bird-head stem and stern ornaments of the Urnfield *Vogelbarke,* we might expect to see other earlier representations of this manner of decorating galleys, yet, of the entire, large corpus of Late Helladic III B–C representations of the Helladic galley type, the Medinet Habu ship representation is unique in this regard. This suggests that this decorative motif and the Urnfield group, for which it had meaning, appeared late in the process.[212]

In this case, those Sea Peoples depicted migrating overland in the terrestrial battle may be identified as elements of the Denyen, Sekels (Thekel), and Peleset.[213] Interestingly, two of these groups—the Peleset/Philistines and the Sekels—indeed eventually settled along the Canaanite coastline.[214] As to the Sherden, they had been active on the Syro-Canaanite littoral for over a century.[215]

Of the Sea Peoples depicted in the Egyptian reliefs, G. A. Wainwright notes that the Peleset, the Sekels, and the Denyen wear "feather" helmets.[216] So did the Weshesh apparently. But what became of them? Within the overall Sea Peoples' story, they appear only in the last invasion and then are never heard from again. Perhaps from this we may conclude that they represented a relatively small and perhaps geographically distant group within the overall milieu of migrating peoples and were acculturated into Egyptian society soon after their invasion.

Miscellaneous

Weights.—The Kenamun scene of Syro-Canaanite merchants arriving in Egypt depicts them interacting with locals employing balance pans.[217] Balance weights are among the most important of personal items that may be used to identify a ship's identity, as demonstrated by those recovered from the Cape Gelidonya and Uluburun shipwrecks.[218] A similar approach is worth pursuing on land sites such as Gurob as well, although the results will remain more ambiguous.

Of fourteen weights found by Petrie at Gurob in one campaign, only six or seven were based on the Egyptian standard.[219] Even among these are anomalies that raise questions as to their Egyptian character. Notes Petrie, "only one of all the weights is properly Egyptian, and two of rounded cubic form are passable as Egyptian in origin; the other eleven are entirely marked off as foreign by the standards, the forms, and the materials."[220] In a summary of weights from Gurob published the next year, Petrie stresses that fewer than half were based on the Egyptian measurement system and that, of those, almost half were made of materials not normally used for Egyptian weights: He notes that only 20 percent of them could be considered normal Egyptian.[221] Some of the weights are of the typical Late Bronze Age Levantine type.[222] While a detailed study of these weights is beyond the scope of the present work, their renewed examination could shed additional light on the question of foreigners at Gurob.

The Gurob "northern" spindle.—Gurob Tomb 11 contained a spindle that may represent the single most eloquent piece of archaeological evidence for a person—presumably a woman—of northern descent residing and working at Gurob (Figs. 4.21, 8 [tomb location]).[223]
Elizabeth J. W. Barber writes the following:[224]

> [W]e have noted that the Egyptian spindles, both in the numerous representations and in the many actual artifacts, had the whorl at the top, along with a groove for hitching the thread. There is one exception. At the New Kingdom site of Gurob in the Faiyum, a spindle turned up with a thread groove (necessarily the top) at one end, and a limestone whorl of about 150 g (5¼ oz.) most of the way towards the other end . . . It is definitely a low-whorl spindle. And a complete anomaly in Egypt. . . .
>
> [There] is some implication that Grave 11 and its spindle belong to a specifically Aegean-related burial. Clearly the spindle is of foreign design, although of local materials: presumably it was made locally for a foreign resident. Hence, the strongest hypothesis is that the form is Aegean, and that the Bronze Age Aegean people—unlike the Egyptians—were using the same sort of low-whorl spindle that their European descendants for the next three millennia have used. Very satisfying. But there is another detail that makes this conclusion even stronger. . . .
>
> One reason why the question of high- and low-whorl spindles is interesting is that it has been suggested that the type of spindle parallels the type of fiber used. We have already seen that we are dealing with at least two different spinning traditions around the Mediterranean, one in the south suited to flax exclusively, and the other to the north and east suited to both wool and flax. Now we see the distribution

Fig. 4.21: Foreign spindle for spinning "Z" yarn, from Gurob, Tomb 11. From Brunton and Engelbach 1927: pl. XIII: 8.

> of high- and low-whorl spindles starting to fall into somewhat different geographic areas: Egypt and the Middle East with one sort, Anatolia and apparently Europe with another sort. There is yet another aspect of the craft of spinning that divides geographically: the direction in which the thread is spun, whether to the left or to the right—"S" or "Z." . . .
>
> An S-*spun* thread . . . is one in which the fibers of the thread lie at a slant similar to that in the middle of the letter "S" when you hold the thread vertically in front of you (i.e., they have a negative slope) [Fig. 4.22]. In a Z-*spun* thread, the fibers slope the other way, like the center part of the letter "Z."
>
> Now, Egyptian linens as they have come down to us from dynastic Egypt are invariably S-spun, whereas the threads in the textile of the other areas around, especially Europe and India, are typically Z-spun, starting with the textiles from around 6000 B.C. at Çatal Hüyük . . . So we can see that the whorl on the Gurob spindle just discussed has *not* accidentally been stuck upside down onto the wrong end of the shaft, because the thread groove itself is cut backwards, from the Egyptian point of view. It is built for hitching Z-spun thread. And that points the spindle once again back to Europe . . .
>
> The only simple explanation for its presence is that some European women were living in Gurob more or less permanently, and doing their own spinning.

At Timna in southern Israel, Z-spun threads, all probably of wool, have been found.[225] At Tell el Amarna they make up only 1–2 percent of the more than thirty-five hundred surviving linen textiles examined.[226] Sparks identifies these as the creation of Syro-Canaanite spinners.[227]

In Papyrus Gurob III.1, right, which dates to Seti II's Year 2, its female author mentions that she taught weaving to foreigners, presumably at the harem of Mi-Wer. She also refers to a previous instance, under Ramses II, in which foreigners were brought to her for their instruction:[228]

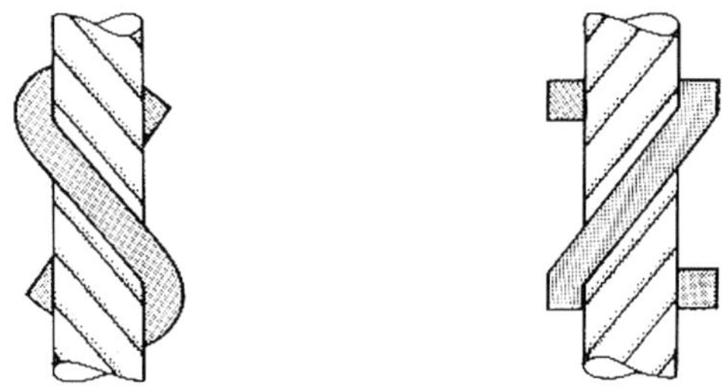

Fig. 4.22: S- and Z-spun thread. From Kemp and Vogelsang-Eastwood 2001: 57 fig. 3.1.

> [*Beginning lost*] Amon [. . .] which I have made, for they are exactly like those which had been made for Pre. I shall have myself boasted about because of them and not let fault be found with me. It is advantageous that my Lord, l.p.h., has had people sent to me to be taught and instructed how to perform this important occupation (of weaving). It is fortunate that my Lord has found someone fit to do that the like of which had not been done for Pre, because those who are here are senior apprentices. It is only such people as are like those people whom my Lord, l.p.h., sent who are capable of functioning and who are capable of receiving my personal instruction, since they are foreigners like those who used to be brought to us in the time of Usermare-setepenre, l.p.h. (Ramses II), the Great God, your good (grand)-father, and who would tell us, "We were quite a number in the households of the officials," and who would receive instruction and so be able to perform whatever was told them.
>
> This is a missive for One's (the king's) information. Year 2, third month of the first season, day 20.

Note that the papyrus does not indicate the ethnic identity of these foreigners. In Thomas's view, in all likelihood they were Asiatic, as female Asiatic servants are regularly found working in the production of linen textiles.[229] Thus, the textual and archaeological evidence converge in pointing to foreign textile workers at Gurob, presumably connected with Mi-Wer. One wonders whether part of the instruction of the foreigners was to teach them the Egyptian S-spun method of spinning.

A roughly contemporaneous parallel to the use of foreign textile workers, presumably in wool in this case, is documented at Pylos.[230] Here Linear B tablets mention women working in various aspects of this industry:[231]

> [A] large proportion of the women have occupational descriptions associating them with the production of woolen or linen cloth: *a-ke-ti-ra₂* 'cloth finishers,' *a-pu-ko-wo-ko* 'headband-makers,' *a-ra-ka-te-ja* 'distaff women,' *i-te-ja* 'weavers,' *pe-ki-ti-ra₂* 'carders,' *ra-pi-ti-ra₂* 'sewing women,' *ri-ne-ja* 'flax workers.' Moreover, some of the other occupational terms recur on the Knossos tablets relating to the textile industry, and it is evident that the production of textiles was an important function of the royal establishments at Pylos too.

Chadwick observes that at least some of these women—bearing ethnics that associate them with foreign locations such as Knidos, Miletos, Lemnos, Kythera, and Chios, as well as possibly Lydia and the Halikarnassos region—found themselves slaves in the Pylian state after the Mycenaeans either captured them in slaving raids or bartered for them at eastern slave markets.[232]

"Roped-reed" burials at Gurob.—In Tomb 605 G. Brunton and R. Engelbach discovered the burial of a young woman wrapped in "reeds, roped together" (Figs. 4.5, 8–9).[233] At the woman's feet the excavators found a casket containing a Mycenaean stirrup jar together with a scarab of Ramses II.[234] The woman wore a necklace that included beads made of red paste, blue glass, carnelian, and a bone spacer, and at her right wrist were two scarabs, five beads of red paste, two of blue glass, and six of carnelian. Near each of the woman's ears lay three penannular jasper rings.[235] Nearby, Tomb 614 included a second burial "roped in reeds."[236]

At Saqqara, the remains found in nine XIXth-Dynasty female burials were swathed in reed mats and wearing amber bead necklaces (Figs. 4.23–25).[237] The beads are reported to be of central European or perhaps Italic design: S. Hood proposes that the deceased might be the womenfolk of foreign mercenaries.[238]

While the Gurob excavators did not record any amber on the female in Tomb 605, red paste, carnelian, and jasper all roughly approximate the color of amber, a material that may have been difficult to acquire in Ramesside Egypt. If Hood is correct in his hypothesis, then the reed burials found in Gurob Tombs 605 and 614—along with the ship-cart model in Tomb 611—may hint that this cemetery was used by northern foreigners.

The Gurob "burnt groups."—Any discussion of foreigners at Gurob must address an unusual phenomenon discovered there by Petrie:[239]

> A very remarkable custom existed in this town, which I believe is unknown as yet elsewhere in ancient Egypt. In many instances the floor of a room has been taken up; a hole about two feet across and a foot deep was dug in the ground. A large quantity of distinctly personal property, such as clothing, a stool, a mirror, necklaces, kohl tubes, and toilet vases of stone and pottery, were thrown in, and then all burnt in the hole. The fire was smothered by potsherds laid flat over it; and lastly the floor was relaid. Such was the arrangement of one instance which I examined in detail; and such is indicated by the state of the things in other finds, and the accounts given by Mr. Hughes-Hughes and by the native diggers. It is evident that the objects buried are such as belong to an individual personally, and not to a household. No bones were ever found with the burnt deposits. These were not therefore funereal pyres. Yet we cannot imagine a general custom of burning and burying valuable property, except on the death of the owner. I conclude therefore that there was a custom among the foreign residents of burying the body in the Egyptian fashion, especially as I found light-haired bodies in the cemetery; and that the personal property which would have been piled on the funereal pyre in the Mediterranean home of the Akhaians, was here sacrificed in the house, and so put out of sight. In most instances Aegean pottery was found in these deposits, an evidence of their belonging to the foreigners.

Petrie dates five of these "burnt groups" to the reigns of Amenhotep III, Tutankhamun, Ramses II (early and late), and Seti II.[240] In his chronological sequencing of the groups, however, Petrie did not take into consideration that the latest artifacts in each group must date it, irrespective of earlier materials in each group.[241] Bell demonstrates that all of the groups are later, belonging to the Ramesside

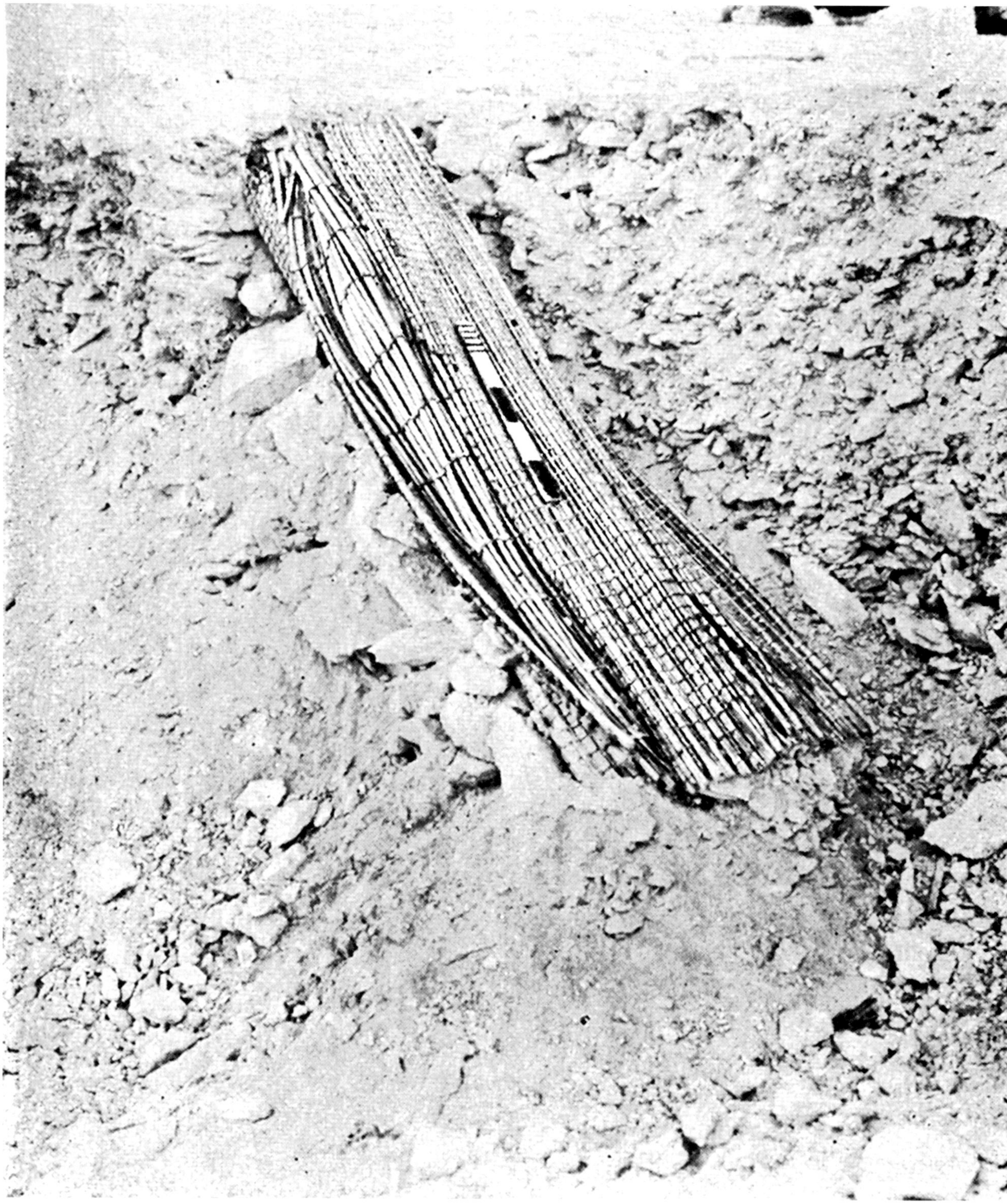

Fig. 4.23: Unopened female roped-reed burial in situ at Saqarra. XIXth Dynasty. From Goneim 1957: LXVII: A.

Fig. 4.24: Female roped-reed burial opened at Saqarra. XIXth Dynasty. From Goneim 1957: pl. LXVII:B.

period despite the fact that they also include older XVIIIth Dynasty artifacts.[242] She discerns a certain progression in the groups, which suggests a continuing history of deposition throughout the XIXth Dynasty and perhaps even later. Ceramics found in contexts later than their production periods, as in the case of the Mycenaean vessels included in the burnt groups, appear to have been kept as heirlooms, perhaps because of their exotic nature.[243] The items found in the burnt groups were unlikely to have come from normal houses or from a temple but rather were probably from an important building or perhaps the harem itself.[244]

Petrie's attribution of the burnt groups as an "Akhaian"—that is, Mycenaean—burial custom is problematic, as the Mycenaeans did not normally practice cremation, nor did they normally burn grave goods.[245] When cremation does arrive in Greece, toward the end of the Late Bronze Age, it appears as an intrusive cultural element.[246]

At Boeotian Thebes, excavations revealed a roundish pit, 1.8 meters in diameter, in which many ivory carvings, sixty Late Helladic IIIA: 2 late ceramics, and four gold beads had been intentionally burnt.[247] Both S. Symeonoglou, the excavator, and R. S. Merrillees note the similarity between the burnt Theban ivory-ceramic hoard and the phenomenon of burnt groups at Gurob.[248]

Fire rituals are known from later periods in Greek history.[249] Evidence of fire, occasionally found in Mycenaean tombs, has been generally interpreted as a component of ritual acts related to the cult of the dead, sacrifices, or purification rituals.[250] Mycenaean pottery, along with many other types of artifacts, has also been found together with burnt human bones at the puzzling thirteenth-century B.C. temple in Amman, Jordan.[251] Finally, note the parallels for the Gurob burnt groups in hearths made from broken sherds covered with clay, found at Xeropolis and possibly also at Tiryns, which may indicate the presence of northern women there.[252]

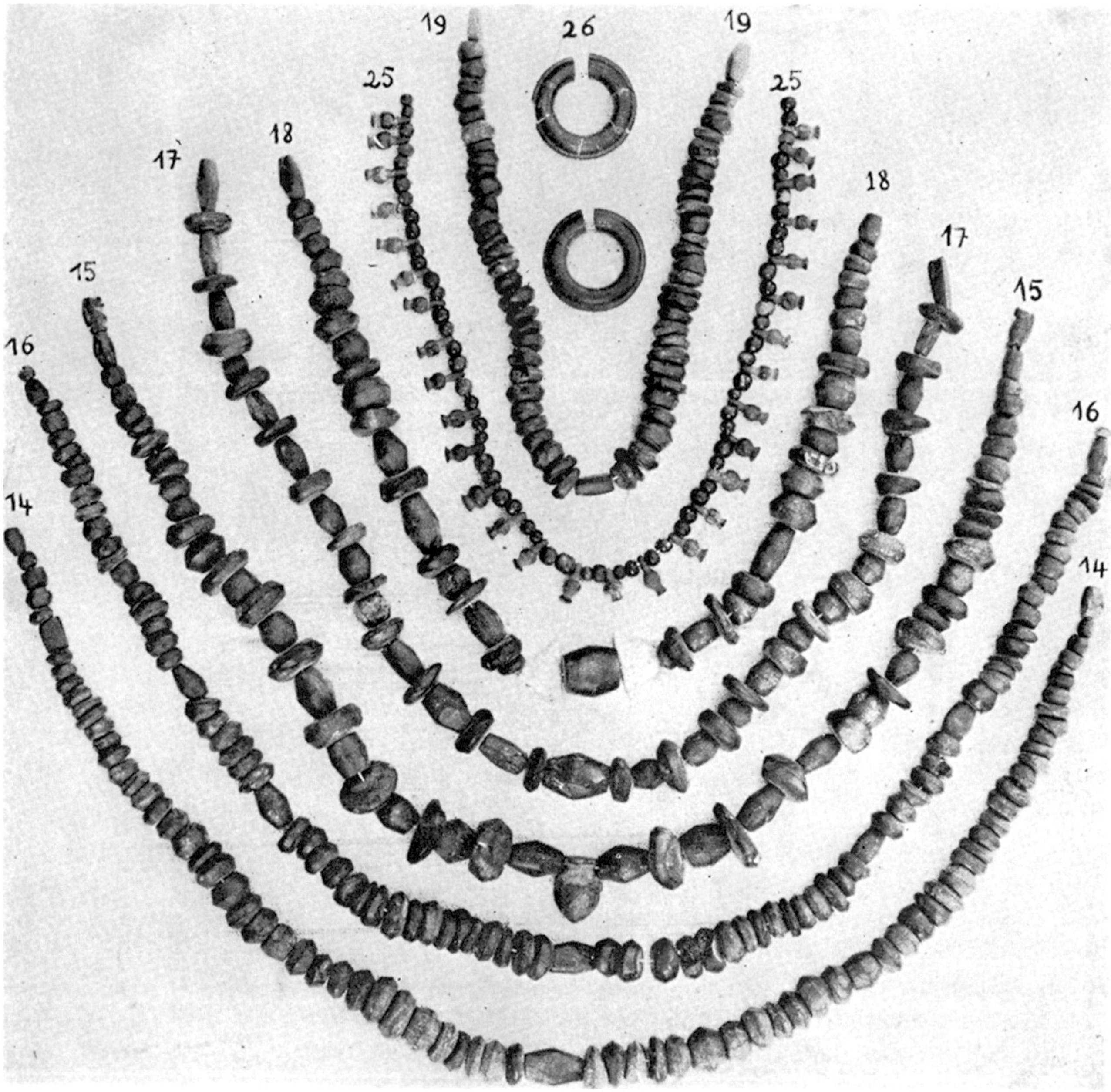

Fig. 4.25: Amber bead necklaces from female roped-reed burials at Saqqara. XIXth Dynasty. From Goneim 1957: pl. LXXI.

Merrillees considers the burnt groups at Gurob as indicative of the use of fire for cultic cleansing.[253] Thomas believes, however, that in reaching this conclusion, Merrillees had studied the material out of context and, furthermore, that he relies too heavily on Petrie's conclusions.[254] Noting that Petrie had given too much consideration to the Mycenaean ceramics contained in the burnt groups, Thomas rejects Merrillees's interpretation.[255] Instead, she prefers to consider the burnt groups as simple trash pits later overbuilt by houses or burnt as the result of a fire. As Bell demonstrates, however, the Mycenaean ceramics were buried long after they had been made. Why would anyone burn as trash exotica that included heirloom Mycenaean pottery as an ongoing custom?

Janet Politi proposes a fascinating explanation for the burnt groups at Gurob.[256] In his Year 34 (1246 B.C.) Ramses II wed Maat-Hor-Neferure, a daughter of the Hittite king Hattusilis III.[257] A fragmentary papyrus from Gurob that refers to garments belonging to Maat-Hor-Neferure has led several scholars to assert that the princess resided at Mi-Wer.[258] Politi builds on this conclusion and adds that Maat-Hor-Neferure must have been installed at Mi-Wer with an entourage of Hittite ladies-in-waiting.[259] She argues that the burnt groups represent the physical evidence of a custom described in Hittite Law:[260]

> If a man takes his wife and leads [her] away to his house, he shall carry her dowry in. If the woman [dies] th[ere] they shall burn the personal possessions of the man (or "he shall burn her personal possessions") and the man shall take her dowry. If she dies in her father's house and she [has] children, the man shall not [take] her dowry.

There are several considerations that argue against this interpretation:

- The Hittite law makes no reference to *burying* the items—only to burning them.
- The law deals with the manner in which the *Hittite husband* of a Hittite wife should act. Assuming that Maat-Hor-Neferure actually did reside at Mi-Wer with her entourage of Hittite ladies-in-waiting—a conclusion that, as we shall see, cannot be supported by the evidence—this would assume that these women arrived in Egypt together with their Hittite husbands to serve in the harem. This seems highly unlikely. In an analogous situation, when Tadu-kheba, the daughter of the Mitannian king Tushratta, journeyed to Egypt to marry Amenhotep III, she left with a retinue of 270 female attendants but only 30 men.[261] The wording "270 women, 30 men, are the dowry-personnel" contains no allusions to any conjugal connections between the two groups.[262]
- As support for her theory, Politi points to the blond hair found by Petrie at Gurob as probably belonging to a Hittite lady-in-waiting.[263] Petrie indeed notes that several of the bodies in the tombs at Gurob showed physical evidence of foreign origins.[264] A body from Tomb 23 had blond hair beneath a black wig (Figs. 4.26, 8–9 [tomb location]).[265] Blond hair also survived on the scalp of a second body from Tomb 24, and Tomb 25 held a male who, although he had dark hair, showed an atypically light complexion. Such racial identifications based on hair and skin color are no longer considered valid.[266]
- Regarding physical evidence of Hittites residing at Gurob, an ushabti from a tomb there bears the name of Sadi-amia, whom Petrie identifies as a Hittite.[267] Bell notes, however, that the most that can be said in this case is that the man was not Egyptian, although Sadi-amia clearly followed typical Egyptian burial customs.[268] The only other artifact at Gurob identified as probably Hittite is a small wooden figure of a harpist with pigtails.[269]
- Crucial to Politi's theory regarding Maat-Hor-Neferure's residence in Mi-Wer is the fragmentary papyrus found at Gurob, which mentions clothing woven for or belonging to her.[270] L. Bell supplies the following translation for Gurob Papyrus, Fragment U:[271]

Fig. 4.26: Blond hair from a tomb at Gurob. UC30139. Copyright of the Petrie Museum of Egyptian Archaeology, UCL.

1. a small (carrying) bag (made) of fine linen[272]

2. [. . .] Queen Maat-Hor-Neferure—may she live!

3. the daughter of the Great Chief of Hatti

4. [. . .] a wrap[273] of 28 cubits 4 palms (in length) and a width of 4 cubits 1

5. [. . .] of 14 cubits 2 palms (in length) and a width of 4 cubits 2

6. [. . .] of X cubits Y palms (in length) and a width of 4 cubits

Simply put, this text does not place Maat-Hor-Neferure at Gurob. The papyrus may be referring to clothing woven for Maat-Hor-Neferure at Mi-Wer and intended for delivery to her. After all, Mi-Wer was a known center of textile production, which shipped its wares to royalty. Indeed, other papyri indicate that garments prepared at the harem were intended for the royal household.[274] As Thomas rightly concludes:[275]

> This does not necessarily indicate that the princess lived in the harem at Gurob, though she may have stayed here occasionally, but some of her clothes were definitely made here. The officials kept detailed records of the garments and of the places to which they were sent.

In evaluating the evidence, or lack thereof, for whether Maat-Hor-Neferure resided at Mi-Wer, it is worthwhile to reflect on the case of another queen whom some scholars have located at Mi-Wer: Amenhotep III's queen Tiye.[276] This assignment is based on the discovery at Gurob of various artifacts related to her and her husband. Yet, as Thomas emphasizes, even this evidence remains tenuous at best and is too limited to allow us to place Tiye at Gurob.[277] How much more questionable is it to place Maat-Hor-Neferure—not to mention a coterie of Hittite ladies-in-waiting together *with their husbands*—at Mi-Wer solely on the basis of the Gurob Papyrus, Fragment U?

If we cannot link the burnt groups to Mycenaeans or to Hittites living at Gurob, then which foreign ethnic group created them? There seems to be general agreement that the burnt groups represent a custom related to the death of the items' owners.[278] Burial methods can adapt when foreigners arrive at a new setting: The particular burial customs that remain will be those that have consequential cultural meaning to the new arrivals.[279] Indeed, ample evidence also indicates that foreigners adopted and amalgamated Egyptian religious practices into their own worship.[280] Thus, if the assumption that the burnt groups are connected in some way with burial is valid, then the logical path of inquiry is to determine which foreign groups that used fire in relation to death and burial can be placed in Egypt at the end of the New Kingdom.

This brings us back to the Sea Peoples' ships at Medinet Habu, the Weshesh, and their Urnfield Culture connections. The central European Urnfield peoples cremated their dead and buried them in urns surrounded by personal items.[281] This burial custom, resulting in vast fields of cremation urns, is one of the most typifying characteristics of the Urnfield Culture, *hence its name.*

I believe that the most likely, and the simplest, explanation for the burnt group phenomenon at Gurob is that they represent physical evidence of an Urnfield element—perhaps the Weshesh—in the midst of rapid acculturation at Gurob. In this scenario, the Weshesh/Urnfelder no longer cremated their dead, having adopted Egyptian burial practices, but still kept alive a memory of their traditions by burning and burying the deceased's personal items. The appearance of Mycenaean pottery, then, can be explained as memories connected to the movement of populations through Mycenaean regions.

5 Conclusions

Sometime, probably during the late thirteenth or early twelfth century, a competent artisan in Egypt constructed a simple model of a ship. Wooden ship models are virtually unknown in Egypt during the New Kingdom apart from those found in royal XVIIIth-Dynasty tombs, and this model is remarkable for several other reasons as well. First, the ship represented in the model is easily identifiable as a galley type first used by Mycenaeans and was perhaps their invention. It had also been readily adopted and adapted by elements of the Sea Peoples, however, as we learn from a study of the Medinet Habu naval battle scene, a similar ship painted on a cremation urn at Hama in Syria, as well as other evidence. Second, the ship was meant to sit on a cart with wheels.

The model is unique in the degree of detail that it supplies regarding the structural and polychromatic aspects of these Helladic galleys. As with all depictions of ships and boats, however, it is important to remember that this is a representation and not the ship itself and that there could be—and, more important, there are—differences between the copy and its prototype.

The model maker may have been an Egyptian artisan commissioned to build it. Despite the foreign prototype ship with which he (I assume that the artisan was a male) dealt, the similarity of the rendering of the forecastle deck to royal XVIIIth-Dynasty wooden ship models is striking, and the exaggerated rockered form of the hull also speaks of a strong Egyptian influence. Unfortunately, we do not have a clear view of the wagon that supported the ship model: Only its four wheels survive.

The most important aspect of the model from a ship-construction point of view is its three-dimensional confirmation of the stanchion system, which in an actual Helladic galley would have served to support the superstructure and centerline deck. This system of stanchions sets the Helladic galley apart from all other oared ship of its age, making it the prototype for all later developments of Greek warships. As the Sea Peoples brought this type of ship to the Levant, it also served as a starting point for later Phoenician galleys.

W. M. F. Petrie reconstructed the model once with the stempost bird-head device facing outboard and once with it reversed and facing inboard, which is its present state, being glued solidly in place. The question of which of the two reconstructions is correct is grounded in the seeming enigma of how the Mycenaean galley tradition survived the Dark Ages. With the fall of the palace-based Mycenaean culture, Greece and the Aegean were largely depopulated. Moreover, while the Dark Ages appear to be shrinking chronologically, it is still difficult to explain how such complex knowledge could have continued under such limiting circumstances. The simplest explanation (Occam's razor) for the continuation of the Mycenaean Greek galley tradition across the Dark Ages is that it survived in Cyprus, where an energetic palace-based Achaean culture prospered during the Late Cypriot III and Cypro-Geometric periods. During the Submycenaean period Cyprus and Greece reestablished communication, and much of the subsequent momentum for Greece's renewal, particularly at Athens, may be attributed to these contacts, which are credited both with the introduction there of iron and with a Cypriot influence on ceramics.

The only evidence for the size of the Gurob model's prototype galley is a row of black daubs on each side of the hull, which must represent oarports. Although most of these dots have been lost, the existent spacing makes a series of twenty-five-dot oarports per side reasonable, suggesting that it represents a *pentekonter.* Given an *interscalmium* of 1 meter per rower, the prototype galley would have required 25 meters to house the rowers and another few meters at bow and at stern, for a total length of about 30 meters. Although some have assumed that these oarsmen on Helladic galleys worked their oars against tholepins, we have in fact no evidence of tholepins on Mycenaean vessels.

The Gurob ship model is also unique for its time in presenting a remarkable display of colors on a Helladic ship representation. Most ship depictions dating from the time about which Homer wrote, as well as the time when he is believed to have lived, are incapable of expressing color as they are, for the most part, monochromatic silhouettes. Only much later, with the appearance of ships on black-figure vases, do we get some idea of their color schemes. At the same time, one must be aware of the limitations of the evidence: The Gurob model's colors are faded today, whereas originally they would have been vibrant.

The model maker used only three colors on the ship model, although his palette contained additional colors: Those he selected are specifically the colors that Homer seems to refer in his ship epithets. Of these the poet most commonly describes ships as "black hulled." This term refers to the use of pitch on the lower parts of the hull as a sealant and as an antifouling agent. The model demonstrates exactly what Homer had in mind. The pitch covers only the bottom of the external side of the hull, which would have been submerged in water. The upper parts of the planking were left free of pitch.

A line of red paint seen below the caprails on the Gurob model is a recurring motif as it reappears on galleys depicted on black-figure vases, indicating that this feature bridged the intervening centuries, for which we lack corroborative evidence. The discovery of red coloring materials in Building P at Kommos in fourteenth- and thirteenth-century-B.C. settings and at Naxos in Sicily inside a mid-fifth-century-B.C. shipshed adds an archaeological context to the iconographic evidence for its appearance on galleys. While it is possible that the "red-cheeked" epithet that Homer employs (rarely) might refer to this red stripe, D. Davis cautions us that we enter here the realm of dactylic hexameter and that ship-related color epithets inserted into the epics may have more to do with the organization of lyrics than with physical reality.

A small, painted, rectangular square with a central hole found with the model sat directly beneath the hull amidships so that a peg attaching the model ship to its wagon transfixed it. To correctly understand the model, one must recognize that this slip of wood is crucial, for it represents the type of base—termed a *pavois* by Egyptologist G. Legrain—to which were attached bars for priestly porters to shoulder as they carried cultic barques overland. The *pavois* is significant for two reasons. First, it indicates that the model represents a cult vessel, and, second, it indicates a syncretism in which a ship foreign to Egypt's shores had taken on indigenous cultic affectations. This is in keeping with other evidence of rapid acculturation by foreigners immersed in the Egyptian mainstream. As the *manner* of transport of cultic ships in general is demonstrably fluid, however, the possibility exists that the Gurob ship model—in addition to being mounted on wheels—could also have been displayed on a *pavois* and that the long pegs found with the model represent the support bars. This theoretical possibility appears as a reconstruction in this book's virtual component.[1]

An examination of the use of wagons with wheels to transport cultic or funerary ship-form catafalques in Egypt reveals that, although the wheel had been known there as early as the Vth Dynasty, its first use in general, as well as for the transport of watercraft replicas specifically, dates to the XVIIth Dynasty. Wheeled conveyances seem to have become relatively common for burial purposes only in Ptolemaic times, to judge from the iconographic evidence.

Although the ship type replicated by the Gurob model is Mycenaean, its wheeled cart is decidedly not. The four-wheeled wagon of the Gurob model belongs to a cultural phenomenon of cultic objects, which first appears in the Aegean world in the twelfth century B.C., apparently arriving there from farther north. The Gurob ship model finds some of its closest parallels in the contemporaneous galley representations from Pyrgos Livanaton (Kynos). It is, therefore, hardly surprising that this particular site also revealed a fragment of a terracotta ship model with a piercing meant to take an axle for a pair of wheels, making it a "ship-cart," also.

The Gurob ship-cart's wheels have a painted radial decoration that does not seem to depict spokes. The wheels are made in two pairs of slightly differing size. Perhaps the larger pair was intended to raise the stern slightly to give additional allowance for the quarter rudder to clear the base. The purpose of the four items that Petrie identified as "awnings" remains unresolved.

It is not possible to determine the model's original purpose prior to its interment in Tomb 611. While it might have been made solely as a funerary offering, the figures holding models on the Dakhla Oasis graffito may more accurately articulate its purpose—that of serving as a processional device to be carried within the context of the event related to the actual ship-cart.

The Gurob ship model must be seen in the context of the only two other known representations of Helladic ships in Egypt: Ramses III's Medinet Habu naval battle relief and the Dakhla Oasis graffito. On the one hand, the model gives a clear three-dimensional rendering of the structural details indicated by the manner in which the bodies of the dead and dying Sea People warriors are interwoven in the capsized ship (Nelson's N.3), now also confirmed in two-dimensional ship representations from Pyrgos Livanaton (Kynos) and Bademgediği Tepe. On the other hand, the Dakhla Oasis graffito demonstrates the use of these ships (or their facsimiles) for cultic purposes. When the Gurob ship model and the Dakhla Oasis graffito are considered together, they exhibit the following features:

- The Gurob model is not so much a model of a ship as a ship-cart model.
- Both representations depict Helladic galleys, specifically of the type with a pointed horizontal forefoot. While at the end of the Late Bronze Age and in the Early Iron Age galleys could have a bow with or without this feature, it was the former type that eventually evolved into the Greek fighting ships of the Geometric, Archaic, Classical, and later periods, as the modest forefoot was transformed into the greatest naval weapon of the ancient world, the waterline ram.
- A reevaluation of the Dakhla Oasis graffito demonstrates that the diagonal member rising from the stern cannot represent part of the vessel's rigging, as it has been erroneously interpreted in the past (including by me); rather, it depicts a solid timber footed in the stern and resting on a tall midships stanchion capable of supporting figures (humans, statues, or both?). The closest iconographic parallels to this are the Dionysian phallus displays on the sixth-century-B.C. Florence cup.
- One of the Dakhla Oasis graffito figures appears to be playing a flute or a trumpet.

A ship cart with a specifically warship prototype as its model, the cult of the phallus, and celebrants playing pipes—at the risk of stating the obvious, all these elements are found later in the Dionysian cult. This raises the question of whether this cult might have penetrated Egypt as part of the cultural baggage carried by Sea People groups on their migrations.

Nonetheless, nearly half a millennium separates the Gurob and Dakhla Oasis ships from the earliest evidence for the Dionysian cult practice of a ship-cart. Is it possible to bridge this long chronological gap?

The Dionysian ship-cart comes into view only on Athenian black-figure cups, and the fact that, when it does, the prototype ship for it is a contemporaneous Archaic ship suggests that if, indeed, the ship-cart was introduced to Athens, then this took place during the Archaic period. We have, however, no reason to assume that the Dionysian ship-cart originated in Athens.

Dionysos already appears in Linear B tablets, confirming his antiquity, and, while one cannot prove that the Kynos ship-cart model fragment was necessarily connected with Dionysian cult, that model does indicate the existence of the concept of a ship-cart in Greece by the Late Helladic IIIC: Furthermore, several classical authors identify Egypt as the source of the phallus in the Dionysian cult.[2] Indeed, when Herodotus visited Egypt, he witnessed a festival of Osiris, whom he identified with Dionysos, and pondered whether the cult originated in Egypt or Greece:[3]

> On the eve of the festival of Dionysos, each Egyptian slaughters a young pig in front of the entrance to his home and then gives it back to the swineherd who sold it to him to take away. The Egyptians celebrate the festival of Dionysos in nearly the same way as the Hellenes do, except they do not have choral dances. And instead of phalluses they have their own invention—marionettes as tall as one and a half feet, which the women carry around through the villages; these marionettes have genitals that move up and down and are not much smaller than their entire bodies. A flute player leads the way, and the women follow, singing praises of Dionysos. There is a sacred story which explains why the genitals are so large and why they are the only part of the marionettes that move.
>
> Now it seems to me that Melampous son of Amythaon was not ignorant of this sacrificial ritual. I think, rather, that he was actually quite familiar with it, for it was Melampous who disclosed the name of Dionysos to the Hellenes, and who taught them how to sacrifice to him and perform his phallic procession. Strictly speaking, he did not reveal everything to them, but the sages who were his descendants completed the revelation. And so it is Melampous who taught the Hellenes the phallic procession for Dionysos which they practice today. What is more, I would argue that Melampous being a clever man, acquired the art of prophecy by himself and then, utilizing information from Egypt, introduced many different rites to the Hellenes, among them those for Dionysos, although along the way he made a few changes. I would certainly not claim that it is by chance that the rite performed for the god in Egypt resembles so closely that carried out in Hellas. If that were true, then this rite would be more similar to other Greek rites in character and would not be considered a recent introduction. Nor would I assert that the Egyptians took this or any other custom or ritual from the Hellenes. But I rather suspect that Melampous learned about Dionysos chiefly from Kadmos of Tyre and those who accompanied him on the journey from Phoenicia to the land now called Boeotia.

When we attempt to determine the Gurob ship-cart model owner's background, it is important to realize the limitations under which we labor. Generally, we have at our disposal three types of evidence: archaeology, iconography, and texts. Each of these forms of information comes with its own inherent limitations.

Under New Kingdom pharaohs, Egypt absorbed many foreigners of varying backgrounds. The reasons for this are varied and numerous. Among the groups absorbed into the fabric of Egyptian society under the Ramessides were the Sea Peoples, who came as invaders. However, some of them eventually made common cause with Egypt, fighting for it against other invaders, including their own kin.

Any attempt to identify the model's owner must begin by understanding the site of Gurob, the settlement where the model was buried. The modern site of Gurob can be identified as the pharaonic harem city of Mi-Wer, located at the entrance to the Fayum. The site appears to have been first settled under Thutmose III and was probably abandoned under Ramses V or soon after his reign, as Egypt spiraled down into the Third Intermediate Period.

The economic importance of the harem at Mi-Wer is indicated both by the lands under its control and by evidence that the harem was an important commercial center and the economic focal point of the surrounding settlement. Remnants of papyri from Gurob indicate various commercial aspects of the harem, with weaving preeminent among them. Textile workers, some of whom were foreigners, received training in weaving at Mi-Wer. Papyri from Gurob indicate that the harem was a clearinghouse for the textiles it manufactured, sending them to a variety of locations and to royalty. Mi-wer was not part of a palace complex but rather an independent institution.

Tomb 611, in which the model was found, belonged to a discrete cemetery located on a rise south of the harem complex. Petrie's assistants examined ten graves here: Those buried in this cemetery appear to have been persons of relatively high status based on the quality of their grave goods.

While the majority of inhabitants at Gurob must have been Egyptian, the town must also have been home to a number of foreign individuals. To a certain degree the evidence allows us to determine which groups can and which cannot be placed at Gurob. We have, for example, both archaeological and textual evidence of the presence of Syro-Canaanites at Gurob, and, indeed, Ramessesemperrēʿ, a high harem official under Merneptah, was a Semite.

From a historical standpoint, large numbers of Libyans likely dwelt in Lower Egypt during Ramesside times. Unfortunately, because theirs was an illiterate society, they remain largely invisible in the textual evidence. Gurob's inhabitants may well have included some Libyans. Ramses III notes that he conquered and resettled Libyans who had been living primarily in the western Delta region. The Wilbour Papyrus identifies Tjuk People living among the Egyptians as a warrior class; however, another group that has been identified as Libyan mercenaries, the Kehek, do not appear there.

Petrie posited Mycenaeans and Cypriots living at Gurob. A review of the evidence, however, which is limited solely to pottery, indicates that it is insufficient to imply the presence of either of these groups at Gurob even though we have abundant evidence of Cypriot traders in Egypt; moreover, a papyrus from Amarna suggests the use of Mycenaean mercenaries during the XVIIIth Dynasty.

Petrie also identified a palace official, Anen-Tursha as a Teresh, a group of the Sea Peoples that joined in the coalition against Merneptah. Whether this worthy is actually a Teresh has been debated, and one notes that the ship model was *not* found in his tomb.

Another important form of information regarding foreigners at Gurob are the balance weights found there, many of which were identified by Petrie as being non-Egyption. These weights would benefit from an updated review of their character and identity.

Egyptian textiles were primarily made of linen. In spinning thread the Egyptians employed an S-spin, and the vast majority of Egyptian textiles that have been examined, such as those at Amarna, show that S-spun threads predominate. For this reason the appearance at Gurob of a spindle used to create Z-spun yarn must indicate the presence of a northern foreign spinner, presumably a woman, at Gurob.

A woman found in Gurob Tomb 605 was buried in an unusual manner, described by the excavators as "reeds, roped together." Nearby, Tomb 614 contained a second "roped-reed" burial. Similar XIXth-Dynasty female burials in reed mats at Saqqara wore necklaces and earrings made of amber, which S. Hood notes may indicate that they were the wives of northern mercenaries. While the female burial in Tomb 605 contained no amber, some of her jewelry was made of materials that replicate its hue. As amber was presumably a rare commodity in Egypt at the time, these items may have been meant to substitute for it. These two tombs, together with Tomb 611, which contained the Gurob ship-cart model, may indicate a preference by northern foreigners for this particular cemetery.

The burnt groups found by Petrie and his assistants at Gurob constitute a unique cultural phenomenon. He discusses five of these groups and dates them in a chronological spectrum from Amenhotep III to Seti II, while ignoring the consideration that a context must be dated by the latest objects found in it. Martha Bell's meticulous restudy of these groups indicates that they were deposited during the XIXth Dynasty or later.

Petrie considered these groups to be Akhaian, that is, Mycenaean, and linked them to the vestigial remnants of a cremation custom. The Mycenaeans did not practice cremation, however, nor did they normally burn grave goods. Scholars have proposed various explanations for the Gurob burnt groups, considering them evidence of cultic cleansing or trash pits or positing that Maat-Hor-Neferure, the Hittite wife of Ramses II, dwelt at Mi-Wer and that the burnt groups represent evidence of a Hittite female death ritual. A review of the evidence suggests that significant problems exist with each of these explanations.

New arrivals in Egypt adopted local practices, but some imported customs may have survived for a time. If the burnt groups are connected with burial, then the simplest explanation for them is that they represent evidence of a group at Gurob who had practiced cremation but is seen here in the immediate process of acculturation, that is, when they have given up cremation but still practiced their burial customs in a vestigial manner by burning the personal items of their dead. The most likely cultural candidate for this phenomenon, then, become migrating elements of the central European Urnfield Culture. This society cremated its dead with their personal possessions, a burial custom that defined the culture: The Medinet Habu naval battle scene places elements of the Urnfield culture in Egypt during the XXth Dynasty.

The naval contingent of the Sea People's attack during Ramses III's Year 8 was limited to the Sherden and the Weshesh. In Papyrus Harris, Ramses III designates only these two groups with the term "of the sea." The Weshesh promptly disappear from history, appearing only here and at Medinet Habu. On the other hand, it is possible to verify at least some of Ramses III's historical statements in Papyrus Harris by measuring them against the evidence for Sherden in Egypt. The Sherden are unique in wearing horned helmets in the monuments, which also facilitates this study. A review indicates a close corroboration between the statements in Papyrus Harris and the reality on the ground as seen in both textual and iconographic evidence.

This raises the question as to what the Egyptians actually meant when, as here, they termed some—but not all—of these northern invaders as "of the sea." If the term meant simply that they were ship based, then all the Sea People groups listed by Merneptah must have reached Libya by ship and should have received this designation.

The Sea People advanced on Egypt during Ramses III's Year 8 by land and sea. I propose that the Egyptians gave the term "of the sea" to only those groups who invaded Egypt by ship during their migrations. This list would include the Sherden under Ramses II, the Ekwesh during the reign of Merneptah, and the Sherden and the Weshesh during the reign of Ramses III. (I do not include here the Teresh, of whom only their chief depicted on the façade of the Migdol entrance at Medinet Habu is given the designation "of the sea" due to the particular character of the relief in which he appears.)

Thus, in this scenario the group appearing in "feather" helmets in the Sea Peoples' ships at Medinet Habu must be identified as Weshesh (and not Peleset), who fought alongside the horn-helmeted Sherden. Furthermore, the ship, which served as the prototype for the five representations of the northern galleys was probably a Weshesh (= Urnfield Culture) and not a Sherden ship. This scenario also argues against the theory that makes Ramses III responsible for the settlement of the Peleset/Philistines in southwestern Canaan, as Ramses III's statement regarding settlement and taxation in Papyrus Harris refers solely to the Sherden and the Weshesh—and not to the Denyen, Sekel, or Peleset, who are said to have met a different fate.

No doubt remnants of other invading Sea Peoples may have ended up living and prospering in Egypt:[4] Yet only two groups—the Sherden and the Weshesh—are specifically mentioned as having been settled and taxed by the pharaohs Ramses II and III. Of all the Sea Peoples who settled in Egypt in Ramesside times, the Sherden appear most prominently. The Wilbour Papyrus confirms this by giving a remarkable snapshot of this group living and prospering in Egypt as a military stratum while in the process of successfully integrating into Egyptian society. This is confirmed by numerous other royal texts, as well as mundane documents dealing with day-to-day life, which witness Sherden sharing in the burdens and privileges of life in Egypt under the Ramessides. Also, the Medinet Habu naval battle relief indicates unequivocally that Sherden participated in a naval invasion of Egypt using ships of the type represented by the Gurob model. These considerations lead to the conclusion that the model was *probably* owned by, and presumably interred, with a person of Sherden ancestry.

Unless and until someone discovers, excavates, and reports on an actual Helladic galley, the Gurob ship model may be the closest we will ever get to this remarkable vessel type, which played such a significant role in changing the course of world history at a critical juncture.

Appendix 1 Lines Drawing of the Gurob Ship Model

Alexis Catsambis

The lines of the Gurob model are included in this volume as an alternate way of conveying its unique form (Fig. App. 1.1).[1] The lines are based on the research conducted by Shelley Wachsmann and the three-dimensional hull reconstruction created by the Institute for the Visualization of History, included with this book.[2]

The model was originally received as a digital mesh, which was then converted to a NURBS object and simplified to a basic polysurface.[3] A framework of sections, waterlines, and buttock lines was created around the model as in conventional representations of ship lines. The digital reconstruction was then bisected along its keel, and thin sections were cut along predetermined intervals (1, 2, 3, C, A, B, C). The model lines were created by using the sections to guide the curves. By activating and deactivating various layers, screenshots were taken that allowed for the model to be accurately depicted in a more traditional format. Minimal fairing was applied in order to maintain and represent the original model's lines.

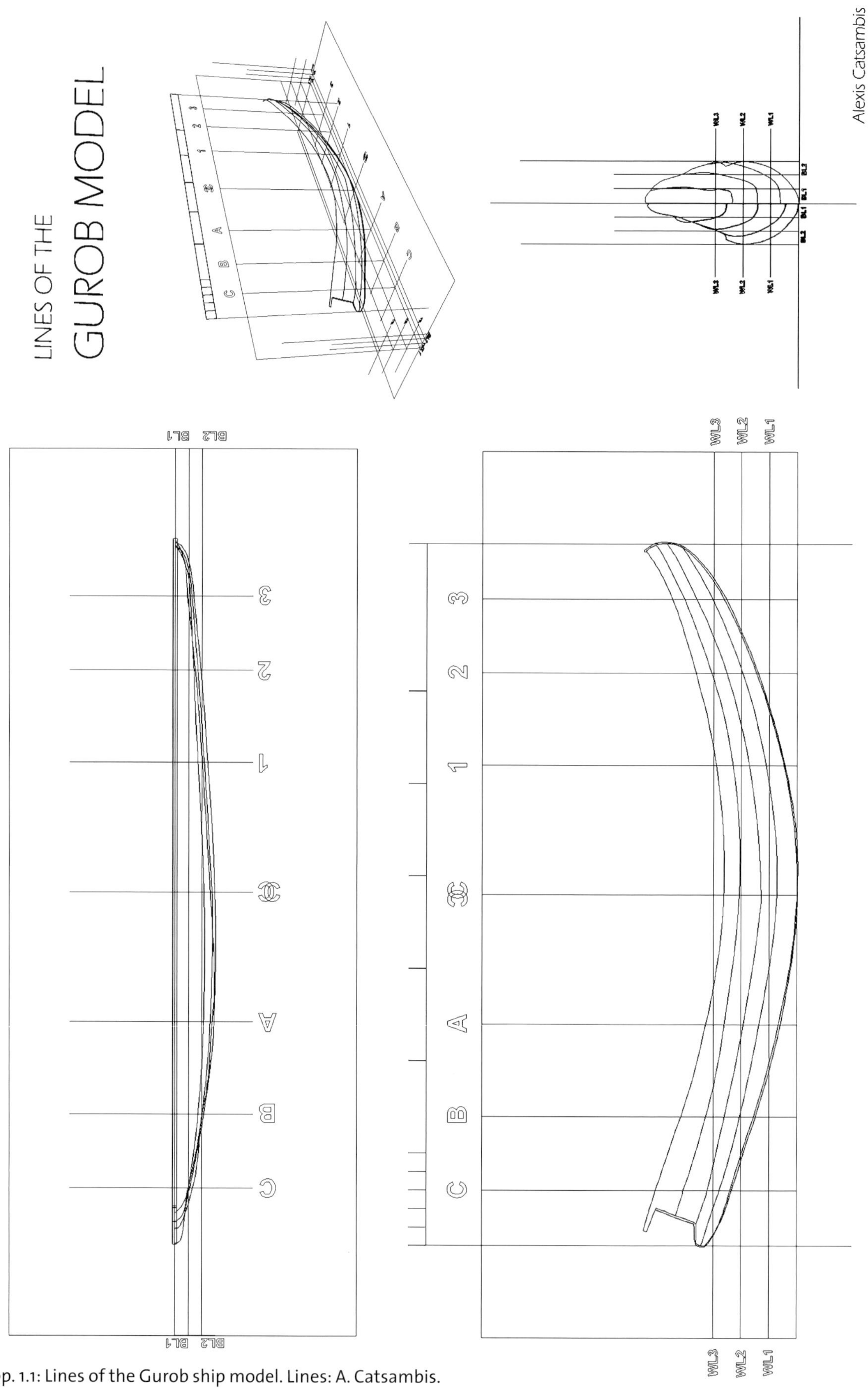

Fig. App. 1.1: Lines of the Gurob ship model. Lines: A. Catsambis.

Appendix 2 Gurob Ship-Cart Model in Virtual Reality

Donald H. Sanders

The field of virtual heritage—the use of interactive 3D computer graphics for the collection, study, teaching, display, and publication of data about cultural heritage—has been a growing influence for more than a decade. Just as the development of photography dramatically changed the practice of fieldwork, subsequent historical analyses, and public outreach throughout the nineteenth and twentieth centuries, so 3D computer modeling is having a similar impact on archaeology today. The first virtual reality re-creations of ancient sites and objects occurred in the early 1990s (with projects created by the Learning Sites team, English Heritage, and Carnegie Mellon University).[1]

Since the early experiments with virtual ancient worlds, computer technologies have improved, software has become easier to use, and virtual heritage has expanded globally to provide visualizations vital to our understanding of the past. These uses have made it clear that the only way to truly understand history is to see it in the same three dimensions in which it unfolded. Hypotheses about cultural change, settlement growth, multiple architectural uses, and object function can be accurately and precisely tested only in three dimensions. Isolated static images (photographs or drawings) cannot offer the same benefits and often lead to erroneous and misleading results.

Although virtual re-creations of historic sites, buildings, and artifacts are currently being built and tested in hundreds of projects around the world, full reports thoroughly integrating the results with other archaeological analyses are still rare. As archaeologists inevitably move toward digital publications, integrated real-time graphics, and linked databases, Shelley Wachsmann and I hope that this hybrid volume will provide researchers with a glimpse into a future encompassing new types of interpretive content and innovative ways to understand historical evidence that will lead to fresh discoveries unobtainable from traditional 2D image sources.

Despite the small size of the Gurob ship-cart model, the use of interactive 3D computer graphics technologies provided Wachsmann with original insights and opportunities for understanding and investigating the object and its many pieces. Working in the digital realm with interactive computer graphics offers other advantages as well. With fragmentary and fragile objects like the Gurob ship, it is unlikely that the Petrie Museum of Egyptology in London (where the fragments are housed) would have allowed him to handle and experiment with the many pieces and possible configurations while trying to understand the original form of the vessel. Also, since the object is stored in a museum quite distant from his office, constant travel to view the object and ponder the uses of its various supplemental fragments was out of the question.

Thus, Wachsmann asked us (the Institute for the Visualization of History, Inc., or VIZIN) to help him study the model more completely, test various theories about its original form and color, try to understand how all the many little pieces might have originally been attached or otherwise used, and provide the reader with a novel ability to examine this remarkable artifact. In the virtual realm, much that is not possible in the real world can be achieved, investigated, and changed. This short review explains some of the steps we at VIZIN took along the way from our first introduction to the wooden model to the fully interactive, virtual-reality package that accompanies this book and can be found online at www.vizin.org/Gurob/Gurob.html. Throughout the various phases of the modeling process, a constant back-and-forth exchange between Wachsmann and us took place. This entailed three stages: (A) re-creating the as-found model from the existing pieces; (B) re-creating two as-built versions of the ship-cart model, extrapolated from the surviving pieces; and (C) refining the final versions to learn more about how the original form may have looked by moving pieces around (in the computer models) and comparing the results with contemporary images of similar vessels.

We began by studying the high-resolution photographs, plans, and detailed descriptions and dimensions of the excavated pieces that Wachsmann provided. These data became the foundation for the as-found digital model of the ancient ship (Fig. App. 2.1). Once we had a good 3D computer model of the various pieces and had reassembled them, we could use the photos as texture maps to ensure that the results depicted the actual remains accurately and in detail. On the project website, the reader can view color versions of the ship-cart model's details that appear here in black and white. From the basic as-found digital model, we could output the files for viewing with virtual reality (VR) software and allow Wachsmann to examine the ship as though he were holding a single object in his hands (Fig. App. 2.2–3). Using the VR version, he could zoom in to view the nuances of the object, rotate it in any direction, and study the vessel from all angles against his color photos. This process opened up new means of researching an artifact well beyond what is normally possible. This VR re-creation of the as-excavated ship is also available online at http://www.vizin.org/Gurob/Gurob.html.

The next step was to work out the form and colors of the little wooden model as originally crafted using what we had learned about the various features from modeling the virtual as-excavated version. Wachsmann's research into ancient ships of the period also helped fill in the gaps and provided suggestive evidence for various details. We first created a computer model of the basic vessel and then began investigating how the remaining loose pieces (such as the stanchions, quarter rudder, and oars) might have fit into the overall design. We next experimented with some of the ship's accessories to see how changing their positions might affect the look of the vessel, which we compared to images of similar ancient ships, and how the parts interacted. We also adjusted the elements based on traces of surviving paint, which might indicate how the parts were originally attached.

Subsequently, we added elements that would have most likely been part of the original design but did not survive. Clues about the existing pieces (such as the forecastle and sterncastle screens and the sternmost, now detached, thwart) indicated possible locations for these missing elements. Remnants of paint along the hull allowed us to complete the oarports and black paint on the lower portion of the hull. We chose a translucent neutral-toned shade to indicate those reconstructed areas for which there are no surviving pieces (Figs. App. 2.4–10). The angle of the quarter rudder could be set by aligning the attachment holes and the angle of the blade's shoe, while the angle of the oars was set to show the model in motion.

Among the last items to have their locations tested were the wheels, the so-called awnings, and other miscellaneous pieces not already used. Based on Wachsmann's suggestions, we tried different locations for the wheels, close together and farther apart (Figs. App. 2.4–9). It immediately became clear that the location and angle of the quarter rudder determined where the wheels could be positioned and that the "awnings" could not be used

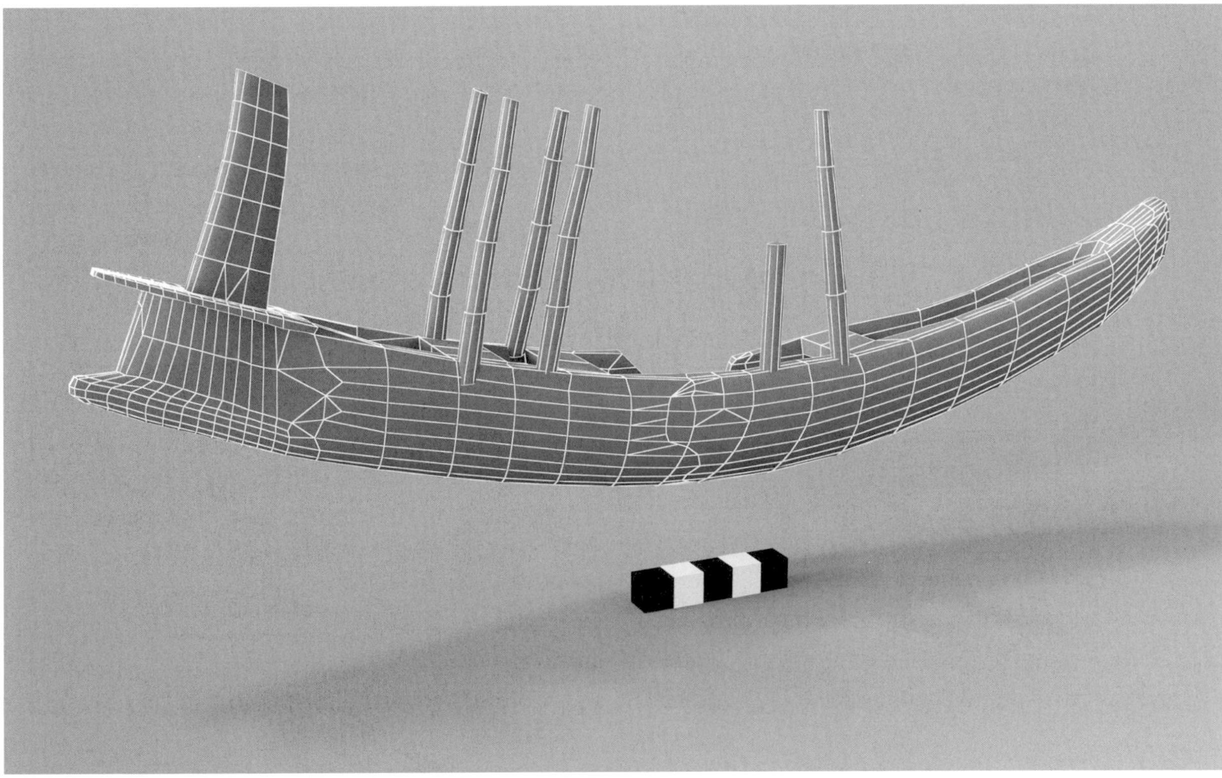

Fig. App. 2.1: Screen grab from the as-found 3D computer model showing the wireframe geometry before the photographic textures were applied. Courtesy Institute for the Visualization of History, Inc.

Fig. App. 2.2: Screen grab from the as-found 3D computer model showing how the photographs of the hull fragments and other elements were used to create realistic textures. Courtesy Institute for the Visualization of History, Inc.

Fig. App. 2.3: Screen grab from the fully textured, as-found 3D computer model of the Gurob ship. Courtesy Institute for the Visualization of History, Inc.

as fenders as they would interfere with the quarter rudder (Fig. App. 2.9). These details would have most likely gone undetected without testing the options in a real-time computer graphics environment. Also, we discovered, by experimenting with wheel placement, that the wheels have slightly different diameters. They could not be placed randomly, as certain arrangements would cause the model to tip oddly (Fig. App. 2.10). This peculiarity was discovered only upon rotating the model in virtual reality and inspecting it from various viewpoints. The differences in the wheels' sizes were not appreciated until that point but clearly reflect how the wheels were originally positioned.

None of these problems would have become evident working only with separate pieces or drawings of presumed configurations. That is, working through the reconstruction process as an interactive 3D computer model has demonstrated that many aspects of the original design could not have been fully appreciated through traditional analytical methods or simply by eye alone. We continued to tweak the positions, colors, and orientation of other elements of the vessel and tested many different arrangements against the archaeological record.

The final step was to create the interactive package that can be found on the associated website built to accompany this volume (http://www.vizin.org/Gurob/Gurob.html). There you will be able to manipulate the computer model and examine the as-found pieces and as-reconstructed original ship as if you were holding them in your hand. You will also be able to click on various elements of the model and study them against high-resolution color photos of the surviving pieces, so you can check the details for yourself.

Fig. App. 2.4: Rendering from one interim as-reconstructed computer model of the ship showing the wheels set close together, the quarter rudder vertical (note how it disappears beneath the groundplane), hypothetical stern castle screen, hypothetical forecastle screen, but no oars and equal-size stanchions (with translucent insets for those elements for which no evidence survives). Courtesy Institute for the Visualization of History, Inc.

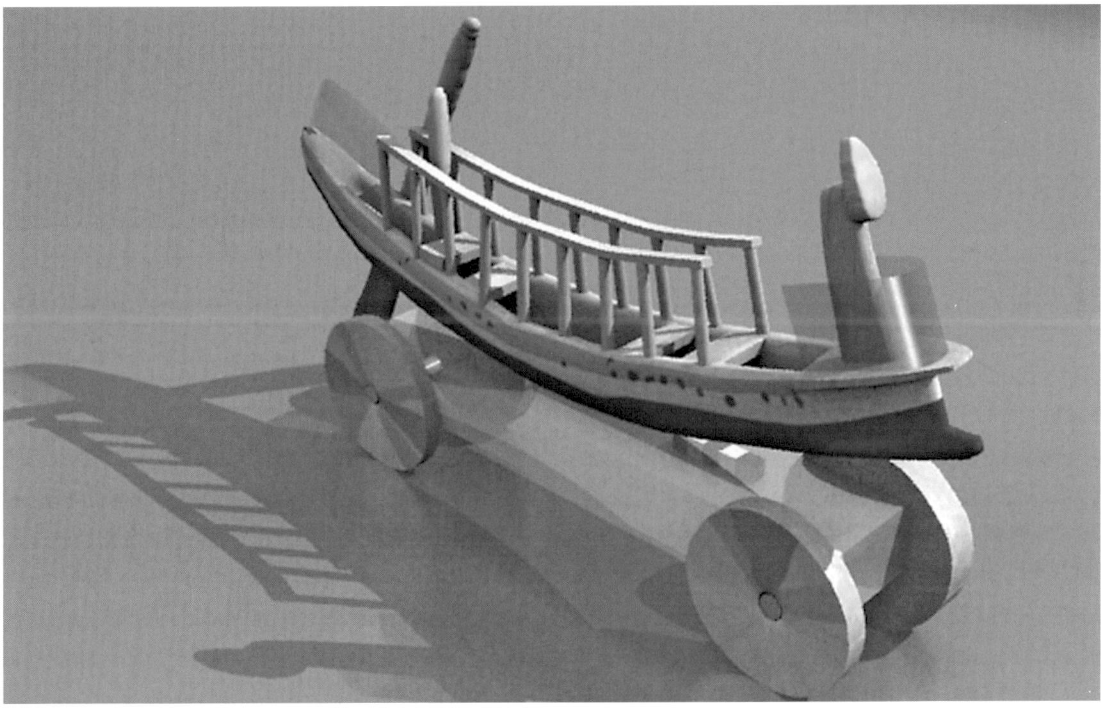

Fig. App. 2.5: Rendering from one interim as-reconstructed computer model of the ship showing the wheels set far apart, the quarter rudder tipped back, hypothetical stern castle screen, hypothetical forecastle screen, horizontal ties between the tops of the stanchions, but no oars (with translucent insets for those elements for which no evidence survives and reconstructed wheel materials and colors). Courtesy Institute for the Visualization of History, Inc.

Fig. App. 2.6: Rendering from one interim as-reconstructed computer model of the ship as a variation of Fig. App. 2.5 but with the wheels set close together. Courtesy Institute for the Visualization of History, Inc.

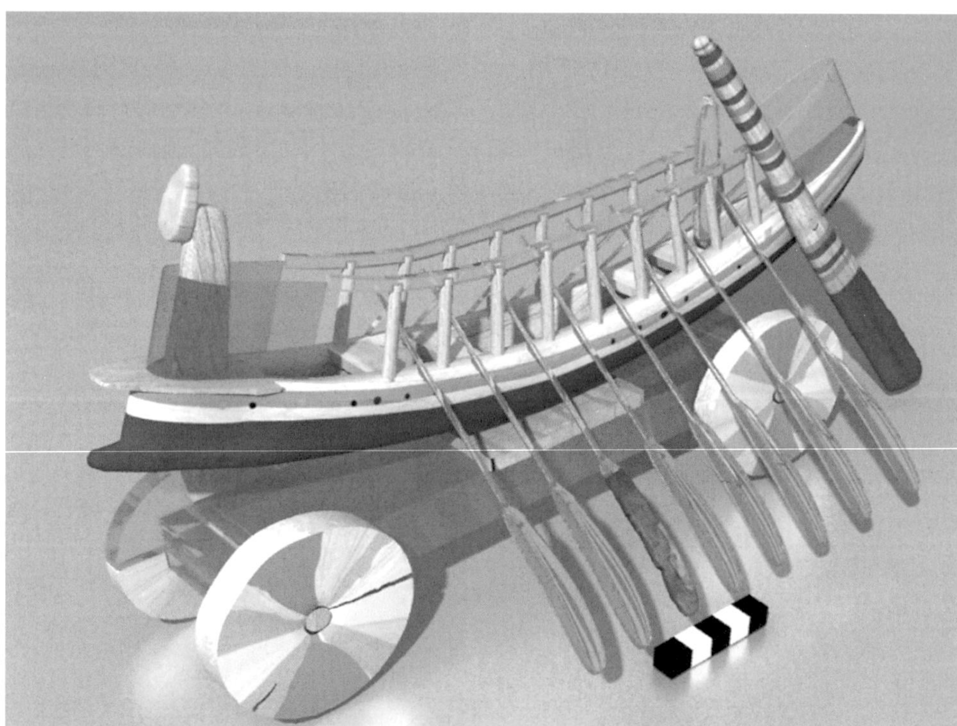

Fig. App. 2.7: Rendering from one more-advanced, interim, as-reconstructed computer model of the ship showing the wheels set far apart, the quarter rudder tipped back, hypothetical stern castle screen, hypothetical forecastle screen, horizontal ties between the tops of the stanchions, and oars (with translucent insets for those elements for which no evidence survives). Courtesy Institute for the Visualization of History, Inc.

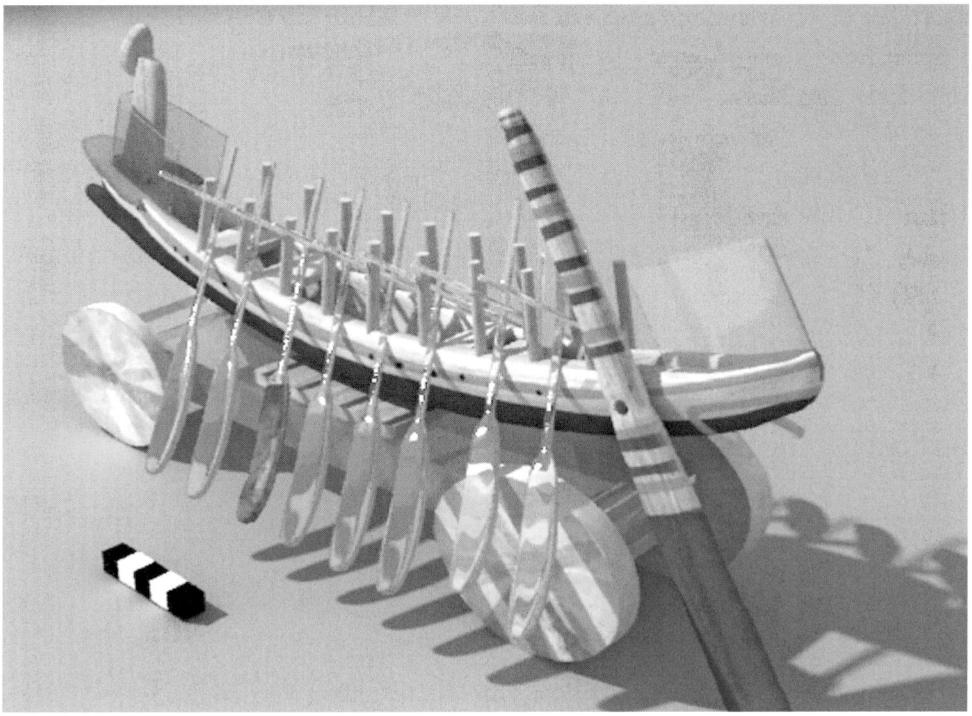

Fig. App. 2.8: Rendering from a variation on the model shown in Fig. App. 2.7, with shorter oars that are tipped a bit more to simulate the ship in motion. Courtesy Institute for the Visualization of History, Inc.

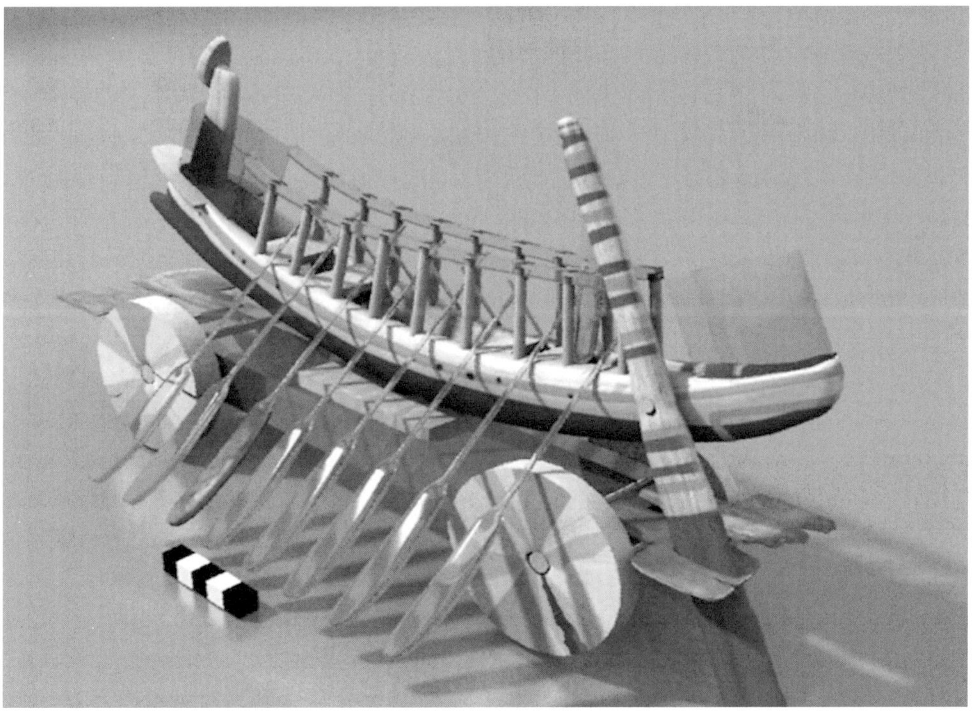

Fig. App. 2.9: A rendering of the Gurob ship 3D computer model as-reconstructed showing a possible configuration of the "awnings." See also Fig. 1.25. Courtesy Institute for the Visualization of History, Inc.

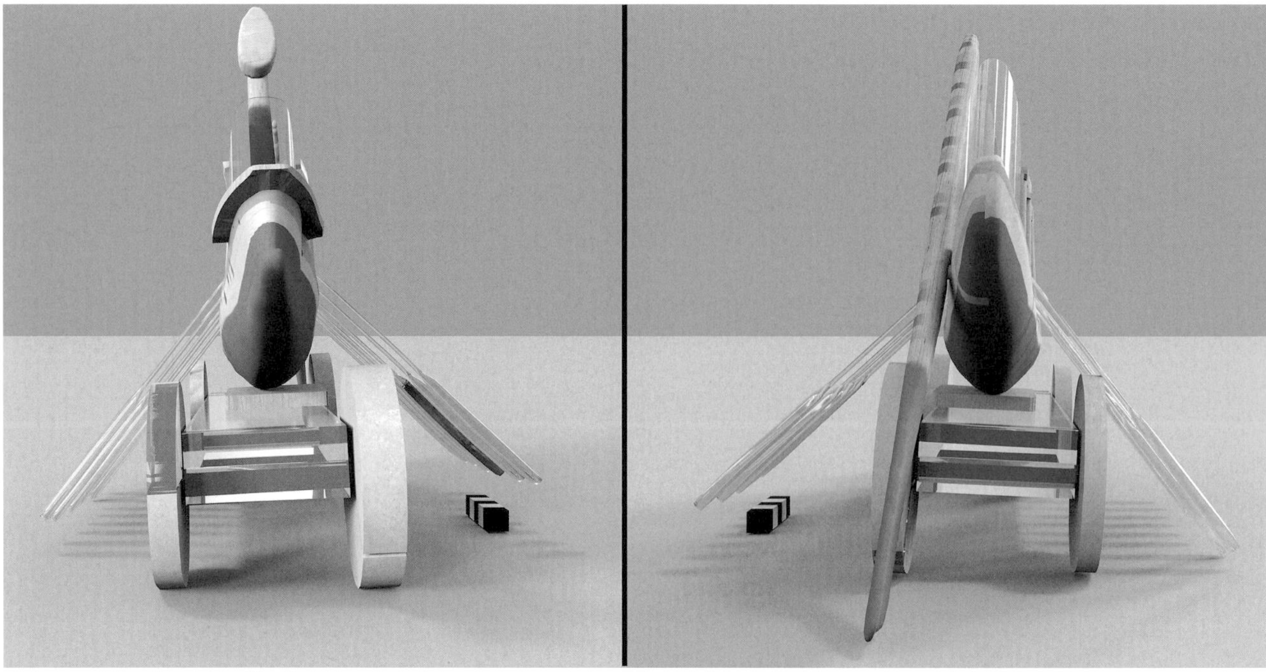

Fig. App. 2.10: Two renderings of the Gurob ship 3D computer model as-reconstructed demonstrating the tilted and wobbly nature of the ship with the different-sized wheels misaligned. Courtesy Institute for the Visualization of History, Inc.

Cautions about Operating Systems

A few caveats and some background history to the development of the interactive online module are needed to explain the differences in the experience for those using PCs versus those using Apple-Mac computers. As originally conceived by Wachsmann in consultation with VIZIN, the online package was to enable all users to simulate a near firsthand experience with the ship-model (in various configurations) using VR software. During the unfolding of this project, that was indeed the case; but recent events have foiled our attempts to allow both sets of users equal degrees of interactivity.

For reasons that go way back to the early days of virtual reality, in the mid 1990s, Apple computer and Windows-based PCs took different approaches to how interactive 3D space would be displayed (within Web browsers, not using stand-alone software or in games). Windows machines could easily display virtual worlds created using VRML (Virtual Reality Markup Language), the international Web standard language for programming virtual reality environments for global distribution across the Internet. Due to limitations in their internal architecture, Apple computers were not originally able to adequately process true 3D spatial models. To compensate, the company developed QTVR (QuickTime Virtual Reality), which is not really virtual reality at all, since there is no actual 3D space. QTVR images are really like bubbles, the interior surfaces of which are coated with flat 2D images and set so that the user is in the center of the bubble. Navigation allows the viewer to simulate moving by zooming in and out and spinning inside the bubble to simulate turning inside the space. Viewers cannot walk behind objects in the scene nor look underneath objects because there is no 3D space to move around in.

QuickTime object movies (unlike standard QTVR panoramas) put individual objects at the center of the bubble and allow users to spin the object and zoom in and out, very closely mimicking actual VR.

With the advent of OS X on Apple computers, the internal architecture and chipset changed, allowing some developers to create software that could indeed view true VRML inside Web browsers, like Safari. However, Apple kept changing the rules over the years, making it very difficult for third-party software to keep up, and native support for and interest in VR quickly waned (while in the Windows world, 3D was becoming the next big wave of change for how information would be viewed and navigated). Apple seems intent on deliberately changing the way third-party applications are distributed and installed. Recent OS X versions actually invalidate the way external software is stored, how shared libraries are accessed, and how appended, but necessary software kits, like Javascript, are accessed. Further, Safari (Apple's preferred Web browser) has completely changed how its plugins work. Thus, many developers (like the primary creator of VRML viewer plugins for OS X, FreeWRL) simply cannot afford to keep pace with the constant changes, effectively locking them out of upgrading their software for Apple-Mac computers.

Although we fully intended to provide the same VR module for everyone, recent circumstances have forced us to reconsider those plans. Windows users will still be able to enjoy all the VR models and their interactivities, but, unfortunately, Apple computer users will need to be content with less interactive QTVR visualizations. It seems odd that this must be, but operating systems and corporate philosophies have conspired to force us into this less-than-ideal result. Should the situation change in the future, we will provide the full VR module to Apple-Mac users.

Instructions for Using the Online VR Module

Installation instructions and navigation tips for exploring the virtual reality (VR) models of the Gurob ship are provided here.

System Specifications

In order to explore the Gurob ship-cart model online, we recommend that you have the following computer configuration:

PC users.—Minimum specs: Pentium 2 processor running at 1 GHz; 512 MB RAM, graphics card with 32 MB RAM, color monitor set at 1024 × 768 pixels, and running Windows XP with the latest version of a Web browser and a VRML viewer installed (see below for further details); recommended specs: Pentium 3 processor running at 2.0 GHz, 1 GB RAM, graphics card with 64 MB RAM, and running WindowsXP.

Apple users.—Minimum specs: OS X, processor running at 500 MHz, 128 MB RAM, graphics card with 32 MB RAM, color monitor set at 1024 × 768 pixels; recommended specs: OS 10.4 or later, processor running at 1 GHz, 512 GB RAM, graphics card with 64 MB RAM with the latest version of Safari Web browser (some older G3 systems with OS 9 may be able to view the QTVR models, but the results may be less than desirable).

Installation

PC users.—Before you can see and explore the VR models, you will need to download and install a virtual reality player and viewer: These are free plugins for your Web browser. We recommend the Cortona 3D VRML viewer originally developed by ParallelGraphics. The executable for this plugin is available from the Cortona 3D website at http://www.cortona3d.com/Products/Viewer/Cortona-3D-viewer.aspx). Be sure to note where the viewer file is being downloaded on your computer; then double-click on the executable file to install it.

There are other VRML viewers available. Although we have not tested them all, theoretically, any one of them is the only virtual reality plugin you will need.

Apple users.—Although the QuickTime viewer comes already installed with Apple-Mac operating system, we recommend that, if your computer supports it, you download the latest version from http://www.apple.com/quicktime/.

Both groups of users will also need to have the latest version of the Adobe Flash® player installed in order to enjoy the satellite overview portion of this interactive module. Both PC and Apple/Macintosh users can download Flash from http://get.adobe.com/flashplayer/.

Once you have installed the latest Flash player and the VR viewing software, you can begin to explore the Gurob digital supplement by visiting the website: http://www.vizin.org/Gurob/Gurob.html.

Navigation Tips

The first screen that appears provides access to the Help file (for plugin installation and virtual world navigation instructions) and to the two (PC and Apple-Mac) versions of the digital supplement. Once you have chosen which version to enter, the next screen that appears allows you to choose which version of the Gurob 3D computer models you wish to view (from among several other options accessible by clicking on a yellow button, including viewing an interactive module showing the location of Gurob).

If you open the satellite (Flash-based) module, then on the first screen you will see a set of images at the right and an information panel on the left that describes how to navigate this module.

If you are interested in one of the virtual reality or QTVR models, then your choices include the as-found version of the ship and two as-reconstructed versions of the model. Each version has links that bring up high-resolution color photographs of the surviving pieces. Once you have made your initial choice, the main navigation screen will appear. It has two divisions: At the top is the VR/QTVR model panel, and below it is the thumbnail image navigation frame (with a set of introductory instructions; and, for Apple/Mac users, the links to the photographs). You will be able to explore, rotate, and zoom into the 3D ship-cart model. Play with it. You cannot break it.

A button in the PC-version of the as-reconstructed virtual world allows you to toggle on and off certain control mechanisms that in turn enable you to change various features of the model as follows: You can change the orientation of the stempost, the location and spacing of the wheels, the orientation of the quarter rudder, and the size of the stanchions. These are some of the details for which some uncertainty exists regarding their original positions. We have suggested one configuration; you may change the model and test other hypothetical solutions. Once you have settled on a configuration, click on the button again to turn off the controls for a clean look at your new creation. Feel free to test as many alternatives as you would like, and be sure to study the linked photographs. The Apple-Mac version of these models, unfortunately, will lack any kind of interactive controls due to the limitations of QTVR.

In general, for specific navigation options and how to use the VRML plugin, refer to the Help files that accompany each downloaded plugin; they each have a different system of controls for moving around in the virtual world (and navigation may differ somewhat depending on the pointing device you choose to use for exploring the models). Some general tips follow:

Navigation Tips for PC Users with ParallelGraphics Cortona

To rotate the model, hold down the left mouse or touchpad button (with the cursor in the model window panel but not on a VRML anchor node) and slide the mouse or your finger (or use the arrow keys).

To zoom into the model, click on the cross-hairs circle and then on to the model; this will bring you very close to the model; you will then need to find a location that is not a hotspot and then mouse click and scroll to zoom away (or use the back-arrow button on the Cortona console to reset the viewpoint).

To access Help—right click in the VR window and select from the pop-up menu.

Navigation Tips For Apple Users with QTVR

To interact with the model, first click in the window holding the image. To rotate the model, use the arrow keys or click and drag using the multi-touch pad. To rapidly rotate the model, click the extra tools on the four edges of the QTVR viewing box. To zoom in and out of the model, use the shift key to move into and the control key to move away from the model. Additional control mechanisms are just barely visible at the very bottom of the QTVR viewing window.

Appendix 3 Ship Colors in the Homeric Poems

Dan Davis

The Gurob ship model is exceptional among representations of Helladic vessels in its preservation of a polychrome paint scheme.[1] Remains of black paint cover the bottom of the hull along its entire length and extend midway up the sides. Above this stretches a single horizontal row of black dots, apparently representing oar ports. Between this row and the caprail is a broad lateral stripe of dark red. At the bow, there are traces of blue paint on the outer edge of the forecastle deck piece, that is, outboard of the footprint of what appears to have been a forecastle screen. The piece was apparently edged in red. Shelley Wachsmann suggests, and reasonably so, that the missing screen piece was painted blue and that blue paint dripped down onto the horizontal piece or was painted together with it. There are also traces of blue paint on the model's port stern quarter.[2]

Based on its form and individual features, the model appears to represent a Helladic galley of the Late Bronze Age.[3] Therefore, any study of the context of the model's paint scheme should include references to colors in roughly contemporary Linear B tablets and the noun-epithet ship descriptors pertaining to colors in the Homeric poems *Iliad* and *Odyssey.* While Linear B includes a variety of color terms and a small corpus of maritime matters, there are no instances of colors used specifically in maritime contexts to shed light on the color schemes of Late Bronze Age ships.[4] Homer, on the other hand, describes ships as, among other things, *black, red-cheeked, purple-cheeked,* and *blue-* or *dark-prowed.* Interestingly, two, perhaps three, out of these four colors appear on the Gurob model. There is, of course, the question of using the Homeric poems as evidence or relevant comparanda of Late Bronze Age ship colors. The *Iliad* and the *Odyssey* were first written down some time in the eighth or early seventh centuries B.C., some four centuries after the collapse of the Mycenaean palaces which served as their setting. What makes the poems relevant to any study of the Late Bronze Age is the oral tradition that characterizes their composition. While many features of the language and heroic setting are quite at home in Homer's time, other features are demonstrably quite ancient and extend back to the Late Bronze Age, especially the formulaic phrases that reached Homer through tradition.[5] Ship epithets, including those involving colors, are highly formulaic and thus may preserve some semblance of a Late Bronze Age reality.[6] Homeric Greek, however, has suffered what some have called a paucity and an ambiguity of colors; one need only recall Homer's confusing but oft-repeated formula the "wine-dark sea" and all of its attempted explanations both in antiquity and today.[7] I will not enter this millennia-old debate but will instead discuss specific uses of colors in both Linear B and Homer to help clarify potential meanings behind the poet's use of ship epithets.

Linear B Tablets and Colors in Homer

Black (μέλας, μέλαινα).—Of all the ship descriptions in Homer, *melas/melaina,* or "black," is the most common. *Melaina naus (μέλαινα ναῦς)* and its morphological variations are found a total of eighty-one times in the *Iliad* and the *Odyssey* and an additional six times in the Homeric *Hymn to Apollo.* Hesiod uses the epithet five times in his *Works and Days.*[8] To these authors, the galleys' dark appearance was apparently the most salient characteristic. The epithet was used occasionally by other authors in later periods but in deliberately archaizing contexts.[9] Some have suggested that the dark color may be attributed simply to the dark appearance of ships as seen on the sea.[10] However, the most common reason adduced to explain its frequency is that the hull planking of ships was covered with pitch (collected from resins of coniferous trees) to waterproof the plank seams and to protect the hull from destructive marine life.[11] While to my knowledge no archaeological evidence of the practice has been found on Late Bronze Age ships, it is well noted in the Classical period and later.[12] The Gurob model appears to be the earliest "physical" evidence for protecting hulls with pitch.

Red-cheeked and purple-cheeked (μιλτοπάρῃος, φοινικοπάρῃος).—Odysseus's ships are called red-cheeked (*Il.* 2.637, *Od.* 9.125) and purple-cheeked (*Od.* 11.124, 23.271). "Red-cheeked" is derived from two Greek words, the color "red," *miltos (μίλτος),* which is made from either cinnabar or red ochre,[13] and "cheek," *paraïon (παρήϊον).* In the Linear B tablets from Knossos, the adjective phrase "painted with *miltos*" *(mi-to-we-sa-e)* is used to describe chariots.[14] *Paraïon,* however, appears to have been flexible in its meaning. In Linear B, *pa-ra-wa-jo* appears to have been confined to the cheek pieces of helmets, as we see listed in tablets from Knossos and Pylos.[15] In Homer *paraïon* is interpreted variously as a literal cheek (as in that of a wolf or lion in *Il.* 16.159 and *Od.* 22.404, respectively), as the cheek strap or cheek piece of a horse's bridle (such as that made with ivory in *Il.* 4.141–142)[16] or the cheek piece of a helmet (such as that made of bronze in *Od.* 24.523).[17]

"Purple-cheeked" is a compound of *phoinix (φοῖνιξ),* meaning either the color purple (red-purple or generally "red") or its dye or pigment, and cheek or cheek piece as described earlier. Several tablets from Knossos list chariots as painted purple *(po-ni-ki-ja).*[18] In Homer, the only other instance of purple in proximity to cheek is *Iliad* 4.141–142, where Menelaus's blood-stained thighs are likened to the staining with purple (color or dye: *φοίνικι)* of the ivory cheek piece of horses (*παρήϊον . . . ἵππων*). Indeed, there is at least one instance in the Knossos tablets of such cheek pieces or straps being decorated with ivory—another instance of Homeric borrowings from the Late Bronze Age.[19]

Miltos with its constituent pigments is a plausible color for encaustic ship paints.[20] *Phoinix,* however, likely is not. The dye, used exclusively to denote royalty throughout antiquity, was extracted from murex shells to color textiles. The murex dye process, as was known in antiquity, resulted in a variety of colors depending on the quantity of secretion and the length of exposure to the sun: yellow, greenish yellow, greenish blue, blue, red blue, purple, and violet.[21] Longer exposures produced the famed Tyrian purple, a very dark shade bordering on crimson. As renowned traffickers in the commodity, the Tyrians eventually earned the name "Phoenicians" as a namesake. The legendary expense of the dye, in addition to the large amount of materials and labor required to produce sufficient quantities, likely precluded its application on something as large as a ship (or twelve ships), even accounting for hyperbole.[22] In any event, Homer employs the term *haliporphyros,* or "sea-purple" (*Od.* 6.53, 306), when he means the murex dye. *Phoinix* may therefore denote an imitation purple or crimson, produced perhaps from vegetable and mineral constituents.[23]

Homer employs these two colors exclusively to describe Odysseus's ships, thus implying their special status to be distinguished from numerous other ship contingents in the *Catalogue of Ships* (*Il.* 2.494–759). Is it possible that

Homer employed "red-cheeked" and "purple-cheeked" as synonyms of *haliporphyros* to evoke the notion of royalty or elite status in his description of Odysseus's ships? Were the chariots from Knossos painted purple for a similar reason? The evidence is too sparse to make even tentative conclusions, but if this reading is correct, the various colors resulting from the murex dye process may also explain Homer's use of *kyanos* in his ship epithet "blue-" or "dark-prowed."[24]

Why are ships red- and purple-*cheeked* in Homer? The term is often assumed to be a metaphorical description of an actual ship feature using an anthropomorphic term. F. E. Wallace considered "-cheeked" to mean the ship's sides, as did M. E. Irwin, who suggested that the sides were "painted like cheeks, since redness is the colour of cheeks."[25] In fact, there are both earlier and later examples of hulls painted or striped in red. The earliest iconographic evidence of red-painted ships is found in the Miniature Frieze of the West House at Akrotiri on Thera (Late Minoan/Late Cycladic I), where we find several vessels colored red in the fresco.[26] Red stripes painted on or near the sheer strake also appear in numerous ships' depictions on black-figure vases of the sixth and fifth centuries B.C., a practice reflected in comments made by Herodotus.[27] These examples demonstrate at least some continuity in striping or painting ships in red between the Late Bronze Age and the Classical period.

T. D. Seymour, however, followed by a host of other scholars,[28] equated "cheeks" with the bow area, presumably because ships painted on Geometric pottery (ca. 900–700 B.C.) commonly had *oculi,* or ships' "eyes" painted near the bow area, either on the hull itself or on the forecastle screen.[29] Homer, however, never mentions *oculi,* although he may have known about them on ships of his own time.

In sum, then, we are left with (1) linkages in the Late Bronze Age of red/purple with chariots and ivory with cheek pieces, (2) Homeric linkages of purple with both ivory and cheek pieces, and (3) some evidence of Minoan/Cycladic ships painted or striped in red, as well as similar examples from the Archaic and Classical periods.

Blue- or dark-prowed (κυανοπρῴροιος).—The epithet blue- or dark-prowed occurs thirteen times in Homer.[30] The compound is formed from either the material *kyanos* *(κύανος)* or the adjective *kyaneos (κυάνεος),* and prow, or *prore (πρῴρη).* The identification of *kyanos* and *kyaneos* has been problematic in both the Linear B tablets and in Homer. The material *kyanos* appears in the Pylos Ta series as *ku-wa-no.*[31] One text lists a stone table inlaid with *kyanos* and silver and gold.[32] In another, a chair is inlaid with phoenixes of both gold and *kyanos,* and a footstool is inlaid with *kyanos* along with silver and gold.[33]

Homer, too, treats *kyanos* as a material. In *Il.* 11.627–628, Hekamede arranged a "fair table with feet of *kyanos,* a well-polished one." Agamemnon's panoply in *Il.* 11.24–44 included a breastplate with ten bands of "dark" *kyanos,* twelve of gold and twenty of tin, with six serpents of *kyanos* creeping upward. On his shield were ten circles of bronze, twenty bosses of tin, and another in the middle of dark *kyanos,* on which a Gorgon was set. The shield's silver strap had a three-headed serpent of *kyanos.* Finally, the palace of Alkinous boasted an interior frieze *(θριγκός)* of *kyanos,* and indeed the Minoan palace at Knossos and the Mycenaean palace at Tiryns boasted interior friezes incorporating blue pigments.[34] Based on the literary and archaeological evidence, *kyanos* has been interpreted as either lapis lazuli (a semiprecious stone composed of lazurite, calcite, pyroxene, and pyrite) or niello (a black metallic alloy of sulfur with silver, copper, or lead). Most now associate the term with imitations of lapis lazuli, either a glass paste or a dark-blue enamel.[35]

The adjective *kyaneos,* on the other hand, is employed in Homer as a color.[36] It seems natural to equate it with dark blue after its noun form, but the contexts of its usage suggest something quite different: Zeus and Hera, for example, are described as having *kyaneos* brows (*Il.* 1.528, 15.102, 17.209); the two Aiantes are a *kyaneos* battalion (*Il.* 4.282); Apollo took Aeneas off in a *kyaneos* cloud; Poseidon is a *kyaneos*-haired god (*Il.* 13.563, 14.390, 15.174, 201, 20.144, *Od.* 3.6, 9.528, 536); the Trojans appear as a *kyaneos* cloud when they descend on the Achaean ships (*Il.* 16.66); a *kyaneos* cloud enfolds Polydorus as he dies (*Il.* 20.417–418); Amphitrite is described as being *kyaneos*-eyed (*Od.* 12.60); Thetis dons a *kyaneos* veil, of which "no other raiment was more black" (*μελάντερον*: *Il.* 24.94); and so on. Although dark blue may fit certain of these contexts, most scholars agree that *kyaneos* generally means "dark" or "black" and can be treated as a synonym of *melas.*[37]

The question is, to which one does Homer refer when he describes ships' prows? The material *kyanos* or the color *kyaneos*? Various solutions have been proposed to determine Homer's meaning. These include the equation of *kyanoproiros* with (1) a "dark" metal, such as bronze or copper, (2) a dark or dark-blue paint at the bow, and (3) a bow ornament in *kyanos*. The first solution, offered initially by W. E. Gladstone in 1877, was dismissed by Wallace, who argued that a bow sheathed in bronze or copper would develop a greenish-blue patina, or verdigris, and would not necessarily appear "dark." [38]

The second solution, that the prow was painted dark blue or black, is prevalent in modern commentaries on Homer and ancient ships.[39] In its simplest reading, it is merely another color added to the palette of red and purple. In terms of color production, we may also recall that blue is one of the colors resulting from the murex dye process; it may have been considered by Homer as nearly synonymous with red and purple.

The third solution, that the bow was decorated with ornaments in *kyanos*, has been a persistent one and warrants some discussion. In his *Lexicon of the Homeric Dialect*, R. J. Cunliffe defines *kyanoproiros* as "having the prow ornamented with designs in κυάνεος," that is, "a glass paste or enamel coloured with a pigment doubtless to be identified with the pigment of brilliant cobalt hue largely used in the palace at Cnossus."[40] S. Mark adopts and expands on Cunliffe's implicit view that the *kyan-* of *kyanoproiros* derives from *kyanos* the material and not *kyaneos* the color. First he deconstructs the prevalent view that *kyanoproiros* referred to blue paint by drawing attention to Homer's distinction of "cheek" and "prow." Because "the bows are painted either red or purple," a translation of *kyanoproiros* as blue- or dark-prowed "seems unlikely." To Mark, Homer's differential use of the term signals a difference in meaning. *Kyanoproiros*, he proposes, likely refers to a "dark or blue eye of inlaid glass on the bow." As evidence he references Homer's description of dark-eyed (*kyanopis*) Amphitrite (*Od.* 12.60), the sea goddess and wife of Poseidon. In his plays *The Persians* (559) and *The Suppliants* (743) the Athenian tragedian Aeschylus (ca. 525–455 B.C.) used *kyanopis* to describe ships. "Consequently, sailors may have used these protective motifs in the hope of acquiring the goodwill of this goddess, or they may have believed that through these eyes she would safely guide their ship."[41]

There are three basic objections to Mark's interpretation. First, as we saw earlier, Homer deliberately avoids anachronizing this feature on the ships of his characters. Second, Amphitrite has no monopoly on the epithet *kyanopis* in Archaic poetry; in Hesiod it is said of Themistonoe (the wife of Cycnus), Clytemnestra, Althaea (mother of Meleager), and Electra.[42] The sixth-century-B.C. lyric poet Anacreon uses it as a general description of nymphs.[43] In any event, Amphitrite was a minor deity and not strongly associated with seafaring cults in antiquity. Finally, as Mark himself points out, fitting rare and costly eye ornaments to the bows of ships seems unlikely, as they may have been "the product of Homer's imagination."[44] Since all of these attempts to explain Homer's use of ship colors seem unsatisfactory, let us seek other possible explanations through a consideration of the nature of traditional oral poetry.

Rhetorical Strategies and Metrical Considerations

Rhetorical strategies and metrical considerations may lie at the heart of Homer's seemingly variable use of ship epithets. Perhaps his most recognizable rhetorical devices are metaphors (unquenchable laughter, shepherd of the people, watery paths, etc.) and similes (like *a* did *x*, so did *b* do *y*). Others include hyperbole, alliteration, hendiadys, and so on. Irwin has suggested that the prow of *kyanoproiros* and the cheeks of *miltoparaios* and *phoinikoparaios* are instances of *metonymy*, a figure of speech in which an attribute of a thing or something related to it is substituted for the thing itself.[45] Instances of modern metonymy include "sweat" for hard labor or "White House" for the president and the president's cabinet. Homeric instances include "Amphitrite" for sea (*Od.* 12.60) and "Hephaestus" for fire (*Il.* 2.426, fr. 18). Perhaps a more accurate diagnosis of the rhetorical device, however, is *synecdoche*, a figure of speech in which a part represents the whole; a species, its genus; or one, many. Modern instances include "hands" for members of the crew or "steel" for sword. In this reading, Homer's *kyanoproiros*, *miltoparaios*, and *phoinikoparaios* are parts that represent the *whole* ship and serve as colorful variations of the more common epithet "black ship" *(melaina naus).*

This still leaves the question of why, when the poems seem heavily formulaic, Homer employed variation in ship epithets. For the uninitiated, a few words on Homeric meter and composition are in order.

Greek epic poetry is written in dactylic hexameter. In its most basic description, one line of verse is divided into six *metra,* or feet. Each foot contains a long vowel *(princeps)* and two short vowels *(biceps):* Together these compose a dactyl. In the first five feet, the two short vowels may be contracted into one long vowel, creating a spondee. The sixth foot is always a spondee, regardless of whether the final vowel is short or long. Words cannot overrun, although sentences can.

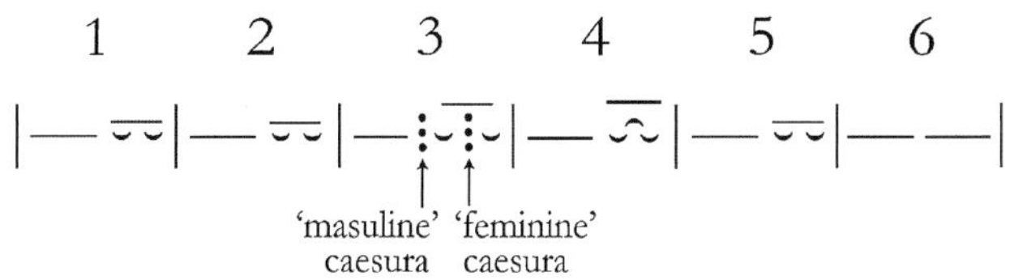

The line is also split roughly into two parts, or *cola.* The opening *colon,* called the *hemiepes,* comprises the first two and a half feet and looks like this:

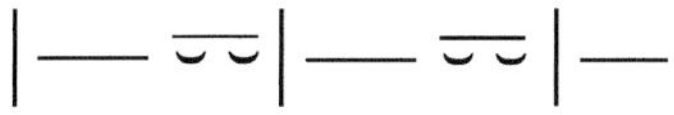

The second colon *(paroemiac)* comprises three and a half feet and has two variants in the third foot, the first with an *anceps* (short or long), the second with a *biceps* (two shorts or one long):

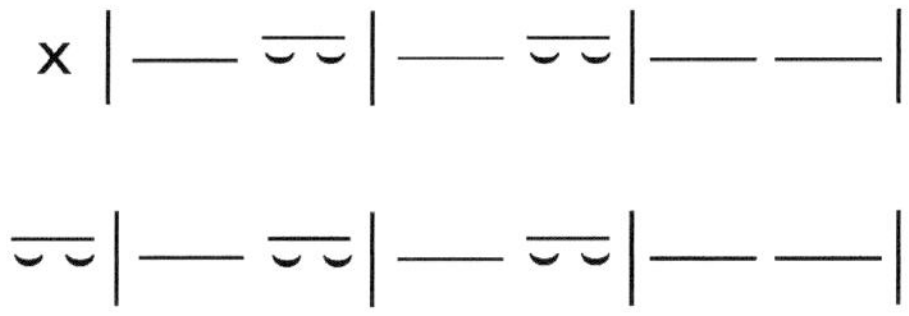

Every verse in Homer has a *caesura,* or pause, in the line. The overwhelming majority have theirs in the third foot, either after the first syllable ("masculine") or between the two shorts ("feminine"). These *caesurae* divide the *cola.*

As a result of the groundwork laid by Milman Parry on Homeric diction and style, most scholars now agree that noun epithets are governed by two general principles: economy and extension.[46] The principle of *economy* states that for each set of formulas (sc. ship epithet) filling a certain segment of hexameter line, there is normally only one expression. In other words, there are rarely two noun epithets that refer to the same thing and have the same metrical value. The principle of *extension* is related to the first, in that for a given idea (sc. ship epithet) there will be a set of expressions of different metrical formation, each filling a different segment of the hexameter line. Well-known examples of this include various expressions for Achilles or wine. Similarly, our various ship epithets stretch to various lengths, some filling an entire colon, others filling just one or two feet. Thus, they offer some degree of flexibility in placement and morphology during composition. If the poet needed to initiate, close, or bridge either of the two cola with a ship epithet, he chose a variety of the highly flexible *melaina naus.*[47] If he needed to fill the last two feet of the line, he simply employed *miltoparaioi.* If he wished to fill the entire second colon with a ship epithet, he chose either *neas phoinikoparaious* or *neos kyanoproroio,* depending on the variant of the colon. Metrical considerations, then, and not just meaning played at least some role in the choice of formulae and ship epithet in traditional oral composition.

So where does this leave us? In my view, there is no need to argue the premise that ships of the Bronze Age, as of later periods, were painted a variety of colors for different purposes, whether practical or symbolic or to suit personal taste. Indeed, there is no evidence to refute the assumption that the paint scheme of the Gurob model is grounded in reality. As Wachsmann has noted, it seems more than a coincidence that the artisan who painted the Gurob model, though having a large palette of colors from which to choose, selected at least two of the four colors (black and red) that Homer employs in his descriptions of ships. However, attempts to find vestiges and the significance of these colors in the Linear B tablets and Homeric poems are fraught with numerous difficulties. We can say with reasonable confidence that the black paint of the Gurob model and the black of Homer's "black ships" almost certainly represent the pitch applied to hulls for waterproofing and protection, a practice well attested in later periods of antiquity. Homer's use of red and purple is more problematic: He may have been rhetorically alluding to some "royal" quality of Odysseus and his ships, as is suggested in the colors employed in Mycenaean chariots. If that is the case, both adjectives were minimally employed, for of Homer's ninety-eight references to ship colors, red and purple occur in only four instances. On the other hand, judging from the appearances of dark red on ships in the Miniature Frieze on Thera, dark-red stripes on ships on black-figure vases and a passing reference in Herodotus, there appears to have been some continuity in painting or striping ships in various hues of dark red between the Late Bronze Age and the Classical period. The Gurob model is simply one more example of the practice, a practice that may have been familiar to Homer and his predecessors.

Homer's blue- or dark-prowed ship is also difficult. If *kyanos* meant some sort of blue material or ornament affixed to the prow, we must concede that there is no physical, textual or iconographic evidence of blue fixtures either on Helladic ships or on those from the Geometric, Orientalizing, or Archaic periods. However, while the prow of the Gurob model was in fact painted blue, Homer's use of the term *kyanos/kyaneos* typically referred to dark or black, thus leaving little or no room for a direct correlation between the model and Homer. The evidence can take us no further than this unless we consider the requirements of Homeric composition. Each of the ship epithets that include a color term has a unique metrical quantity and fits various frames within the hexameter line, thus meeting the requirements of economy and extension that typically govern Homeric formulae. This strongly suggests that compositional requirements and rhetorical devices played a central role in Homer's choice of color terms for ships. In this reading, "cheeks" and "prow" may simply have served as instances of synecdoche and were employed to make the formulae fit within their required metrical shapes.

As P. G. Maxwell-Stuart cautioned in his study of Greek color terminology, it is a mistake to base discussions of color on poetic material "since verse has its own way of subduing or altering words to its own purposes."[48] We can only speculate whether Homer and his predecessors had in mind the color scheme of ships represented by the Gurob model as they sang of ships in the *Iliad* and the *Odyssey.*

Appendix 4 Sherden and Tjuk-People in the Wilbour Papyrus

The Wilbour Papyrus supplies a truly remarkable view of Sherden living in Egypt during the reign of Ramses V. This appendix lists the appearances of Sherden and Tjuk-People in the Wilbour Papyrus within the general contexts and regions in which they appear in Gardiner's translation of the document (Figs. 4.10–11).[1]

SHERDEN

Text A

§3. ll. x+16–17 (p. 4)
MEASUREMENT made . . .
The *Sherden*[2] . . .

§17. 15: ll. 1, 6 (p. 15)
[MEASURE]MENT made to the south of Gerg:
. . .
The *Sherden* Sebknakhte .10——.

§17. 15: ll. 10, 12 (p. 15)
MEASUREMENT made in the Lake of Gold:
. . .
The *retainer of the Sherden* Penḥasi .20——.

§31. 17: ll. 7–8, 14 (p. 17)
Herbage of the House of Amen-Rē', <King> of the Gods.

Measurement made to the north-east of Pi-Ōḳer:

. . .

The *retainer of the Sherden* Amennakhte 3 ar. . $\frac{1}{4}$, mc. $I\frac{2}{4}$

§32. 17: ll. 18–19, 40–43; 18: ll. 3, 5–6 (pp. 17–18)

Herbage of the Mansion of Usima‘rē‘-setpenrē‘ in the House of Amūn.

Measurement made to the north of the Great Byre:

. . .

The *Sherden* Tjarobu .10 __| 5. $\frac{1}{2}$, mc. $I\frac{2}{4}$

Another measurement (made) for him land-cubits 5.45

Another measurement (made) for him .5.45

Another measurement (made) for him 50 resting

. . .

[The *Sher*]*den* Ḳeny .3.8

. . .

[Measur]ement made to the south-east of Pi-Ōḳer:

The *Sherden* Yō‘, together with his brethren .20 being dry

§33. 18: ll. 27–28, 32 (pp. 18–19)

Herbage of the Mansion of Ra‘messe-miamūn, Beloved like Rē‘

Measurement made to the north of Sasa (?):

...

The *Sherden* Tjarobu .5——[.]

§54. 23: ll. 16, 20, 24 (p. 24)

T Measurement made in the Island [of A]mūn, Manifold-of-Brave-Deeds:

. . .

The *standard-bearer of the Sherden* Pe[n]ernūte .20.80

. . .

The *Sherden* Mahar[y] .20.80

§55. 24: ll. 13–14, 31 (pp. 24–25)

Apportioning domain [of this house] (in) the Nome of Ninsu.

T Measurement made in the *p*[*a‘ĕ*-land east] of the Village of Djasasati:

. . .

The *Sher*[*den*] Pra‘(ḥi)wenm[af] ....

§59. 26: ll. 1, 34, 37, 39; 27: ll. 1, 3, 9, 12, 19, 24, 27, 40, 43; 28: ll. 6–7, 24, 29 (pp. 26–29)

Apportioning [domain] of this house (in) the Tract of Ḥardai.

. . .

T Measurement made to the west of the Byre of Horus:

. . .

The *retainer of the Sherden* Pkhōre .3ar..*I*, mc. $I\frac{2}{4}$

. . .

The *Sherden* Pḳaruiu .5. *1*, mc. $1\frac{2}{4}$

...

T MEASUREMENT made (to) the south-west [of the Village of Inroyshes]:

...

The *Sherden* Setnakhte ...., mc. $1\frac{2}{4}$

...

The *Sherden* P'onninsu [land-cubits (?) 5 (?)] it [was not seen (?)]

...

T MEASUREMENT made in the Mound of..... [nūf?]e:

...

The *retainer* [*of the Sher*]*den* Pentwēre, who is dead, (cultivated) by the hand of his children. 5. $\frac{1}{4}$, mc. $1\frac{2}{4}$

...

MEASUREMENT made to the north-east of this [place] in P-u- ḥer-nekheb-'o:

...

The *Sherden* Mahar[y] .5. $\frac{1}{2}$, mc. $1\frac{2}{4}$

...

T MEASUREMENT made to the west of this place:

...

The *standard-bearer of Sherden* Tenroy 10 ___| 5. $\frac{1}{4}$, mc. $1\frac{2}{4}$

...

T MEASUREMENT made to the west of Pi-Medjwe west of the Keep of Sha':
The *Sherden* Tja'o .5. $\frac{1}{2}$, mc. $1\frac{2}{4}$

...

T MEASUREMENT made in the island of Amūn Uniting-himself-with-Eternity east of P-[Kenroy]:

...

The *Sherden* Ḥanofre, (cultivated) by his hand land-cubits 100 resting

§67. 30: ll. 10–12 (p. 31)
DOMAIN of this house (administered) by the hand of the controller Merōn.
T MEASUREMENT made in P-mi-meki:
Land cultivated by the hand of the *Sherden* Ptaḥ(em)maini 5, mc. 5, mc. 2[5]

§70. 32: ll. 42–43, 47–49 (p. 34)
The house of Ḥaremḥab in the House of Amūn.
T MEASUREMENT made in the riparian land east of the Village of Inroyshes:

...

The *Sherden* Set(em)ḥab, which the cultivator Kha'empopĕ acquired for himself (?) 36 resting
Another [measure]ment (made) [for him] land cubits 12 resting
Another [measurement (made) for him] in 3(?)] places .12 resting

§76. 33: ll. 19–20, 35–36, 39–40 (pp. 34–35)
THE MANSION of Ra'messe-miamūn in the House of [Rē']:
APPORTIONING DOMAIN of this house in the Tract of [Ḥar]dai.

. . .
T Measurement made to the north of the Keep of Wadjmosĕ:
The *retainer of the Sherden* Pkhōre 10 __| 5. *I*, mc. *I* $\frac{2}{4}$
. . .
T Measurement made to the north-west of this place:
The *retainer of the Sherden* Khensemwēse 3 ar.. *I*, mc. *I* $\frac{2}{4}$

§86. 36: ll. 34–35, 40, 43–44, 46–47; 37: ll. 3, 8 (pp. 38–39)
Fields of Pharaoh (in) this domain.
Measurement made to the south of Ḥ-saḥto:
. . .
Measurement made to the north-west of the Village of Djasa(sa)ti:
. . .
The *standard-bearer of Sherden* Setkhaʻ .5 ——.
Measurement made to the north of Pi-Wayna:
. . .
The *Sherden* Amenkhaʻ .10 ....
The *retainer of the Sherden* Raʻmosĕ ....
. . .
Measurement made to the south-west of the Village of Djasa(sa)ti:
. . .
The *retainer of the Sherden Sady* 5 being dry

§87. 37: ll. 30, 34–36 (p. 39)
The House of Osiris, lord of Abydus, the Great God, Ruler of Eternity.
. . .
Measurement made in P-mi-meki:
The *standard-bearer of Sherden* Ptaḥ(em)maini in apportionment of land cultivated for The Mansion in the House of Amūn, | (in the) domain (under the authority of) the controller Merōn arouras 5. *I* $\frac{1}{4}$, mc. *I* $\frac{2}{4}$

§101. 40: ll. 21–23 (p. 42)
The Sunshade of Rēʻ-Ḥarakhti, which is (in) the Keep of <ʻOnayna>.
Measurement made in the riparian land west of the Keep of ʻOnayna:
Land cultivated by the hand of the *Sherden* Pḥamnūte *10*, mc. *5*, mc. *50*

§104. 40: ll. 42, 49–50 (p. 43)
Herbage of the House of Amen-Rēʻ, King of the Gods.
. . .
Measurement made in the New land of (T-)Ḳaha:
[The *Sher*]*den* Menthḥikhḥopsh[*ef*] 10 __| 5. *I*, mc. *I* $\frac{2}{4}$

§104. 40: ll. 42, 51; 41: ll. 3, 7 (p. 43)
Herbage of the House of Amen-Rēʻ, King of the Gods.
. . .

Measurement made [to] the north-east of (the) Village of Dja[sasati]:

. . .

The *Sherden* Merysēt .5 ——.

. . .

The *Sherden* Inwau .5 ——.

§106. 41: l. 34; 42: ll. 18, 21, 27 (p. 44)

Herbage of the Mansion of Usima'rē'-setpenrē' in the House of Amūn.

. . .

Measurement made to the south-west of Sekh-(en-)W'ab-yeb:

. . .

The *Sherden* Ḳenḥikhopshef .10 ___| 5. $\frac{1}{4}$, mc. *1*$\frac{2}{4}$

. . .

The deputy Iia, (cultivated) by the hand of the *Sherden* | Nakht'o .10 ___| 5. $\frac{1}{2}$ mc. *1*$\frac{2}{4}$

§118. 44: ll. 6, 11–13, 17, 20–22, 24, 31–32 (p. 46)

Apportioning domain of this house (in) the Tract of Ḥardai.

. . .

T Measurement made (to) the east of this place:

The *standard-bearer of Sherden* Anḥer[rekh] .10 ___| 5. *1*, mc. *1*$\frac{2}{4}$

T Measurement made in the riparian land north[-east] of this place.

. . .

The *Sherden* Didi .10[.]40

. . .

T Measurement made (to) the east of Pen-Kenroy:

The *retainer of the Sherden* Dḥutemḥab .3 ar. . *1*, mc. *1*$\frac{2}{4}$

T Measurement made to the south of the [Dy]ke of Spermeru:

. . .

The scribe Setnakhte, (cultivated) by [the hand of the *Sher*]*den* Setkha' 10 ___| 5. *1*, mc. *1*$\frac{2}{4}$

. . .

T Measurement made to the north of the [Pen-shō]s (in) the Lake of Iryut:

The *retainer of the Sherden* Pḳaha (?), together with his brethren 10 ___| 5. 2 AR., mc. *1*$\frac{2}{4}$

§119 45: ll. 9, 17–18 (p. 47)

Apportioning domain of this house (in) the Nome of Ninsu.

. . .

T Measurement made to the north-east of Entōre:

Apportioned for Amūn of the Granary, (cultivated) by the hand of the *Sherden* Pkhōre 5. *1*, mc. *1*$\frac{2}{4}$

§123. 46: ll. 1–2, 6: 47: ll. 12–13, 44, 47–48; 48: ll. 2, 6, 17, 24–25, 27–28, 43, 45; 49: l. 4 (pp. 48–51)

Apportioning domain of this house (in) the tract of Ḥardai.

T Measurement made in the Basin (?) (*ẖnm*) of Sharopĕ (in) the Sycomore-lake:

. . .

The *Sherden* Ḳens'ankh .10 ___| 5. *1*, mc. *1*$\frac{2}{4}$

. . .

T MEASUREMENT made to the north of this place:

The *standard-bearer of the Sherden* Pnūdjem .5. *I*, mc. $1\frac{2}{4}$

. . .

T MEASUREMENT made to the north of Pen-Ro-n-yeb:

. . .

The *Sherden* Dḥutemḥab [.]5. *I*, mc. $1\frac{2}{4}$

The *Sherden* Nefer'abĕ .5. *I*, mc. $1\frac{2}{4}$

. . .

T MEASUREMENT made to the east of Pen-Shōs:

The *Sherden* Yō' .3 ar. . *I*, mc. $1\frac{2}{4}$

. . .

T MEASUREMENT made to the south-west of this place and to the north of Pen-Wedḥu:

. . .

The *Sherden* 'Adjd .5. *I*, mc. $1\frac{2}{4}$

T MEASUREMENT made to the north-east of Spermeru:

. . .

The *retainer of the Sherden* Pabuemḥab .3 ar. . $\frac{1}{2}$, mc. $1\frac{2}{4}$

The *Sherden* Neb'anyeb .3 ar. . $\frac{1}{2}$, mc. $1\frac{2}{4}$

. . .

T MEASUREMENT made to the north-west of Pen-Ōn:

. . .

His Majesty's charioteer Merenptaḥ, (cultivated) by the hand of | the *Sherden* Siptaḥ arouras 20. *I*, mc. $1\frac{2}{4}$

. . .

The *Sherden* Piuiu (in) herdsman's territory | of the Mansion of Ḥeḳma're'-setpenamūn arouras 50 ___| 20. *I*, mc. $1\frac{2}{4}$

§134. 51: ll. 45–49; 52: ll. 18, 24–25; 53: ll. 6, 17, (pp. 54–56)

APPORTIONING DOMAIN of this house (in) the Tract of Ḥardai.

T MEASUREMENT made to the east of Yaya:

The stable-master of the Residence Psiūr (cultivated) by the hand of | in ploughing of the *Sherden* Pḥurŏpu .3 ar. . *I*, mc. $1\frac{2}{4}$

Another measurement (made) for him, (cultivated by the hand of) the *Sherden* Ptaḥmosĕ .3 ar. . *I*, mc. $1\frac{2}{4}$

. . .

T MEASUREMENT made to the east of Yaya:

. . .

The *Sherden* Pwa'ru .10 ___| 5. $\frac{1}{2}$, mc. $1\frac{2}{4}$

The *standard-bearer of Sherden* Tenroy .10 ___| 5. $\frac{1}{2}$, mc. $1\frac{2}{4}$

. . .

T MEASUREMENT made to the east of Sharopĕ | (in) the Lower Nome:

. . .

The Sherden Ḳens'ankh .5. $\frac{1}{2}$, mc. $1\frac{2}{4}$

§135. 53: ll. 26, 30–31, 34–35 (p. 56)

DOMAINS of this house under the authority of officials:

. . .

T Measurement made to the north of the House of Pḥati'a (in) the waters of (?) the basin (?) (*ḥnm*):

Land cultivated by the hand of the *standard-bearer of Sherden* E'ḥowte'o 3 ar., mc. 5, mc. *15*

. . .

T Measurement made to the south-west of Sapa:

Land cultivated by the hand of the *standard-bearer of Sherden* Setemḥab *8*, mc. 5, mc. *40*

§139. 54: ll. 25–26, 28 (p. 57)

The Mansion of King 'Akheperen(?)rē' in the House of Amūn.

T Measurement made to the north-east of Pen-Iḳarya:

. . .

The *Sherden* Penyebnūdjem .5. $\frac{1}{2}$, mc. $1\frac{2}{4}$

§143. 55: ll. 7–9 (p. 58)

The Tabernacle of Usima'rē'-miamun under the authority of the *standard-bearer of Sherden* Usima'ré-nakhte.

T Measurement made to the west of Pi-Khay:

Land cultivated by him *40*, mc. 5, mc. *200*

§145. 55: ll. 28–29; 56: ll. 10–11, 15, 20; 57: ll. 19, 26, 42–45 (pp. 58–60)

The Mansion of Ra'messe-miamūn in the House of Rē':

Apportioning domain of this house (in) the Tract of Ḥardai.

. . .

T Measurement made to the north-east of the Village of Irkak:

The *Sherden* Nakhtkēme .5. *1*, mc. $1\frac{2}{4}$

. . .

The soldier Sady, (cultivated) by the hand of the *Sherden* Setkha' .3 ar. . $\frac{1}{2}$, mc. $1\frac{2}{4}$

. . .

The *retainer of the Sherden* Penbuwa .5. *1*, mc. $1\frac{2}{4}$

. . .

T Measurement made to the east of this place:

. . .

The *Sherden* Setkha' .5. $\frac{1}{2}$, mc. $1\frac{2}{4}$

. . .

T Measurement made [to the sou]th-east of the House of Meryrē:

The *Sherden* Penyebnūdjem in apportionment of land cultivated for the House of [the King's] Wife arouras 20. 5, mc. $1\frac{2}{4}$

T Measurement made to the north of the Mound of Naḥiḥu:

The *Sherden* Penyebnūdjem in apportionment of land cultivated for the House of [the King's] Wife arouras 20. 5, mc. $1\frac{2}{4}$

§150. 58: l. 37; 59: ll. 8–9, 21, 25, 27, 31–32 (pp. 61–62)

Apportioning domain of this house (in) the Tract of Ḥardai.

. . .

MEASUREMENT made (in) the *meshru* north of Spermeru:

The *standard-bearer of Sherden* Ptaḥemḥab, who is dead, (cultivated) by the hand of <his> children 10 ___| 5. *I*, mc. $1\frac{2}{4}$

. . .

MEASUREMENT made to the south-west of Entōre:

. . .

The *retainer of the Sherden* Mesman, (cultivated) by the hand of <his> children 10 ___| 5. *I*, mc. $1\frac{2}{4}$

. . .

The *Sherden* Seti .5. *I*, mc. $1\frac{2}{4}$

. . .

MEASUREMENT made to the east of Pen-n-Nḥasy:

The *Sherden* Setnakhte 20 ___| 5. *1*, mc. $1\frac{2}{4}$

§151. 59: ll. 42–44 (p. 63)

The House of Ra'messe-miamūn, Repeater of *Sed*-festivals.

MEASUREMENT made to the north of the Village of (Ir)kak

The *Sherden* Nakhtkēme . 5. *I*, mc. $1\frac{2}{4}$

§158. 61: ll. 35–36, 42, 44 (p. 65)

THE MANSION The heart of King Menma'rē' is pleased in Abydus:

APPORTIONING DOMAIN of this house.

. . .

T MEASUREMENT made in the clearing (?) north of Spermeru:

. . .

The *standard-bearer of Sherden* Pentwēre 5. $\frac{1}{2}$, mc. $1\frac{2}{4}$

§172. 64: ll. 38–41 (p. 68)

THE HOUSE of the King's Great Wife Ḥenw'ōte:

DOMAIN of this house which was in the seventeenth [hou]se.

MEASUREMENT made to the north-west of the Mound of Naḥiḥu:

Land cultivated by the hand of the *Sherden* Penyeb[nūdjem] *20*, mc. *5*, mc. *100*

§179. 67: ll. 16–18 (p. 70)

HERBAGE of the House of Rē'-Ḥarakhti.

MEASUREMENT made to the south-east of the flooded land of P-Deben:

The *herdsman of the Sherden* Rowey 5. $\frac{1}{2}$, mc. $1\frac{2}{4}$

§180. 67: ll. 36–37, 41 (p. 71)

HERBAGE of the Mansion of Ra'messe-miamūn in the House of Rē'.

MEASUREMENT made to the north of [the] House of Meryrē':

. . .

The *Sherden* Set[em]uia .5. $\frac{1}{2}$, mc. $1\frac{2}{4}$

§181. 67: l. 42; 68: ll. 4, 9, 13–14 (p. 71)
HERBAGE of the Mansion of Ra‘messe-miamūn, Beloved like Rē‘.

. . .

MEASUREMENT made to the east of this place:

. . .

The *Sherden* Setha .5. $\frac{1}{2}$, mc. $1\frac{2}{4}$

. . .

MEASUREMENT made to the north of the Pylons of (the) Temple of Seth:
The *Sherden* Setkha‘, together with his brethren .5. $\frac{1}{2}$, mc. $1\frac{2}{4}$

§188. 70: ll. 15, 35, 38, 49, 51–52 (pp. 73–74)
Food for white goats of the Mansion of Usima‘rē‘-setpenrē‘ in the House of Amūn.

. . .

MEASUREMENT made to the north of the House formerly of Tjawati:

. . .

The *Sherden* Pḥa‘py .5. $\frac{1}{2}$, mc. $1\frac{2}{4}$

. . .

MEASUREMENT made to the east of Maniu-Merōn:

. . .

The *retainer of the Sherden* Seten‘amu 5. *1*/*2*, mc. $1\frac{2}{4}$
The *Sherden* Sety .10. $\frac{1}{2}$, mc. $1\frac{2}{4}$

§192. 71: ll. 16–17, 19–20, 33 (p. 75)
Food for white goats of the Treasury of Pharaoh.
MEASUREMENT made to the west of Pen-medja(ry?):

. . .

The steward Usima‘rē‘nakhte, (cultivated) by the hand of the *Sherden* Siptaḥ 5. $\frac{1}{2}$, mc. $1\frac{2}{4}$
MEASUREMENT made to the west of the House of Psiūr

. . .

The *Sherden* Piu .5. $\frac{1}{2}$, mc. $1\frac{2}{4}$

§211. 75: ll. 18, 27, 37 (p. 79)
APPORTIONING DOMAIN of this house (in) the Tract of Ḥardai.

. . .

MEASUREMENT made to the north of Na-Amūn:

. . .

The *retainer of Sherden* Yeb .3 ar. . ———.

§218. 77: ll. 1, 14, 16, 38–39, 47 (pp. 81–82)
APPORTIONING DOMAIN of this house (in) the Tract of Ḥardai.

…

T MEASUREMENT made in the high ground north of this place:

. . .

The *Sherden* Sunero .5. $\frac{1}{4}$, mc. $1\frac{2}{4}$

. . .

The *Sherden* Bataemḥab .3 ar. . $\frac{1}{4}$, mc. $1\frac{2}{4}$

The *Sherden* Ḥōri, son of Ḥōri .3 ar. . $\frac{1}{4}$, mc. $1\frac{2}{4}$

. . .

The *Sherden* Setnakhte .5. $\frac{1}{4}$, mc. $1\frac{2}{4}$

§228. 80: l. 19; 81: ll. 31–32 (pp. 85–86)

Apportioning domain of this house (in) the Tract of Ḥardai.

. . .

T Measurement made to the south-west of this place:

The *tender of crocodiles (?) of Sherden* Imnē 5 ___| 1 ar. . $\frac{1}{4}$, mc. $1\frac{2}{4}$

§234. 83: ll. 23–25 (p. 88)

The Mansion of King Menma'rē' in Abydus.

T Measurement made in [the] New land of Neby east of Sakō:

The scribe Setnakhte, (cultivated) by the hand of the *Sherden* Tja'o .10. $\frac{1}{4}$, mc. $1\frac{2}{4}$

§236. 83: ll. 30–31; 84: ll. 32, 39; 85: ll. 5, 10; 86: ll. 3, 12 (pp. 88–91)

The Mansion of Ra'messe-miamūn in the House of Rē':

Apportioning domain of this house (in) the Tract of Ḥardai.

. . .

T Measurement made to the north-west of this place:

. . .

The *standard-bearer of Sherden* Ḥori 30. $\frac{1}{4}$, mc. $1\frac{2}{4}$

. . .

T Measurement made to the south-west of the Sycomores of | Irkak:

. . .

The *Sherden* Setnakhte .5. $\frac{1}{4}$, mc. $1\frac{2}{4}$

. . .

T Measurement made in the basin (?) (*ḥnm*) south-east of Men'onkh:

. . .

The *retainer of Sherden* Pel'o .10.14

§241. 89: ll. 11, 26, 38 (p. 94)

The Landing-place of Pharaoh in Ḥardai.

. . .

Measurement made to the north of Sakō:

. . .

The *Sherden* Mahary .3 ar. . $\frac{1}{4}$, mc. $1\frac{2}{4}$

§245. 91: ll. 1, 22, 24, 28 (p. 96)

Herbage of the Mansion of Ra'messe-miamūn, Beloved like Rē'.

. . .

MEASUREMENT made to the east of this place:

. . .

The *Sherden* Wennofrĕ .20. $\frac{1}{4}$, mc. *1$\frac{2}{4}$*

. . .

Again the *Sherden* Wennofrĕ .5. $\frac{1}{4}$, mc. *1$\frac{2}{4}$*

Text B

§2. 2: ll. A, x + 1–4; 3: ll. 7, 21, 26, 29; 5: l. 14 (pp. 109–111)

… …. BALANCE (?) *10*

… …. BALANCE, ARABLE LAND, *5*

… …. BALANCE, ARABLE LAND, *5*

… … … … … … … … … … … … IN CULTIVATION (?) BY THE *SHERDEN* KHORY

BALANCE *10*

. . .

Reg[ion] . . . Heḳy (??) on [fields of the] Mansion in the House of Amūn arable land, 80 BALANCE. . . . . THE [*SHERD*]*EN* RA'MOSĔ

. . .

[Region] … … … … … Pen-Sonbe on fields of this house arable [land], … … … … … ….
[IN CULTIVATION (?) BY THE *SHERDEN*] SETI, THE *SHERDEN* TJAUMĒĒSHE' AND THE *SHERDEN* MEREMO[PĔ]

. . .

[Region of] . . . the lake of the Pharaoh south of Nayroti [on fiel]ds of this house [arable land] . . . . .
[IN CULTIVATION (?) BY THE *SHERD*]*EN* SETKHA' *9*

. . .

… … … … … … … … … … … … ..THE *SHERDEN* (?) PENMEN… …..

. . .

[TOTAL, ARABLE LAND, AROURAS (?) *214*], MAKING AROURAS 107. . . . .*9*, BALANCE *62* (IN) CULTIVATION BY THE SHERDEN, AROURAS *18$\frac{1}{2}$*

§4. 6: ll. 3, 20; 7: l. 23 (pp. 112–114)
KHATO-LAND of Pharaoh under his authority (in) the … … … … ….

. . .

Region south- . . . . . of this place (on) fields of the Stretch of Pharaoh arable land, arouras 20, BALANCE 50 (IN) CULTIVATION (?) BY THE *SHERDEN* BES

. . .

• • TOTAL, ARABLE LAND, AROURAS *90*, MAKING AROURAS 45 (IN) CULTIVATION (?) BY THE *SHERDEN*, AROURAS 115, MAKING AROURAS *57$\frac{1}{2}$*

§5. 7: l. 24; 8: ll. 6, 11, 26, 34; 9: ll. 5, 9–10 (pp. 114–115)
KHATO-LAND of Pharaoh under his authority starting from the backland of Heracleopolis in Memphis (on) the East Bank—(administered) BY THE HAND OF THE DEPUTY ḤŌRI.

. . .

Region south-east of the Castle of Meryrē' (on) fields of this house arable land, arouras 20, BALANCE 9
. . . WITH SEED OF (?) . . . [THE *SHER*]*DEN* P. . . . . TOGETHER WITH HIS BRETHREN, *19* . . . BALANCE *25*, BALANCE *18*

. . .

west

• Region south-east of the Village of I-Rēʿ [on fields of the House of] Rēʿ arable land arouras 10,
BALANCE 5, IN CULTIVATION (?) BY [THE *SHER*]*DEN* KHORY

. . .

REGION NORTH OF THIS PLACE ON FIELDS OF PHARAOH ARABLE LAND, *10* IN CULTIVATION (?) BY THE *SHERDEN* KAIA

. . .

• • • • • TOTAL … … …. AROURAS *67*, BALANCE 52 (IN) CULTIVATION (?) BY *SHERDEN*, ARABLE LAND 185 (?)

BALANCE 25

… … …. ARABLE LAND, *20*, MAKING *12* $\frac{1}{2}$

. . .

• • Region of the Village of Waro (on) fields of the House of Rēʿ arable land, arouras 340, BALANCE 20,
WITH SEED [OF] *SHERD*[*EN*]

. . .

Region north-west of the *meshru* (on) fields of the House of Rēʿ arable land, arouras 5,

BALANCE *300*(*?*) BALANCE 52

Total, arable [land] .….*90* .….making arouras (?) … [40] … (in) cultivation (?)

BALANCE *108* $\frac{1}{2}$

BY THE *SHERDEN* ḤAITBI (?) … … ….*227* (?), MAKING AROURAS *113* $\frac{1}{2}$

§6. 9: ll. 11, 30; 10: l. 1 (pp. 115–116)

KHATO-LAND of Pharaoh under his authority starting from P-tjesy-[Ḥaur] (in) Me[mphis?] … … ….

. . .

THE *SHER*[*DEN*] … … ….[RAʿMES]SENAKHTE *10* .….*9*, TOTAL *29*, MAKING AROURAS *1*[*4* $\frac{1}{2}$]

. . .

CULTIVATION (?) BY *SHERDEN*, ARABLE LAND, *300* + … … *11*(?)*3* $\frac{1}{2}$ 3

§7. 10: 2–4, 4A (p. 116)

KHATO-LAND of Pharaoh under the authority of the standard-bearer of the Residence Mernptaḥ.

ITS FIELDS: Region north of [the] Wall of P-tjesy-Ḥaur (on) fields of the House of Rēʿ arable land, arouras 100 (?) . . .
(10, 3 A) IN CULTIVATION (?) BY THE *SHERDEN* … … ….

• Region of the riparian land north of this place (on) fields of this house arable land, arouras 10,
BALANCE *1* AR. (10, 4A) ~~IN CULTIVATION (?) BY THE *SHERDEN*~~ … … ….ḤUY ~~… … …. 9 (?)~~

§17. 17: ll. 26–28; 18: 4 (pp. 124–125)

KHATO-LAND of Pharaoh under the authority of Pkatjana, the overseer of cattle of The Mansion in the House of Ptaḥ (sacks, ——)

ITS FIELDS: Region west of the Temple (on) fields of the House of Ptaḥ arable land, 40, BALANCE 15,

BALANCE, TIRED LAND (?), *15*

IN CULTIVATION (?) BY THE *SHERDEN* SEBKNAKHTE, TOGETHER WITH HIS BRETHREN

balance 40

Region of the *meshru* north of Spermeru (on) fields of this house arable land, arouras 60, BALANCE 40,
IN CULTIVATION (?) BY THE *SHERDEN* EʿḤOWTEʿO

balance, arable land, 10

. . .

TOTAL, ARABLE LAND, *13*, MAKING AROURAS $6\frac{1}{2}$ (IN) CULTIVATION (?) BY *SHERDEN*, MAKING AROURAS $65/32\frac{1}{2}$

THE TJUK-PEOPLE

Text A

§123. 46: ll. 1, 8, 28 (p. 48)
APPORTIONING DOMAIN of this house (in) the Tract of Ḥardai.
. . .
T MEASUREMENT made in the pond east of Sharopĕ:
. . .
The *standard-bearer of the Tjuk-people* Nebwaʿ .5 it was not seen (?)

§150. 58: ll. 37–38, 43 (p. 61)
APPORTIONING DOMAIN of this house (in) the Tract of Ḥardai.
MEASUREMENT made to the south-east of the Castle of Iōt:
. . .
The *Tjuk* Shaukatja .3 ar. . $\frac{1}{2}$, mc. $1\frac{2}{4}$

§187. 70: ll. 9–11 (p. 73)
Food for white goats of the Mansion of Millions (of Years) of Raʿmesse-Amenḥikhopshef-miamūn.
MEASUREMENT made to the south of Lake of Dīme:
The *Tjuk* Pkatja 5., mc.

§218. 77: 1, 14, 45–46, 48 (pp. 81–82)
APPORTIONING DOMAIN of this house (in) the Tract of Ḥardai.
T MEASUREMENT made in the high ground north of this place:
. . .
The *Tjuk* Bataḥotpe 5. $\frac{1}{4}$, mc. $1\frac{2}{4}$
The *Tjuk* Penemḥēye 5. $\frac{1}{4}$, mc $1\frac{2}{4}$
. . .
Again the *Tjuk* Penemḥēye .5. $\frac{1}{4}$, mc $1\frac{2}{4}$

§ 241. 89. ll. 11–12, 17 (p. 94)
THE LANDING-PLACE of Pharaoh in Ḥardai.
MEASUREMENT made in Imy-tay-m-t-nē:
. . .
The *Tjuk* Karoiay 5 __| 1 ar. . $\frac{1}{4}$, mc. *12/4*

Appendix 5 Radiocarbon Age Analysis of the Gurob Ship-Cart Model

Christine A. Prior

Samples were removed from the ship-cart model at the Petrie Museum of Egyptian Archaeology, University College London, on September 10, 2010, by Susanna Pancaldo (UCL) and Mike Sim (GNS Science). One sample was removed from the broken center section in the body of the ship, and a second sample was removed from a broken wheel (Fig. App. 5.1).

The samples were wrapped in aluminum foil, labeled, and sent to the Rafter Radiocarbon Laboratory, National Isotope Centre, GNS Science, New Zealand, for analysis.

Upon receipt, the two samples were assigned the following laboratory identification numbers:

- Ship 01 was assigned laboratory number R32559/1. It was identified on the submission form as a fragment of wood from the ship.
- Wheel 01 was assigned laboratory number R32559/2. It was identified on the submission form as wood from one of the wheels.

Pretreatment

Both samples were chemically pretreated to remove contaminants before being combusted to CO_2 and converted to graphite for radiocarbon dating by accelerator mass spectrometry (AMS) at the National Isotope Centre, GNS Science.

Pretreatment description for sample R32559/1.—Sample consisted of fine fragments of wood wrapped in aluminum foil. Microscopic examination revealed the wood fragments have visible vascular structure but a degraded powdery appearance. Noted presence of some white cobweb-looking material, a few minerals, and a green synthetic fiber. Removed white substance and fiber and removed larger minerals. No further grinding or other surface cleaning applied to wood. Cleaned in soxhlet extractor with organic solvent washes of hexane, isopropanol, and acetone. No color change or weight loss noted. Dried in vacuum oven. Treated cleaned wood with cellulose extraction process: acid wash, followed with alkali pulping, then oxidation in hydrogen peroxide, and final acid wash. Resulting cellulose was yellow. Dried in vacuum oven.

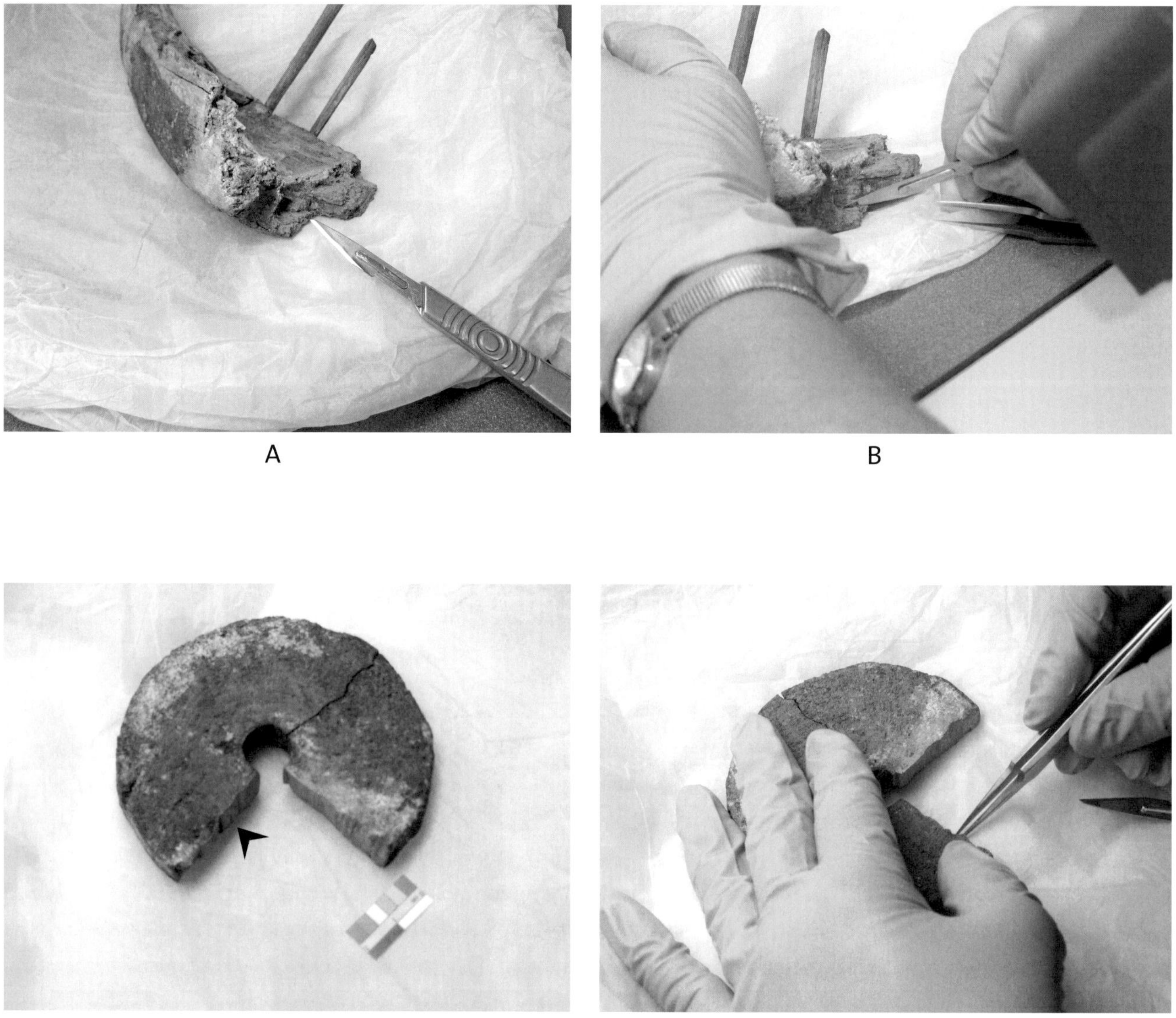

Fig. App. 5.1: (A) Location of Sample Ship 01; (B) Removing Sample Ship 01; (C) Location of Sample Wheel 01; (D) Removing Sample Wheel 01. Courtesy Radiocarbon Lab, National Isotope Centre, GNS Science.

Pretreatment description for sample 32559/2.—Sample consisted of small fragments of wood wrapped in aluminum foil. Microscopic examination revealed fragments of wood have vascular structure but a degraded powdery appearance. Observed some white material over surfaces and presence of white fibers. Scraped outer surfaces to remove white material and removed fibers with tweezers. Cleaned wood fragments in soxhlet extractor with organic solvent washes of hexane, isopropanol, and acetone. No color change and only a small weight loss noted. Dried in vacuum oven. Treated cleaned wood with cellulose extraction process: acid wash, followed with alkali pulping, oxidation in hydrogen peroxide, and final acid wash. Resulting cellulose was yellow. Dried in vacuum oven.

Rafter ID Number	NZA Number	Description	Sample ID	$\delta^{13}C$, ‰	Radiocarbon Age, Years BP	Calibrated Age Range, 2σ
32559/1	35075	wood from Gurob ship model	Ship 01	-25.4	2941 ± 20	1256 to 1054 B.C.
32559/2	35076	wood from Gurob ship model	Wheel 01	-24.3	3027 ± 25	1386 to 1211 B.C.

Table 1. Radiocarbon Age Results

Results

Full AMS results sheets and calibration reports have been supplied for these analyses, but the radiocarbon age results are summarized in table 1:

The conventional radiocarbon age is reported in radiocarbon years BP as defined by international convention among radiocarbon researchers.[1] Because radiocarbon "years" are not the same as calendar years, a conventional radiocarbon age must be converted to a calendar age range through a process known as calibration. These conventional radiocarbon ages were calibrated using WinsCal, a GNS calibration program, and the most recent international data set from IntCa109.[2] The two-sigma calendar age range cited represents the 95 percent confidence interval (Figs. App. 5.2–3).

Discussion

Two radiocarbon ages taken from different areas of the ship model indicate that the wood used to make it grew before 1054 B.C. The wood sample that was taken from the wheel appears to be about a century older than the wood sampled from the body of the ship.

This can be explained as an example of what is known in radiocarbon dating as the "old wood problem." Trees form annual growth rings, which reproduce the radiocarbon content of the atmosphere during the year of growth. Thus, tree rings from the center of the tree, when it was a sapling, have older radiocarbon ages than the outermost growth rings. For some long-lived tree species (e.g., oak, beeches, conifers) it is not unusual for the outer tree rings to be several hundred years younger than the inner tree rings. Other types of trees may live for 50–100 years. In addition, a gap between the time a tree was felled and when the wood was used to make an object is common. Wood preserves well in a dry climate, and old wood may be recycled and reused.

The wheel was likely carved at the same time as the body of the ship. Since it is possible that the model was constructed from wood that came from different trees and since the part of the wood could come from heartwood that was already several decades old by the time the tree was cut down, it is not an inconsistency that the two wood samples produced different radiocarbon ages.

The object cannot be older than the youngest material used to make it, but it can be substantially younger than the oldest. While the wood used to make the wheel was possibly alive as early as 1386 B.C., the wood made into the ship was growing at 1256 B.C. This would establish 1256 B.C. as the earliest probable year that the ship was carved since it cannot have been made before the tree was cut down.

The opposite argument cannot be made; that is, that because the wood from the wheel dates to 1386–1211 B.C., the object could not have been made later than 1211. Old wood, either through reuse or in-built age, could have been used later in time. So, for that reason, I believe that the full calibrated age range for the younger of the two measurements is the possible time of manufacture. Therefore, the ship model dates to sometime between 1256 B.C. and 1054 B.C.

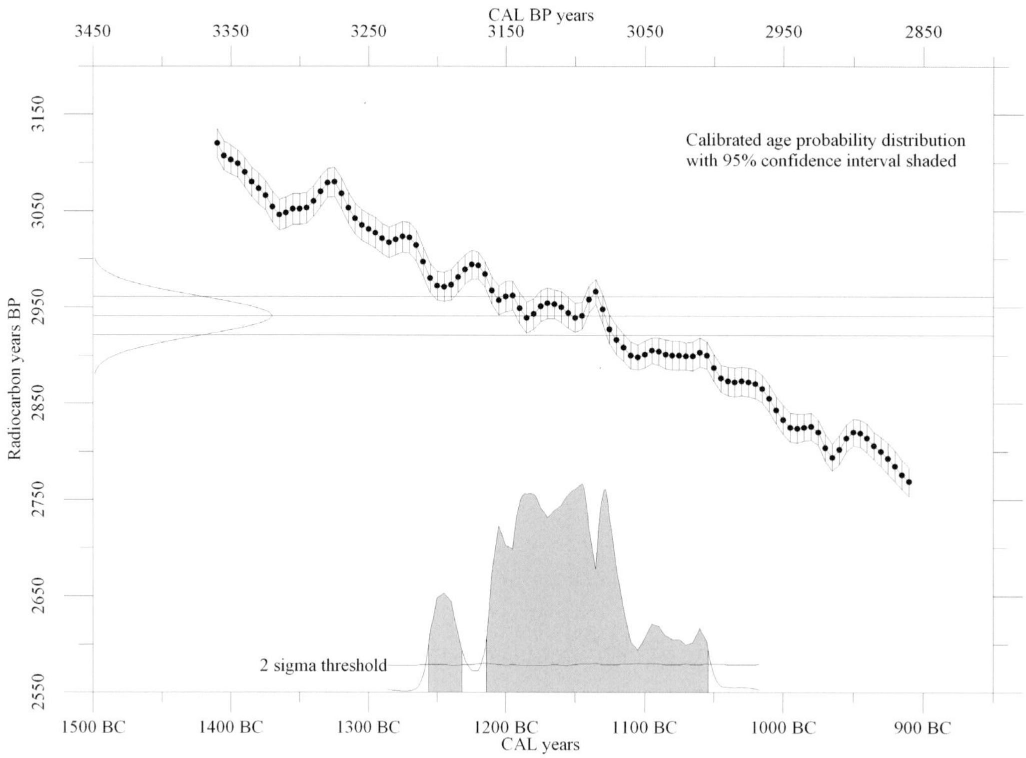

Fig. App. 5.2: Calibration of Sample Ship 01. Courtesy Radiocarbon Lab, National Isotope Centre, GNS Science.

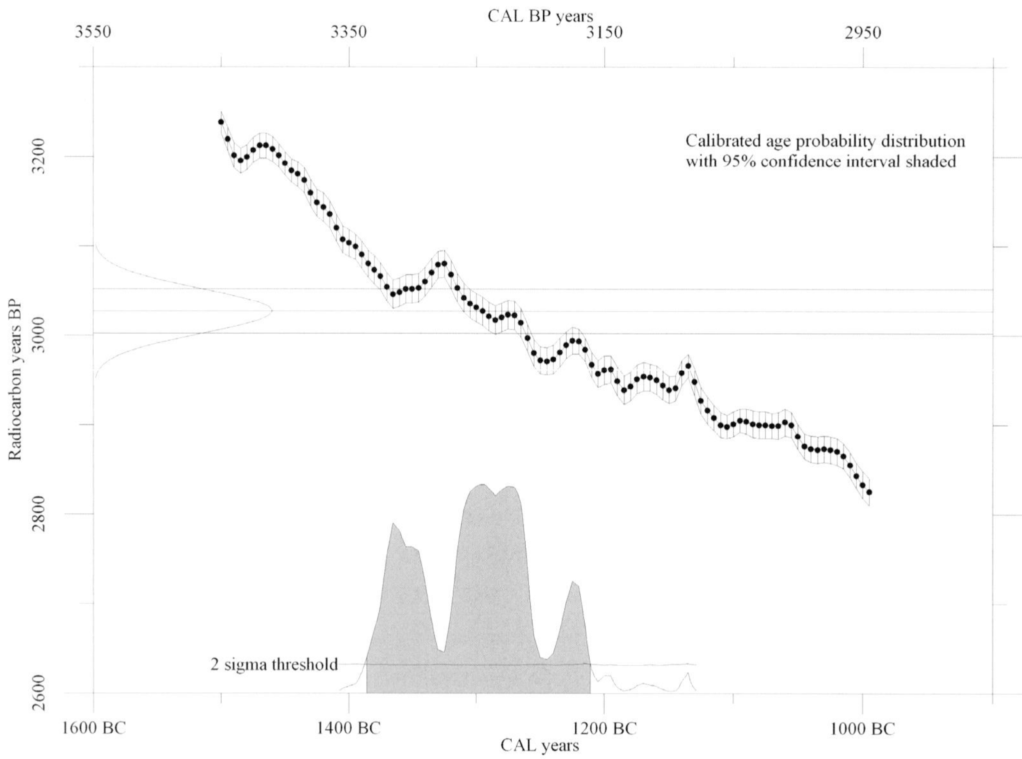

Fig. App. 5.3: Calibration of Sample Wheel 01. Courtesy Radiocarbon Lab, National Isotope Centre, GNS Science.

Appendix 6 Analysis of Pigments from the Gurob Ship-Cart Model

Ruth Siddall

The stratigraphy of paint layers is clearly seen from examination with a hand lens. The wheels of the boat are painted over with a white ground, and then spokes or segments are overpainted in red, yellow, and blue. The red and/or the yellow paint were applied first, and then the blue paint was applied last. The hull of the boat is also underpainted with a white ground. The lower part of the hull is painted black, and traces of red, yellow, and blue pigment decorate the upper sections.

Samples of white, red, blue, and yellow pigments were obtained from the most damaged wheel of the model boat. Two samples of yellow and of red were taken, one of blue, and one sample of the white ground. The white ground was also incorporated in samples of the overpainted colors. A sample of black paint was taken from the hull of the boat in the damaged area amidships (Figs. App. 6.1–2).

Samples were collected using a scalpel and then mounted on glass microscope slides in Meltmount with an RI = 1.662. The sample was covered with a glass cover slip, and the particles were dispersed with an application of light pressure. Slides were examined, and pigments identified using optical polarized light microscopy, employing magnifications of ×400 and ×1000; higher magnifications were obtained with a ×1000 oil objective. Analysis of the materials was carried out on a Leitz Orthoplan-Pol optical microscope. Observations were made in both plane-polarized light (PPL) and under crossed-polars (XPL). Samples and pigment identifications are summarized in table 1.

The Pigments

White.—The white paint covers the entire planar surface of the wheel. It is overpainted with the other colors used and also used alone.

Microscopic examination of the pigment showed the particles to be colorless and transparent. Grain shape was generally of anhedral flakes with a platelike habit. However, a significant proportion of the grains were euhedral rhombic. The grains were well sorted, with a mean particle size of 4 μm, with a few larger particles up to 10 μm in diameter. Particles were not well dispersed, clumping together in crumblike aggregates. When observed in plane-polarized light and with the stage rotated, the relief of the particles was observed to vary from very low to high.

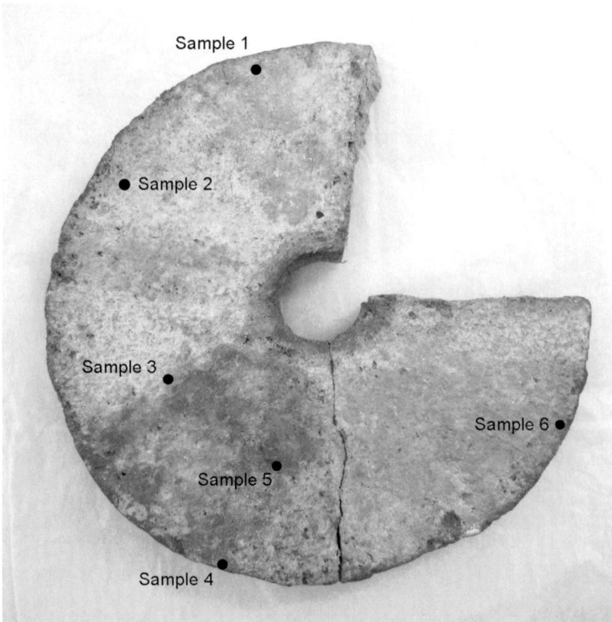

Fig. App. 6.1: Locations of paint samples collected from the wheel. Courtesy of Ruth Siddall.

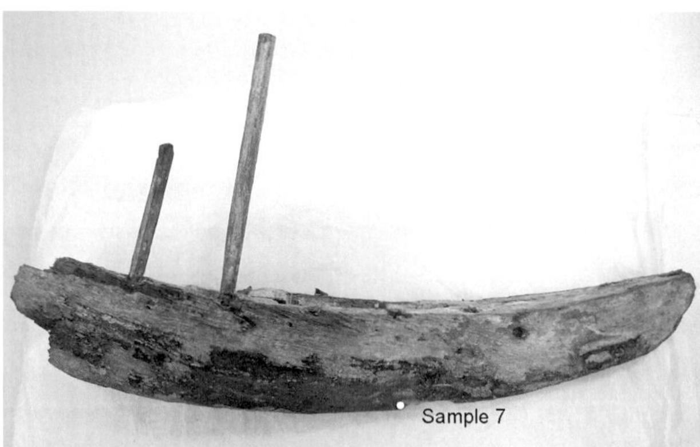

Fig. App. 6.2: Locations of paint sample collected from the hull. Courtesy of Ruth Siddall.

Under cross-polarized light, the particles showed high birefringence on larger particles ($\partial >$ fifth order on Newton's scale) and second- to third-order on particles < 10 μm. The sample was homogenous with no detrital grains, fossil fragments, or microfossils present.

These optical properties are diagnostic of carbonate minerals; the two most common are the calcium carbonate calcite and the magnesium carbonate dolomite. These compounds form end members of a solid-solution series. Both phases exhibit very similar optical properties and, when euhedral, form rhombic crystal habits. However, the lack of an obvious fizzing reaction when treated with acid indicates that this is likely to be a more magnesium-rich form.

The high purity and fine, euhedral particle size of this pigment are consistent with the optical properties of the magnesium-calcium carbonate mineral huntite. Huntite ($Mg_3Ca[CO_3]_4$) has been recognized on a wide range of artifacts and architectural pieces, dating from as early as the Old Kingdom, although primarily from the New Kingdom. Its occurrence and optical properties have been described by Heywood, as well as by Eastaugh and his colleagues.[1] Particles are typically finely fibrous or rhombic (only a few microns in diameter). Huntite forms monomineralic, soft, friable rocks of very high purity, and their bright, clean whiteness would have been attractive to pigment prospectors. Although rare, sources are known from Egypt in association with dolomite and magnesite. Huntite is capable of forming an opaque, matt, pure white pigment and has been used to this effect, for example, in

Sample	Color	Paint Finish	Pigment
1	yellow	opaque, matte	saffron + goethite
2	white	opaque, matte	huntite
3	red	translucent, matte	red ocher (hematite)
4	red	translucent, matte	red ocher (hematite)
5	blue	opaque, impasto, matte	Egyptian blue
6	yellow	opaque, matte	saffron + goethite
7	black	opaque, matte	asphalt + quartz

Table 1. Samples and Pigments Identified on the Gurob Ship-Cart Model

highlighting white figures against a background painted using calcium carbonate white.[2]

Blue.—The blue paint is an impasto (thickly applied) decoration using a coarsely ground pigment; individual particles are discernable with a hand lens. This paint was the last layer to be applied; it overlies the red and yellow paints used on the wheel.

The pigment contains subhedral to anhedral, angular grains of blue, pleochroic crystals, with a mean grain size of 40 µm. The particles are translucent and relatively free of inclusions or impurities (Fig. App. 6.3: D). Under XPL, the grains are highly birefringent, and interference colors are masked by the blue body color. Also present are colorless particles of angular glass (isotropic in XPL) and a few angular quartz grains.

The optical properties of the pigment particles is consistent with an identification of Egyptian blue. Egyptian blue is synthetic calcium copper tetrasilicate, analogous to the rare mineral cuprorivaite, $CaCuSi_4O_{10}$. This pigment was in very common use in Egyptian painting and more widely in the Mediterranean region from the XIVth Dynasty until Late Antiquity. Its manufacture, occurrence, optical and physical properties, and identification have been described previously.[3]

Red.—The red paint is applied to show radial, fan-shaped spokes on the wheel (Fig. App. 6.3: A–B). In appearance the paint layer is translucent; when observed through a hand lens it appears almost as a stain on the white ground surface. Under examination using PPL, the pigment was observed to be composed of a very fine-grained, red pigment with particle size below the resolution of the optical microscope (< 1 µm). Under XPL, the particles were highly birefringent, with the red body color masking the interference colors. These occurred as individual grains but more commonly as crumblike aggregates. Also present were a few grains of subrounded quartz that were coated with the red pigment.

This sample is composed of very pure and very fine particles of the red iron oxide, hematite (Fe_2O_3), and hematite-coated quartz grains. It is typical of a red ocher, an earth pigment. Such materials would have been widespread and abundant. A source cannot be identified, but any area associated with iron-rich deposits would suffice, including many localities in the Eastern Desert or at the iron deposits such as those of El Gedida in the Western Desert.

Yellow.—The yellow paint is applied in a manner similar to the red, as fan-shaped spokes of the wheel (Fig. App. 6.3: E–F). It forms an opaque paint layer.

Under the optical microscope, the yellow pigment was observed to be dominated by strong, orange-yellow colored fragments of cellular plant material. These were in finely chopped segments a few tens of microns across and ~40–250 µm in length. Elongated, nonwoody plant cell structures (parenchyma) are commonly observed and well defined in many of the particles, whereas in others dense bundles of fine fibers were observed. Under cross-polarized light, these particles were weakly birefringent and showed straight extinction.

Mineral (inorganic) yellow is also present in minor amounts. This occurs as finely particulate (< 1 µm) grains and also as rounded, translucent, yellow particles, c. 40 µm, which are shown to have high birefringence and also to be finely polycrystalline in XPL.

The yellow pigment appears to be composed of a mixture of a mineral yellow and an organic, plant-derived yellow. The former is goethite, iron oxide hydroxide ($FeO[OH]$), the prime constituent of yellow ochers. The organic component appears to be a plant-derived direct dye. The cell structure observed strongly resembles that described by Winton for saffron stamens (*Crocus sativus*), and, though degraded, the sample compared well with a modern preparation of ground saffron.[4] Alternatively, this could be safflower (*Carthamus tinctorius*), which show a similar cell structure. A yellow dye may also be extracted from safflower; however, it is less likely to be used as a direct dye or pigment. Further organic chemical analysis would be required to confirm this identification.

Both saffron and safflower were known to be available in the eastern Mediterranean region. Saffron may have been locally derived or traded with Minoan settlements in the Aegean.[5] Safflower seeds, oils, and dyes have been identified in numerous Egyptian contexts, and the plant is cultivated in Egypt to this day.[6]

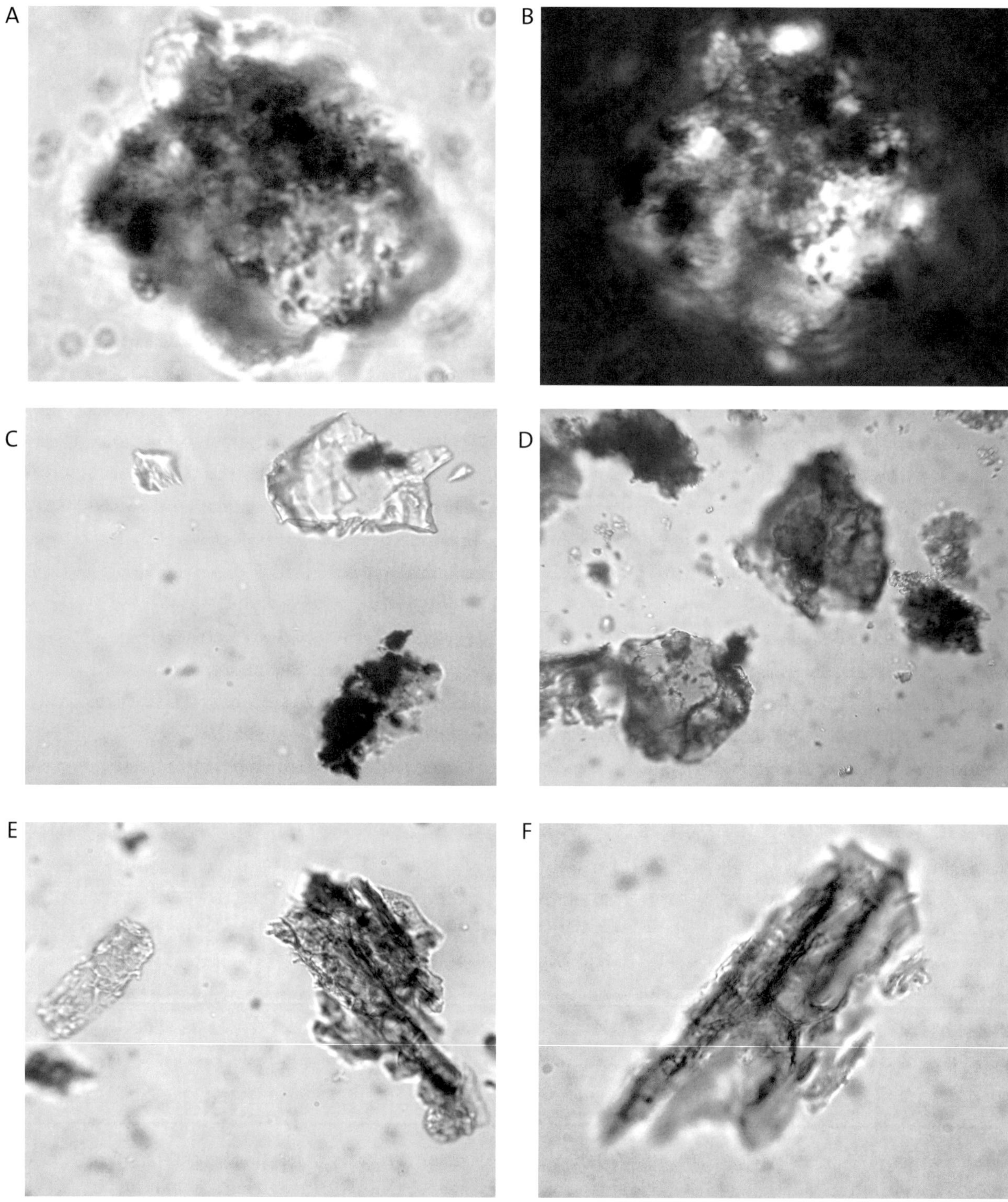

Fig. App. 6.3: (A) Red paint: red ocher, a quartz grain coated by very fine particles of hematite (PPL). Image taken at ×1000 magnification, field of view (FOV) is 20 μm. (B) The same grain under crossed-polars. (C) Black paint: dark brown particles of asphalt with quartz (colorless) in PPL. Image taken at ×400 magnification, FOV 200 μm. (D). Blue paint: particles of Egyptian blue (PPL). Image taken at ×400 magnification, FOV 200 μm. (E). Yellow paint: organic direct dye composed of saffron. Canal-like parenchyma cells are visible. To the left of the image is a contaminant grain of Egyptian blue. Image taken at ×400 magnification, FOV 200 μm. (F). Yellow paint: organic direct dye particle composed of saffron or a similar plant. Canal-like parenchyma cells are visible. Image taken at ×400 magnification, FOV 100 μm. Courtesy of Ruth Siddall.

Black.—The black pigment particles are angular, often demonstrating curved, "conchoidal" fractures along their faces (Fig. App. 6. 3: C). In plane polarized light they appear opaque black or very dark brown. At high magnification, the margins of the particles appear weakly translucent and brown. Under crossed-polars the grains are isotropic and black, although some internal reflections cause the translucent, brown grain boundaries to glow. Particles are 5–10 μm in diameter.

Quartz is also present and abundant in this sample as a relatively coarse-grained fraction, with particles 10–100 μm. Particles are highly angular, and their surfaces show a strong conchoidal fracture. Shards and splinters are present, indicating that the quartz has been manually pulverized. Additionally, macroimages of the black-painted hull show that quartz is also present on the hull in large (0.1–0.5 mm), rounded grains. This may be secondary; sand grains may have adhered to the stick black paint at a later date.

Black pigments are notoriously difficult to identify using microscopy because they are largely opaque and/or isotropic in XPL. However, the brown coloration observed here, coupled with the angular and conchoidal fractures demonstrated by the particles, are typical of asphalt or a related solid hydrocarbon.[7] Mineral blacks and carbon blacks derived from bone chars, plant chars, and soot do not show this brown coloration. The quartz was probably added to both extend the paint and to make it more workable, although some coarse-grained, sand-sized, quartz grains may have adhered to the boat hull at a later date; the asphalt paint would have been sticky and taken a long time to dry.

The asphalt-based paint used on the hull of the boat is now matt, but it is likely that the original effect would have been smooth and glossy.

The use of asphalt as an adhesive, a waterproofing agent, and a modeling material has been known from the eastern Mediterranean Neolithic.[8] Asphalt-based pigments are infrequently recognized on archaeological artifacts, but they likely were in common usage and have been poorly reported. They are known from a number of Romano-Egyptian artifacts that I examined and were used either as extenders (for darkening the shade of other pigments) or in their own right as black pigments.[9] Evidence indicates that these materials would have been in ubiquitous use as pigments from earlier periods, particularly as their sources are geologically abundant in the Middle East region. They could have been derived from natural hydrocarbon seeps associated with the Dead Sea (*Lacus Asphaltites*) and many other locations within that region. Hydrocarbon black pigments are also derived from humic earth deposits (peat and lignites), which occur in the northern Mediterranean, but these are a less likely source of this pigment from southeastern Mediterranean sources. Their dark brown or black body color and isotropic properties makes these particles easy to miss when using PLM and, since they are composed of hydrocarbons, they would be detected only with specialist chemical analytical techniques. Languri has written on the identification and characterization of these materials using mass spectrometric techniques.[10]

Conclusions

The palette used here is largely typical of Egyptian painting of the New Kingdom. Egyptian blue and varieties of red and yellow ocher are frequently encountered in Egyptian painting, and their occurrence here is unsurprising. Similarly, huntite is well recorded, and usage suggests that it was well known and intentionally selected by craftsmen to produce a strong, opaque white. The presence of asphalt as a black pigment has been less frequently recognized: Nevertheless, it is likely that its use was far more widespread than the assumptions made from a survey of the literature. Asphalt was also used for waterproofing ships, and the presence of this material on the hull of the Gurob ship-cart model is arguably more than an artistic choice. Organic yellow pigments have been infrequently identified in Egyptian painting, and identification has largely been as glazes and varnishes on coffins.[11] Inorganic yellows derived from the minerals goethite, jarosite, and orpiment are far more widespread. Therefore, the presence of saffron or safflower here is of note. Further analysis is required to determine the exact plant species used in this context.

Appendix 7 Wood Identification

Caroline Cartwright

The British Museum's Conservation and Scientific Research Department identified samples of wood from a loose fragment (item #12) and from one of the wheels (item #30) found with the Gurob model.[1] Both samples are of sycomore fig *(Ficus sycomorus)*. This taxon is frequently misspelled and misused. *Ficus sycomorus* is a fig tree from the *Moraceae* family and is unrelated to the European sycamore, also known as field maple, which belongs to the genus *Acer,* from the *Sapindaceae* family. Editors frequently mistakenly "correct" "*sycomorus*/sycomore" to read "*sycamorus*/sycamore." This is botanically wrong.

Ficus sycomorus are large trees indigenous to Egypt that prefer to grow along the banks of the Nile. Their light, porous wood is not of the highest quality and is often prone to insect attack. Despite these drawbacks, this wood was much favored in ancient Egypt for coffin manufacture, particularly during the Middle Kingdom, and was also widely used for small domestic and funerary objects, which were often painted. This wood had the advantages of being readily obtainable and easy to work. Furthermore, other factors may have influenced its selection for the construction of the Gurob ship-cart model. The sycomore fig tree, apart from bearing edible fruits, held a highly significant place in ceremonial and religious belief, and both the tree and its fruits were frequently depicted, (for example, in a wall painting from the Tomb of Nebamun).[2]

Glossary of Nautical Terms

As some readers of this volume may not be familiar with nautical terms, I have tried to keep them to a minimum in this book. Some terminology is unavoidable however, when dealing with details of ship and boat construction.[1]

Aft: Toward the stern.

Amidship(s): In the vessel's central area.

Anchor: An object normally used to prevent a vessel from drifting by securing it to the floor of the waterway.

Anchor cable: A strong rope for towing or securing a vessel. Also termed a hawser.

Aphlaston: Recurving sternpost ornament common to Classical period warships.

Astern: Behind the vessel.

Axial rudder: An oar-shaped rudder maneuvered with a tiller placed over the stern and secured to a stanchion.

Bank: A row of oarsmen.

Beam: (a) The width of a vessel; (b) a transverse timber.

Bireme: A Phoenician galley with two banks of superimposed oars.

Bow: The forward end of a vessel, beginning where the hull starts curving inward toward the stem.

Bow anchor: Anchor stationed at the bow of a vessel.

Bowsprit: A spar jutting forward from the bow.

Bowstick (bowsprit): Device seen in the bows of representations of Egyptian Middle Kingdom vessels. Their purpose is not clear. They may have served as a fairlead for a bow anchor hawser.

Brailed sail: A rig that employs brailing lines, also known as brails, to control the amount of sail exposed to the wind and to determine the sail's geometry.

Brailing lines (brails): Lines attached at intervals to the foot of the sail and brought up vertically through "brailing rings" or "brailing fairleads" sewn to the sail. The lines are carried up over the yard and then brought astern in a bunch. The sail is controlled in a manner similar to modern Venetian blinds.

Breadth (beam): The width of the hull.

Buttock lines: Lines drawing projections that define vertically oriented longitudinal hull shapes.

Caprail: Timber attached to the tops of a vessel's frames, normally representing the uppermost edge of a vessel along its sides.

Cleat: A timber used to regulate the lines that control a sailing vessel.

Crutch: Stanchion used to bear a mast, yard, and or boom when unstepped or not in use.

Dieres (GK.): A Greek galley with two superimposed banks of oars.

Figurehead: A carved figure situated at the bow of a vessel.

Fore: Concerning the region of the bow of a vessel.

Forecastle: A raised deck in the bow or the structure constructed to enclose it.

Furl: To secure a sail or a flag.

Galley: A seagoing vessel propelled primarily by oar. Galleys usually carried sail, also.

Galley, double-banked: A galley with two banks of oars.

Galley, single-banked: A galley with a single bank of oars.

Halyards: Lines used to position the yard.

Hawser: An anchor cable.

Helmsman: The person who steers a vessel.

Inboard: Located or facing inside a vessel.

Interscalmium: The Latin term for the distance required by a rower to work his oar, approximately 1 m.

Keel: The primary longitudinal central timber of most vessels.

Lines: (a) Cordage used aboard a vessel, often to carry out a specific function; (b) a hull's shape as defined by a series of geometric projections, usually presented in three views.

Loom: (a) The part of an oar between its blade and handle; (b) the staff of an oar-shaped quarter or axial rudder above the blade.

Mast: A vertical timber that supports a sail.

Mast partner: A construction supporting the mast at deck level.

Masthead: The top of a mast, above the yard(s).

Maststep: A mortise cut into the keel—or into one or more timbers placed above the keel—to receive the heel of the mast.

Merchantman: A cargo ship.

Midship: The central area of a vessel.

Molded (dimension): The measurements of timbers as viewed from the sheer (longitudinal perspective), or the body, views of a vessel's lines drawing.

Moor: To make fast a vessel by means of anchoring and/or securing to land.

Mortise and tenon: A type of joinery used in early ship construction for attaching planks or other timbers to each other.

Oarport: Aperture in the side of a vessel's hull through which the loom of an oar passed.

Oarsman: A man who rows a ship; a rower.

Oculus (oculi): Latin term for a ship's "eye(s)." Commonly appearing as apotropaic and/or decorative devices on vessels.

Ophthalmos (ophtholamoi): Greek term for a ship's "eye(s)." Commonly appearing as apotropaic and/or decorative devices on vessels.

Outboard: Located or facing outside the vessel.

Outrigger: A rowing box or frame (Gk. *parexeiresia*) on Greek triremes and later ancient galleys to facilitate superimposed banks of rowers.

Papyriform: In the shape of a papyrus umbel. A form of stem and/or sternpost decoration appearing on depictions of some types of Egyptian vessels.

Pavois: French term proposed by G. Legrain for the support structure common to Egyptian terrestrial cultic boats. Poles could be secured to the *pavois* to allow porters to shoulder these vessels when transporting them overland.

Pentekonter: Greek term for a fifty-oared galley.

Poop: Deck covering the stern.

Port: Left side when facing the bow.

Proembolion: Greek term for a subsidiary ram found on some ancient warships located above the waterline ram.

Prow: The forward part of the vessel; the bow.

Quarter rudder: An oar-shaped rudder attached to the side of the hull at the stern.

Raking (mast): The slant of a mast toward either extremity of a vessel.

Ram: Waterline projection at a galley's bow normally sheathed in metal and used to impact other vessels. The encased bow timbers were designed to transmit the energy of a collision throughout the hull.

Rigging: The network of cordage employed to support and manage a vessel's sail system.

Rockered keel: A curving keel lower amidships than at the vessel's extremities.

Scarf: A joint employed to connect parallel timbers.

Sewing: The joining of a vessel's planking and other members with ligatures, often used in conjunction with mortise-and-tenon joints or dowels. Also termed lashing or stitching.

Sheer plan: Side view of a hull.

Sheer strake: The highest uninterrupted run of planking from bow to stern.

Sheets: Lines attached to a square sail at its lower corners (clews) and employed to trim the sail.

Sided (dimension): The measurements of an unmolded surface (i.e., the upper surface of a keel).

Stanchion: A load-bearing vertical post.

Starboard: Right side when facing the bow.

Stays: Lines that run from the bow and the stern to the masthead and support it.

Steering oar: An oar used to maneuver a vessel; the steering oar lacks a tiller and is less efficient than a quarter or an axial rudder.

Stem (stempost): One or more upright timbers attached by means of a scarf to the keel/keel plank at the bow and to which the hull planking is joined.

Step: To place a mast in its maststep; hence, to raise it.

Stern: The back of a vessel.

Sterncastle: A raised deck in the stern or the structure constructed to enclose it.

Sternpost: One or more upright timbers attached by means of a scarf to the keel/keel plank at the stern and to which the hull planking is joined.

Strake: An uninterrupted run of planking stretching from bow to stern.

Stylis: An identification device consisting of a pole with a crosspiece carried in the sterns of Classical ships. The *stylis* bore the name or attributes of the ship's patron deity.

Thole (tholepin): One or two pins set upright in the side of a hull to serve as an oar's fulcrum.

Throughbeam: A beam that extends through the planking and protrudes outside the hull on either side.

Thwart: A beam placed from side to side, across a hull, which may be used as a rower's seat.

Tiller: Lever attached to the loom of a rudder to facilitate ease of steering.

Trireme (Gk. *Trieres*): A three-banked war galley in which a single rower pulls each oar.

Wale: A particularly thick strake that contributes to the longitudinal stiffening of a hull.

Waterline: An imaginary horizontal line on the hull indicating the level of the water in which the vessel floats.

Yard: A spar used to spread a square sail on a mast.

Notes

Preface

1. Wachsmann 1998.

2. Monroe 1990: pl. 74. See p. 5, Fig. 1.6: A in this volume.

3. Wachsmann 2008.

4. See esp. p. 56, Fig 2.25, p. 63, Fig. 2.34 in this volumn.

5. Legrain 1917.

6. Item no. 1. See p. 20, Fig. 1.20 in this volume.

7. See chapter 3, Wheels, Wagons, and the Transport of Ships Overland, pp. 85–162.

8. Hammacher 1985: 25, 27–28, Figs. 33–34, 29–30; Sylvester 1992: 168–170.

Acknowledgments

1. See pp. 43–48, Figs. 2.10–18 in this volume.

2. Coleman and Morrow (forthcoming).

Chapter 1: The Gurob Ship-Cart Model

1. Brunton and Engelbach 1927. On the site of Gurob, see pp. 167–180 in this volume. For maps and satellite photos of the site, see the virtual-reality package that accompanies this book online at http://www.vizin.org/Gurob/Gurob.html. This book follows K. A. Kitchen's chronology of ancient Egypt. For Dynasties XI–XX, see Kitchen 1987: 49–52. For Dynasties XXI–XXVI, see Kitchen 1973: 465–466, 588–589.

2. Petrie 1931: 91; Thomas 1981A: 3.

3. Brunton and Engelbach 1927: pl. LII. For a digital avatar of the model in its present state, see the Gurob Ship-Cart Model Digital Supplement/Gurob ship model as excavated. For color photos of the model and the loose parts, see the Gurob Ship-Cart Model Digital Supplement/Image catalogue accessed at http://www.vizin.org/Gurob/Gurob.html .

4. Stephen Quirke, curator of the Petrie Museum, notes (pers. comm.): "I doubt that more ever existed than the meager account in the publication and on the tomb-card. If the tomb-card was not present here [at the Petrie Museum], we might hope for some future discovery of some Brunton archives, but, since the Petrie tomb-card does survive here, I would be surprised if there was anything else in the way of documentation from the time of the excavation itself. Informal records might be more rewarding but are of course much harder to harvest, as they extend to all autobiographical material relating to that year and all correspondence. Perhaps the fragments were removed from the tomb by the digger without record and presented as its contents to the dig supervisor, Petrie, without information on precise location(s) within the tomb-shaft or chamber. Perhaps Petrie was so fascinated by the object itself that his focus of attention was moved instantly from the recording of the circumstances of the find to the task of reconstruction."

5. Brunton and Engelbach 1927: 17.

6. Brunton and Engelbach 1927: 17.

7. Petrie 1933B: 74.

8. Thomas 1981A: 21, 86: 747; Thomas 1981B: pl. 56: 747.

9. Thomas 1981A: 86: 747.

10. This conservation effort, as described on the unsigned artifact card, included removing the gapfill by first softening the bond using

swabs soaked in a mixture of distilled water with industrial methylated spirits, a minute amount of a nonionic detergent and sodium hexmataphosphate. The model's fragile parts were then strengthened by an adhesive consisting of a 10 percent solution of Paraloid B72 (crylic copolymer of ethyl methacrylate and methyl acrylate).

11. Littauer and Crouwel 1979A: 113 n. 25.

12. M. R. Bell 1985: 62; Monroe 1990: fig. 74; Creasman and Doyle 2010: 22 fig. 19.

13. Merriman 2011: 82 (quote), 226 no. 222.

14. The destruction, or ritual killing, of items found in burials is a common cross-cultural phenomenon that transcends space and time. There are numerous and varied explanations for it (Grinsell 1961: 476–478). Although the breaking of objects has a long history in pharaonic Egypt, this trend seems to have been abandoned in the New Kingdom (480–482). Fragments of drinking cups found in Mycenaean burials, particularly in the Argolide, suggest intentional breakage after a libation or the drinking of a toast as part of the burial ritual (482 and additional bibliography there).

15. See below, Appendix 1: Lines Drawing of the Gurob Ship Model, pp. 207–208, in this volume.

16. See Appendix 7: Wood Identification, p. 249, in this volume. Also, Jones 1990: 7.

17. Regarding stem/bow configurations on contemporaneous Aegean galleys, see pp. 73–74 in this volume.

18. Item no. 6. See p. 11, Fig. 1.11 (D), p. 30 in this volume.

19. Early Egyptian wooden models are normally treated in this manner, but from the First Intermediate Period on, such models were normally made of a single block of wood with structural details defined in paint, although they have slightly raised sheer strakes. For examples of wooden (and ivory) models with hollowed hulls, see Reisner 1913: 20 fig. 88, 161, pl. VI (nos. 4814–4815), 53 fig. 190, 54 figs. 192–193, pl. XIII (no. 4882), 55 fig. 195 (no. 4883), 56 figs. 199–200, pl. XIV (no. 4886), 57 fig. 204, 58 fig. 205, pl. XIV (no. 4887), 59 fig. 209, pl. XV (no. 4888); Petrie 1933A; 1933B: 65 fig. 57, 71 figs. 73–74; Landström 1970: 14 fig. 20, 23 fig. 66, 24 figs. 70–72, 40 fig. 106, 50 figs. 140–142, 54 figs. 147–150. For examples of "block" models, see Reisner 1913: 1 figs. 1–3, 3 fig. 14, pl. I (no. 4798), 4 figs. 15–17, pl. I (no. 4799), 5 fig. 20, pl. 1 (no. 4800), 6 figs. 21, 23, pl. II (no. 4801), 7 fig. 26, pl. II (no. 4802), 8 figs. 32–35, pl. III (no. 4803), 9 figs. 37–39, pl. III (no. 4804), 11 figs. 45, 47, pl. IV (no. 4805), 12 fig. 52, pl. IV (no. 4807), 14 fig. 60, pl. IV (no. 4808), 17 fig. 76, pl. V (no. 4811), 30 fig. 120, pl. VII (no. 44841), 32 fig. 125, pl. VIII (no. 4844), 33 figs. 126, 130–131, pl. VIII (no. 4845), pl. IX (no. 4846), 35 figs. 132–133, 36 fig. 134, pl. IX (no. 4847), 38 fig. 138, 39 figs. 140, 143, pl. X (no. 4851), 12 figs. 149, 151 pl. X (no. 4859), 43 fig. 152, 44 fig. 153 (no. 4860), 45 fig. 158, pl. XI (no. 4861), 48 fig. 170, pl. XI (no. 4869), 49 fig. 173, 50 fig. 174, pl. XII (no. 4872), 51 fig. 179, 52 fig. 185, pl. XII (no. 4880), 63 fig. 226, 64 fig. 231, pl. XV (no. 4909), 65 fig. 234, pl. XVI (no. 4910), 67 fig. 243, pl. XVI (no. 4911), 68 fig. 251 (no. 4912), 69 fig. 254, pl. XVII (no. 4913), 70 fig. 259, 71 fig. 260, pl. XVII (no. 4915), pl. XVII (no. 4916), 74 fig. 275, pl. XVIII (no. 4917), 75 fig. 280, 78 fig. 290, pl. XVIII (no. 4918), 88 fig. 327, 90 fig. 328, pl. XIX (no. 4929), 93 fig. 340 (no. 4931), 96 figs. 348–349 (no. 4944), 98 fig. 354 (no. 4946), 99 figs. 355, 357, pl. XXI (no. 4947), 100 figs. 359–360, pl. XXII (no. 4948), 102 figs. 362–363, pl. XXII (no. 4949), 103 figs. 365–366, pl. XXIII (no. 4951), 105 fig. 369, pl. XXII (no. 4952), 106 fig. 372, pl. XXIV (no. 4953), 108 fig. 377, pl. XXIV (no. 4955); Petrie 1933A; 1933B; Winlock 1955: pls. 70, 72–76, 78–82, 84: R, 85: U; Landström 1970: 65 figs. 194–195, 71 figs. 205, 207, 74 figs. 216, 219–221, 75 figs. 222–224, 78 fig. 234, 79 fig. 238, 82 figs. 247–248, 83 fig. 251, 88 fig. 270, 89 fig. 271, 91 fig. 276, 92 fig. 287, 99 fig. 313, 102 figs. 323–324, 108 fig. 337, 117 figs. 359–360, 118 figs. 363, 366; Jones 1990: pls. X–XII, XIII (top), XIV–XXIII, XXV–XXXV; 1995: 16 fig. 5, 25 fig. 16, 30 figs. 22–23, 31 figs. 24–25, 33 fig. 29, 44 fig. 39, pl. VII.

20. Gesso, a mixture of whiting and glue, acted as a base for paint on Tutankhamun's ship models and served to hide the sometimes poor quality of the wood (Jones 1990: 7). It served similar purposes on the Gurob model. See pp. 243–247 in this volume

21. See pp. 210, 213–216, Figs. App. 2.4–10 in this volume.

22. See p. 32, item 26: B, in this volume.

23. Egyptian wooden models that carry masts vary in the detail in which the base is represented. In many cases the mast is simply inserted into a hole amidships (Reisner 1913: 1 figs. 1–2, 3 fig. 14 (no. 4798), 7 fig. 26 (no. 4802), 8 fig. 35, 9 (no. 4803), 9 figs. 37, 39 (no. 4804), 11 figs. 46–47 (no. 4805), 19 fig. 86 (no. 4813), 30 fig. 120 (no. 4841), 33 figs. 126, 130, 34 (no. 4845), 45 fig. 158 (no. 4861), 48 fig. 170 (no. 4869), 50 fig. 174 (no. 4872), 63 fig. 226, 64 fig. 231 (no. 4909), 68 fig. 251, 69 (no. 4912), 69 fig. 254, 70 (no. 4913), 96 figs. 348–349 (no. 4944), 99 fig. 357 (no. 4947), 100 fig. 360 (no. 4948), 103 fig. 366, 104 (no. 4951), 105 fig. 369 (no. 4952). In others, such as the models of Meket-Re (XIth Dynasty), the artisan(s) included detailed replications of the mast support structure (Winlock 1955: pls. 33, 34, 36, 44–45, 51, 55, 70, 72–76, 78–82, 85: T, V). See also Reisner 1913: 4 fig. 15, 5 fig. 18 (no. 4799), 14 fig. 60, 15 figs. 70–72, pl. IV (no. 4808), 32 (no. 4844), 65 fig. 234, 66 fig. 239 (no. 4910), 70 fig. 258 (no. 4914), 75 figs. 278, 280, 78 figs. 290, 292 (no. 4918). The maststep or partner may also be indicated by color (Reisner 1913: 5 fig. 20, [no. 4800]).

24. See p. 31, items 13–16, in this volume.

25. See pp. 102, 104 Fig. 3.19, Fig. 3.20, in this volume.

26. See p. 7 n. 11 in this volume.

27. See Appendix 2: The Gurob Ship-Cart Model in Virtual Reality, pp. 209–218 in this volume.

28. For a reconstructed virtual model as it might have originally appeared, which also allows the viewer (on the Windows version) to manipulate the possible distance between the wheels, the angle of the quarter rudder, and other details, see the virtual-reality package that accompanies this book online at htp://www.vizin.org/Gurob/Gurob.html.

29. See pp. 1, 32 item nos. 18–21 in this volume.

30. Compare, for example, the castle screens on the following ship depictions: p. 40 Fig. 2.7, p. 63 Fig. 2.34, p. 71 Fig. 2.42 in this volume.

31. See pp. 86–97 in this volume.

32. Reisner 1913: IV, [Thutmose III] 132–133 (nos. 5127–5131), 137 (nos. 5156–5157), 145–146 (nos. 5193–5200); [Amenhotep II] 96–99 (nos. 4944–4946), 113–143 (nos. 5034–5126, 5132–5155, 5158–5192); [Tutankhamun] Jones 1990. See also Landström 1970: 98–110. The exceptions are a model from the Sépulture des prêtres d'Amun, which may date to the Middle or New Kingdoms (Reisner 1913: 88, 90–91, pl. XIX [no. 4929]; Landström 1970: 118 fig. 363 and n. 3; Tooley 1995: 56 fig. 58) and a XIXth-Dynasty model of the cult vessel of the Egyptian deity Anuket from Elephantine exhibited in the Louvre (E 12710 and AF 9756).

33. Reisner 1913: XXVII–XXVIII.

34. Reisner 1913: III, XVIII–XXI. Type IV models represent papyriform craft dating from the Old to the Middle Kingdoms.

35. Reisner 1913: [Thutmose III] 145 (no. 5193); [Amenhotep II] 96 figs. 348–349 (no. 4944), 98 fig. 354 (no. 4946), 138–139 (no. 5164), 139 (no. 5165), 139–140 (no. 5167), 140 (nos. 5168–5170); [Tutankhamun] Jones 1990: pls. XVI–XXIII. See also Landström 1970: 107 figs. 321–334, 108 figs. 337–338, 109 fig. 339, 110 fig. 340.

36. Reisner 1913: 139 (no. 5164), 139 (no. 5165), 140 (nos. 5168–5170), 145 (no. 5193).

37. See Appendix 7: Wood Identification, p. 249 in this volume.

38. See chapter 2, The Iconographic Evidence, pp. 33–84 in this volume.

39. Basch 1987: 140–201; Casson 1995: 59–60, 62, 64–68, 70–74, 77; Wachsmann 1998: 130–153, 166–176.

40. See pp. 33–35 in this volume.

41. Stieglitz 1972–1975; Göttlicher 1978: 31 Taf. 7 (no. 107); Wachsmann 1998: 152, fig. 7.51.

42. Jones 1990: 7; El Goresy 2000: 51.

43. On the materials used to create the colors on the model and its wheels, see Appendix 6, Analysis of Pigments from the Gurob Ship-Cart Model, pp. 243–247 in this volume.

44. For a discussion on Homer's use of colors in describing ships, see Appendix 3, Ship Colors in the Homeric Poems, pp. 219–224.

45. Casson 1995: 43.

46. Lucas and Harris 1962: 338–351; Baines 1985: 286–289; El Goresy 2000 and additional bibliography there.

47. See p. 25 n. 32 in this volume.

48. Reisner 1913: [Thutmose III] 132 (nos. 5128–5130), 133 (no. 5131), 134 (nos. 5136, 5138–5140), 145 (nos. 5196–5197), 146 (no. 5200); [Amenhotep II] 97 (no. 4944), 98 (no. 4946), 114 (nos. 5034–5039), 115 (nos. 5040–5044), 116 (nos. 5045, 5047–5049), 117 (nos. 5051–5054), 118–119 (no. 5059–5063), 120 (nos. 5064–5067), 121 (nos. 5068–5071), 122 (nos. 5073–5077), 123 (nos. 5078–5082), 124 (nos. 5083–5087), 125–126 (nos. 5088–5092), 126 (nos. 5093–5096), 127 (nos. 5097–5101), 128 (nos. 5102–5106), 129 (nos. 5107–5108, 5112), 130 (nos. 5113–5115, 5117), 131 (nos. 5119–5124), 132 (nos. 5125, 5126), 133 (no. 5135), 134 (nos. 5136–5140), 135 (nos. 5141, 5145), 136 (nos. 5147–5151), 137 (nos. 5152, 5154–5155), 138–139 (nos. 5160, 5162–5164), 139 (nos. 5165–5167), 140 (nos. 5168–5170), 141 (nos. 5172–5175), 142 (nos. 5177–5180), 143 (nos. 5181–5185), 144 (nos. 5186, 5188, 5191); [Priests of Amun] 90–91 (no. 4929); [Tutankhamun] Jones 1990: 7.

49. Seymour 1914: 307; Casson 1995: 45 n. 15; Mark 2005: 99.

50. Seymour 1914: 307; Casson 1995: 45 n. 18; Mark 2005: 115.

51. For color images of some of these, see Spathari 1995: 81 fig. 90, 84–85 fig. 94, 88 fig. 98, 93 fig. 106, 94–95 fig. 107, 96 fig. 108, 100 figs. 115–116.

52. Blackman and Lentini 2006: 193, 196, 197 fig. 7; 2009: 74 fig. 58; 2010: 41, 42 n. 4 and fig. 11.

53. See pp. 202, 220–221 in this volume.

54. Spathari 1995: 93 fig. 106. Note, however, that black-figure painters worked with a limited palette of colors, so the red strip on the upper portion of the forecastle screen in this representation may have been intended to indicate only that it differed in hue from the screen's bottom.

55. Note that blue color was found in a mid-fifth century B.C. context in a Naxos shipshed (Lentini and Blackman 2009: 74 fig. 59, 75; 2010: 42 fig. 12). Mark (2005: 101–102) prefers to connect *blue-* or *dark-prowed* with blue-colored *oculi,* but convincing evidence for this is lacking. At present the only eyes on known Helladic ship representations are those of their bird/bird-head stem or stern ornaments with the single notable exception of a Late Helladic terracotta ship model from Philakopi (Marinatos 1933: pl. 15: 26; Basch 1987: 141, 142 fig. 296; Wachsmann 1998: 150 fig. 7.42). Furthermore, none of the classical-period marble eyes known to date have traces of blue paint, although this might be due to their states of preservation: T. Nowak writes (pers. comm.): "There is no conclusive archaeological evidence for the use of the color blue on any of the marble eyes. The possible presence of blue

on one example was suggested with hesitation by Blümel. A marble eye from Zea was donated to the antiquities collection of the National Museum in Berlin by Lolling in 1881. It has a provenance similar to the other eyes from Zea. Blümel (1964: 19) describes the coloration of its iris in a catalog entry: 'Remnants of color survive. The wide innermost ring could have been yellow, where it was not covered by the head of the nail. This was followed by a narrow red ring and then by a wide ring that was either white or blue, these rings were further followed by a wider red ring and finally by a narrow yellow ring.' Blümel's description is the only possible evidence for the presence of the color blue. None of the other eyes from Zea retain traces of blue coloration. Lolling's (1880: 384–387) original note tells us that in general the middle rings of the irises represented on the eyes he studied retained no traces of pigment in the middle ring, ie. the position of the possible blue coloration noted by Blümel—but we have no way of knowing which specific examples were in his possession when he wrote his article." See also Nowak 2006.

56. See pp. 221–222, 224 in this volume.

57. Doumas 1992: 58 fig. 26, 62–63 fig. 29, 68–70 fig. 35, 71–74 fig. 36, 75–77 fig. 37, 78–79 fig. 38, 80 fig. 39, 81 fig. 40, 82 fig. 42, 83 fig. 43. On Minoan/Cycladic ships see Basch 1987: 93–132; Wachsmann 1998: 83–122, and additional bibliography there.

58. Doumas 1992: 68–70 fig. 35, 71–73 fig. 36, 77 fig. 37, 78 fig. 38, 80 fig. 39, 81 fig. 40, 86 figs. 49–50, 87 fig. 51, 88 fig. 52, 89 figs. 53–54, 90 fig. 55, 91 fig. 56, 92 fig. 57, 93 fig. 58, 94 figs. 59–60, 95 figs. 61–62, 95 figs. 61–62.

59. Doumas 1992: 85 fig. 48.

60. Fitzgerald 1994: 170–178; Steffy 1994: 41, 43, 49, 51–52, 54, 59, 258–259; Asouti 2003: 473–475, 480; Hillman and Liphschitz 2004.

61. Doumas 1992: 71–73 fig. 36. Excavations at Building P at Kommos revealed red hematite, presumably for the painting of hulls (Shaw and Shaw 2006: 52–53, 78, 82).

62. Doumas 1992: 71–72 fig. 36.

63. Doumas 1992: 70 fig. 35, 78–79 fig. 38.

64. Red splotches discerned on the Kinneret Boat (from Israel, first centuries B.C.–A.D.) during its excavation proved to be a coating of pine resin, presumably applied to prevent rot (Wachsmann, Raveh, and Amos 1990: 12–13, color pl. B; White 1990; Wachsmann 2009: 255).

65. Doumas 1992: 71–72 fig. 36.

66. Doumas 1992: 76 fig. 37.

67. Doumas 1992: 58 fig. 26, 62–63 fig. 29.

68. See Appendix 5: Radiocarbon Age Analysis of the Gurob Ship-Cart Model pp. 239–242 in this volume.

69. See chapter 2, The Iconographic Evidence, pp.33–84.

70. See chapter 4, Foreigners at Gurob, pp. 163–199.

71. See pp. 102–105 Figs. 3.18–3.21, 107 Fig. 3.24, 202 in this volume.

72. In the dimensions of the pavois, "Th." refers to "thickness."

Chapter 2: The Iconographic Evidence

1. For a discussion of the attributes of Helladic galleys see pp. 65–84 in this volume. For summaries of New Kingdom and Late Period Egyptian vessels, see Reisner 1913: 96–99 nos. 4944–4946; Landström 1970: 98–141; Jones 1990; 1995: 49–61, color pl. VIII; Casson 1995: 21–22, 36–37; Vinson 1994: 37–49; Partridge 1996: 60–72; Doyle 1998; Wachsmann 1998: 17–32, 37–38, 52–60.

2. Regarding the transfer of technology of galley construction during Greece's "Dark Age," see pp. 80–83 in this volume.

3. Wachsmann 1981; 1982; 1997; 1998: 163–204; 2000; Basch 1987: 66–69.

4. See pp. 33–40 in this volume.

5. See p. 183 n. 133 in this volume.

6. *MH* I: pls. 32, 34 (land battle), 37, 39 (naval battle); Murnane 1980: 13–18.

7. Gardiner 1920: 110; O'Conner 2000: 100; Redford 2000: 13; Rainey and Notley 2006: 109. For an overview of this region, see Oren 1984, and additional bibliography there.

8. *BAR* IV: §77; Nelson 1943: 46; Casson 1995: 38.

9. Nelson 1929: 22–23.

10. Nelson 1929: 22. On the Tjemhu (Libyans), see pp. 57, 59, 178, 181–182 in this volume.

11. On the Egyptian ships taking part in the battle, see Landström 1970: 111–113; Casson 1995: 36–38; Wachsmann 1998: 29–32.

12. *BAR* IV: §65; Edgerton and Wilson 1936: 54; Nelson 1943: 44.

13. Basch 1987: 140–151; Wachsmann 1998: 131–153.

14. See Holland 1929: 198–203.

15. Edgerton and Wilson 1936: 45 n. 19a; Wainwright 1961: 74–75; O'Conner 2000: 85.

16. *MH* VIII: 5, pl. 600; Murnane 1980: 6, 7 fig. 4. Unfortunately, the chief of the Peleset/Philistines in the same tableau now lies buried behind a Graeco-Roman wall. On the question of the Sekel material culture at Dor, see Gilboa 2005.

17. Nelson 1943.

18. Casson 1995: 38; Wachsmann 1998: 317–319. On the use of the grapnel in classical naval warfare, see Diod. Sic., *Library,* 13.16.1, 13.50.5, 13.67.2.

19. Wachsmann 1998: 317–319.

20. Nelson 1943: 48–50, 53–54.

21. For a convenient listing of the ethnic groups that took part in the various attacks, see Redford 1992: 246 table 1.

22. Rainey and Notley 2006: 109.

23. *BAR* IV: §64; Edgerton and Wilson 1936: 53. Sherden fighting *for* Egypt also appear prominently in the Year 8 battle scenes and the surrounding tableaus.

24. *BAR* IV: §403.

25. On the vocalization of the Šikala, see Rainey 1982: 134; Wachsmann 1998: 164 n. 10.

26. *BAR* IV: §402; Lesko 1980: 85. On the Kehek, see pp. 181, 185–187, 205 in this volume.

27. On the Sherden, see pp. 35, 52–53, 57, 182–190, 206, 225–237 in this volume.

28. Redford 2000: 11–12.

29. *MH* II: pl. 127; Nims 1976.

30. Lesko 1980: 86; 1992: 153–155.

31. Redford 2000: 12. See also *MH* I: pls. 29–44, 45: B, 46, 49, 50, 51: E–G, 52, 53: B–D; *MH* II: pls. 65: C, 98–100, 125, 127; Edgerton and Wilson 1936: 35–58; Redford 1992: 250–256; Yurco 1997; O'Conner 2000; Rainey and Notley 2006: 108–109.

32. Lesko's proposed reconstruction, along with questionable iconographic comparisons, has led Michal Artzy (2001: 35) to question the accuracy of the Sea Peoples' ship representations in the naval battle scene. Similarly, A. Raban (1989: 167, 171) states that the Sea Peoples' ship representations at Medinet Habu signify a "new" vessel type, which he constructs from a patchwork of the Helladic galley type, the pictograph of a ship on the Phaistos Disk, and some Late Bronze Age Canaanite ship components. The evidence is overwhelmingly at odds with these conclusions.

33. Nelson 1943: 53; Casson 1995: 38; Wachsmann 1998: 317–319.

34. Naville 1898: pls. LXIX–LXX, LXXII–LXXV; N. d. G. Davies 1930: 34–35; Wreszinski 1988 (II): pls. 26–33; Bodenheimer 1949: 257, 258 fig. 32, 259; 1960: 167–168; 1972: fig. 32; Danelius and Steinitz 1967; Beaux 1990; Bianchi 1997; Wachsmann 1987: 5–6, pl. II; 1998: 17 fig. 2.11, 18 figs. 2.12–13, 19 fig. 2.14, 22, 169.

35. Note that ship models held by figures in the Dakhla Oasis ship graffito have sternposts capped with inward-facing bird heads. See pp. 41, 43 Fig. 2.10, 44 Fig. 2.12 in this volume. On nautical ship decorations, see Svoronos 1914; on their spiritual significance, see Hornell 1943: 271–289; Wachsmann 1998: 191, 193–196.

36. Wachsmann 1998: 177 fig. 8.23. On bird and bird-head devices on Helladic and later Mediterranean ships, see Wachsmann 1996 and pp. 78–83 in this volume.

37. Wachsmann 2000: 122–123.

38. Hencken 1968B: 514–531, 537, 568–570, 625–628; Bouzek 1985: 177 fig. 7, 178; Wachsmann 1997: 349–354; 1998: 178–183; 2000: 121–133.

39. Hencken 1968B: 515; Bouzek 1985: 178. Bouzek (1994: 217) notes that the dates for the Br D may be earlier than previously thought.

40. Kossack 1954: 122 Taf. 11: 5; Hencken 1968B: 515, 516 fig. 478: f; Wachsmann 1998: 178, 179 fig. 8.28: A.

41. Bouzek 1985: 177 fig. 88: 6, 178; Wachsmann 1998: 178, 181 fig. 8.32, 183.

42. See pp. 188–190 in this volume.

43. Winkler 1939: 1–2, 78–79, and foldout map: site 69. The negative resides in the archives of the Egypt Exploration Society, London, registered as Winkler-Mond-Site 69, negative M916.

44. Červiček 1986: 68 fig. 504 (negative M916).

45. Excerpted from the Winkler diaries, translated and transposed by Ursula Murphy, for Coleman and Morrow, forthcoming.

46. Basch 1994; 1997; Wachsmann 1998: 203 fig. 8A.3, 204.

47. For a discussion on the possible identity of this figurehead, see p. 78 in this volume.

48. On the use of models in tandem with cultic ships moved over land, see pp. 116 Fig. 3.35, 117 Fig. 3.36, 152 in this volume.

49. Červiček 1986: 67–68, negatives M898, M907.

50. Basch (1994: 22; 1997: 21) suggests that the diagonal line might be a shroud (Fr. *hauban*) or a backstay (Fr. *étai arrière*). In the past, I also mistakenly identified it as rigging—a halyard (Wachsmann 1998: 203).

51. Khnumhotep: Petrie 1952: pl. 17; Landström 1970: 43 fig. 117; Kenamun: Davies and Faulkner 1947: pl. 8; Basch 1987: 64 fig. 113; Casson 1995: fig. 57; Wachsmann 1998: 43 figs. 3.3–4. A relief now in Berlin includes a crew member shown lying prone on the yard while working on the sail (Vinson 1993: 136 fig. 4a, 137 fig. 4b).

52. Landström 1970: 42 fig. 113, 47 fig. 125, 50 figs. 137, 139, 51 fig. 144, 54 figs. 147, 151, 60 figs. 178–180, 62 fig. 185, 63 fig. 187, 64–65 fig. 191, 65 fig. 193, 82 fig. 249, 82–83 fig. 246, 83 fig. 252, 114 fig. 353, 135 fig. 396; Wachsmann 1998: 13 figs. 2.2–3, 16 fig. 2.9, 35 figs. 2.48, 2.51, 36 fig. 2.55, 37 fig. 2.56. Rigging on ship models should be approached with caution as unattached pieces could be reconstructed in the past at the whim of the archaeologist. See p. 91 n. 27 in this volume.

53. Gaballa and Kitchen 1969: 6, tab. I–II. Nefertem was the god of the primeval lotus blossom.

54. Gaballa and Kitchen 1969.

55. Otago Museum E 44.446. Brunton 1934: 150 fig. 3, 153-155 (item no. 3). This ship is accepted as authentic by Fabre (2004: 89). On sail in the Predynastic period, see Le Baron Bowen 1960A; 1960B; Basch 1987: 48–50; Wachsmann 1998: 248 fig. 11.2.

56. UC 15343. Petrie 1921: pl. XXXIV: 46: K; 1974B: 20 no. 42, pl. XXI: 46K; Brunton 1934: 149, 150 fig. 2, 153-155 (item no. 2); Crowfoot, Kaczmarcyk and Fleming 1977: 5 n. 2, 8: D, 9-10, 12. Crowfoot and his colleagues conclude (p. 9) also that a single hand perpetrated a group of the forgeries, including UC 15343, and that, "It is likely that other examples of work by this forger exist both in museums and in private collections."

57. Boardman 1958: 7 n. 9 (Florence 3897); Webster 1953–1954: pl. 2; 1970: 30, pl. 2; Deubner 1956: pl. 22; Green 2007: 110 fig. 40, 111 fig. 41.

58. Boardman 1958: 7.

59. Boardman 1958: 7 n. 10. See Basch 1987: 242–243, 245 figs. 513–516, 246 figs. 518–518 *bis*. On the *ex-voto* ship models from Samos, see Ohly 1953: 111–118, Beil. 34–35; Kopcke 1967: 145–148, Beil. 82–83; Kyrieleis 1980: 89–94, Taf. 18–20; Basch 1987: 243, 244 figs. 513–517, 245 figs. 518–518 *bis*, 246; Johnston 1985: 46–47, 50–51, 54–64 [Arch. 3–24].

60. Webster 1970: 30; Green 2007: 110–111.

61. Csapo 1997.

62. Ath. V: 196a–203b; Rice and Kallixeinos 1983: 21, 110–111, 180.

63. Ath. V: 201e, quoted from C. B. Gulick, trans., *Athenaeus* 1967: 413. See also Rice and Kallixeinos 1983: 21, 110–111, 181. In modern Greece during the "Cooking of the Bourani" in Tyrnavo, Thessaly, phalli are still paraded during the celebrations, which occur on the first day of Lent (Tomkinson 2003: 50, 53). Farther afield, the cult of the phallus continues in the Japanese Hounen Matsuri fertility festival, which takes place each year on March 15 at the Shinto shrine of Tagata Hounen, at Komaki in Aichi prefecture (Anonymous, n.d., Hounen Matsouri, Tagata Jinja; Plutschow and O'Neill 1996: 176–177, 262–263 no. 8, pls. 9–10; Ellmo and Lidén 2005: 13–14). During this festival a 2.5-meter-long cypress-wood phallus is carried on a mobile altar between temples.

64. I follow here O'Conner (1990: 29–31) in preferring "Tjemhu" as a general and inclusive term for the various ancient Libyan ethnic groups encountered by Egyptians during the New Kingdom and Third Intermediate Period.

65. Parkinson and Schofield 1993A; 1993B; 1995; Schofield and Parkinson 1994.

66. Hencken 1971: 8, 17–18, 19 fig. 3, 20, 21 fig. 4, 22 fig. 5: a.

67. Dakoronia 2002: 290 fig. 8; 2006: 27 fig. 5a.

68. Marinatos 1984: 35, 36 fig. 18, 37; Wachsmann 1998: 115.

69. Artzy 2001: 39.

70. Judges 14:3, 15:18; 1 Samuel 14:6, 17:26, 36, 31:4; 2 Samuel 1:20; 1 Chronicles 10:4; Gollaher 2000: 10; Maeir 2007.

71. *BAR* III: §588, 601; Rainey and Notley 2006: 108. The removal of phalli from dead uncircumcised enemy, in this case Philistines, has a biblical parallel (I Samuel 18:24–27, II Samuel 3:14; Gollaher 2000: 12). On the date of Merneptah's Tjemhu War, see Kitchen 1990: 19.

72. Ekwesh: *BAR* III: §588 n. a, 601; Sherden: *BAR* IV: §129: 4; Teresh: *BAR* IV: §129: 6; Weshesh: *BAR* IV: §403. See Wainwright 1961: 75.

73. Redford 1983: 482–483; 1992: 248 n. 34.

74. W.-D. Niemeier 1998: 46 and additional bibliography there; Wiener 2009; Winnicki 2009: 80.

75. W.-D. Niemeier 1998: 46. See also Hölscher 1955: 46 n. 7; Basch 1994: 23 n. 16.

76. Gollaher 2000: 13–14.

77. Marinatos 1984: 36 fig. 18; Doumas 1992: 52 fig. 19, 63 fig. 29, 147 fig. 111. Note that two of the three figures are immature males. This presents another difficulty as both ancient and modern customs indicate that circumcision may be performed at an older age (Bryk 1934: 59, 61, 73–74, 76, 79, 84, 87–88; Pritchard 1969A: [206] and [327] fig. 629; Gollaher 2000: 1–6, 46–47).

78. Astour 1972: 456.

79. Basch 1994: 23–25; 1997: 27–28.

80. *BAR* III: §572–617 (Merneptah), IV: §405 (Ramses III and his predecessors). Edgerton and Wilson 1936: 24–30 (First Tjemhu War), 74, 76–94 (Second Tjemhu War); *BAR* IV: §405 (Papyrus Harris). For summaries of the historical evidence, see Wilson 1935; Redford 1992: 248–250; Rainey and Notley 2006: 108.

81. Habachi 1980; Kitchen 1982: 71–72; O'Conner 1990: 31, 87.

82. *BAR* III: §80, 83–86, pp. 39 fig. 1, 44 fig. 2; Gardiner 1920; Faulkner 1947: 35–36; *BRKS* 1986: 2–4, 13–22, pls. 2, 5–6, 7: C; Oren 1987; Rainey and Notley 2006: 92–93, 119, and additional bibliography there; Eisenberg 2007.

83. Petrie 1888: 26 no. 78 lines 14–15; *BAR* III: §488, 491; Wainwright 1962: 93; Habachi 1980: 27–29; Wachsmann 1981: 187–188; Kitchen 1982: 40–41; 1990: 17–19; Sandars 1985: 50 n. 14.

84. *MH* I: pls. 22–23; Edgerton and Wilson 1936: 13–15.

85. Edgerton and Wilson 1936: 14 n. 24a, 15 n. 6–30a.

86. On the Karkisha, see *BAR* IV: §405; Huxley 1960: 20–21; Astour 1972: 456; Redford 1992: 249 n. 41. John A. Wilson (1935: 77) apparently considers them a Tjemhu tribe. Note, in this regard, that the inscription of the Year 5 First Tjemhu War conflates events of the Year 8 attack by the Peleset, Sikilu, and possibly one other ethnic group (Edgerton and Wilson 1936: 19, 30 n. 51b, 31–32; Rainey and Notley 2006: 108).

87. *MH* I: pl. 19.

88. Hölscher (1955: 46) concludes that some of the Meshwesh practiced circumcision, while G. A. Wainwright (1962: 92–93) deduces

that they did not. Bates (1914: 140–141) notes that much later Herodotus (*Hdt.* II: 36–37, 104) does not list the Tjemhu among those cultures who practice circumcision.

89. *BAR* IV: §405; Wilson 1935: 81–82; Edgerton and Wilson 1936: 27, 79, 87, 94; Yoyotte 1951; Kitchen 1990: 19–22; O'Conner 1990: 86; Winnicki 2009: 74–76. Note also that, in regard to changing customs, as opposed to Merneptah's removal of the hands rather than the phalli of the Ekwesh due to their circumcision, by Ramses III's time, mores may have changed, and this was no longer a consideration. Alternately, see O'Conner's interpretation, quoted below.

90. O'Conner 1990: 53.

91. *MH* II: (sidelocks) pl. 129: A, C–F; (sidelocks and phallic sheaths) pls. 68, 72, 74–75, 77–78, 98–100, 102, 113–114, 120–122, 125; *MH* VIII: 600, 625–626, 628; Bates 1914: (phallic sheaths) 122, 123 figs. 17–19, 124 figs. 20–23, 125–126; (sidelock) 43 figs. 3–4, 134 fig. 43, 135 fig. 45, 136–137, pls. II: figs. 1–3, 8, VI: figs. 1–3, VII; Wilson 1935: 73; Pritchard 1969A: [250] n. 8; O'Conner 1990: (phallic sheaths) 47, 49 fig. 1, 50, 51 fig. 2, 52–54, 69, 73. On phallic sheaths in general, see Ucko 1969.

92. Bates 1914: 129–130, pl. V; O'Conner 1990: 55.

93. Bates 1914: pl. V: figs. 2–3.

94. Pritchard 1969A: [4] and [250] n. 8; Carter 2001: pls. XVIII–XIX.

95. For reviews of the evidence, see Basch 1987: 140–151; Casson 1995: 37–42; Spathari 1995: 45–55; Wachsmann 1998: 130–153, 166–176.

96. Ingholt 1940: 71, pl. XXII: 2; Riis 1948: 48 fig. 25, 97 fig. 130: B: 112, 105–106, pl. 12C (no. G VIII, 551 [5B902]); Hencken 1968B: 627; Wachsmann 1998: 174 fig. 8.19, 175–176; Romey 2003: 62–63.

97. This detail may appear in paint on a Late Helladic terra-cotta model from Tiryns (p. 67 Fig. 2.39:D).

98. Basch 1987: 144, 145 fig. 303, 146 figs. 304, 306; Wachsmann 1998: 136 fig. 7.19, 137–138.

99. Ingholt 1940: 69–84, pls. XXI–XXVI; Riis 1948: 46–52, 202.

100. Ingholt 1940: 82, pl. XXV: 3–4; Riis 1948: 120 fig. 136, 121–122, 131–134, 202–203; Kimmig 1964: 239, 243.

101. Albright 1951: 106.

102. Bieńkowski 1982: 82. See also Kimmig 1964: 245; Romey 2003: 71.

103. On the Medinet Habu evidence, see p. 40 in this volume.

104. Wreschner 1971: 218, pl. 47: C.

105. Artzy 1994A: 9, 10 figs. 2–3; 1994B; 2001: 39–40; 2003: 232, 234, 235 figs. 3–5, 236 fig. 6; Wachsmann 1998: 202 fig. 8A.2, 203.

106. Grant 1932: 21, 82–83, pl. XLVIII: no. 1107; Keel 1994: 33 n. 39, 34 fig. 20; Keel, Shuval, and Uehlinger 1990: 386 no. 30, 389 fig. 98; Wachsmann 2000: 135 fig. 6.30.

107. Dothan and Zuckerman 2004: 33 fig. 35: 10, 41; Mountjoy 2005: 425, pl. XCVIII; 2006B.

108. Marinatos 1933: 172 no. 13, 218, 219 fig. 10, pl. XIII: 13, 16; Basch 1987: 146, 147 fig. 307; Wachsmann 1998: 139 fig. 7.23, 140 fig. 7.24.

109. Wachsmann 1998: 131 fig. 7.7 (Late Minoan), 135 fig. 7.16: A, 139 fig. 7.21 (Late Helladic IIIC), 147 fig. 7.36, 151 fig. 7.48: A (Late Cypriot III), 138 fig. 7.20: A (Late Geometric). This detail may also appear on the Bademgediği galley representation, pp. 74–75, 77, Fig. 2.47 in this volume.

110. Wachsmann 1998: frontispiece, 201 Fig. 8A.1; 2000: 134 fig. 6.29.

111. D. Masters (pers. comm.); Mountjoy 2011.

112. Compare this to the curved items held by the two figures on the Beith Shemesh seal discussed earlier, p. 64 Fig. 2.35 in this volume.

113. See Appendix 1, Lines Drawings of the Gurob Ship Model pp. 207–208 in this volume.

114. Many, if not most, Egyptian models have exaggerated sheers. See, for example, Reisner 1913: pls. II: no. 4802, III: no. 4803, IV: no. 4807, V: no. 4811, VII: no. 4841, VIII: no. 4851, IX: no. 4847, X: no. 4851, XI: no. 4869, XII: no. 4872, XIV: 4886, XIX: 4929, XXII: 4948, XXIII nos. 4951–4952, XXIV: no. 4953; Landström 1970: 59, 100.

115. On Egyptian and Syro-Canaanite seagoing vessels as portrayed in Egyptian art, see Wachsmann 1998: 11–32, 42–47, and additional bibliography there.

116. See also Wachsmann 1998: 149 figs. 7.39–41, 150 fig. 7.42, 151 fig. 7.47.

117. Landström 1970: 141 fig. 411; Basch 1987: 335 figs. 718–720; Spathari 1995: 26–27 fig. 18.

118. See, for example, Petrie (1933B: 74), quoted earlier; Morrison and Williams 1968: 7, 9 BA.2, 11 BA.6, pl. I:b; Wachsmann 1998: 139, Fig. 7.22, 140.

119. Kirk 1949: 117 fig. 5, 118; Morrison and Williams 1968: 10, pl. I: c, BA3.

120. Bass 1972: 22 fig. 20; 1987: 146–147 fig. 309; Casson 1995: 32 fig. 29. This is a reversal for Basch, who prior to this had thought the ship to be sailing to the left (Basch 1975: 201–202, fig. 2).

121. Basch and Artzy 1985: 325–326, 332 fig. 8: B–C.

122. Wachsmann 1998: 147 fig. 7.36.

123. Casson 1995: 38, 42 n. 4; Wachsmann 1998: 157–158.

124. Steffy 1991.

125. Johnston 1985: 29–30 no. and fig. BA 21; Wachsmann 1998: 152 fig. 7.49.

126. Palaima 1991: 285–287; Wachsmann 1998: 128–130, 163–164, 166; Hoftijzer and van Soldt 1998: 343 RS 34.129.

127. *Od.* 13: 113–115; Doumas 1992: 58 pl. 26, 62–63 pl. 29; Wachsmann 1998: 89 fig. 6.7, 91 fig. 6.11.

128. Kirk 1949: 125–127.

129. Hornell 1970: 202.

130. J. R. Steffy (pers. comm.).

131. Casson 1995: 34–35, 74 n. 16, 331. Compare, for example, the Galilee Boat (first centuries B.C.–A.D.) with the Madregue de Giens Wreck (first century B.C.), both of which had cutwater bows (Pomey 1982: 142 fig. 6, 143 fig. 7, 144–145; Steffy 1990: 30, 31 fig. 5.3, 38, foldout 2; Rival 1991: 155 pl. 27, 222 pl. 70).

132. Kirk 1949: 125–126; van Doorninck 1982: 283–285; Casson 1995: 49; Basch 1987: 150–151; Wachsmann 1998: 157–158; Mark 2005: 104–114, 185–186; 2008.

133. Casson 1995: 58.

134. On the continuity of the Helladic galley design through the Greek Dark Age, see pp. 80, 82–83 in this volume. Regarding Phoenician galleys, see Basch 1969A; 1969B; 1987: 303–335; Casson 1995: 55–60, 94–96; 1970. These vessels evolved through the Phoenician bireme and the Greek *dieres* to eventually culminate in divergent types of triremes.

135. The eastern coast of the Mediterranean—the ancient geographical equivalent of modern Syria, Lebanon, and Israel—appears to have been a center of maritime innovation throughout antiquity, including the following developments: (a) the invention of the brailed rig, (b) the invention of pegged mortise-and-tenon joinery used in conjunction with shell-based construction, first seen on the Uluburun shipwreck, which became the normal form of ship construction soon after the Classical and during the Roman periods in the Mediterranean, (c) the development of the bireme, and (d) a "root" in the transformation from shell-based to frame-based Mediterranean construction (Casson 1995: 55–58; Wachsmann 1995; 1998: 239–241, 251–253; Wachsmann and Kahanov 1997: 3–8; Kahanov, Royal, and Hall 2004; Barkai and Kahanov 2007; Pomey, Kahanov, and Reith 2012).

136. DeVries and Katzev 1972: 43, 56 fig. 9. Note that the metal sheeting is attached with nails, the heads of which are visible. Biremes on orthostats from Sennacherib's southwest palace at Ninveh (Kuyunjik) show similar metallic coverings and fittings (Basch 1987: 311 fig. 655, 312 fig. 656, 315 fig. 662, 316 figs. 664–665; Barnett, Bleibtreu, and Turner 1998: 38, 134–135).

137. Hdt. 1: 166.

138. See pp. 7–8 in this volume. Homer termed the castles in the bow and stern *ikria* (Seymour 1914: 309; Casson 1995: 44 n. 4, 179 n. 58, 218; Mark 2005: 115–116).

139. *MH* I: pl. 39; Wachsmann 1998: 29 fig. 2.35, 30 fig. 2.36, 31 figs. 2.37, 40, 32 figs. 2.41–42.

140. Basch 1987: 166 figs. 334–336, 174 figs. 360–361, 184 figs. 387–388, 192 fig. 404, 200 fig. 420.

141. Naville 1898: pls. LXXII–LXXV; Wachsmann 1998: 17 fig. 2.11, 20 figs. 2.15–16, 21 figs. 2.17–18. For forecastles on Egyptian XVIIIth-Dynasty wooden models, see p. 25 n. 35 in this volume.

142. See p. 25 in this volume.

143. See p. 8 in this volume.

144. See also Wachsmann 1998: 131 fig. 7.8A, 134 fig. 7.15, 135 figs. 7.16–17, 136 figs. 7.18–19, 139 fig. 7.21, 142 fig. 7.29, 143 fig. 7.30: A, 150 fig. 7.42, 151 fig. 7.48, 174 fig. 8.19, 177 fig. 8.23.

145. See also Wachsmann 1998: 131 fig. 7.7, 139 fig. 7.23.

146. Wachsmann 1998: 149 figs. 7.39–41, 150 fig. 7.42.

147. Wachsmann 1981.

148. Redford (1992: 255 n. 70) and O'Conner (2000: 85, 99 n. 3) misconstrue these Sea People galleys to be sailing craft solely on the basis of their *seeming* lack of oars. O'Conner takes the crew members intertwined in ship N.3's hull as an indication that one sees the ship in the process of breaking apart. In offering this interpretation he seems not to have given sufficient weight to the following considerations: (a) the Sea Peoples' ships being caught at anchor before their crews could run out their oars (see p. 33n8 in this volume), (b) the overwhelming evidence for an open rowers' gallery on the Helladic ship type, and (c) the fact that, for various reasons, artists often depict galleys without their oars (Figs. 2.33, 37: A, 38: A[?], 39: B–C, 40: A–B, 42: A, 44).

Scenes of battles on ships in Geometric art rarely show oarsmen pulling at their oars, perhaps because the rowers were also the fighters (Basch 1987: 177 fig. 373; Casson 1995: 51). For the most part, however, rowers and their oars are conspicuously absent in most scenes of combat on galleys (Basch 1987: 166 figs. 333–336, 167 figs. 337, 340, 171 figs. 349–352, 174 fig. 360, 177 fig. 372, 178 fig. 374, 179 fig. 376, 192 figs. 403–404, 193 figs. 407, 411; Casson 1995: figs. 30, 65–68). Oarsmen make an appearance in Geometric art normally when galleys are shown leaving, arriving, or voyaging on the open sea, sometimes carrying the bodies of deceased warriors (Basch 1987: 164 figs. 328–329, 172 figs. 353–356, 173 figs. 357–359, 174 fig. 363, 181 fig. 381, 182 fig. 382, 184 figs. 387–388).

149. Stanchions present, no crew: Basch 1987: 166 figs. 333–336, 167 figs. 337, 339, 171 figs. 349–352, 172 figs. 353, 182 fig. 382. Crew rowing from the open rowers' gallery, no stanchions: Basch 1987: 172 figs. 354–356, 173 figs. 357–359.

150. On rowers mentioned in the Linear B tablets from Pylos, see Ventris and Chadwick 1973: 183–188, 431–432; Killen 1983; Palaima 1991: 285–286; Palmer 1998: 69–71, 118, 124, 129–132, 161–162, 217, 222, 239; Wachsmann 1998: 123–127, 159–161; 1999.

151. Dakoronia 1990; 1993; 1995; 1996A; 1996B; 1999; 2001; 2002; 2006; Wachsmann 1998: 131–135, 137.

152. Dakoronia 2006: 24, 25 fig. 1, 26. For a detailed discussion, see Wachsmann 1998: 131 fig. 7.8, 132 figs. 7.9–10, 134.

153. Wachsmann 1998: 131–134.

154. Dakoronia 1990: 122 fig. 1 (figure reversed); 1996A: 171 fig. 9; 2006: 28 fig. 8; Wachsmann 1998: 135 fig. 7.16.

155. Meriç and Mountjoy 2002: 90, 91 fig. 5 no. 29, 92–93; Mountjoy 2005; 2006A: 110, 111 fig. 4, 112, and P. Mountjoy (pers. comm.).

156. On these protuberances, see p. 79 n. 182 in this volume.

157. Compare Mountjoy 2005: pl. XCVII: b with Boardman 1967: 73 fig. 6: 21 (= Wachsmann 1998: 138 fig. 7.20: A).

158. Morricone 1975: 360 figs. 357: a, 358; Sandars 1985: 135 fig. 92, 137; Wachsmann 1998: 140 fig. 7.26: A, 176 fig. 8.22: A; Mountjoy 2005: 424–426, pls. XCVI, XCVII: a, c.

159. Güntner 2000: 33 (Motiv Mensch 17), pl. 12.6; 2006: 179 fig. 5.

160. See also Basch 1987: (Geometric): 166 figs. 333–334, 172 fig. 353, 175 fig. 362, 177 fig. 372; Wachsmann 1998: (Late Helladic): 152 figs. 7.49, 51(?).

161. Westerberg 1983: 91 fig. 19; Wachsmann 1998: 149 fig. 7.40.

162. Güntner 2000: 34 (Motiv Mensch 18), pl. 12.7; 2006: 179, 180 fig. 6.

163. Gjerstad et al. 1934A: 484 no. 262; 1934B: pl. LXXVII: top row center; Sjöqvist 1940: fig. 20: 3; Furumark 1941: 335 and 333, fig. 56: 40: 1; Karageorghis 1960: 146 pl. X: VII; Vermeule 1964: pl. XXXII: A; Casson 1995: fig. 59; Basch 1987: 147, 148 fig. 311; Wachsmann 1998: 141 fig. 7.28, 142–143.

164. Vermeule 1964: 258; Morrison and Williams 1968: 8, 11 BA.8; Casson 1995: 36, fig. 59; Basch 1987: 147, 148 fig. 311; Wedde 2000: 98.

165. Kirk 1949: 116.

166. Furumark 1941: 237, 238 fig. 25, 239–242, 444 (9), 446.

167. Wachsmann 1981: 200.

168. Dikaios 1971: 593 no. 5549/3; 1969: pl. 72: 8 (5549/3); Mountjoy 2005: 424 n. 6, pl. XCVII: d.

169. See p. 138 n. 245 in this volume.

170. See p. 79n.182 in this volume.

171. See pp. 82–83 in this volume.

172. Morricone 1975: 360 fig. 358; Wachsmann 1998: 140 fig. 7.26: A.

173. Morricone 1975: 360 fig. 357: b; Wachsmann 1998: 176 fig. 8.22: B.

174. Casson 1995: 53–57.

175. Egyptians sometimes decorated small craft with bird-head stem ornaments. Such craft appear in Tutankhamun's Opet procession in the forecourt of Luxor Temple, as well as on a papyrus reed raft in the tomb of Ipy (TT 217, Ramses II) (Davies 1927: pl. 30; *FPO*: pls. 76–78, 80; Wachsmann 1998: 181 fig. 8.33, 183).

176. Wedde (2000: 324 n. 643, fig. 643; 2002: 838 n. 6: 2) states categorically that the bird head at the bow of the Tragana galley caps the stempost and identifies the bird figure, of which only the posterior survives, as representing a live bird standing on the forecastle deck (Fig. 2.42: B). The crucial sherd is missing, however, making Wedde's reconstruction pure conjecture. In my view, the Enkomi krater—the only existent Helladic galley depiction that shows a bow with a bird figure and a stempost—indicates that, during the Late Helladic period, a bird figure could stand in place of the stempost bird-head finial, with the stem ending in other decorative devices (Fig. 2.38: A). I am unaware of any Late Helladic ship depiction in which a bird *and* a bird-head stem finial appear together in the manner in which Wedde reconstructs the Tragana galley.

In the Geometric period, what I take to be bird statues are seen attached to the galleys' stems and sterns. One krater, now in the Metropolitan Museum of Art, depicts the same galley at two different moments as a battle rages around it (Basch 1987: 178 fig. 374; Casson 1995: figs. 65–66; Wachsmann 1998: 184 fig. 8.42: D). The sternpost on which the bird sits in both representations is an abstract bird's beak.

177. Wachsmann 1998: 177–197.

178. Basch 1987: 122 fig. 239; Doumas 1992: 73 pl. 36; Wachsmann 1998: 92 fig. 6.12.

179. Basch 1987: 94, 107 figs. 192–193.

180. Betts 1968: 333 figs. 11, 13, 15, 334; 1973; Onassoglou 1985: pls. XI: 4, XII: 1–14, 18–20, XIII: 21–23, 28–29, 31–33.

181. Wachsmann 1998: 190–192, 193 fig. 8.61.

182. Wachsmann 1998: 191, 193–197.

183. Wachsmann 1998: 190, 191 fig. 8.57.

184. On the use of bird heads and birds on Helladic and, later, Mediterranean watercraft, see Wachsmann 1996; 1998: 177–197.

185. Furumark 1941: 250–252, 253 fig. 30, 254, 255 fig. 31 (nos. 36–52); Benson 1961; 1975; Dothan 1982A: 198–200, 201–202 figs. 61–63, 203; Meiberg 2011: 47–51, 54–69, 81–90, 99–112, 126–144, 162–165, 182–187. Note, however, that the vertical bird beak, which is seen commonly on bird/bird-head bow/stempost ornaments of ship representations, appears to be an anomaly, as it does not normally feature on other representations of these birds in contemporaneous Aegean and Philistine art.

186. Karageorghis and Des Gagniers 1974A: 38; 1974B: 122–123 nos. XI: 1–3.

187. Basch 1987: 171 figs. 349–352, 172 fig. 353, 173 fig. 357, 174 figs.

360–361, 175 fig. 362, 176 fig. 367, 177 figs. 370–372, 178 fig. 374, 188 fig. 395, 189 figs. 396–397, 192 figs. 405–406, 193 figs. 407–408, 410–411, 194 fig. 414.

188. Homer twice likens the form of his ships to the horns of cattle (*Il.* 18.3, 19.344; Morrison and Williams 1968: 45; Basch 1987: 159; Casson 1995: 45 n. 16; Mark 2005: 98, 102, 104). Some have taken this to mean that these curving appendages on Geometric galleys represent the *actual* horns of cattle. Wedde (2002: 838), in arguing that Geometric stem and stern finials represent horns, writes as follows of my identification of them as abstract bird heads: "Such a reading possesses an undeniable attractiveness: its economy as an explanatory model allows it to be formulated in a single sentence with no exceptions. In addition, it would furnish support for an argument in favor of seeing a fundamental continuity in Greek galley architecture from the Bronze Age down through the classical and Hellenistic periods." First, such a fundamental continuity exists, as discussed by Wedde himself (2006). Second, he seems to be arguing, counterintuitively, that being part of a well-defined and copiously documented cultural continuum somehow argues *against* the identification of Geometric galley stem and stern devices as abstract bird heads. Instead, he prefers a hypothesis (horns on Geometric galleys), which would have horns appearing and then just as suddenly and mysteriously vanishing, outside of any cultural continuity.

189. Lenz 1998.

190. Wachsmann 1998: 186; Mark 2005: 97–98.

191. See pp. 8, 10 in this volume.

192. Morrison and Williams 1968: 12 Geom 1, pl. 1:d; van Doorninck 1982; Basch 1987: 158, 159 fig. 320, 160 figs. 321, 323, 161; Casson 1995: 36 fig. 60; Wachsmann 1998: 138 fig. 7.20.

193. Basch 1987: 159; Casson 1995: 43; Wachsmann 1998: 174; Wedde 1999: 471–472; 2006.

194. For overviews of this period, see Desborough 1964, 1972; Thomas and Conant 1999; Snodgrass 2000; Tandy 2000; Coldstream 2006; Dickinson 2006.

195. Wedde 2006: 269.

196. Karageorghis 1982: 86–127, and additional bibliography there.

197. Wachsmann 1981: 205 fig. 21, 206; 1998: 142 fig. 7.29, 143–144, 145 fig. 7.32, 146 fig. 7.33, 147 figs. 7.34–36, 148 figs. 7.37–38, 149 fig. 7.40, 151 fig. 7.48; Basch 1987: 149 fig. 313, 150 figs. 314–316.

198. On the probable introduction of iron from Cyprus to Greece, see Desborough 1964: 25–26; 1972: 78, 315–316, 340–341; Muhly 1980: 51; Snodgrass 1980: 340–347; 2000: 219, 221–222, 326, 336; Karageorghis 1982: 113; Thomas and Conant 1999: 70. For the most recent discussion on the role of Cyprus in the introduction of iron production, see Muhly and Kassianidou 2012: 134–135.

199. Desborough 1972: 340–341.

200. Sjöqvist 1940: 209; Catling 1964: 52; Yon 1971: 51–52.

201. Basch 1987: 150 fig. 314.

202. Karageorghis 1963: 277 figs. 17–18, 278 fig. 19; 1967: 343 fig. 148, 344.

203. Courbin 1957: 370, 371 figs. 55–57, 372–373, 374 figs. 58–62, 375, 376 figs. 63–65, 377–385.

204. *Il.* 16.170; *Od.* 8.37; Seymour 1914: 311; Casson 1995: 46 n. 22; Mark 2005: 122.

205. Basch 1987: 164 figs. 328–329, 165 fig. 330, 166 figs. 333–336, 338, 167 figs. 337, 339–340.

206. Melena 1975; Palaima 1991: 295–296.

207. *Od.* 4.782, 8.53; Seymour 1914: 311; Casson 1995: 46 n. 23; Mark 2005: 122.

208. On the size of Helladic galleys, see Wachsmann 1998: 157.

209. On ropes made of flax *(Linum usitatissimum),* see Casson 1995: 231 n. 27; Charlton 1996: 9–10, 19–24, 27–28, 31–36, 145.

210. Mark 1991; 1996; Casson 1992; 1995: 10 n. 27, 203. Of particular importance in this regard is the evidence from the Pabuç Burnu shipwreck (Polzer 2004; 2005; in press).

211. Pomey 1981; 1985; 1996: 140; 1997A; 1997B: 90–93; Kahanov 1998; 2004; Kahanov and Pomey 2004: 11–26; Mark 2005: 25–69.

212. Ventris and Chadwick 1973: 298 (Na 568), 562; Palaima 1991: 287–288.

213. Williams 1986.

214. Wedde 2000: 98.

215. Dakoronia 2002; 2006: 27 fig. 7.

216. See pp. 80, 82–83 in this volume.

217. See pp. 10, 14, 210, 212, 213–215 Figs. App.2.4–9 in this volume.

218. Winlock 1955: pls. 33, 35–37, 40–43, 70, 72–76, 84; Landström 1970: 71 figs. 205–206, 74 fig. 221, 78 figs. 234, 236, 79 fig. 238, 82–83 figs. 246, 249–252, 86 fig. 260, 99 fig. 313, 101 fig. 321, 115 figs. 354–355; Jones 1990: 4, 16: A1, 17–22, pls. X nos. 375, 460, 462–463, 481, 491, 556, 581, 609, 610, 612, 617, XII no. 334, XIII no. 352, XV no. 437, XXX no. 375, XXXI no. 334.

219. See, for example, Hatshepsut's Punt ships, the Syro-Canaanite ships in the tomb of Kenamun (Fig. 37), and the determinative of the *mnš* ship (Daressy 1895: pls. XIV–XV; Naville 1898: pls. 72–75; Säve-Söderbergh 1946: 58 fig. 12; Davies and Faulkner 1947: pl. 8; Basch 1987: 63 figs. 111–112, 64 figs. 113–114, 65 fig. 116; Wachsmann 1998: 20 figs. 2.15–16, 21 figs. 2.17–18, 42 fig. 3.2, 43 fig. 3.3, 44 figs. 3.5–6, 47 fig. 3.10).

220. Casson 1995: 46.

221. Basch 1987: (single quarter rudder) 176 fig. 369, 177 fig. 372, 178 fig. 374(?), 182 fig. 382, 187 fig. 393, 188 figs. 394–395, 190 fig. 399, 192

figs. 403–406, 193 figs. 407–412, 194 fig. 413; (pairs of quarter rudders) 164 fig. 328, 167 fig. 340, 171 fig. 352, 172 fig. 355, 173 fig. 357, 174 fig. 360, 176 fig. 368, 184 fig. 388, 191 fig. 401.

Chapter 3: Wheels, Wagons, and the Transport of Ships Overland

1. See pp. 20–21, 32 item nos. 27–30 in this volume.

2. Jones 1995: 9–11.

3. Yadin 1963: 44, 147; Pritchard 1969A: [101] fig. 311; Partridge 1996: 76, 77 fig. 63.

4. Partridge 1996: 76.

5. Winlock 1947: 153–157, 170; McLeod 1958: 397 n. 5 (and additional bibliography there); Yadin 1963: 86; Littauer and Crouwel 1979B: 56, 76; Hayes 1990: 193.

6. Wachsmann 2010. On Ahhotep, see Vandersleyen 1971: 129–130, 175–196 and additional bibliography there.

7. Winlock 1924: 252–253.

8. Von Bissing 1900: 19–21.

9. Maspero 1892: 321 fig. 297; Vernier 1927A: 219 (no. 52668); 1927B: pl. XLIX.

10. Landström 1970: 141 fig. 411; Basch 1987: 335 figs. 719–720; Spathari 1995: 26, 27 fig. 18.

11. Reisner 1913: IV; Landström 1970: 98 figs. 311–312; Jones 1995: 30. Ship models were often interred in tombs with the deceased. In his pioneering publication, *Models of Ships and Boats* (1913), G. A. Reisner defined the different type of watercraft depicted by these models. The earliest models date to the Predynastic period and are mainly made of ceramic. Wooden ship models become common in the VIth Dynasty and continue to appear till the XIIth Dynasty, which is the period in which they are the most abundant. Perhaps the best-known collection of ship models found in a nonroyal tomb is the little flotilla from the tomb of Meket-Re, which dates to the XIIth Dynasty (Winlock 1955): Fifty-eight wooden boat models have been reported from the XIIth-Dynasty tomb of Djehutynakht at Deir el-Bersha (Freed, Berman, and Doxey 2009: 166–177). During the New Kingdom, ship models fell out of style, with the notable exception of some royal tombs and two nonroyal ship models. On Egyptian ship models, see Reisner 1913; Landström 1970; Vinson 1994; Jones 1995: 26–33; Tooley 1995: 51–56.

12. Van Walsem 1997: 226–231; Creasman and Doyle 2010.

13. Landström 1970: 110 figs. 340–341, 118 figs. 363–364.

14. Compare Landström 1970: 98; Jones 1995: 32.

15. Maspero 1915: 428 no. 4030.

16. Length: 38.5 cm; breadth: maximum 6.7 cm, narrowing at its ends to 1.5 cm; weight: 372 gms. I base the following description primarily on the commentaries by von Bissing (1900: 19) and Vernier (1927A: 217–218 no. 52667).

17. For Middle Kingdom–New Kingdom quarter and axial rudder arrangements, see Landström 1970: 78 figs. 234, 236, 79 fig. 238, 82–83 fig. 246, 82 fig. 249, 83 figs. 250–252, 86 fig. 260, 89 fig. 271, 90 fig. 275, 92 figs. 283, 287–288, 93 fig. 293, 99 figs. 313, 316, 101 figs. 319, 321, 102 figs. 322, 324, 106 figs. 327–330, 107 figs. 331–334, 114 fig. 352, 115 figs. 354, 356, 116 figs. 357–358, 117 figs. 361–362, 118 figs. 364–365, 119 figs. 368–369, 121 fig. 371, 122–123 fig. 372, 125 fig. 375, 128 figs. 381–382, 130 fig. 383, 134 figs. 389–391, 393, 135 figs. 394, 396, 136 fig. 399, 138 figs. 404–405; Jones 1990: pls. V, XVI–XXIII, XXV, XXVIII–XXXV.

18. The steering oar held by the helmsman has received modern attention (Von Bissing 1900: 19). The two surviving parts of the oar have been welded together by a goldsmith in Cairo and received a sulfur-induced patination at the museum.

19. Edgerton 1926–1927: 257.

20. Edgerton 1926–1927: 258 figs. 2–3.

21. Landström 1970: 95 fig. 297, 119 fig. 368.

22. Doyle 1998: 90 fig. 6–12, 97 fig. 6–19, 105 fig. 6–31, 108 fig. 6–37, 112 fig. 6–43.

23. Compare Biers and Terry 2004: 51 no. 28. Note, however, a bronze kohl stick from Assasif, which is contemporary in date to Ahhotep but is dissimilar to the item held by the standing figure in the silver model. The item was deaccessed from the Metropolitan Museum (MMA 16.10.447) and is now in the Museum of Natural History. The kohl stick is 9.7 cm long and has "a small round 'spoon' rather than the elongated oval 'paddle' " (Christine Lilyquist, pers. comm.).

24. Landström 1970: 98 fig. 312.

25. Wachsmann 1995: 10.

26. Reisner 1913: XVI.

27. Reisner 1913: XVI, 6 n. 4; Doyle 1998: 137–139.

28. Von Bissing 1900: 19.

29. Landström 1970: 99 figs. 313, 316, 101 fig. 321, 115 figs. 354, 356; Jones 1990: pls. XXX–XXXI; Doyle 1998: 128–132.

30. Reisner 1913: [no. 4798] 3 fig. 14 and pl. 1, [no. 4835] 27 and pl. 30; Landström 1970: 76, 77 fig. 226, 82 figs. 247–248, 83 fig. 251.

31. Wachsmann 1998: 257–258, 259 fig. 12.5.

32. Daressy 1895: pls. XIV–XV; Säve-Söderbergh 1946: 56 fig. 10, 57 fig. 11; Davies and Faulkner 1947: pl. VIII; Casson 1995: 35–36, fig. 57; Basch 1987: 63 figs. 110–112, 64 figs. 113–114; Wachsmann 1998: 42 fig. 3.2, 43 figs. 3.3–4, 44 figs. 3.5–6, 45.

33. Wachsmann 1998: 42 fig. 3.1, 46 figs. 3.7–8, 47 figs. 3.9–10, 50 fig. 3.14.

34. Wachsmann 1998: 42, 44–45.

35. Dunand 1954: 337–338 nos. 10089–10092.

36. Dunand 1950: pl. 69 no. 10089; Wachsmann 1998: 54, 55 fig. 3.21. On representations of vessels derived from models rather than the actual prototypes, see Reich 1991; Wachsmann 1998: 52–54.

37. Wachsmann 1998: 63–64, 65 figs. 4.5–6, 66 figs. 4.7–9, and additional bibliography there.

38. On ship models from Cyprus, see Westerberg 1983; Monloup 1984: 145–160; Basch 1987: 70–74, 148–151, 249–262; Wachsmann 1998: 61–67. Later iron firedogs in the shape of Geometric-period galleys have been found at Paleopaphos and Salamis in Cyprus (Karageorghis 1963: 277 figs. 17–18, 278 fig. 19, 292–294; 1967: 343 fig. 148, 344; Basch 1987: 188, 189 figs. 396–397, 258, 260 fig. 562). Similar firedogs have been found at Argos (Courbin 1957: 369 fig. 54, 370, 371 figs. 55–57, 372–373, 374 figs. 58–62, 375, 376 figs. 63–65, 377–385; Göttlicher 1978: 64 and Taf. 25 [nos. 338–339]; Wachsmann 1998: 186, 188 fig. 8.50: A). Karageorghis (1963: 292) believes that those found on Cyprus arrived there from Greece.

39. Doumas 1992: 71–72 fig. 36; Spathari 1995: 44 fig. 44.

40. *PM* IV: II: 953 fig. 923; Alexiou 1969: 114 fig. 56.

41. Wachsmann 1998: 93 fig. 6.17, 94.

42. Casson 1975: 7; Doumas 1992: 63 fig. 29 [partial], 71–74 fig. 36, 75–77 fig. 37, 80 fig. 39, 81 fig. 40; Wachsmann 1998: 93 fig. 6.14, 94 fig. 6.19, 99 fig. 6.27.

43. Renfrew 1967: 5, 18, pls. 1: 12, 3: 12–14; Basch 1987: 78, 79 figs. 153–156; Wachsmann 1998: 69, 70 fig. 5.1.

44. Caskey 1964: 327, pl. 56: C; Long 1974: 48, pl. 24 fig. 69; Göttlicher 1978: [no. 335] 64, Taf. 25; Johnston 1985; 26–27 (BA17); Wachsmann 1998: 102, figs. 6.34–35, 104. There has been some confusion as to whether one or two metal models were found at Keos. There is indeed only one model (Wedde 2000: 309 no. 310).

45. Basch 1986: 421 fig. 5, 422 fig. 7; Wachsmann 1998: 77, 80, 82 fig. 5.24: A.

46. Casson 1995: 41–42, 445–446; Basch 1987: 98 figs. B1–2, 99 figs. B3–4, 6–7, 102 figs. D1–2, 4, 6, 103 figs. D7–9, 106 fig. G3; Wachsmann 1998: 100 fig. 6.29: A–C, G–K.

47. Wachsmann 1998: 92 fig. 6.13, 93 fig. 6.14, 106 and additional bibliography there.

48. Wachsmann 1987: 60–61, pls. XXVII: B, XXIX: 3, XXXVI: A: 5, LV: 6, LVI: 5. For discussions on the items brought by Minoans in the Theban tombs, see Vercoutter 1954: 121–127, 134–135, 153–156; 1956: 305–366, pls. XXV–LXVII; Wachsmann 1987: 49–92, pls. LIV–LVIII.

49. Warren 1995: 5.

50. *BAR* II: §1–12; Vandersleyen 1971: 30–40; Redford 1992: 115, 120–122, 125–129; 1997: 13–16 docs. 68–70; Rainey and Notley 2006: 63–64.

51. Habachi 1972: 36–37 ll. 12–15. Translation from Redford 1997: 14 no. 69 ll. 12–15.

52. Bietak 1995; 1996: 73–81, color pls. III–VIII, pl. 33; Bietak, Marinatos, and Palyvou 2000; Bietak et al. 2007; Morgan 1995.

53. Bietak 1992: 26–28; 1996: 68.

54. Cline 1998; Niemeier and Niemeier 2000: 764–765; Bietak 2000.

55. Niemeier 1991; Niemeier and Niemeier 2000: 765–767.

56. *PM* I: 18, 26, 297, 319, 380, 418, 419 fig. 304: b, 420–422, 553; II: I: 220, 303, 357 n. 1, 360; III: 9; IV: I: 130; Redford 1992: 120 n. 120; Warren 1995: 3.

57. The *interscalmium* in the classical world was the distance measured between tholepins, believed to be about 1 m (Vitr. *De arch* 1.2.4; Morrison, Coates, and Rankov 2000: 133, 245–246).

58. Glanville 1932: 22 no. 56; Bietak 2005: 17.

59. On this subject, see Save-Söderbergh 1946: 43–50; Vercoutter 1954: 165–166; 1956: 53–55; Heltzer 1988; Wachsmann 1998: 51–52. Bietak (2005) identifies *Prw-nfr* with the harbor of Avaris/Tell el Da'ba. Barber (1998: 15) offers a unique interpretation of Keftiu ships in which she identifies them as vessels "that use colorful fabrics on a frame cabin to shield passengers from the elements during the voyage," thus connecting the name to the Aegean patterned cloth covering used on some Nile ships.

60. Yadin 1963: 186–189, 191–194, 200, 202. See p. 88 n. 5 in this volume.

61. Yadin 1963: 86–90, 190, 192–193, 200, 206, 210, 211–217, 220–221.

62. Canney 1938: 132–133. Schäfer (1974: 239 fig. 252) identifies a funerary ship being dragged by men on a wavy surface in a VIth-Dynasty scene from Dahshur as representing a vessel on the water. J. A. Wilson (1944: 207, 211, pls. XIII–XIV) believes the vessel to be depicted being pulled over a rocky surface. For the burial ship dragged on a sledge, see Jones 1995: pl. II. The tomb of Ipy (TT 217) contains a scene of such a vessel under construction on its sledge (Wreszinski 1988[I]: 369; Davies 1927: pl. 34; Wachsmann 1998: 237 figs. 10.26–27, 238). See also Creasman and Doyle 2010 and additional bibliography there.

63. Jones 1995: 18–19.

64. Davies 1948: 36; Jones 1995: 19 fig. 8; Porter and Moss 1960: 408.

65. Dittman 1941: 67 (e), Abb. 6.

66. Exodus 37:3–5; Deuteronomy 10:8.

67. I Samuel 5, 6:1–12. Recent excavations at Ashkelon and at Tell

es-Safi have revealed bronze *situlae* and ceramic objects in the shape of uncircumcised phalli, some of which appear in cultic contexts, indicating the existence of a Philistine cult of the phallus. This has led A. Maeir to propose that the plague was one of impotence and that the five gold *'opalim'* sent by the Philistines to the Israelites along with the Ark, were phalli (Stager 1996: 68*–70*, fig. 13; Maeir 2007, 2008). Evidence from the Dionysian cult suggests that, rather than impotence, the Philistines may have suffered from priapism (Csapo 1997: 266–268).

68. Tylor 1896: pl. II. Recently, Sobeknakht's tomb gained attention when archaeologists who were cleaning it discovered a text describing the deceased's deeds in repulsing a Kushite invasion of southern Egypt (Anonymous 2003).

69. Porter and Moss 1937: 184–185, 178 Tomb 10.

70. For an early and rare image of an oxcart with spoked wheels dating to the mid-XVIIIth Dynasty, see Hayes 1959: 164 fig. 90, 165.

71. Davies 1925; 1926. See also Wilson 1944: 212; Schäfer 1974: 239 n. 216, 260 n. 4, pl. 40; Porter and Moss 1960: 26–27.

72. Davies 1925: 17, pl. V. An alternative translation (17 n. 2) reads, "The assistant casts water and milk for . . ."

73. Taylor 2001: 187, 188 fig. 131; Creasman and Doyle 2010: 17 fig. 9.

74. Marucchi 1927: 38–39, 40 fig. 11 [no. 216]; Creasman and Doyle 2010: 17 fig. 8.

75. Creasman and Doyle 2010: 17 figs. 7–8. See also Wachsmann 1998: 236 figs. 10.26–27, 238.

76. Lefebvre 1924: 129–130; 1923: pls. XXX, XXXIV; Suys 1927: 85; Cherpion, Cortegianai, and Gout 2007: 134–135 (Scène 92, 6 [GL 81]).

77. Suys 1927; Picard 1931; Stevenson Smith 1949: 25; Terrace 1963: 273.

78. Cherpion, Cortegianai, and Gout 2007: 134 (Scène 92, 5 [GL 81]).

79. Petosiris's eight-spoked wheels with studs may have been influenced by those used on Achaemenian chariots, which continued the development of Assyrian wheeled vehicles (Yadin 1963: 299–302, 413, 420–421, 430–433, 442–443, 452, 454, 459; Littauer and Crouwel 1979B: 106 figs. 55–56, 80–82, 108). Eight-spoked wheels appeared in Assyria on both chariots and carts as early as the reign of Tiglat-Pileser III (744–727 B.C.), while studded wheels came into use during the reign of Sennacherib (704–681 B.C).

80. Fakhri 1944: 144 fig. 24, 145.

81. Fakhri 1944: 134.

82. Fakhri 1944: 138 fig. 19, 140 fig. 20, 146 fig. 26.

83. Compare Littauer and Crouwel 1979A: 108 fig. 1 (section).

84. Fakhri 1944: 145 fig. 25.

85. Wilkinson 1837: 341 no. 243; Dittman 1941: 68 (i), 69 Abb. 8; Littauer and Crouwel 1979A: 113 no. 3; Göttlicher 1992: 69 Abb. 40, 71. Wilkinson notes that the mummy wrapping belongs to S. D'Athanasi. For the date, see Dittman 1941: 68 Abb. 8, 69.

86. Dunand and Zivie-Coche 2004: 116–119.

87. Kemp 1991: 185, 188; Jones 1995: 20–22.

88. Legrain 1917: 1–5. He compares it (1 n. 1) to an Italian Catholic *portantina,* a construction with poles used to transport reliquaries and statues. For a three-dimensional image of a *pavois* on an XVIIIth-Dynasty granite cult ship statue from Karnak, see Shaw and Nicholson 1995: 48.

89. For an example of Egyptian sledges see Reisner 1913: [no. 4928] 88, 89 fig. 326; Hayes 1990: 193 fig. 118; Carter 2000: pl. V; Reeves 1990: 119–121; Creasman 2005: 122, 123 fig. 57. On the use of sledges in Egypt, see Partridge 1996: 131–137; Creasman and Doyle 2010. Note also a Ptolemaic-period trolley on four small wheels from Medinet Madi (Fig. 3.23; Dittmann 1941; Littauer and Crouwel 1979A: 112).

90. See pp. 20, 28, item no. 1 in this volume.

91. Kemp 1991: 185.

92. Kitchen 1974; 1982: 172.

93. Kitchen 1974: 169 translation, l. 7, 171.

94. Canney 1936: 50–51; 1938: 133–135, 145–146. The most famous of these sacred river ships was Amun's yacht, the *Amunuserhet,* used in the Opet Festival, as well as for the Beautiful Festival of the Valley (Foucart 1921–1922; Landström 1970: 119–121; Kemp 1991: 185; Jones 1995: 22–25). The Opet Festival, as well as the vessels used in it, was most completely documented during the reign of Tutankhamun in the forecourt of the Luxor Temple (*FPO*). Legends of golden *dahabiehs* sailing at night on the sacred lakes of Amun and Mut may be memories of these craft (Wilbour 1936: 52; Peck 1994–1995, and additional bibliography there).

95. Kemp 1991: 188, 205–206; Myśliwiec 2000: 165.

96. Van Siclen 1986.

97. Kemp 1991: 202.

98. Vos 1993. Regarding the date, see p. 7.

99. Vos 1993: 40 no. 7, 52 §19: 15–18.

100. Vos 1993: 40.

101. Vos (1993: 40) notes that the boat was placed "on a wheeled carriage."

102. Göttlicher 1992: 73, 74 fig. 41; Hornig 2002: 13 nos. 13–16, 14 nos. 17–19, and additional bibliography there.

103. See p. 102 n. 89 in this volume.

104. Hdt. II, 59, 63, translation by G. Rawlinson (1997: 153–155).

105. Dattari 1901: 77 and pl. XXX (no. 1158), 236, pl. XXVII

(no. 3557); Weber 1914: 256 Abb. 127. The coins date to the reigns of Hadrian (no. 1158) and Antoninus Pius (no. 3557). On the Isiac cult in Alexandria and the Nile Delta, see Dunand 1973: 110–122.

106. On the Ploiaphesia/Navigium Isidis, see pp. 155–158 in this volume.

107. Weber 1914: 255, 256 no. 468, Abb. 128. For an Isiac ship-shaped lamp, see Spathari 1995: 137 fig. 167. For a lamp in the form of a ship-cart on wheels bearing the figure of Harpocrates, see Hornig 2002: 12 no. 10, 13 Abb. 13.

108. Michaud 1974: 587, 590 fig. 42 (#12776), 591 fig. 43 (#12775); Johnston 1985: 144 Rep. 6A; Mandel-Elzinga 1988: 257 Abb. 5–6, 258; Werner 1997: 110 fig. 21.

109. On the gazelle *(Gazella dorcas)* see Bodenheimer 1960: 49–50, 127; 1972: 36 fig. 21, 2: C–D. Egyptian religious iconography associates the gazelle with the goddess Anuket (Simpson 1953: 88).

110. The cult ship of Sokar is capped with the head of a stern-facing oryx with straight horns and a forward-facing bull's head (Fig. 2.20: B), but the recurving horns of the Benaki ship's animal figurehead argue against this identification.

111. Moll 1929: 63, pl. A IV h: no. 484.

112. L. Bell (pers. comm.).

113. Černý 1962; Taylor 1996: 49–51; Dunand and Zivie-Coche 2004: 119–121. Taharka's shrine of Amun at Karnak, which appears in a vignette in a papyrus concerning an oracle delivered by Amun at Thebes in 651 B.C. is remarkable in that, unlike almost all other such scenes, the shrine is *not* shown on a boat palanquin (Parker 1962: 4, pl. 1).

114. Černý 1962: 35–36.

115. *BAR* IV: §669–687; A. M. Blackman 1941: 84; Myśliwiec 2000: 41–42.

116. Krutchten 1986: 122–189; Myśliwiec 2000: 40.

117. Fakhri 1944: 42 n. 4; *BAR* IV: §650–658.

118. Černý 1962: 39.

119. Krutchten 1986: 337–354.

120. Černý 1962: 36.

121. Černý 1930: 496.

122. Taylor 1996: 50–51. For a rather humorous case of a Rameside writer of an oracular request hinting at such priestly manipulations, see Barnes 1949.

123. Černý 1930; 1935: 56–58; 1962: 43–45.

124. Fakhri 1944: 41–44.

125. Fakhri 1944: 33, 35–40; Černý 1962: 47; Parke 1967: 194–241.

126. Černý 1962: 40.

127. Gardiner 1933; Černý 1962: 40.

128. Blackman 1926; Černý 1935; 1941.

129. Dawson 1925; Blackman 1925; Černý 1962: 40–41.

130. Černý 1962: 41, 42 fig. 9, 43.

131. Legrain 1914: 75; Canney 1936: 53; 1938: 143–144, 146; Hornell 1938; Seligmann 1966: 452–454, fig. 2; Wickett 1990; 2009; Wachsmann 2002A; 2002B, and additional bibliography there.

132. On the significance of this date, see McPherson 1941: 306.

133. In recent memory only the *moulid* of Abd el Rahim el Qanawi at Qena shared the distinction of the transport of boats during the festival, but J. W. McPherson (1941: 5, fig. facing 5) notes that the custom had been discontinued in his day. This *moulid* also takes place on the fourteenth day of Shaaban. On Egyptian *moulids,* see Lane 1973: 239; McPherson 1941; Atia 1999.

134. Otto 1967: pl. 28.

135. See pp. 102, 125 in this volume.

136. Legrain 1914: 84; A. M. Blackman 1923: 78–79; Hornell 1938: 145; Seligman 1966: 452–455; Otto 1967: 130–132; Peck 1994–1995: 72. For information on Abu el Haggag and his festival, see Legrain 1914: 44–91.

137. Wachsmann 2002B: 826.

138. The route of the *dura* is said to be based on a trick played by Abu el Haggag on the female Christian head of Luxor (Otto 1967: 132; Wachsmann 2002B: 823–824). The ruse is reminiscent of the manner in which Elissa (Dido) acquired the hill of Byrsa when she founded Carthage (Verg. *Aen.* I: 365–368), perhaps not surprising given Abu Haggag's Tunisian origins. On the specific route of the *dura,* which begins in the Corniche, see Wachsmann 2002B: 825.

139. Hornell 1938: 146.

140. Duff Gordon 1969: 117.

141. Wickett 1990; Hornell 1938: pl. I–J.

142. Kamil 1996: 30 (first published 1983).

143. *FPO*: key plan, pls. 18–19, 23, 27–28, 31.

144. Long 1974: 46, 48–49, pl. 19 fig. 52. Regarding the date of the sarcophagus, see pp. 11–14 in Long.

145. On the Karphi chariot, see Pendlebury, Pendlebury, and Money-Coutts 1937–1938 (1940): 81 Cat. no. 24: D, pl. XXXV: 4; Seiradaki 1960: 28 n. 3, pl. 13; Hutchinson 1962: pl. 21; Gesell 1985: 210, pl. 159: a–b; Sakellarakis 2006: 95, 97 no. 11046; Mohen and Eluère 2000: 128. The peculiar three-wheeled arrangement topped by animal figures/protomes of the chariot suggests a foreign element introduced into Crete at this time. Its unique configuration points to a distinctly European source: Three-wheeled cult objects appear in central Europe, from the better preserved of the two Dupljaja bird chariots from Serbia (Dubovac-Zuto Brdo culture, ca. 1600–1300 B.C.) to later Urnfield cult objects that combine three wheels with the ubiquitous waterbird that

was so predominant to their belief system (Kossack 1954: 10–12, 28, 53, 59, 79 Taf. 1, 3: 1b; Schauer 1987: 17 Abb. 13; Pare 1989: 85 fig. 4; Mohen and Eluère 2000: 96; Vasić 2004; Bouzek 2005: 27–28). Thus, the Karphi chariot may be best understood as displaying a syncretic combination of the Cretan local sacred—bulls—with the central European motif of a three-wheeled cultic object, in this case a chariot. Another possible link to the Dupljaja bird chariot is the parasol that appeared, apparently as a foreign element, on some Mycenaean depictions of chariots (Wardle 1973: 328, 331 fig. 19; Crouwel 1973; French 1973).

The Karphi bull chariot is not unique in demonstrating syncretism in Early Iron Age Crete. Large Late Minoan IIIC vessels could be decorated with local cultic symbology and used as cremation urns (Popham 1986; 2001: 284–285). At this time the foreign custom of cremation was becoming more common in Crete: The earliest known cremations on the island date to the Late Minoan IIIA2 (Rehak and Younger 2001: 464 n. 541).

146. Wachsmann 1998: 131–135, 137.

147. Dakoronia 2002: 283–284, 289 figs. 1–4. The internal keel is indicated by a painted red line inside the hull.

148. Dakoronia 2002: 283–284.

149. Dakoronia 2002: 284.

150. Frickenhaus 1912; Boardman 1958; Kerényi 1976, ills. 56–57, 59: A–B; Wilson 1987–1988: 135 fig. 36; Göttlicher 1992: 103, 104 fig. 59, 105 figs. 60–61, 106–108.

151. Simon 1983: 94 fig. 12; Robertson 1985: 291; Basch 1987: 227, 228 figs. 475 [Bologna Museum no. 130]; British Museum B 79; Athens Acropolis 281.

152. Basch 1987: 207 fig. 428, 209 fig. 430, 209 figs. 431–432, 434, 210 figs. 435–436, 211 fig. 438, 212 figs. 439–440, 213 figs. 441–444, 214 figs. 445–447; 215 figs. 448, 450, 217 figs. 452–453, 218 figs. 454–455, 219 figs. 456–457, 220 figs. 458–459, 221 figs. 460, 460 *bis,* 222 figs. 462–463, 225 fig. 469, 226 figs. 470: B, 471, 227 figs. 472–473, 234 figs. 484–485, 235 fig. 486, 238 figs. 497, 499, 242 fig. 510; Casson and Linder 1991: 68, 70 fig. 5–4.

153. Von Bothmer 1985: 64, 182 [Cat. 47]: right and front, 184 fig. 100:A.

154. On bird-head stem and stern decorations, see Wachsmann 1998: 177–197.

155. If the latter interpretation is correct, then this setup finds parallels in later, Byzantine galleys, which have been reconstructed with a broad, transomlike stern (Pryor 1995: 102–104, 109).

156. R. J. A. Wilson (1987–1988: 135 fig. 36) raises the possibility that the lead sheet might be a forgery. I believe that the consideration that it *differs* from the ship-carts on the *skyphoi,* discussed earlier, argues against this view. If it is a valid artifact, then a sixth–fifth-century B.C. date seems possible given the 1983 excavation of a rich necropolis of that date at Montagna di Marzo. See also R. J. A. Wilson 1981–1982: 85. I thank Dan Davis for bringing this representation to my attention.

157. Nautical authors writing on the subject tend to identify these objects as fabrics or draperies (Johnston 1985: 141–143 [Reps. 2–4]; Basch 1987: 227; Göttlicher 2008). Mary B. Moore suggests that it may represent a wicker basket or wicker work (pers. comm., compare Bobart 1936: 36 fig. 20, 37 fig. 21, 49 fig. 29, 50 fig. 30; Amyx and Pritchett 1958: pl. 51: d; Von Bothmer 1985: 61 [Cat. 19], 113 no. 19 Side A, 117 fig. 72; Beazley 1986: pl. 55: 2). I thank Mary B. Moore for her comments.

158. Tarquinia Museum no. 678. Basch 1987: 225, 226 fig. 470: A.

159. Basch 1987: 207 fig. 428, 208 fig. 430, 209 figs. 431, 434: A–B, 210 figs. 434: C–E, 435, 436: A–C, 211 fig. 438, 212 fig. 440, 213 figs. 442–443, 444(?), 214 figs. 445, 447, 215 figs. 448, 450(?), 217 fig. 452, 218 figs. 454–455, 219 fig. 457, 220 fig. 459, 224 fig. 486, 226 figs. 470–471; Spathari 1995: 84–85 fig. 94, 87 fig. 95, 88 fig. 99, 93 fig. 106, 97 fig. 109.

160. Watzinger 1924: 31 no. 53, Taf. 15: D53; Göttlicher 1992: 105 Abb. 60.

161. Basch 1987: 428, 432–435 figs. 936–941; Spathari 1995: 136–137 fig. 165; Göttlicher 2004; 2008.

162. Basch 1987: 237, 238 fig. 497, 270 fig. 574; Spathari 1995: 105 fig. 121.

163. O. Höckmann (pers. comm.). See Basch 1987: 321 figs. 675–676, 678, 322 fig. 681, 323 fig. 682, 324 fig. 685, 325 figs. 687, 692–693, 329 figs. 703–704, 330 figs. 712–715. I thank Olaf Höckmann for bringing this issue to my attention.

164. Ashmolean Sherds 1924 264; Boardman 1958; 1999: 137–138; Csapo 1997: 277–278, pl. 8.

165. Boardman 1958: 5 fig. 1.

166. Csapo 1997: 277, pl. 8: A.

167. Boardman 1958: 6, 7 n. 8.

168. Csapo 1997: 277–278, pl. 8: B.

169. Boardman 1999: 137. On the Opet Festival, see pp. 102 n. 94, 113, 119 Figs. 3.39–40 in this volume.

170. Richard 1992.

171. Hamilton 1978: 386–387. See also Dugas 1936: 226–227, pl. III:1.

172. Basch 1987: 221 fig. 460 *bis,* 222 figs. 462–464, 226 fig. 471, 227 fig. 472, 238 fig. 497, 272 fig. 576, 273 fig. 579, 274 fig. 580; Spathari 1995: 94 fig. 107, 96 fig. 108, 97 fig. 109, 107 fig. 122, 108 fig. 123, 109 fig. 124.

173. D. Davis (pers. comm.) notes that ship departure scenes were popular themes in Archaic and Classical art, and one wonders whether the stage provided the inspiration or vice versa (Spathari 1995: 78–79 fig. 87, 82–83 fig. 92, 107 fig. 122, 108–109 fig. 123, 109 fig. 124).

174. Robertson 1985: 292–293.

175. Pickard-Cambridge, Gould, and Lewis 1968: 11–12; Parke 1977: 111. This *hieros gamos* aspect of the Anthesteria may explain the choice of wheels with two parallel spokes, normally seen on bridal wagons, which appear on a few of the Archaic Dionysian ship cart representations. See p. 121 n. 153 in this volume.

176. Bremmer 1994: 19.

177. Ar. *Ach.* 238–264; Boardman 1958: 7; Simon 1983: 101–102; Csapo 1997.

178. Philostr. *VS* I.25.531. Translation by W. C. Wright from Philostratus and Eunapius 1921: 107. Regarding the date of compilations, see there p. xii.

179. Philostratus's use of the term *trireme* is curious at this late date. The term continues in use until the fourth century A.D. but may be employed here simply for a vessel resembling a warship (W. Murray, pers. comm.).

180. *Hymn. Hom. Bacch.*, translation by Evelyn-White 1998: 429, 431, 433. See also Isler-Kerényi 2007: 185.

181. Beazley 1986: pl. 67; Basch 1987: 226 fig. 471; Spathari 1995: 97 fig. 109; Yacoub 2007: 172, 173 fig. 86, 174.

182. Harrison 1885. See also Gardner 1881A; 1881B.

183. Morrison and Williams 1968: 92–95, 100, 102–105, 107–108, 113–114, pls. 14, 16–18, 21 (Arch. 51–56, 61, 65–67, 69, 74, 74 bis, 79, 81, 81 bis, 84, 90–93); Casson 1995: figs. 88–89; Basch 1987: 207 fig. 428, 208 fig. 430, 209 figs. 431-434, 210 figs. 435-436, 211 fig. 438, 212 figs. 439-440, 213 figs. 441-444, 214 figs. 445-447, 215 figs. 448-450, 217 fig. 452, 218 figs. 454-455, 219 figs. 456-457, 220 figs. 458-459; Spathari 1995: 84–85 fig. 94, 86–87 fig. 95, 87 figs. 96–97, 88 figs. 98–99, 89 fig. 100.

184. Davies 1978: 72–74.

185. On ships' "eyes" *(ophthalmoi/oculi)*, see Hornell 1970: 51, 88, 249, 272–273, 285–289; Wachsmann 1998: 14, 52, 149, 195; Nowak 2006; Carlson 2009. Most recently *ophthalmoi* have been found on a fifth-century-B.C. merchantman that sank at Tektaş Burnu, Turkey, and at Yavneh Yam on the Israeli coast (Nowak 1999; 2001; Israel Antiquities Authority 2008).

186. Paus. II, 35, 1. Translation from Gardner 1881B: 315.

187. J. Hale (pers. comm.).

188. Deubner 1956: 22–35; Parke 1977: 33–40; Burkert 1983: 154–158; Simon 1983: 55, 58–72; Pinney 1988; Neils 1992: 13; Robertson 1996: 56–65; Hurwit 1999: 30–31; Shear 2001: 29–38, 42; Hale 2009: 136–137.

189. Mikalson 1975: 34; Norman 1983: 41; Simon 1983: 55; Neils 1992; Hurwit 1999: 44–47; Shear 2001.

190. Davison 1958: 24–26, 31; Brommer 1979: 33; Simon 1983: 58–72; Robertson 1996: 58; Neils 1996; 2001: 21–24, 49–71, 125–201; 2005: 201–223; Neils and Tracy 2003: 16–17, 27–29. Regarding the *peplos,* see Mansfield 1985: 51–68, 77–78; Barber 1992A.

191. Nagy 1980; Mansfield 1985: 51, 58, 69; Göttlicher 1992: 108–110; Hurwit 1999: 45, 286; Wachsmann 2012.

192. Deubner 1956: 33; Parke 1977: 40; Norman 1983: 43; Mansfield 1985: 18; Hurwit 1999: 226. Regarding lacunae in the iconographic record of the Panathenaia, see Pinney 1988: 465; Hurwit 2004: 232–234.

193. The Panathenaia probably had its origins in the Bronze Age (Robertson 1996: 57–58; Simon 1996: 23). On the mythical origins of the Panathenaia, see Parke 1977: 33; Robertson 1985; 1996: 61–65; Neils 1992: 21; Boegehold 1996: 95–96; Hurwit 1999: 44 n. 53.

194. Davies 1967: 36–37; Jordan 1972: 154; Parke 1977: 37; Simon 1983: 55; Neils 1992: 13, 15, 16 fig. 1: ll. 78–81, 97; Robertson 1992: 91–93; Boegehold 1996: 96–97; Kyle 1996; Parker 1996: 75–76, 89–92; Parker 2005: 262; *Panathenaic Games;* Hale 2009: 136. The earliest evidence for Panathenaic ship races is late, however (Kyle 1992: 97). On evidence for ship races primarily, but not exclusively, dedicated to Dionysos, see Gardner 1881A; 1881B; Harrison 1885; Davies 1967: 36–37, 39; Davies 1978.

195. Parke 1977: 33; Norman 1983: 41; Neils 2007.

196. Parke 1977: 38; Norman 1983: 41–43. According to Mansfield (1985: 2–18, 54–58), two *peploi* were woven, one by aristocratic women and one by professional male weavers. See also Barber 1992A: 112–117; Hurwit 1999: 45.

197. Norman 1983: 42 n. 13 and additional bibliography there.

198. Robertson 1985: 276, 279–280, 289–291.

199. Hurwit 1999: 45.

200. Gardner 1914; Mansfield 1985: 76–77; Göttlicher 1992: 108–109; Robertson 1992: 93–119; Martin 1999: 143, 144 Abb. 11–12, 145; Neils 2001: 49, 51 fig. 38.

201. Philostr. *VS* 2: 550. According to Jennifer Tobin (1993: 81), this Panathenaia is the one dating to 143/4 A.D.

202. Philostr. *VS* 2: 550. Translation by W. C. Wright from Philostratus and Eunapius 1921: 147. On Herodes Atticus's propulsion system, see Mansfield 1985: 75 n. 90; Göttlicher 1992: 109.

203. Tobin 1991; 1993.

204. Welch 1998: 136–145; Rife 2008: 109–111.

205. Mansfield 1985: 46–50, and additional bibliography there. On Athena's possible connections to navigation and ship construction, see Detienne 1970.

206. Deubner 1956: 33–34; Parke 1977: 39.

207. Mansfield 1985: 52–53, 68; Barber 1992A: 114; Shapiro 1996: 217; Hurwit 1999: 45.

208. Norman 1983.

209. On the late date for the appearance of the Panathenaic ship, see also Mansfield 1985: 69; Parker 2005: 262 ns. 38–39.

210. Norman 1983: 45.

211. For a list of textual references to the Panathenaic ship, see Mansfield 1985: 70–78; Göttlicher 1992: 108–110; Martin 1999: 142.

212. Svoronos 1899; Basch 1987: 346, 347 fig. 732.

213. Deubner 1956: 248; Norman 1983: 43 n. 19, 44; Simon 1983: 6; Basch 1987: 346, 347 fig. 732; Palagia 2008: 233–234, 237.

214. Deubner 1956: 248–254, pls. 34–40; Pickard-Cambridge, Gould, and Lewis 1968: 51; Norman 1983: 43–44; Simon 1983: 6–8, 16, 18, 21, 25, 54, 59, 76–77, 81, 83, 90, 100–102, pls. 1–3; Palagia 2008: 233–234.

215. Deubner 1956: 248; Simon 1983: 6; Palagia 2008: 217, 224.

216. Deubner 1956: pls. 34–35, 40; Simon 1983: 6, pls. 1–2.

217. On depictions of Archaic ships in general, see Morrison and Williams 1968: 81–117 pls. 8–21; Basch 1987: 202–264; Spathari 1995: 74–102.

218. For the possible three-dimensional profile of the ram, see pp. 144 Fig. 3.61, 145 Fig. 3.62:B in this volume (Erechtheion model). For images of Archaic ships with boar-head rams, see Morrison and Williams 1968: pls. 9: a, c, 10: a, c, 11: b, d, 12: b, d, f, 13, 14: a–g, 15: a–b, 16: a–d, 17: a–e, 18: a–b, 19, 20: a–d; Basch 1987: 206 fig. 427, 207 fig. 428, 208 fig. 430, 209 figs. 431–432, 434: A–B, 210 figs. 434: C–E, 435, 436: A, C; 211 fig. 438, 212 fig. 440: A–C, 213 figs. 441–444, 214 figs. 445, 447, 215 fig. 448, 217 figs. 452: A–B, 453, 218 fig. 454, 219 figs. 456–457, 220 fig. 459: A–B, 221 figs. 460, 460 *bis*, 222 figs. 462–463, 224 fig. 468, 225 fig. 469, 226 figs. 470: B, 471, 227 figs. 472–473, 233 fig. 482, 234 fig. 485, 235 fig. 486: 2–3, 238 figs. 497–499, 242 fig. 510: A, 243 fig. 511; Spathari 1995: 74 fig. 79, 82 fig. 91, 84–85 fig. 94, 87 figs. 95, 97, 88 figs. 98–99, 89 fig. 100, 90 figs. 101–102, 91 fig. 104, 93 fig. 106, 95 fig. 107, 97 fig. 109, 99 figs. 113–114.

219. Casson 1995: 85; Steffy 1991: 17–25.

220. Casson 1995: 85, fig. 107; Basch 1987: 241 figs. 506–507, 274 fig. 582, 275 figs. 583–585, 299 figs. 632–634, 300 fig. 636, 301 fig. 640; Spathari 1995: 113 fig. 131, 115 figs. 134–135, 118 fig. 137, 125 fig. 142, 127 fig. 147, 128 fig. 148, 129 figs. 152–155.

221. Basch 1987: 208 fig. 430, 210 figs. 435–436, 215 fig. 450, 217 fig. 453, 219 fig. 456, 224 fig. 468; Spathari 1995: 88 fig. 98, 89 fig. 100, 93 fig. 107.

222. Neils 2001: 88–93; Brinkmann 2008; Østergaard 2008; Brinkmann et al. 2007.

223. This raised element extends below the other crosses also, indicating that it is not an element related to the ship's architecture or its conveyance. For images of the other crosses, see Simon 1983: pls. 1 fig. 1, 2 fig. 1.

224. Simon 1983: 6 n. 14; Mansfield 1985: 70 n. 70; Parker 1996: 91.

225. Mansfield 1985: 58 n. 26, 59 n. 30.

226. Mansfield 1985: 70; Werner 1997: 110 fig. 18.

227. Basch 1987: 207 fig. 428, 208 fig. 430, 209 figs. 431–432, 434, 210 figs. 435–436, 214 fig. 446, 215 figs. 449–450, 217 figs. 452–453, 219 figs. 456–457, 220 fig. 458, 226 fig. 471, 227 fig. 472; Spathari 1995: 88 figs. 98–99, 90 fig. 102, 92 fig. 105, 94–95 fig. 107, 97 fig. 109. On the brailed sail, see Casson 1995: 37, 47–48, 70; Cariolou 1997; Wachsmann 1998: 251–254.

228. Philostr. *VS* 2.550. Translation by W. C. Wright from Philostratus and Eunapius 1921: 147; italics added.

229. Mansfield 1985: 70 n. 69.

230. See pp. 120–132 in this volume.

231. See p. 105 in this volume.

232. Pittakis 1862: 94; Martin 1999: 146, 147 Abb. 15. Basch (1972: 44) comes close to this conclusion, noting that the Erechtheion model might represent "some still surviving type of archaic ship," but he does not equate the model with the Panathenaic ship. See also Saglio, Pottier, and Lafaye 1877: 1325 fig. 4587 and n. 14; de Ridder 1896: 139, 140 fig. 95 (425), 141; Staïs 1910: 276 (with photograph), 277 [no. 7083]; Paton et al. 1927: 571–572 fig. 229; Moll 1929: B VII: 1; Morrison and Williams 1968: 179, pl. 27: B [Clas. 20]; Basch 1969c: 442–443, 444 fig. 5:A; 1987: 228, 229 fig. 477, 230 fig. 478, 231; Lloyd 1975: 48 n. 25; Göttlicher 1978: 68, pl. 27 [no. 362]; Johnston 1985: 79–81 [Clas. 2]; Spathari 1995: 110–111 fig. 128.

233. Boetticher 1863: 194; Paton et al. 1927: 571–572 fig. 229. The Archaic wooden galley models found in the German excavations of the Heraion on Samos serve as a Greek cultural parallel for the dedication of *ex-voto* ship models to a temple (see p. 50 n. 59 in this volume).

234. Morrison and Williams (1968: 179 [Clas. 20]) citing R. P. Austin. I thank Harry Kritzas (pers. comm.) for this translation and his comments on the inscription.

235. D. Blackman (pers. comm.).

236. Paton et al. 1927: 571–572 fig. 229; Morrison and Williams 1968: 179 [Clas. 20].

237. Basch 1987: 228.

238. Morrison and Williams 1968: 179 [Clas. 20]. Note that Morrison includes the model with his Classical ship representations, apparently due to the date of the inscription. See also Basch 1972: 44.

239. Martin 1999: 146.

240. Compare particularly Basch 1987: 209 fig. 432, 215 fig. 450, 217 fig. 453, 222 figs. 462–463, 224 fig. 468, 227 fig. 472 (upper photo, galley at

left), but see also 207 fig. 428: C, 210 figs. 435, 436: B–C, 212 fig. 440: A, 215 figs. 449: B, 450; 217 fig. 452: A (galley on the right); Spathari 1995: 87 fig. 97 (galley at left), 89 fig. 100, 91 fig. 104, 93 fig. 106, 97 fig. 109, 99 fig. 114.

241. On the *hypozoma,* see Casson 1995: 91–92, 147, 211 n. 45, 265 n. 3; Kennedy 1976.

242. Basch 1987: 228, 229 fig. 479, 231 figs. 480–481. Kennedy (1976: 162) identifies the rope motif as a fender.

243. Basch 1987: 209 figs. 432–433, 212 fig. 440, 214 fig. 446, 217 figs. 452–453, 218 figs. 454–455, 219 fig. 457, 227 fig. 472, 252 fig. 536, 253 figs. 537–539; Spathari 1995: 80 fig. 88, 84 fig. 94, 86 fig. 95, 87 figs. 96–97, 89 fig. 100, 92 fig. 105, 94 fig. 107, 99 fig. 114.

244. Spathari 1995: 97 fig. 109, 110 fig. 125; Cariolou 1997: 86, 88 fig. 3.

245. Basch 1987: 205 fig. 425, 210 fig. 435, 226 fig. 471, 237 fig. 493–494 (abstract), 243 fig. 511; Spathari 1995: 82–83 fig. 92, 87 fig. 96, 94 fig. 107, 97 fig. 109; Wachsmann 1998: 189, 191 figs. 8.56–57, 192 fig. 8.58: A–E.

246. Lehmann and Lehmann 1973: 179–190; Basch 1987: 354–362.

247. Martin 1999: 145.

248. Basch (1987: 202–227) includes numerous illustrations of Archaic galleys in which the artists are at pains to indicate the long, narrow lines of galleys.

249. See p. 121 n. 154 in this volume.

250. The chisel-shaped ram is not unique to these representations, however (Basch 1987: 270 fig. 574 [British Museum stamnos E440], 428, 432 figs. 936–938, 433 figs. 939, 943 [Sparta model]).

251. Plut. *Vit. Thes.* 23.1. Translation by Bernadotte Perrin from Plutarch 1914: 49. Mansfield (1985: 68, 74–75) enumerates a total of four documented iterations of the Panathenaic ship, but we should keep in mind that not all reconstructions and repairs may have reached us or even have been originally documented.

252. On the various iterations of the *Amunuserhet,* see Foucart 1921–1922.

253. Pritchard 1969B: 25–29. Herihor had a representation, apparently of this new *Amuneserhet,* carved in the temple of Khonsu at Karnak, but the chronology suggests that the depiction was prepared before Wenamun could have returned with the timber for the new vessel's construction (Kitchen 1973: 252 n. 45; *SKH:* xiv: pls. 19, 21; Egberts 1991).

More recent analogies are HMS *Victory,* Nelson's flagship at the Battle of Trafalgar and the *Olympias.* Although HMS *Victory* still exists in all its glory, very few of her timbers are original. The modern Greek trireme replica *Olympias,* launched in 1987, underwent major repairs for the 2004 Olympics so that it could deliver the Olympic flame to Piraeus. There are currently plans to refit the vessel so that it can take part in New York's July 4, 2012, celebrations (Hollander 2011).

254. Pittakis 1862: 91–92; de Ridder 1896: 140 fig. 95 (425); Moll 1929: no. B VII: 1; Johnston 1985: 81; Martin 1999: 146.

255. Morrison and Williams 1968: 179 (Clas. 20). On the *histopede,* see Casson 1995: 47 nn. 32–33.

256. For lamps with central stalks, see Walters 1914: 18 nos. 105 (pl. VIII), 106 fig. 11, 110 no. 734 fig. 129. For a lamp in the form of a ship hung from a central element, in this case a female figure, see Basch 1987: 449, 452 fig. 997. One is reminded of the bronze Sardinian ship models (ca. 11–10th to 6th centuries B.C.), which have various constructions to allow them to be hung from amidships (Lilliu 1966: 389 no. 271, figs. 533, 534 *bis,* 390 no. 273, fig. 537, 392 nos. 275–276, figs. 541–543, 544, 546, 393 nos. 277–278, figs. 547–548, 396 no. 281, fig. 552, 397 no. 282, fig. 553, 398 no. 284, figs. 555–556, 399 nos. 285–286, figs. 557–558, 400 no. 287, fig. 559, 401 no. 288, fig. 560, 402 no. 289, fig. 562, 403 no. 290, fig. 563, 405 nos. 291–292, figs. 564–565, 406 no. 293, fig. 566, 407 no. 294, figs. 567–568, 408 no. 295, fig. 569, 410 no. 297, figs. 571–572, 411 no. 298, fig. 573, 412 no. 299, figs. 574–575, 422 no. 314, fig. 593, 423 nos. 315–316, figs. 594–595, 425 no. 318, fig. 597, 426 no. 319, figs. 598–600, 427 no. 320, figs. 601–602, 431 no. 321, figs. 603, 605–606, 436 nos. 323–324, figs. 608–610, 437 nos. 325–326, figs. 611–612, 438 no. 327, figs. 612 *bis,* 613, 439 no. 328, fig. 614, 441 nos. 329–331, figs. 615–617). The closest parallel to the Erechtheion model in shape, if not in size, appears to be a bronze ship/lamp model now in the Cincinnati Art Museum, which also has a central stalk (Adams 1965; Cincinnati Art Museum 1970: 34 no. 1962.392; Johnston 1985: 81–82 [Clas. 3]). Note also an apparently Roman-period bronze ship/lamp, found in the sea (Basch 1987: 449, 452 fig. 996).

257. Martin 1999.

258. The one-and-a-half rows of oar ports raises the question as to whether the Kerameikos model might represent a type of galley termed *hemiolia.* Casson (1958; 1995: 62 n. 95, 128–132, fig. 117) demonstrates, however, that the name, which implies a "one-and-a-half" arrangement, derives from the practice of pirate ships to have half the oarsmen on one of the complete rows ship their oars in order to lower sail, mast, and rigging as the *hemiolia* approached a victim. A black-figure cup eloquently depicts this procedure. This vessel type started out as a pirate ship, but by the fourth century B.C. had been adopted by war fleets. In general, given its heritage, the *hemiolia* would seem an odd prototype for a Panathenaic ship.

259. Martin 1999: 148 Abb. 17, 149.

260. Martin 1999: 138 Abb. 1–2, 139 Abbs. 4–5, 140, 141 Abb. 10, 149.

261. Martin 1999: 140.

262. Martin 1999: 148.

263. Pers. comm. On the ship of Theseus, see p. 142 in this volume.

264. Martin 1999: 138.

265. Martin 1999: 141 Abb. 9.

266. Martin 1999: 148 Abb. 17.

267. Basch 1972: 44, 45 fig. 30.

268. Casson 1995: 86, 344–345.

269. See pp. 134–146 in this volume.

270. Spetsieri-Choremi 2000. On the City Eleusinion, see Thompson and Wycherley 1972: 150–155, pls. 77–78, Miles 1998 and additional bibliography there.

271. Spetsieri-Choremi 2000: 14.

272. Baika 2002.

273. Mansfield 1985: 76 n. 98. Pausanias (1:29:1) notes, however, that in his day (second century A.D.) the Panathenaic ship was stored near the Areopagus, where it could be viewed. This is apparently not the same ship as the one built by Herodes Atticus (Gardner 1914; Tobin 1993: 89; Hurwit 1999: 277 n. 85).

274. On Athens's sacred ships, see Jordan 1972: 153–184.

275. Mentioned in a presentation at Tropis X (Tzalas 2008) and pers. comm. See also communications there by D. J. Blackman (2008) and Trakadas (2008). Regarding the ritual offering basket *(kanoun)*, see Roccos 1995: 642 fig. 1, 647 fig. 2, 650 figs. 5–6, 651 fig. 7, 652 figs. 8–9, 653 fig. 10, 654 figs. 11–12, 656 figs. 14, 15 no. 49, 658 fig. 16, 659 fig. 17, 661 fig. 20; Neils 2001: 93, 149 fig. 111, 157, foldout: East Frieze: E49.

276. See pp. 41, 117 in this volume.

277. See quote from Plut. *Vit. Thes.* 23.1, p. 142 in this volume; Dyer 1891: 377; Jordan 1972: 158, 160–161; Casson 1975: 7 n. 11; Simon 1996: 12; Hale 2009: 86, 88.

278. W. M. Murray (pers. comm.).

279. Carlson 2009: 347–349, 359–363.

280. The bows of the Kerameikos model and the Eleusinion relief are missing and, therefore, cannot be of help in this regard.

281. Casson 1995: 270–273; Davis 2000: 31–40; 2007: 90.

282. Wenamun, for example, sailed in early January 1075 B.C. from Egypt to Byblos and arrived there in May (Egberts 1991: 59–61, 67). The Ahikar text from Elephantine, which contains an Egyptian custom account from 475 B.C., indicates a long sailing season that lasted from February/March to November/December (Porten and Yardeni 1993: xx–xxi; Yardeni 1994: 69–70). At Crete Paul (Acts 27:9–10) argued against sailing to a better harbor in which to winter, specifically noting that it was late in the sailing season, which was defined in Midrash as stretching from Shvuot to Sukkot, roughly from May to September (Sperber 1986: 99–100). This custom of limiting maritime traffic to the summer months evolved into legislation during the medieval period, when sailing outside the season became unlawful (Ashburner 1909: cxlii–cxliii).

283. Morgan 1988: 144–145, 162; Wachsmann 1998: 105–122.

284. Palmer 1955: 10–12; 1988: 248, 254–255, 265; Bennett and Olivier 1973: 233; Chadwick 1976: 90, 179, 192.

285. Göttlicher 1992: 114–128; Alföldi (1937: 47–51) proposes that the Navigium Isidis merged with the renewal of vows to the emperor, which fell on January 3. Contra this view, see Brady 1938: 89–90.

286. Brady 1938: 89–90; Levi 1942: 33; Barb 1971: 154; Bruneau 1974: 340–341; Witt 1997: 178.

287. Bruneau 1974: 335–341.

288. Bruneau 1961: 442–445.

289. The earliest appearance of Isis Pelegia on a coin is from Byblos and dates to the reign of Antiochus IV Epiphanes (175–164 B.C.). See Alföldi 1937: 23, 46, pl. XVIII: 1–39; Bruneau 1961: 438–442; Witt 1997: pls. 60–63; Spathari 1995: 142, 143 fig. 175.

290. Apul. *Met.* 11. Regarding the date, see Griffiths 1975: 9–10.

291. Griffiths 1975: 6, 15; Witt 1997: 166.

292. Apul. *Met.* 11, 16. Translation from Griffiths 1975: 89, 91. Griffiths (46–47) considers the ship of Isis described by Apuleius as a fully functional trader with a crew onboard.

293. Moll 1929: 28, 67, Taf. B XIb: 97; Stern 1981: 440–441, pl. VII no. 23; Göttlicher 1992: 119–120.

294. Museo Nazionale Romano, Palazzo Altemps Inv. 77255.

295. Levi 1942: 32–35, pl. I: 2; 1947: 164–165.

296. Compare this with scenes of harbors identified by Alföldi (1937: 48–49, pl. XI: 1, 3) as representing the Navigium Isidis.

297. See comments by Griffiths (1975: 44) regarding the question of this scene representing the Navigium Isidis.

298. Alföldi 1937: 22, 48, pl. XVI: 19–25.

299. Laing 1918: 250–251; Alföldi 1937: 30–58; Brady 1938: 89–90.

300. For the traditional view, see OED: s.v. carnival. For the connection to antiquity, see Burckhardt and Hohl 1924 (first ed. 1852): 193, 194 n. 1; Alföldi 1937: 57–58, Brady 1938: 90; Levi 1942: 33; Barb 1971: 154–155; Lehmann and Lehmann 1973: 213; Parke 1977: 109–110; 1977. Contra this view, see Wittmann 1938: 213 n. 476; Griffiths 1975: 47 no. 6; Göttlicher 1992: 128.

301. Laing 1918: 249; Alföldi 1937: 58; Barb 1971: 154; *PM* I: 299; II: I: 251–252; II: II: 843.

302. Monloup 1984: 158–160, pl. 29. Perhaps this is a Phoenician version of a Dionysian ship cart? On possible Semitic connections to Semele/Dionysos, see Astour 1965: 169–194. I thank V. Karageorghis for bringing this model to my attention.

303. See pp. 133 n. 219, 137 Fig. 3.54:B in this volume.

304. Monloup 1984: 159. Compare DeVries and Katzev 1972: 56 fig. 9; Basch 1987: 310 fig. 654, 311 fig. 655, 312 figs. 656–657, 313 figs. 658–659, 314 figs. 660–661, 315 figs. 662–663, 316 figs. 664–666, 318 fig. 669, 321 figs. 675–678, 322 figs. 679–681, 323 figs. 682–683, 324 figs. 685–686, 325 figs. 687–694, 326 figs. 695–697, 329 figs. 703–711, 330 figs. 712–715, 331 fig. 717, 395 figs. 719–720.

305. Monloup 1984: 166–169 nos. 615–643.

306. De Graeve 1981: 73 and pl. XLV (no. 103).

307. Wachsmann 1998: 239; 2003.

308. Vercoutter 1964, 1965: 68–69; Creasman and Doyle 2010.

309. Newberry 1942: 65.

310. *BAR* I: §429–430, 432.

311. Sayed 1977: 170.

312. Mersa Gawasis: Fattovich 2005; Fattovich and Bard 2006; Delgado 2008: 307–310; Veldmeijer and Zazzaro 2008; Ward and Zazzaro 2010. Ayn Soukhna: Abd el-Raziq, Castel, and Tallet 2004: 19–21; 2006; 2007; 2010.

313. Pritchard 1969B: 240.

314. See references in Newberry 1942: 64;Wachsmann 1998: 239.

315. Wadi Tumilat: Redmount 1994, and additional bibliography there; Wachsmann 2003. Athos: Isserlin 1991; Isserlin et al. 1994; 1996; Karastathis and Papamarinopoulos 1994; 1997; Karastathis, Papamarinopoulos, and Jones 2001; Jones et al. 2000.

316. Reinach 1909: 327 fig. 134, 328 fig. 135; Moll 1929: 66, Taf. B VIII: 27; Caprino et al. 1955: 115 (Scene CXI) Tav. LXVI fig. 132, Tav. LXVII fig. 133, Tav. R; Scheid and Huet 2000: (Scene 111) 412 fig. 131, 413 fig. 132.

317. Caprino et al., 1955: 72–73.

318. Caprino et al. 1955: (as water transport): Tavs. III fig. 7 (Scene II), IV fig. 8 (Scene II), XVIII fig. 36 (Scene XXVI), XXII fig. 44 (Scene XXXIV), XXXVII fig. 74 (Scene LIX), LI fig. 102 (Scene LXXXI), Q (Scene LXXXI), (as pontoons for bridges): Tavs. IV fig. 9 (Scene III), LII fig. 104 (Scene LXXXV), LXIV fig. 127 (Scene CVIII), LXIX fig. 138 (Scene CXV), LXX fig. 139 (Scene CXV), Q (Scenes CVIII, LXXXIV), T (Scene LXXXIV); Scheid and Huet 2000: (as water transport): 313 figs. 9–10 (Scene 2), 339 fig. 45 (Scene 27), 345 fig. 53 (Scene 33), 427 fig. 149 (Scene 3), 431 fig. 157 (Scene 33); (as pontoons for bridges): 314 figs. 11–12 and 315 fig. 13 (Scene 3), 382 fig. 97 and 383 fig. 98 (Scene 78), 407 fig. 126 and 408 fig. 127 (Scene 108), 416 fig. 135 (Scene 115), 433 fig. 161 (Scene 78).

319. A similar concept is seen in the construction of supergalleys from the fifth century B.C. and later, which Casson (1995: 107–112, 114 n. 57) identifies as catamarans. The *schedia* built to the orders of Vespasian prior to the battle of Migdal (A.D. 67) also may have been pairs of local boats connected to form catamarans (Wachsmann 1990: 121–122; 2009: 180, 330–332).

320. Caprino et al. 1955: (Scene XXXVIII) Tavs. XXIII fig. 47, XXIV fig. 48; (Scene XCIII) Tav. LVI fig. 111; Scheid and Huet 2000: 348 fig. 56 (Scene 37), 428 fig. 152 (Scene 25).

321. Caprino et al. 1955: Tavs. LVI figs. 111–112 (Scene XCIII), R (Scene XCIII); Scheid and Huet 2000: 339 fig. 29 (Scene 29).

322. Joseph. *BJ* 7, 121–157.

323. Joseph. *BJ* 7, 147.

324. Joseph. *BJ* 3, 462–505, 522–542; Wachsmann 1990: 121–122; 2009: 169–194.

325. Wachsmann 2009: 191–193.

326. Joseph. *BJ* 3, 414–427; 4, 439.

327. Runciman 1993: 436–437.

328. Scandurra 1972: 209–210, 220–221 figs. 10–18.

329. Wiseman 1978: 44–50; Werner 1997: 114–116; Pettegrew 2011.

330. MacDonald 1986; Strassler 1996: 486 ill. 8.8; Werner 1997, and additional bibliography there.

331. Thuc. 1, 13, 5.

332. MacDonald 1986: 193.

333. Werner 1997: 100.

334. Werner 1997: 112–114. For a list of ancient authors, see Lohmann 2008.

335. MacDonald 1986: 193–195.

336. Werner 1997: 111.

Chapter 4: Foreigners at Gurob

1. On the model *pavois* found with the Gurob ship-cart model, see pp. 20 Fig. 1.20, 28 item no. 1, 102, 103 Fig. 3.18, 104 Figs 3.19–20, 105 Fig. 3.21 in this volume.

2. Götze 1957: 80–81; Mellink 1966: 457–458; Veenhof 1995: 861.

3. Sparks 2004: 28–29, 47–48. For an example of such a phenomenon at Gurob, see pp. 193, 195, 197–199 in this volume.

4. Sparks 2004: 48.

5. Spiegelberg and Erman 1898; Ward 1994: 62; Sparks 2004: 34 fig. 3.4: A, 35. For a color photo of the stele, see Russman 2008: 265 fig. 165; Homan 2010: 48. On Asiatics in New Kingdom Egypt, see Redford 1992: 214–237. On Syro-Canaanite clothing, see Davies and Faulkner 1947: 44–45; Pritchard 1957; Wachsmann 1987: 45–46.

6. Enigmatic figures, possibly Minoans, also appear earlier carrying metal ingots on a Middle Kingdom relief (*PM* II: 177 fig. 90; Davies 1932: 61 fig. 13; Wachsmann 1987: 50, pl. LII: A).

7. Vercoutter 1956; Wachsmann 1987.

8. Wachsmann 1987: 33–37.

9. Wachsmann 1987: 127–129.

10. Wachsmann 1987: 4–26, 37–40.

11. See 97 n. 52 in this volume.

12. See 52 n. 65 in this volume. Of unclear significance is C. Doumas's (2008: 125 n. 4) observation that children portrayed in the tomb of Inkherkhau (TT 359) at Deir el Medina have tonsures reminiscent of those seen on youths at Thera, suggesting a possible Egyptian adoption of an Aegean trait (Meskell 2002: 86 fig. 3.11, 87; Weeks 2005: 507). Additionally, the ceiling of a chamber of Inkherkhau's tomb bears Aegean-style decorations, including running spirals, *bucrania,* and other motifs reminiscent of those appearing on Minoan textiles in earlier tombs (Weeks 2005: 502–503; Barber 1992B:10 n. 1, 338, 339 table 15.1, 340, 342, 348–351, 396). On the influence of Minoan textiles on Egyptian art motifs, see Kantor 1947: 20–21, 26, 29, 43, 58–59, 99; Barber 1992B: 311–357; 1998.

13. Thomas 1981A: 6; Bell 1991: 254; Leahy 2000; Sparks 2004: 26–28. See the case of the Semite Ramessesemperrē', pp. 178, 187, 205 in this volume.

14. S. Jones 1997: 56–83; Sherratt 2005; Woudhuizen 2006: 15–28; Yasur-Landau 2010: 9–33.

15. Woudhuizen 2006: 21 fig. 1:b.

16. Woudhuizen 2006: 15.

17. Lehmann 1979; Rainey 1982: 133–134; Wachsmann 1998: 164; Hoftijzer and van Soldt 1998: 343.

18. Niemeier 1998; Wachsmann 1998: 129.

19. EA 38: 7–22. Translation from Moran 1987. See also Wachsmann 1998:130.

20. Hoftijzer and van Soldt 1998: 343 (RS 20.18: 7–13).

21. Sweeney and Yasur-Landau 1999.

22. Ward 1994.

23. Ward 1989: 290.

24. Ward 1989: 295–299.

25. Ward 1989: 288.

26. Shisha-Halevy 1978.

27. Shisha-Halevy 1978: 162 line 3.

28. Kitchen 1982: 88–89; Bell 2007: 110 nn. 40–41. On Maat-Hor-Neferure, see pp. 166, 197–199 in this volume.

29. Yon 1992: 119.

30. PN = personal name.

31. Ward 1994: 62.

32. Gaballa 1972: 129–132, 133 n. (k); 1973: 109, 110 (c); Kitchen 1982: 40, 88, 99–100, 110.

33. Ward 1989: 289–291; 1994: 63–64.

34. Ward 1994: 66.

35. Ward 1989: 289.

36. Ward 1994: 61–62; Leahy 2000; Sparks 2004. For the situation in the early New Kingdom, see Panagiotopoulos 2006.

37. Sparks 2004: 45–47.

38. Davies and Faulkner 1947: pl. VIII; Merrillees 1973: 179; Wachsmann 1998: 42 fig. 3.2, 313, 314 fig. 14.6.

39. Davies and Faulkner 1947: 46; Wachsmann 1998: 313.

40. Thomas 1981A: 5.

41. Petrie 1890: 40–45 nos.79–86; 1891: 16–20 nos. 36–42; Petrie 1931: 101–102; Bell 1991: 254–255; Thomas 1981A: 5; David 1986: 98, 192.

42. Thomas 1981A: 5.

43. Bell 1991: 255.

44. For maps and satellite photos of the site, see the virtual-reality package that accompanies this book online at http://www.vizin.org/Gurob/Gurob.html.

45. Thomas 1981A: 1.

46. David 1986: 40 42.

47. Thomas 1981A: 1–4; David 1986: 89–98.

48. Kemp 1978: 124–125; Lacovara 1997: 297.

49. Anonymous 2008; Shaw 2007; 2008A; 2008B; 2010; 2011A; 2011B. Annual field reports are available on the project website (http://www.gurob.org.uk/).

50. Petrie 1890: 5 no. 2, 32–33 nos. 64–65 (Merneptah); 1891: 16 no. 35 (Ramses III); Thomas 1981A: 2, 4–5.

51. On the Wilbour Papyrus, see p. 176 Fig. 4.10, 177 Fig. 4.11, 187 in this volume.

52. Thomas 1981A: 4.

53. Thomas 1981A: 5; Redford 1992: 290; Trigger et al. 1996: 222–232.

54. Tombs 499, 601, 605–606, and 609–614. Brunton and Engelbach 1927: pl. II (lower); Thomas 1981A: 21. Bell (1985: 62 n. 8) notes correctly that Brunton and Englebach (1927: pl. XVIII) list the tombs under point H, but this is curious, as they place this point to the north of the wadi that separates it from this tomb cluster.

55. For digital versions of the maps and QuickBird satellite images, see the virtual-reality package that accompanies this book online (http://www.vizin.org/Gurob/Gurob.html). The original photo was taken on November 12, 2002, and has a resolution of 60 cm.

56. Thomas 1981A: 21; Bell 1985: 63.

57. Brunton and Engelbach discuss in summary form only six of the most important of these tombs vis-à-vis chronology (1927: 4, 16–17 pls. XXIV, XXIX, XXXI [Tombs 601, 605, 606, 609, 611, and 613]). The tomb register, however, lists the remaining tombs (pl. XVIII). See also Thomas 1981A: 21.

58. Bell 1985: 63.

59. Bell 1985. On Late Helladic IIIB synchronisms see also Hankey and Warren 1974: 148–149. On a possibly foreign female burial in Tomb 605, see p. 193 in this volume.

60. Griffith 1898A: 95; 1898B: pl. XXXIX [l. 33]; Gardiner and Bell 1943: 43; *WP* II: 44–47; Kemp 1978; Thomas 1981A: 1, 5; David 1986: 42.

61. *WP* II: 45 n. 5, 69.

62. Gardiner 1953: 145. This followed the establishment of a harem at nearby Lahun during the XIIth Dynasty (*WP* II: 45 n. 8).

63. *WP* II: 30, 44, 47.

64. Gardiner and Bell 1943. The phenomenon of naming a lake after an important settlement on or near its shores is paralleled by the various names given to Israel's Sea of Galilee throughout history. Over time, the lake received its name from major cities on its shores or within its region (Galilee) (Wachsmann 2009: 39–41).

65. Gardiner and Bell 1943.

66. *WP* II: 44, 46.

67. *WP* II: 18, 30, 44–47, 74, 129 (§39), 141–142 (§111–112), 157 (§278–279), 192 (§12). The manner of assessing the entries dealing with harem lands in the Wilbour Papyrus is different from that used to assess those owned by other institutions (*WP* II: 55).

68. *WP* II: 18.

69. *WP* II: 45. On Ptolemais Harmos, see Gardiner and Bell 1943: 41, 43, 45–46.

70. Memphis represents the only other such stand-alone harem complex known from the late New Kingdom period (Kemp 1978: 132–133).

71. Lacovara 1997: 300, 304, 305 fig. 5.

72. Davies and Davies 1973: 20–25, pl. I; Lacovara 1997: 304, 306 fig. 6.

73. Gardiner 1968: X–XII, 14–35.

74. Thomas 1981A: 9–11, 47 (Cat. no. 215); Bell 1991: 251–252.

75. Petrie 1890: 35 no. 69, pl. XVIII: 11, 17, 29, 31; Hayes 1959: 266, 267 fig. 161; Thomas 1981A: 2 no. 2; Trope, Quirk, and Lacovara 2005: 83 fig. no. 64.

76. Borchardt 1911: Blatt. 1–4; Thomas 1981A: 3 no. 4; Fay 1986: 50–51; Kozloff et al. 1992: 192 pl. 17, 209–210.

77. Petrie 1890: 34 no. 68; Thomas 1981A: 16 nos. 45–46; Bell 1991: 251 n. 6. See pp. 191–193 in this volume. In 1924 W. S. Blackman reported that weaving continued to be an important source of income at nearby Lahun into the modern era (David 1986: 42).

78. Barber 1992B; 1994. There is, however, some evidence for men weaving on vertical looms in Egypt during the New Kingdom (Barber 1994: 259–260, 261 fig. 11.2, 262 fig. 11.3, 263).

79. See pp. 192, 197–198 in this volume.

80. Thomas 1981A: 7–8; Bell 1991: 251.

81. Thomas 1981A: 5.

82. Petrie 1891: 19 no. 39, pl. XXII: 1–3; 1974A: 6 no. 12: 13–14, pl. II: 13–14; Thomas 1981A: 87 no.754; 1981B: pl. 40 no. 754. On Syro-Canaanite toggle pins, see Henschel-Simon 1937; Biers and Terry 2004: 59 no. 37.

83. Sparks 2004: 34 fig. 3.3: A–B, 35.

84. EA 17: 41–45. Note also the many pairs of decorated toggle pins sent by Tushratta, presumably as part of his daughter's (Tadu-Kheba) dowry (EA 25: [i] 22–32; Cochavi-Rainey and Lilyquist 1999: 102–105). On the background to Amenhotep III's diplomatic marriages to two Mitannian princes, see Rainey and Notley 2006: 77, 79.

85. Petrie 1891: 19 no. 39, pl. XXII: 3. For gold and silver toggle pins, see Hennessy 1966: 161, pl. XXV: A.

86. Schulman 1976; Bell 1991: 255.

87. Brunton and Engelbach 1927: pl. XLIII: 44: X, Z; Sparks 2004: 38, 39 fig. 3.6: A, C.

88. Petrie 1890: 41 no. 80, pl. XVIII: 4.

89. *WP* II: 80, 81 n. 1; *WP* III: §123: 46, 28; Faulkner 1941: 18; Caminos and Gardiner 1954: 176–177, 180; Kitchen 1973: 295 n. 291; Hulin 2011. See Appendix 4: Sherden and Tjuk-People in the Wilbour Papyrus, pp. 225–237 in this volume. On the Libyan character of the Third Intermediate Period in general, see Ritner 2009: 1–8.

90. II Chronicles 12:3. On the campaign, see Kitchen 1973: 294–300; Rainey and Notley 2006: 185–189.

91. The Kehek may have been a Tjemhu tribe, although this remains unclear (Bates 1914: 47; Kitchen 1990: 15 n. 3; O'Conner 1990: 44–45, 76, 96, 104, 107; Winnicki 2009: 76–77). See also pp. 39, 185–187, 205 in this volume.

92. Gardiner 1947: 194*; Cavillier 2005: 12.

93. *BAR* III: §588.

94. *BAR* IV §: 402, 410. See texts quoted on pp. 186–187 in this volume.

95. Edgerton and Wilson 1936: 27 l. 40; see also 27 nn. 40a, 42. For a scene of the branding of feather-helmeted Sea Peoples, see *MH* I: pl. 42.

96. *BAR* IV §: 405. On the probable locations of sites mentioned, see pp. 201 n. g, 202 nn. a, c.

97. Wilson 1935; Faulkner 1941: 18; 1953: 45; Wainwright 1962; Kitchen 1990; O'Conner 1990; Redford 1992: 102, 247–250; Trigger et al. 1996: 252–253, 271–272, 273 fig. 3.25, 274–278; Leahy 2000.

98. Ward 1989: 289; 1994: 65.

99. See, for example, Peet 1972.

100. Ward 1994: 65–66.

101. Petrie 1890: 42–43 no. 82, pl. XXVIII: 1–3, 5–9, 11–14, 16–18; 1891: 17–18, pls. XVII: 3, 28, 40, XVIII: 52 (local imitation), XIX: 12 and 17 (local imitations); 1931: 101–102; Brunton and Engelbach 1927: 12 (glazed), 13, pls. XII: 4, XIV: nos. 6, 37, XV: no 217 (glazed "Aegean"), XVI: no. 245, XVII: no. 475, XVIII: no. 605, XXV no. 4 (glazed "Aegean"), XXVII: no. 19, XXIX: no. 39, XXXIX: nos. 95: G–H, K, P, 97: A–B; 98: B; Merrillees 1968B: 48–57; Merrillees and Winter 1972: 116–117; Thomas 1981A: 5, 43 no. 171, 44 nos. 172, 175–176, 46 nos. 201–207, 47 nos. 208 (local imitation); Bell 1985; 1991: 255–277.

102. Petrie 1890: 43 no. 82. See also Hassler 2011.

103. Petrie 1890:43 no. 82; Säve-Söderbergh 1946: 66 n. 4. J. M. Kelder (2009), the latest to espouse this view, proposes that Mycenaean ceramic finds at Amarna indicate a Mycenaean presence there to fit a perceived political reality.

104. Bass 1967: 165–167; 1991: 69–70; 1997: 76–77; Merrillees 1968B: 66; 1973: 180.

105. Basch 1972: 50; Wachsmann 1998: 206, 208, 211–212.

106. Bass 1967: 163–165; 1976: 57–59; 1991: 69–71. Other important indicators of the origins of a ship include stone anchors, weight sets, and, on occasion, figurines found on board (Pulak 1996; 2008: 300–302; Wachsmann 1998: 211–212. Note that, in the case of ships, a vessel might not have been built at its homeport (Wachsmann 1998: 310–313; Hoftijzer and van Soldt 1998: 337–338 [KTU 4.338]). For example, C. Pulak's latest research suggests that the ship that sank at Uluburun hailed from Israel's Carmel Coast, but the ship's hull was constructed of Lebanese cedar, which does not grow locally (Pulak 2008: 299, 302–303; see also 1998: 213; 2001: 29, 30 fig. 1, 31 fig. 2, 32–33, 35). Alternately, the hull could have been laid on the Camel coast from timber imported from the Lebanon. The Galilee Boat's builders built that vessel on the shores of the Sea of Galilee some thirteen centuries later using imported Lebanese cedar for most, although not all, of its hull planking (Werker 1990: 65–69; 2005: 233–236).

107. On the evidence for Europeans on board the ship, see Bass 1986: 274, 283–284, 285 ill. 20, 296, pl. 17: fig. 2; 1987: 722, 726; Pulak 1991: 8–9; 1992: 11; 1997: 253–256; 2001: 45–48; 2005B; 2008: 301–302; Bachhuber 2003; 2006.

108. On the Amarna papyrus, see pp. 52 n. 65, 164 in this volume.

109. The guard of Akhenaten consisted of Egyptians, Syrians, and Nubians (Yoyotte 1949: 68; Wreszinski 1988 [II]: 11–14). On Ramses II's Sherden guard, see Breasted 1906: 2 fig. 1, 3 fig. 2, 4.

110. Astour 1964: 193–194; 1965: 353–355; 1973: 25. Two Akkadian texts from the House of Urtenu in Ugarit—RS 94.2530 and RS 94.2523—mention Hiyawa-men to whom rations are to be sent (Lackenbacher and Malbran-Labat 2005: 236–238; Singer 2006). Hiyawa has been identified as the Akkadian equivalent of Hittite Ahhiyawa, but even in this case, those referred to in the texts are clearly located in Lukka (usually identified as Lycia in Asia Minor) and not Ugarit. One should not be dogmatic about the absence of Mycenaeans at Ugarit, however, as the Mycenaean passengers onboard the Uluburun shipwreck indicate that high-ranking Mycenaean agents were presumably reaching Cyprus and possibly the Levantine coast as well by the end of the fourteenth century B.C. On the Ahhiyawa, see p. 53 n. 74 in this volume; on the Europeans onboard this ship, which sank at Uluburun, see p. 182 n. 107 in this volume.

111. Heltzer 1988. Earlier, during the latter part of the reign of Zimri-Lim, an Ugaritian dragoman appears in Mari assisting Caphtorite (Minoan) traders in the acquisition of tin ingots (Malamat 1971; 1998: 34–35).

112. On the Alashia/Cyprus equation, see Holmes 1971; Hellbing 1979; Wachsmann 1986; 2006; Goren, Finkelstein, and Na'aman 2002: 197–198; 2004: 48–75; Muhly 2006. Against this identification, see most recently Strange 1980: 168–184, 209 App. II: map; Bass 1991: 75–76; Merrillees 1987; 2006.

113. EA 33–40.

114. Holmes 1973.

115. EA 35: 30–34.

116. EA 33: 19–32; 35: 40–42; 37: 13–29; 40: 16–20, 24–28.

117. EA 39: 14–20, 40: 24–28.

118. Ward 1989: 297–298.

119. Simpson 1972: 154–155.

120. Basch 1978: 118, 119 figs. 32–33, 120–121; Frost 1979: 146, 155 fig. 4, 156–157; Wachsmann 1998: 62, 274, 281 fig. 12.44.

121. Petrie 1890: 36 no. 69, 40 no. 79, pl. XIX; 1931: 101; Wainwright 1959: 198–199; Thomas 1981A: 6. On Sadi-amia, see p. 197 n. 267 in this volume.

122. Petrie 1890: 40–41 no. 80.

123. Anen-Tursha's ear piercings are identical to those on the golden mask of Tutankhamun (Carter 2001: pls. LXXI, LXXIII; Reeves 1990: 111, 115).

124. Petrie 1890: 36 no. 69.

125. Bell 1991: 254.

126. Thomas 1981B: 6.

127. Bell 1991: 254.

128. *BAR* III: §588, 601; Redford 1992: 246, 248–249; Rainey and Notley 2006: 99, 108; Winnicki 2009: 84.

129. *BAR* III: §588. It is not clear why the number of bodies and single hands do not tally: The Athribis Stele gives the number of 722 [+ *X*] men (*BAR* III: §588 n. d, 601).

130. Wainwright 1959: 201 n. 1; Murnane 1980: 6, 7 fig. 4.

131. *BAR* III: §588, 601 (Merneptah); IV: §129 (Ramses III).

132. Hdt. I: 94; Niemeier 1998: 46, and additional bibliography there. The material culture of the Etruscan culture represents a continuation of the Proto-Villanovan and Villanovan cultures, which derive from Apennine branches of the Urnfield culture (Hencken 1968A; 1968B; Tykot 1994: 61 n. 8). Nevertheless, Herodotus's (I: 94) claim that the Etruscans originated in Asia Minor appears to have received some support from recent mitochondrial DNA studies (Achilli et al. 2007; Pellecchia et al. 2007).

133. Research on the Philistines is truly immense. Significant reference works include, but are not limited to, Dothan 1982A; Dothan and Dothan 1992; Brug 1985; Sandars 1985: 164–170; Mazar 1985; Stager 1991; 1995; Howard Jr. 1994; Gitin, Mazar, and Stern 1998; Sweeney and Yasur-Landau 1999; *Sea Peoples*; Killebrew 2005: 197–251; 2006–2007; 2008; Rainey and Notley 2006: 104–105, 107–110; Yasur-Landau 2010.

134. Loretz 1995: 126. On the historical and iconographic sources of the Sherden in relation to Egypt, see particularly Gardiner 1947: 194*–199*; Yoyotte 1949: 68–69; Mertens 1960: 71–77; Kahl 1995; Cavillier 2005; Winnicki 2009: 81–83.

135. Roberts 2009: 62–63.

136. Gardiner 1947: 195*; *WP* II: 80; Mertens 1960: 81.

137. Bouzek 1994: 232–233; 1996.

138. EA 81: 14–24.

139. EA 122: 31–43; 123: 9–28; Loretz 1995: 127–128, and additional bibliography there; Rainey 2003: 194*–196*. Moran (1987: 201–202) had previously read Sherden in the plural.

140. Rainey 2003: 195*.

141. Yoyotte 1949:68.

142. On Sherden at Ugarit, see Rainey 1965: 26; 1967: 90; Dietrich and Loretz 1972; Liverani 1977; Heltzer 1979; 1982: 40, 106–107, 125–127; Tykot 1994: 62–63; Loretz 1995; Kahl 1995. The Onomasticon of Amenope, which dates to the late twelfth century B.C., locates Sherden along the modern Israeli coast, probably north of Dor (Aharoni 1979: 270; Dothan 1986; Tykot 1994: 67; Killebrew 2005: 204–205; Rainey and Notley 2006: 110).

143. Translation from Gardiner 1947: 195*. See also Petrie 1888: 26 no. 78, lines 14–16; *BAR* III: §491; Yoyotte 1949: 68; Kitchen 1982: 40–41.

144. *BAR* III: §479; Gardiner 1947: 195*; Cavillier 2005: 12.

145. Wachsmann 1981: 187–188.

146. Hdt. II: 151–152, 154, 161; Boardman 1980: 114–115.

147. *BAR* IV: §307; Redford 1992: 225, 227, 243; Cavillier 2005: 12–14.

148. Cavillier 2005: 15–16.

149. *BAR* III: 307. See also Gardiner 1947: 194*–195*.

150. Gardiner 1947: 194*; Cavillier 2005: 12.

151. See pp. 39, 52–53, 57, 181, 183, 189, 205–206 in this volume.

152. *BAR* III: §574, 579, 588, 600; Gardiner 1947: 196*; Redford 1992: 246–250.

153. Caminos and Gardiner 1954: 45.

154. Caminos and Gardiner 1954: 64. See also Gardiner 1947: 194*.

155. On the term "Great Green," which means "sea," and thus "Sherden of the sea," see Kitchen 1978: 170; 1983: 78; Schaden 1979: 147–151.

156. Gardiner 1947: 196*. For Sherden in the Medinet Habu reliefs, see *MH* I: 36–41, 50: A, C–D, 51: F–G, 52: A, 53: D. See p. 39 Table 1 in this volume.

157. *MH* I: pls. 37, 41.

158. *BAR* IV: §129: 4; Gardiner 1947: 196*–197*; Wainwright 1959: 201 n. 1; *MH* VIII: pl. 600; Murnane 1980: 6, 7 fig. 4.

159. Gardiner 1947: 196*; Sandars 1985: 125 fig. 79: right; Wreszinski 1988 (II): 160: A–B; Murnane 1980: 6. For a detailed close-up watercolor reproduction of a Sherden warrior's head and helmet where the paint had survived, see *MH* II: pl. 65: C (detail of pl. 62: upper left, second register, first Sherden on left). This figure sports a beard *painted* in black. Apparently the Sherden were bearded, but the Egyptian artists included this detail only in paint, which is now for the most part lost. The Sherden leader has a *carved* thick beard and wears a horned helmet topped with a projecting spike ending in a ball or disk. Similar helmets with disks are worn by some Sherden (Breasted 1906: 2 fig. 1, 3 fig. 2, 4, 30; Cavillier 2005: 13–15) while on others the ball/disk is missing.

160. See p. 190 n. 207 in this volume.

161. Wainwright 1961: 74. On the possible identity of the feather-helmeted Sea Peoples' group manning the other three ships seen in the Medinet Habu naval battle, see pp. 188–190 in this volume.

162. *MH* I: pl. 32.

163. *BAR* IV: §397–412. This is the longest papyrus known from ancient Egypt, and it has survived in excellent condition (*BAR* IV §: 152). See also pp. 39, 181, 183, 188–190, 206 in this volume.

164. *BAR* IV: §397.

165. *BAR* IV: §402.

166. *BAR* IV: §403. Gardiner 1947: 196*. See pp. 39 Table 1 in this volume.

167. *BAR* IV: §410.

168. *WP* II: 1, 9. The Wilbour Papyrus is eclipsed only by the Harris Papyrus at 45 meters and the Ebers Medical Papyrus at 20 meters.

169. *WP* I–IV; Katary 1976.

170. *WP* II: 9.

171. *WP* II: 12.

172. Schulman 1976; Bell 1991: 255.

173. Schulman 1976. On Ramessesemperrē', see also pp. 178, 187, 205 in this volume.

174. *WP* II: 80; Thomas 1981B: 6. See Appendix 4: Sherden and Tjuk-People in the Wilbour Papyrus, pp. 225–237 in this volume.

175. Faulkner 1953: 45.

176. *WP* II: 76.

177. Cavillier 2005: 31.

178. *WP* II: 80; Faulkner 1941: 18.

179. *WP* II: 60–61.

180. *WP* II: 80.

181. See p. 185 n. 136 in this volume.

182. *WP* II: 82.

183. *WP* II: 83.

184. *WP* II: 88.

185. *WP* II: 80; III: 49 §123, 47, 19.

186. Regarding the date of the papyrus, see Gardiner 1941: 43; Janssen 1996.

187. Gardiner 1941: 37–56, pl. VII; Janssen 2004.

188. Gardiner 1941: 40 (lines 4, 10), 41 (lines 5, 4), 46; Janssen 2004: 22, 25.

189. Gardiner 1940: 23.

190. Gardiner 1940: 24–25; Cruz-Uribe 1988: 222.

191. Wente 1990: 171–201. Remains of Butehamon's house, located inside Medinet Habu, still survive (Hölscher 1954: 4, 5 fig. 3, pl. 5).

192. Wente 1967: 11–12, 38 (no. 9, British Museum 10326), 60 (no. 28, British Museum 10375), 83 n. *k* (no. 50, Papyrus Turin, 2026); 1990: 190 no. 312, 192 no. 313, 194 no. 315.

193. Petrie 1905: 22 no. 34, pl. XXVII: 1; *WP* II: 80.

194. Daressy 1915: 142; Cavillier 2005: 33.

195. Petrie 1905: 22 no. 34, fig. XXVII: 2.

196. For the chronology of the XXnd Dynasty, see Kitchen 1973: 588.

197. Kitchen 1973: 295 nn. 290–292.

198. For an overview of acculturation as it applies to the Philistine phenomenon, see Stone 1995.

199. Gardiner 1947: 196*; Winnicki 2009: 84–85.

200. *BAR* IV: §64.

201. In their translation of this text Edgerton and Wilson (1936: 53 n. 17a) propose to identify this location as Azarwa rather than Arvad.

202. *BAR* IV: §403.

203. Alt 1944; Albright 1975: 509, 511; Dothan 1982A: 3; Dothan and Dothan 1992: 26–28; Singer 1985: 109–111, 114; 1985: 109–114; 1988: 1, 4–6; 1992; Wood 1991; Bietak 1993: 292–294; Finkelstein 1995: 213–214; 2000: 159–161; Stager 1995: 340–344; Killebrew 2005: 204.

204. See pp. 183–188 in this volume.

205. *BAR* III: §588 n. a; 601.

206. *MH* I: pl. 32.

207. On past identifications of the feather-helmeted Sea People in the naval battle scene as Philistines, see, for example, Yadin 1963: 251, 340–341; Dothan 1982A: 5, 7, 10 fig. 7, 11; 1982B: 24–26; Stager 1991: 32–33, 35; Wood 1991: 44–45; Barako 2001: 137–138, 281 ill. 2; 2003: 28–29. Y. Yadin (1963: 249, 345) believes that the decorative devices on the diadems surrounding the "feather" helmets depicted at Medinet Habu may represent tribal or clan markings. T. Dothan (1982A: 13) suggests that the decorations may be indicative either of military status or of internal divisions within groups of Sea Peoples. Most recently, R. G. Roberts (2009: 65–67) reviewed the evidence for the varied helmet decorations and concluded that these were not meant to ascribe ethnic identification to the feather-helmeted enemies. He notes that the helmets "have little, if any, interpretive value" (67). I concur with this evaluation, with the exception of the horned-helmets, which are worn exclusively by the Sherden. The specific identity of the feather-helmeted groups simply cannot be established based solely on their helmet designs.

208. See p. 35 n. 15–16 in this volume.

209. On the source for the five representations of a single Sea Peoples' ship, see pp. 39–40 in this volume.

210. Woudhuizen (2006: 115–116, 119) came to the same conclusion but for somewhat different reasons. He equates the Weshesh with the Italic Oscans, a connection first suggested by Chabas (1873: 250).

211. See pp. 52 n. 65, 164, 183, 185 in this volume.

212. A Late Helladic IIIC sherd from Tiryns probably bears a representation of an actual *Vogelbarke* and not a ship. See p. 40 n. 41, Fig. 2.9: C in this volume.

213. *MH* I: pl. 32.

214. It would be tempting to include the Denyen as well in the category of Sea People groups that subsequently settled in Canaan, as per Yadin's (1968) proposal to equate the Denyen with the biblical tribe of Dan, but there are serious objections to such a reconstruction (Rainey and Notley 2006: 110).

215. See p. 190 n. 211 in this volume.

216. Wainwright 1961: 74.

217. Davies and Faulkner 1947: 45–46.

218. Bass 1967: 135–142, 164; 1973: 34; 1986: 292 fig. 31; Bass et al. 1989: 7, 8. fig. 14, 9, 20; Pulak 1988A: 4–5, 25, 30 fig. 37, 31 fig. 38, 32 fig. 39, 33; 1988B: 14–15; 1989: 5, 6 fig. 4, 8; 1990: 9–10; 1991: 6–7, 8 fig. 7; 1992: 7–8, 11; Pulak 1993: 5, 9; 1994: 10, 12, 16 fig. 15; 1995: 52 Abb. 19, 56; 1996; 2000; 2005A: 87 Abb. 43, 88 Abb. 44; 2008: 300, 369–370 no. 235. See p. 182 n. 106 in this volume.

219. Petrie 1890: 41–42 no. 81.

220. Petrie 1890: 41 no. 81.

221. Petrie 1891: 20–21 no. 42.

222. Sparks 2004: 34 fig. 3.4: D.

223. Brunton and Engelbach 1927: pls. XIII: 8, XIV: T. 11; Barber 1992B: 64, 65 fig. 2.32 (left), 66, 299; 1994: 193 fig. 8.3: c, 267; Hassler 2011: 132.

224. Barber 1992B: 63–66, 351. See also Barber 1994: 193 fig. 8.3: C.

225. Sheffer and Tidhar 1988: 226 Text. Cat. nos. 8, 11–12, 228.

226. Kemp and Vogelsang-Eastwood 2001: 58–60.

227. Sparks 2004: 42.

228. Translation from Wente 1990: 36 no. 34. Griffith 1898A: 95; 1898B: pl. XXXIX; Thomas 1981B: 6, 9; Bell 1991: 254. The Metropolitan Museum has a wooden statuette of a woman head of weavers named Teye from Gurob (Hayes 1959: 266, 267 fig. 161; Thomas 1981A: 2, 9 n. 101, 16 no. 45). A headrest from Gurob also gives its owner's title as "Head of Spinners" (Thomas 1981A: 16 no. 46).

229. Thomas 1981A: 6; Hayes 1955: 105 n. 380.

230. Chadwick 1976: 150–152; 1988: 71, 78–79, 82–84, 90–93.

231. Chadwick 1988: 90.

232. Chadwick 1976: 80–81; 1988: 80–81, 83–84, 89, 91–93.

233. Brunton and Engelbach 1927: 16.

234. On the chronological significance of Gurob Tomb 605, see pp. 167 n. 59 in this volume.

235. Brunton and Engelbach 1927: 16–17, pl. XLV: 96: C, E, G, J; Bell 1985: 67–68, pl. II. On penannular rings, see Petrie 1974A: 22 nos. 45–47, pls. XVIII: 1–36; XXVI: 11.

236. Brunton and Engelbach 1927: 17 no. 613, pl. XVIII: no. 614; Bell 1985: 62.

237. Goneim 1957: 6, 27–28, pl. LXXI. My thanks to David Jeffreys for bringing this reference to my attention.

238. Hood 1993: 233. Of particular interest in this regard are two amber seals bearing Linear B inscriptions found at Bernstorf, in the vicinity of Freising in Upper Bavaria (Stahl 2006: 340 Taf. 7.b). I thank Richard Janks for bringing these seals to my attention.

239. Petrie 1891: 16 no. 36.

240. Petrie 1891: 16–19 nos. 37–38; 1931: 102, 111; Thomas 1981A: 13–14; Hassler 2011: 125–129, 132–133.

241. The initial date for Mycenaean IIIB has as its basis the Gurob "burnt groups" and the "palace dump" at Tell el Amarna (Hankey 1981; Bell 1985: 61 n. 1). Petrie first defined the Mycenaean period as synchronous with the XVIIIth and XIXth Dynasties at Gurob, but he did not seem to realize that these objects gave only a *terminus post quem* date to the burnt groups, a problem already noted at the time by Torr (1892; Bell 1991: 256).

242. Bell 1991: 255–257.

243. Merrillees 1973: 180–182; Thomas 1981A: 5–6; Bell 1991: 264–265; Wachsmann 1987: 115.

244. Thomas 1981A: 14.

245. Symeonoglou 1973: 18 n. 37; Cavanaugh and Mee 1998: 71–72, 74–76, 92–97, 108, 112–113, 135–136; Thomas 1981A: 13; Bell 1991: 255. See also pp. 120 n. 145 in this volume.

246. Symeonoglou 1973: 18 n. 38; Cavanaugh and Mee 1998: 71–72, 74–76, 92–97, 108, 112–113, 135–136; Thomas and Conant 1999: 105; Popham 1986; 2001: 284–285, 290; Dickinson 2006: 69, 73, 179–181.

247. Symeonoglou 1973: 16–18, figs. 7–8.

248. Symeonoglou 1973: 18 n. 49; Merrillees 1973: 180. S. Symeonoglou relates the "burnt pit" there to building B–C, which was part of the Mycenaean royal palace at Thebes and speculates that the items were discarded following an earthquake and burnt in a purification ritual meant to prevent malevolent spirits from causing additional harm. In searching for a link to fire rituals performed specifically at Thebes, Symeonoglou emphasizes that the cult of Dionysos and his mother, Semele, at Delphi utilized an element of fire purification, as "the resurrection of the two gods was connected with fire, and Dionysos himself lay next to the Immortal Fire" (Delcourt 1965: 113–115; Symeonoglou 1973: 18.)

249. Nilsson 1923; Delcourt 1965; Symeonoglou 1973: 18.

250. Symeonoglou 1973: 18 n. 47.

251. Hennessy 1966: 159, 162; 1985; Herr 1983; Mazar 1990: 255–256.

252. Popham 2001: 295.

253. Merrillees 1973: 180.

254. Thomas 1981A: 5 n. 47.

255. Thomas 1981A: 13.

256. Politi 2001.

257. *BAR* III: §415–424; Kitchen 1982: 83–89; Thomas 1981A: 17; Bell 1991: 254; Bell 2007: 110–112. Maat-Hor-Neferure's story was revived as propaganda by the priests of Khons-the-Plan-Maker-in-Thebes ca. 300 B.C. (*BAR* III: §429–447; Kitchen 1982: 229).

258. *WP* II: 45; Kitchen 1982: 89, 110.

259. Politi 2001: 111; Hassler 2011: 132–133.

260. Politi 2001: 111.

261. EA 25: (iii) 64–67, (iv) 64. See Moran 1987: 82 n. 29, 83 n. 46.

262. EA 25: (iv) 64.

263. Politi 2001: 111 (UC 30139).

264. Petrie 1890: 41 no. 80 Hassler 2011:130.

265. For color images of light human hair from Gurob (UC 30317–30319), see the Gurob Ship-Cart Model Digital Supplement/ Image Catalogue/Human hair from Tombs 23–25 at Gurob (UC 30317–30319).

266. Thomas 1981A: 6. This hair, which is on exhibit at the Petrie Egyptological Museum, may be worthy of DNA studies to better determine its identity.

267. Petrie 1890: 38 no. 75, 40 no. 79. Sadi-amia's ushabti comes from the same tomb as Anen-Tursha, whom Petrie thus identified as a Teresh, one of the Sea People groups. See p. 183 in this volume.

268. Bell 1991: 254.

269. Petrie 1890: 41 no. 80, pl. XVIII: 38; Trope, Quirk, and Lacovara 2005: 83 fig. no. 64.

270. Gardiner 1953: 149; 1968: X–XI: VI ("Gurob" Fragments, U), 22–24; David 1986: 42.

271. I thank L. Bell for this translation.

272. Lesko 2004: 230, 122–123.

273. Probably an outer garment, such as a cloak or mantle.

274. Gardiner 1953: 149; 1968: X: IV ("Gurob" Fragments, G), V ("Gurob" Fragments, T), 20–22; Thomas 1981A: 17.

275. Thomas 1981A: 17.

276. Thomas 1981A: 17 for a list of scholars.

277. Thomas 1981A: 17 ns. 147–150.

278. Petrie 1891: 16 no. 36 (quoted earlier); Symeonoglou 1973: 18 n. 47; Politi 2001: 111 Hassler 2011: 132.

279. Sparks 2004: 40.

280. See, for example, the case for the syncretism of Syro-Canaanite cults practiced in Egypt (Sparks 2004: 38).

281. Harding 2000: 111–118, 395, 400–404; Mohen and Eluère 2000: 114–116; Vandkilde 2007: 144–153.

Chapter 5: Conclusions

1. For the virtual-reality reconstruction, visit online (http://www.vizin.org/Gurob/Gurob.html).

2. Pylos: Xa 102, Xa 1419; Khania: Gq 5. See Ventris and Chadwick 1973: 127 411; Chadwick 1976: 85, 87, 99–100; Baumbach 1979: 146–147; Ruipérez 1983: Hallager, Vlasakis, and Hallager 1992: 76–80, 86, pl. 1. On clasical authors identifying an Egyptian source for the phallic cult of Dionysos, see Scherf 2008.

3. Hdt. II, 48–49. Translation by A. L. Purvis from Herodotus, Strassler, and Purvis 2007: 139. Regarding the significance of the puppets mentioned by Herodotus, see Csapo 1997: 269–274.

4. A Late Helladic IIIC jug found in a shrine at Mendes is particularly evocative in this regard (Redford 2012: 92, 94 fig. 6.26).

Appendix 1: Lines Drawing of the Gurob Ship Model

1. I would like to thank Shelley Wachsmann for inviting me to participate in this volume and F. Castro and D. Sanders for their advice throughout various stages involved in creating the Gurob model lines.

2. For those unfamiliar with lines drawings, see Landström 1970: 144–145; Steffy 1994: 10–20.

3. NURBS stands for "nonuniform rational Bézier spline," which is a mathematical model commonly used in computer graphics for generating and representing curves and surfaces through the use of control points.

Appendix 2: The Gurob Ship-Cart Model in Virtual Reality

1. Learning Sites, Inc., helped create the Fortress of Buhen virtual world in 1993 and 1994 (see http://www.learningsites.com/EarlyWork/buhen-2.htm); English Heritage worked on a virtual Stonehenge (no longer available); and researchers at Carnegie Mellon University, under Carl Loeffler, developed a virtual Egyptian temple (no longer available); these and other virtual worlds were presented at the First Annual Virtual Heritage Conference, Bath, England, in November 1995.

Appendix 3: Ship Colors in the Homeric Poems

1. I would like to extend my gratitude to Shelley Wachsmann for inviting me to contribute this appendix and to Sarah James, who made helpful comments and criticisms on an earlier version. For ancient authors I use the abbreviation standard of *OCD*[3].

2. The black, dark-red, and blue pigments have doubtlessly undergone shifts in color over the past three millennia. On the composition

of the pigments, see Appendix 6, Analysis of Pigments from the Gurob Ship-Cart Model, pp. 243–247 in this volume. For color images of the polychromatic paint on the model, see the virtual-reality package that accompanies this book online at http://www.vizin.org/Gurob/Gurob.html.

3. See chapter 2, The Iconographic Evidence, pp. 31–84. On the distinct differences between Minoan/Cycladic ships and Helladic or Mycenaean ships, see Wachsmann 1998: 83–158 (esp. 95, 130–31).

4. On color terms in Linear B and Homeric correlates see Blakolmer 2000; on maritime matters in the Linear B tablets see Palaima 1991.

5. See, for example, West 1997: 233–237.

6. Cf. Hoekstra 1969: 124–130.

7. On attempts to account for the apparent deficiency of color terms in early Greek using linguistic models, see Moonwomon 1994. On Homer's troublesome "wine-dark sea," see Maxwell-Stuart 1981: 1–11.

8. "Black ship" *(melaina naus)* in Homer: *Il.* 1.141, 300, 329, 433, 485, 2.170, 358, 524, 534, 545, 556, 568, 630, 644, 652, 710, 737, 747, 759, 5.550, 700, 8.222, 528, 9.235, 654, 10.74, 11.5, 824, 828, 12.126, 13.267, 15.387, 423, 16.304, 17.383, 639, 19.331, 24.780; *Od.* 2.430, 3.61, 360, 365, 423, 4.646, 731, 781, 6.268, 8.34, 51, 52, 445, 9.322, 10.95, 169, 244, 272, 332, 502, 571, 11.3, 58, 12.186, 264, 276, 418, 13.425, 14.308, 15.218, 258, 269, 416, 503, 16.325, 348, 359, 17.249, 18.84, 21.39, 307, 23.320, 24.152; *Hymn Hom. Ap.* 397, 405, 457, 459, 497, 511; in Hes. *Op.* 636, fr. 96. I.19, 72, fr. 606.15.

9. For example, Alc. frag. 326.4 and 34a.12; Hermipp. 63.3.

10. Suggested by Kober 1932: 31.

11. Said of Noah's Ark in *Genesis* 6:14; Kober 1932: 30; Torr 1964: 34–35, n. 88; Kurt 1979: 33; Steffy 1994: 277; Casson 1995: 211–212.

12. The Uluburun ship from ca. 1300 B.C., the oldest Mediterranean seagoing ship from which we have any substantial remains of hull timbers, preserved no obvious traces of pitch (C. Pulak, pers. comm.). On the practice of applying pitch to hulls in the Classical period, see Casson 1995: 211–212 and notes 46–50. See also p. 27 n. 49 in this volume.

13. Cinnabar is a thick, reddish mercuric sulfide (HgS) and a natural mineral of mercury. Red ochre is composed of hydrated oxides of iron of various concentrations. The red of the Gurob boat model is derived from red iron oxide and hematite, both of which are typical components of red ochre (see Appendix 6, Analysis of Pigments from the Gurob Ship-Cart Model, pp. 243–247 in this volume). *Miltos* and *melas* (black) share the same Pre-Indo-European root (Moonwomon 1994: 45–46).

14. Ventris and Chadwick 1973: 367 (KN Sd 4404); see also Chantraine 1983: 702, s.v. *miltos.* Aura Jorro and Adrados 1985: 455, s.v. *mi-to-we-sa* and *mi-to-we-sa-e.*

15. *Pa-ra-wa-jo:* Ventris and Chadwick 1973: 379 (PY Sh 737, 740), 380 (KN Sk 789), and 524 (KN Sk 8100). For the cheek straps or cheek pieces of horses, another word was preferred, *o-po-qo:* Ventris and Chadwick 1973: 365–368 (KN Sd 4401, 4403, 4404, 4409, 4413, Sf 4428). See also Aura Jorro and Adrados 1985: s.v. *o-po-qo.*

16. Hom. *Il.* 4.141–142: *as when a woman of Maeonia or Caria stains ivory with purple to make the cheek-piece of horses (Ὡς δ' ὅτε τίς τ' ἐλέφαντα γυνὴ φοίνικι μιήνῃ | Μῃονίς ἠέ Κάειρα, παρήιον ἔμμεναι ἵππων).* See Heubeck and Hoekstra 1989: 22.

17. Hom. *Od.* 24.523: *on a sudden he poised and hurled his spear and struck Eupeithes through his helmet with bronze cheek-pieces (αἶψα μάλ' ἀμπεπαλὼν προΐει δολιχόσκιον ἔγχος, | καὶ βάλεν Εὐπείθεα κόρυθος διὰ χαλκοπαρῄου).*

18. Ventris and Chadwick 1973: 366–368 (KN Sd 4401, 4402, 4409, 4413, Sf 4428); Chantraine 1983: 1218, s.v. *phoinix;* Aura Jorro and Adrados 1993: 135, s.v. *po-ni-ki-ja.* Moonwomon (1994: 45–46, 58–59) interprets *phoinix* simply as "red" and ignores the evidence indicating its proximity to purple and crimson.

19. Ventris and Chadwick 1973: 365–366 (KN Sd 4403).

20. Wallace 1927: 9–10, 43; Torr 1964: 35.

21. Wilson 1938: 7–8.

22. As pointed out by Wallace (1927: 9).

23. Wilson 1938: 9 and n. 32.

24. Moonwomon (1994: 50) states that *phoinix* and *miltos* in these contexts "seem to mean exactly the same thing." Synonymous usage of the two adjectives in Homer (sc., epic poetry) is plausible in isolation, but the distinction in Linear B (a palace inventory) must have been real and suggests a similar distinction in Homer.

25. Wallace 1927: 9, 63; Irwin 1974: 92.

26. See p. 28 nn. 61–63 in this volume.

27. Hdt. 3.58: *τὸ δὲ παλαιὸν ἅπασαι αἱ νέες ἦσαν μιλτηλιφέες.* For color images of numerous examples of painted ships (galleys and sailing vessels), see Spathari 1995: figs. 98, 103, 93, 107, and 108. See also the early fifth-century-B.C. painting of an Etruscan merchantman from the Tomba della Nave at Tarquinia, which shows a wide, dark-red stripe down the entire length of the hull (Moretti 1961; color image in Pomey 1997B: 81).

28. Seymour 1914: 307; Kober 1932: 95; Torr 1964: 36–37; Casson 1995: 45 and n. 18 (see also 212 and n. 49, where he also calls these painted areas "bow patches and decorative effects"); Kirk 1985: 221–222; Heubeck and Hoekstra 1989: 22; Mark 2005: 99–102.

29. Depictions of Geometric ships can be found in Morrison and Williams 1968: pls. 1–10; Casson 1995: figs. 65–72; Basch 1987: 166–181.

30. *Il.* 15.693, 23.852, 878; *Od.* 3.299, 9.482, 539, 10.127, 11.6, 12.100, 148, 354, 14.311, 22.465.

31. Aura Jorro and Adrados 1985: 415–416, s.v. *ku-wa-no.*

32. Ventris and Chadwick 1973: 339 (PY Ta 642).

33. Ventris and Chadwick 1973: 344 (PY Ta 714).

34. Knossos: *PM* I: 533–534; Tiryns: Demangel 1944–1945.

35. Halleux 1969. The blue pigment used on the model is typical of Egyptian Blue, which is a synthetic calcium copper tetrasilicate (see Appendix 6, Analysis of Pigments from the Gurob Ship-Cart Model, pp. 243–247 in this volume).

36. Chantraine 1983: 593–594, s.v. *kyanos.*

37. Irwin 1974: 89–90; Heubeck, West, and Hainsworth 1988: 179.

38. Gladstone 1877: 378–379; Wallace 1927: 11 (cf. Kober 1932: 73). Mention should be made of a galley with a yellow ram, apparently representing a nailed sheath of copper or bronze, depicted on a fresco from Til Barsip on the upper Euphrates and dated to the second half of the eighth century B.C. (see Thureau-Dangin et al. 1936: frontispiece; Basch 1987: 306–308, fig. 649; for a color photo of the mosaic, see DeVries and Katzev 1972: 56, fig. 9).

39. See, for example, Wallace 1927: 11; Casson 1995: 45 and n. 18 and 212 and n. 49; Irwin 1974: 91–92.

40. Cunliffe 1988: s.v. *κυανοπρῴροιος* and *κύανος*.

41. Mark 2005: 101.

42. Themistonoe: Hes. [*Sc.*] 356; Clytemnestra: frags. 23a14, 27; Althaea: fr. 25.14; Electra: *Schol. Pind. Nem.* ii.17.

43. Anac. fr. 12.2.

44. Mark 2005: 102. See also p. 28 n. 55 in this volume.

45. Irwin 1974: 92.

46. Parry 1971: 6–7 [1928, 7], 16–17 [19–20]; 276 [1930, 86]. For a discussion of Parry's work and subsequent contributions, see Bakker 1995.

47. Kurt 1979: 32–33. In fact, various declensions of *melaina naus* are distributed throughout each foot in a variety of combinations: The term occurs in the last two feet (5–6) a total of 56 percent (combining dative and genitive forms); 2 percent for 1–2–3; 6 percent for 2–3; 4 percent for 2–3–4–5; 15 percent for 3–4–5; and 17 percent for 3–4–5–6. Also, bucolic diaeresis (word ending between the fourth and the fifth foot) are also frequent in verses with *melaina naus,* accounting for some 60 out of the 81 verses.

48. Maxwell-Stuart 1981: 4.

Appendix 4: Sherden and Tjuk-People in the Wilbour Papyrus

1. Translations from *WP* III.

2. Italics added for the words "Sherden," "Tjuk," and their vocational descriptions.

3. See Gardiner's (*WP* III: 116) comments on this lacuna.

Appendix 5: Radiocarbon Age Analysis of the Gurob Ship-Cart Model

1. Stuiver and Polach 1977.

2. Reimer et al. 2009.

Appendix 6: Analysis of Pigments from the Gurob Ship-Cart Model

1. Heywood 2001; Eastaugh et al., 2004A; 2004B.

2. Heywood 2001.

3. Tite, Bimson, and Cowell 1987; Riederer 1997; Scott 2002; Eastaugh et al., 2004A; 2004B.

4. Winton 1906.

5. Tzachili 2005.

6. Serpico and White 2000.

7. Eastaugh et al. 2004B; Bothe 2007.

8. Bothe 2007.

9. See, for example, Rowe, Siddall, and Stacey 2010.

10. Languri 2004.

11. Serpico and White 2001.

Appendix 7: Wood Identification

1. See pp. 22 Fig. 122: B (item no. 30), 31 Fig. 1.30: D(item no. 12) in this volume.

2. Cartwright 2008A; 2008B.

Glossary

1. This glossary derives from definitions from the following sources of nautical terms: Steffy 1994: 266–298 (includes an illustrated glossary); Casson 1995: 389–402 (Greek and Latin nautical terms); Hocker 1998; Wachsmann 2009: 397–403; British Columbia Ministry of Culture and the Arts 2010.

Bibliography

Abbreviations

AA = *Archäologischer Anzeiger*

AAA = *Archaiologika Analekta ex Athenon (Athens Annals of Archaeology)*

ABSA = *Annual of the British School at Athens*

AEANE = *Ancient Egypt, the Aegean, and the Near East: Studies in Honor of Martha Rhoads Bell*, ed. J. Phillips. San Antonio: Van Siclen, 1997.

Age of the Galley = R. Gardiner, ed., *The Age of the Galley: Mediterranean Oared Vessels since Pre-Classical Times.* Conway's History of the Ship series. London: Conway Maritime Press, 1995.

Age of Iron = T. A. Wertime and J. D. Muhly, eds., *The Coming of the Age of Iron.* New Haven, Conn.: Yale University Press, 1980.

Agora = *The Athenian Agora: Results of the Excavations Conducted by the American School of Classical Studies at Athens Princeton.*

AHL = *Archaeology and History in Lebanon*

AJA = *American Journal of Archaeology*

AJBL = *Australian Journal of Biblical Archaeology*

AJSL = *American Journal of Semitic Languages and Literatures*

Alc. = Alcaeus

AM = *Mitteilungen des Deutschen Archäologischen Instituts, Athenische Abteilung*

AmAnth = *American Anthropologist*, n.s.

Anac. fr. = Anacreon fragment

AnatSt = *Anatolian Studies: Journal of the British Institute of Archaeology at Ankara*

AncEg = *Ancient Egypt*

AntCl = *L'Antiquité Classique*

AO = *Archaeology Odyssey*

AOAT = *Alter Orient und Altes Testament*

AP = *Archaeological Prospection*

Apul. *Met.* = Apuleius, *The Golden Ass (Metamorphoses)*

AR = *Archaeological Reports* (supplement to JHS)

Ar. *Ach.* = Aristophanes, *Acharnians*

ArcheologiaWar = *Archeologia: Rocznik Instytutu historii kultury materialnej Polskiej akademii nauk* (Warsaw)

ArchEph = *Archaiologike Ephemeris*

ArchNews = *Archaeological News*

ASAE = *Annales du Service des Antiquités de l'Égypt*

ASAtene = *Annuario della Scuola archeologica di Atene e delle Missioni italiane in Oriente*

ASCIM = *Atti e memorie del secondo Congresso internazionale di micenologia, Roma-Napoli, 14–20 ottobre 1991,* vol. 3, *Archaeologia.* Incunabula Graeca 98, ed. E. De Miro, L. Godart, and A. Sacconi. Rome: Gruppo editoriale internazional, 1996.

Athlit Ram = *The Athlit Ram.* L. Casson and J. R. Steffy, eds. College Station: Texas A&M University Press, 1991.

BAR = J. H. Breasted, ed., *Ancient Records of Egypt: Historical Documents from the Earliest Times to the Persian Conquest,* 5 vols. London: Histories and Mysteries of Man, 1988 (reprint).

BARIS = British Archaeological Series, International Series

BASOR = *Bulletin of the American Schools of Oriental Research*

BCH = *Bulletin de correspondance hellénique*

Beyond Babylon = J. Aruz, K. Benzel, and J. M. Evans, eds., *Beyond Babylon: Art, Trade, and Diplomacy in the Second Millennium* B.C. New York: Metropolitan Museum of Art, 2008.

BibAR = *Biblical Archaeology Review*
BICS = *Bulletin of the Institute of Classical Studies of the University of London*
BIFAO = *Bulletin de l'institut français d'archéologie orientale de Caire*
BJRL = *Bulletin of the John Rylands Library*
BMFA = *Bulletin of the Museum of Fine Arts* (Boston)
BMMA = *Bulletin of the Metropolitan Museum of Art*
BRKS = Epigraphic Survey, The Battle Reliefs of King Sety I: Reliefs and Inscriptions at Karnak 4. (OIP 107). Chicago: Oriental Institute of the University of Chicago, 1986.
BSA = *Annual of the British School at Athens*
BSFE = *Bulletin de la Société française d'égyptologie*
CAH = *Cambridge Ancient History*
Cd'É = *Chronique d'Égypt*
CJ = *Classical Journal*
Color of Life = R. Panzanelli, E. D. Schmidt, and K. D. S. Lapatin, eds., *The Color of Life: Polychromy in Sculpture from Antiquity to the Present.* Los Angeles: Getty Museum, 2008.
CQ = *Classical Quarterly*
CR = *Classical Review*
CRAI = *Comptes rendus des séances de l'Académie des inscriptions et belles-lettres* (Paris)
Crisis Years = W. A. Ward and M. Sharp Joukowsky, eds., *The Crisis Years: The 12th Century* B.C. *from beyond the Danube to the Tigris.* Dubuque: Kendall/Hunt, 1992.
CVA Bologna 2 [Italy 7] = Laurenzi, L., and G. B. Montanari, eds., *Corpus vasorum antiquorum: Italia, Museo Civico di Bologna.* Milan: Bestetti e Tumminelli, 1929.
EA = El Amarna
EABSG = S. Wachsmann, ed., *The Excavations of an Ancient Boat in the Sea of Galilee (Lake Kinneret)* ('Antiqot 19). Jerusalem: Israel Antiquities Authority.
EAL = W. V. Davies and L. Schofield, eds., *Egypt, the Aegean, and the Levant: Interconnections in the Second Millennium* BC. London: British Museum Press, 1995.
EgyArch = *Egyptian Archaeology*
Eirene = *Eirene: Studia graeca et latina*
Emporia = R. Laffineur and E. Greco, eds., *Emporia: Aegeans in the Central and Eastern Mediterranean. Proceedings of the 10th International Aegean Conference/10e Rencontre égéenne internationale, Athens Italian School of Archaeology, April 14–18, 2004*, vol. 1. Liège: University of Liège, Histoire de l'art et archéologie de la Grèce antique, and University of Texas at Austin, Program in Aegean Scripts and Prehistory, 2005.
EtrStud = *Etruscan Studies: Journal of the Etruscan Foundation*
FPO = Epigraphic Survey, The Festival Procession of Opet in the Colonnade Hall: With Translations of Texts, Commentary, and Glossary, vol. 1 (pls. 1–128). Chicago: Oriental Institute of the University of Chicago, 1994.
fr. = fragment
GM = *Göttinger Miszellen*
Goddess and Polis = J. Neils, ed., *Goddess and Polis: The Panathenaic Festival in Ancient Athens.* Princeton, N.J.: Princeton University Press, 1992.
GP = *Geophysical Prospecting*
Hdt. = Herodotus
Hermipp. = Hermippus
Hes. [*Sc.*] = Hesiod, *Shield of Herakles*
Hes. *Op.* = Hesiod, *Works and Days* (*Opus Diesque,* the Latin version)
Hom. *Il.* = Homer, *Iliad*
Hom. *Od.* = Homer, *Odyssey*
HSBUA = G. F. Bass, ed., *A History of Seafaring Based on Underwater Archaeology.* New York: Walker, 1972.
HSCP = *Harvard Studies in Classical Philology*
Hyksos = E. D. Oren, ed., *The Hyksos: New Historical and Archaeological Perspectives.* University Museum Monograph 96, University Museum Symposium Series 8. Philadelphia: University Museum, University of Pennsylvania, 1997.
Hymn Hom. Ap. = *Homeric Hymn to Apollo (Hymnus Homericus ad Apollinem)*
Hymn Hom. Bacch. = *Homeric Hymn to Dionysos (Hymnus Homericus ad Bacchum)*
IEJ = *Israel Exploration Journal*
IJNA = *International Journal of Nautical Archaeology*
Il. = *Iliad*
INAN = *INA Newsletter*
INAQ = *INA Quarterly*
IOS = *Israel Oriental Studies*
IstMitt = *Istanbuler Mitteilungen*
JAEI = *Journal of Ancient Egyptian Interconnections*
JAG = *Journal of Applied Geophysics*
JAOS = *Journal of the American Oriental Society*
JARCE = *Journal of the American Research Center in Egypt*
JDI = *Jahrbuch des Deutschen Archäologischen Instituts*
JEA = *Journal of Egyptian Archaeology*
JHS = *Journal of Hellenic Studies*
JIAN = *Journal international d'archéologie numismatique*
JIES = *Journal of Indo-European Studies*

JMAA = *Journal of Mediterranean Anthropology and Archaeology*

JNES = *Journal of Near Eastern Studies*

Joseph. *BJ* = Josephus, *The Jewish War (Bellum Judaicum)*

JRA = *Journal of Roman Archaeology*

JRS = *Journal of Roman Studies*

JSOT = *Journal for the Study of the Old Testament*

Kush = *Kush, Journal of the Sudan Antiquities Service*

LA = *Liber Annuus*

Libya and Egypt = A. Leahy, ed., *Libya and Egypt c1300–750 BC*. London: Centre of Near and Middle Eastern Studies (School of Oriental and African Studies, University of London) and the Society for Libyan Studies, 1990.

MDIK = *Mitteilungen des Deutschen Archäologischen Instituts, Abteilung Kairo*

MEM = P. Dikaios, ed., *Acts of the International Archaeological Symposium, "The Mycenaeans in the Eastern Mediterranean" (Nicosia, 27th March–2nd April 1972)*. Nicosia: Department of Antiquities, Cyprus, 1973.

MH I = Epigraphic Survey, *Medinet Habu*, vol. 1, *Earlier Historical Records of Ramses III*. Chicago: University of Chicago Press, 1930.

MH II = Epigraphic Survey, *Medinet Habu*, vol. 2, *Earlier Historical Records of Ramses III*. Chicago: University of Chicago Press, 1932.

MH VIII = Epigraphic Survey, *Medinet Habu*, vol. 8, *The Eastern High Gate with Translation of Texts*. Chicago: University of Chicago Press, 1970.

MM = *Mariner's Mirror*

MonPiot = *Monuments et mémoires*. Fondation Eugène Piot

MW = *Miscellanea Wilbouraniana*

NEA = *Near Eastern Archaeology*

NEGS = *Newsletter*. European Geophysical Society

NGM = *National Geographic Magazine*

NTS = Not to scale

OCD[3] = Hornblower, S., and A. Spawforth, eds., *Oxford Classical Dictionary*.[3] Oxford: Oxford University Press, 1996.

OCPM = *Oxford Classical and Philosophical Monographs*

Od. = *Odyssey*

OED = *The Oxford English Dictionary*, vol. 2, *BBC–Chalysography*. J. A. Simpson and E. S. C. Weiner, eds. Oxford: Clarendon, 1967.

OIC = *Oriental Institute Communications*

OIP = *Oriental Institute Publications*

OJA = *Oxford Journal of Archaeology*

OLA = Orientalia Lovaniensia Analecta

OpAth = *Opuscula Atheniensia*

Panathenaic Games = O. Palagia and A. Spetsieri-Choremi, eds., *The Panathenaic Games: Proceedings of an International Conference Held at the University of Athens, May 11–12, 2004*. Oxford: Oxbow, 2007.

Paus. = Pausanias

Pays d'Ougarit = M. Yon, M. Sznycer, and P. Bordreuil, eds., *Le pays d'Ougarit autour de 1200 av. J.-C. histoire et archéologie: actes du colloque international, Paris, 28 juin–1er juillet 1993. Ras Shamra-Ougarit*. Paris: Éditions Recherche sur les civilizations, 1995.

Peoples in Transition = S. Gitin, A. Mazar, and E. Stern, eds., *Mediterranean Peoples in Transition: Thirteenth to Early Tenth Centuries BCE. In Honor of Trude Dothan*. Jerusalem: Israel Exploration Society, 1998.

PEQ = *Palestine Exploration Quarterly*

Philostr. *VS* = Philostratus, *Lives of the Sophists*

Pictorial Pursuit = E. Rystedt and B. Wells, eds., *Pictorial Pursuits: Figurative Painting on Mycenaean and Geometric Pottery. Papers from Two Seminars at the Swedish Institute at Athens in 1999 and 2001*. Stockholm: Svenska Institutet i Athen, 2006.

Plut. *Vit. Thes.* = Plutarch, *Lives (Theseus)*

PM = A. Evans, *The Palace of Minos*, 4 vols. London: Biblo and Tannen, 1964.

PN = Personal name

Polemos = R. Laffineur, ed., *Polemos: Le contexte guerrier en Égée à l'age du bronze. (Actes de la 7e Recontre égéene internationale, Université de Liège, 14–17 avril 1998*, vol. 2). Liège: Université de Liège and the University of Texas at Austin, 1999.

PRAI = *Proceedings of the Royal Anthropological Institute of Great Britain and Ireland*

QDAP = *Quarterly of the Department of Antiquities in Palestine*

RA = *Revue Archéologique*

RÉg = *Revue d'égyptologie*

RÉG = *Revue des études grecques*

Res Maritimae = S. Swiny, R. L. Hohlfelder, and H. W. Swiny, eds., *Res Maritimae: Cyprus and the Eastern Mediterranean from Prehistory to Late Antiquity: Proceedings of the Second International Symposium: "Cities on the Sea," Nicosia, Cyprus, October 18–22, 1994*. Cyprus American Archaeological Research Institute Monograph Series. Atlanta: Scholars Press, 1994.

RhM = *Rheinisches Museum für Philologie*

SAOC = Oriental Institute of the University of Chicago, Studies in Ancient Oriental Civilization

Schol. Pind. *Nem.* = Scholiast to Pindar's *Nemea*

Sea Peoples = E. D. Oren, ed., *The Sea Peoples and Their World: A Reassessment*. University Museum Monograph 108: University

Museum Symposium Series 11. Philadelphia: University Museum, 2000.

SicArch = *Sicilia Archeologica*

SIMA = *Studies in Mediterranean Archaeology*

SKH = Epigraphic Survey, *Scenes of King Herihor in the Court with Translations of Texts: The Temple of Khonsu*, vol. 1 (pls. 1–110). University of Chicago, Oriental Institute Publications 100. Chicago: Chicago University Press, 1979.

Skyllis = *Skyllis Zeitschrift für Unterwasserarchäeologie*

SM = *Scripta Mediterranea*

SMEA = *Studi micenei ed egeo-anatolici*

TAPS = *Transactions of the American Philosophical Society*

Tel Aviv = *Tel Aviv.* Journal of the Institute of Archaeology, Tel Aviv University

Thalassa = R. Laffineur and L. Basch, eds., *Thalassa: L'Égée préhistorique et la mer. Actes de la troisième Recontre égéenne internationale de l'Université de Liège, Station de recherches sous-marines et océanographiques [StaReSO], Clavi, Corse [23–25 avril 1990].* Liege: Université de Liège, 1991.

Thera I = C. Doumas, ed., *Thera and the Aegean World*, vol. 1. *Papers Presented at the Second International Scientific Congress, Santorini, Greece, August 1978.* London: Thera and the Aegean World, 1978.

Thuc. = Thucydides

Tropis I = H. Tzalas, ed., *Tropis* I: *Proceedings of the 1st International Symposium on Ship Construction in Antiquity (Piraeus, August 30–September 1, 1985).* Athens: Hellenic Institute for the Preservation of Nautical Tradition, 1989.

Tropis II = H. Tzalas, ed., *Tropis* II: *Proceedings of the 2nd International Symposium on Ship Construction in Antiquity (Delphi, 27–29 August 1987).* Athens: Hellenic Institute for the Preservation of Ship Construction in Antiquity, 1990.

Tropis III = H. Tzalas, ed., *Tropis* III: *Proceedings of the 3rd International Symposium on Ship Construction in Antiquity (Athens, 24–27 August 1989).* Athens: Hellenic Institute for the Preservation of Ship Construction in Antiquity, 1995.

Tropis IV = H. Tzalas, ed., *Tropis* IV: *Proceedings of the 2nd International Symposium on Ship Construction in Antiquity (Athens, 28–31 August 1991).* Athens: Hellenic Institute for the Preservation of Ship Construction in Antiquity, 1996.

Tropis V = H. Tzalas, ed., *Tropis* V: *Proceedings of the 5th International Symposium on Ship Construction in Antiquity (Nauplia, 26–28 August 1993).* Athens: Hellenic Institute for the Preservation of Ship Construction in Antiquity, 1999.

Tropis VI = H. Tzalas, ed., *Tropis* VI: *Proceedings of the 6th International Symposium on Ship Construction in Antiquity (Lamia, 28–30 August 1996).* Athens: Hellenic Institute for the Preservation of Ship Construction in Antiquity, 2001.

Tropis VII = H. Tzalas, ed., *Tropis* VII: *Proceedings of the 7th International Symposium on Ship Construction in Antiquity (Lamia, 28–30 August 1996)*, 2 vols. Athens: Hellenic Institute for the Preservation of Ship Construction in Antiquity, 2002.

Tropis IX = H. Tzalas, ed., *Tropis* IX: *Proceedings of the 9th International Symposium on Ship Construction in Antiquity (Agia Napa, Cyprus, 25–30 August 2005).* Athens: Hellenic Institute for the Preservation of Ship Construction in Antiquity. In press.

Tropis X *Abstracts* = H. E. Tzalas, ed., *10th International Symposium on Ship Construction in Antiquity (Hydra, 27 August–3 September 2008): Program and Abstracts*, 2008.

TT = Theban Tomb

UF = *Ugarit-Forschungen: Internationales Jahrbuch für die Altertumskunde Syrien-Palästinas*

Verg. *Aen.* = Virgil, *Aeneid*

Vitr. *De arch.* = Vitruvius, *On Architecture*

VT = *Vetus Testamentum*

Worshipping Athena = J. Neils, *Worshipping Athena: Panathenaia and Parthenon.* Wisconsin Studies in Classics. Madison: University of Wisconsin Press, 1996.

WP I = A. Gardiner, *The Wilbour Papyrus*, vol. 1, *Plates.* Oxford: Brooklyn Museum at the Oxford University Press, 1941.

WP II = A. Gardiner, *The Wilbour Papyrus*, vol. 2, *Commentary.* London: Brooklyn Museum at the Oxford University Press, 1948.

WP III = A. Gardiner, *The Wilbour Papyrus*, vol. 3, *Translation.* London: Brooklyn Museum at the Oxford University Press, 1948.

WP IV = R. O. Faulkner, *The Wilbour Papyrus*, vol. 4, *Index.* London: Brooklyn Museum at the Oxford University Press, 1952.

WPT = S. Sherratt, ed., *Proceedings of the First International Symposium: The Wall Paintings of Thera (Petros M. Nomikos Conference Centre, Thera, Hellas, 30 August–4 September 1997).* 2 vols. Athens: Thera Foundation, 2000.

ZÄSA = *Zeitschrift für ägyptische Sprachen und Altertumskunde*

ZDPV = *Zeitschrift des deutschen Palästina-Vereins*

References

Abd el-Raziq, M., G. Castel, and P. Tallet. 2004. Dans le golfe de Suez, les mines de cuivre d'Ayn Soukhna. *Archéologia* 414: 10–21.

———. 2006. Ayn Soukhna et la mer Rouge. *Égypte, Afrique, et Orient* 41: 3–6.

———. 2007. L'exploration archéologique du site d'Ayn Soukhna (2001–2004). In *Proceedings of the Ninth International Congress of Egyptologists/Actes du Neuvième Congrès International des Égyptologues, Grenoble, 6–12 Septembre 2004,* vol. 1. J.-C. Goyon, ed. Leuven: Uitgeverij Peeters en Departement Oosterse Studies: 61–68.

———. 2010. Ayn Soukhna. June 21. http://www.ifao.egnet.net/archeologie/ayn-soukhna/.

Achilli, A., A. Olivieri, M. Pala, E. Metspalu, S. Fornarino, V. Battaglia, M. Accetturo, et al. 2007. Mitochondrial DNA Variation of Modern Tuscans Supports the Near Eastern Origin of Etruscans. *American Journal of Human Genetics* 80: 759–768.

Adams, P. R. 1965. Annual Report: Transmittal and Summary for the Year 1962–1963 to the Shareholders of the Cincinnati Museum Association. *Cincinnati Art Museum Bulletin* 7(3–4): (unpaginated).

Aharoni, Y. 1979. *The Land of the Bible: A Historical Geography.* A. F. Rainey, trans. London: Burnes and Oates.

Albright, W. F. 1951. Review: *Ḥamā: Fouilles et Recherches de la Fondation Carlsberg, 1931–1938. Les cimetières à crémation.* By P. J. Riis. *AJA* 55: 106–107.

———. 1975. Syria, the Philistines, and Phoenicia. In *CAH*³, vol. 2, part 2. Cambridge: Cambridge University Press: 507–534.

Alföldi, A. 1937. *A Festival of Isis in Rome under the Christian Emperors of the Fourth Century.* (*Dissertationes Pannonicae,* vol. 2, 7). Budapest: Institute of Numismatics and Archaeology of Pázmány University.

Alt, A. 1944. Ägyptische Tempel in Palästina und die Landnahme der Philister. *ZDPV* 67: 1–20.

Amyx, D. A., and W. K. Pritchett. 1958. The Attic Stelai. Part III, Vases and Other Containers. *Hesperia* 27: 255–310.

Anonymous. 2003. Elkab's Hidden Treasure. *Al-Ahram* 649, August 6. http://weekly.ahram.org.eg/2003/649/he1.htm.

———. 2008. The Egypt Exploration Society: Survey at Medinet el-Gurob. The Egypt Exploration Society. November 23. http://egyptexplorationsociety.tumblr.com/post/35309104/survey-at-medinet-el-gurob.

———. n.d. Hounen Matsouri, Tagata Jinja. http://www.yamasa.org/japan/english/ destinations/aichi/tagata_jinja.html#1.

Arnaud, D. 1987. *Recherches au Pays d'Astata. Emar* IV. Paris: Éditions Recherche sur les Civilisations.

Artzy, M. 1994A. Nomads of the Sea. In *Res Maritimae:* 1–16.

———. 1994B. On Boats, on Rocks, and on "Nomads of the Sea." *C.M.S. News* 21.

———. 2001. The Medinet Habu Boat Depictions: Can We Trust Ramses III? In *Tropis* VI: 35–44.

———. 2003. Mariners and Their Boats at the End of the Late Bronze and the Beginning of the Iron Age in the Eastern Mediterranean. *Tel Aviv* 30: 232–246.

Ashburner, W. 1909. *The Rhodian Sea Law.* Oxford: Clarendon.

Asouti, E. 2003. Wood Charcoal from Santorini (Thera): New Evidence for Climate, Vegetation, and Timber Imports in the Aegean Bronze Age. *Antiquity* 77: 471–484.

Astour, M. C. 1964. Greek Names in the Semitic World and Semitic Names in the Greek World. *JNES* 23: 193–201.

———. 1965. *Hellenosemitica.* Leiden: Brill.

———. 1972. Review: Some Recent Works on Ancient Syria and the Sea People. *JAOS* 92: 447–459.

———. 1973. Ugarit and the Aegean: A Brief Summary of Archaeological and Epigraphic Evidence. In *Orient and Occident: Essays Presented to Cyrus H. Gordon on the Occasion of His Sixty-Fifth Birthday.* AOAT, Bd. 22. H. A. Hoffner Jr., ed. Kevelaer: Butzon & Bercker: 17–27.

Athenaeus. 1967. *The Deipnosophists.* C. B. Gulick, trans. London: Heinemann and Harvard University Press.

Atia, T. 1999. *Mulid! Carnival of Faith.* Cairo: American University in Cairo Press.

Aura Jorro, F., and F. R. Adrados. 1985–1993. *Diccionario Micénico,* 2 vols. Madrid: Instituto de Filología.

Bachhuber, C. 2003. Aspects of Late Helladic Sea Trade. (Master's thesis, Texas A&M University.)

———. 2006. Aegean Interest on the Uluburun Ship. *AJA* 110: 345–363.

Baika, K. 2002. Dispositif du halage des hangars navals antiques: Études ethno-archéologiques. In *Tropis* VII: 43–83.

Baines, J. 1985. Color Terminology and Color Classification: Ancient Egyptian Color Terminology and Polychromy. *AmAnth* 87: 282–297.

Bakker, E. J. 1995. Noun-Epithet Formulas, Milman Parry, and the Grammar of Poetry. In *Homeric Questions: Essays in Philology, Ancient History, and Archaeology, Including the Papers of a Conference Organized by the Netherlands Institute at Athens (15 May 1993).* J. P. Crielaard, ed. Amsterdam: Gieben: 97–125.

Barako, T. J. 2001. The Seaborne Migration of the Philistines. (PhD diss., Harvard University.)

———. 2003. How Did the Philistines Get to Canaan? One: By Sea . . . *BibAR* 29(2) (March/April): 26–33, 64.

Barb, A. A. 1971. Mystery, Myth, and Magic. In *The Legacy of Egypt.* J. R. Harris, ed. 2nd ed. Oxford: Clarendon: 138–169.

Barber, E. J. W. 1992A. The Peplos of Athena. In *Goddess and Polis: The Panathenaic Festival in Ancient Athens.* J. Neils, ed. Princeton, N.J.: Princeton University Press: 103–117.

———.1992B. *Prehistoric Textiles: The Development of Cloth in the Neolithic and Bronze Ages with Special Reference to the Aegean.* Princeton, N.J.: Princeton University Press.

———. 1994. *Women's Work: The First 20,000 Years: Women, Cloth, and Society in Early Times.* New York: Norton.

———. 1998. Aegean Ornaments and Designs in Egypt. In *The Aegean and the Orient in the Second Millennium. Proceedings of the 50th Anniversary Symposium, Cincinnati, 18–20 April 1997. Aegaeum* 18. E. H. Cline and D. Harris-Cline, eds. Liège: Université de Liège: 13–17.

Barbieri, A., and G. Purpura. 1977. Un giacimento archeolgico in acque profonde nel Canale di Sicilia. *SicArch* 10(34): 54–62.

Barkai, O., and Y. Kahanov. 2007. The Tantura F Shipwreck, Israel. *IJNA* 36: 21–31.

Barnes, J. 1949. The Nevill Papyrus: A Late Ramesside Letter to an Oracle. *JEA* 35: 69–71.

Barnett, R. D., E. Bleibtreu, and G. Turner. 1998. *Sculptures from the Southwest Palace of Sennacherib at Nineveh: Plates.* London: British Museum Press.

Basch, L. 1969A. Phoenician Oared Ships: Part I. *MM* 55: 139–162.

———. 1969B. Phoenician Oared Ships: Part II. *MM* 55: 227–245.

———. 1969C. Trois modéles de navires en marbre au Musée de Sparte. *AntCl* 38: 430–445, pls. I–III.

———. 1972. Ancient Wrecks and the Archaeology of Ships. *IJNA* 1: 1–58.

———. 1975. Another Punic Wreck in Sicily: Its Ram. 1. A Typological Sketch. *IJNA* 4: 201–219.

———. 1978. Le navire *mnš* et autres notes de voyage in Égypte. *MM* 64: 99–123.

———. 1986. The Aegina Pirate Ships of c. B.C. 1700. *MM* 72: 415–438.

———. 1987. *Le musée imaginaire de la marine antique.* Athens: Hellenic Institute for the Preservation of Nautical Tradition.

———. 1994. Un navire grec en Égypte à l'époque de Ulysse. *Neptunia* 195: 19–26.

———. 1997. Une représentation de navire de type égéen dans l'oasis de Dakhleh (Égypte) vers 1200 av. J.-C. In *Res Maritimae: 17–29.*

———, and M. Artzy. 1985. App. 2. Ship Graffiti at Kition. In *Excavations at Kition* V: I: *The Pre-Phoenician Levels, Areas I and II.* V. Karageorghis and M. Demas, eds. Nicosia: Department of Antiquities, Cyprus: 322–336.

Bass, G. F., ed. 1967. *Cape Gelidonya: A Bronze Age Shipwreck. TAPS,* n.s. 57: 8. Philadelphia: American Philosophical Society.

———. 1972. The Earliest Seafarers in the Mediterranean and the Near East. In *HSBUA: 11–36.*

———. 1973. Cape Gelidonya and Bronze Age Maritime Trade. *AOAT* 22: 29–38.

———. 1976. *Archaeology beneath the Sea.* New York: Harper Colophone.

———. 1986. A Bronze Age Shipwreck at Ulu Burun (Kaş): 1984 Campaign. *AJA* 90: 269–296, pl. 17.

———. 1987. Oldest Known Shipwreck Reveals Bronze Age Splendors. *NGM* 172: 693–733.

———. 1991. Evidence of Trade from Bronze Age Shipwrecks. In *Bronze Age Trade in the Mediterranean.* SIMA 90. N. H. Gale, ed. Jonsered: Åström: 69–82.

———. 1997. Beneath the Wine Dark Sea: Nautical Archaeology and Phoenicians of the *Odyssey.* In *Greeks and Barbarians: Essays on the Interactions between Greeks and Non-Greeks in Antiquity and the Consequences for Eurocentrism.* J. E. Coleman and C. A. Walz, eds. Bethesda: CDL Press: 71–101.

———, C. Pulak, D. Collon, and J. Weinstein. 1989. The Bronze Age Shipwreck at Uluburun: 1986 Campaign. *AJA* 93: 1–29.

Bates, O. 1914. *The Eastern Libyans: An Essay.* London: Macmillan.

Baumbach, L. 1979. The Mycenaean Contributions to the Study of Greek Religion in the Bronze Age. *SMEA* 20: 143–160.

Beaux, N. 1990. *Le cabinet de curiosités de Thoutmosis III.* OLA 36. Louvain: Dep. Oriëntalistiek.

Beazley, J. D. 1986. *The Development of Attic Black-Figure.* Berkeley: University of California Press.

Bell, L. 2007. Conflict and Reconciliation in the Ancient Middle East: The Clash of Egyptian and Hittite Chariots in Syria, and the World's First Peace Treaty between "Superpowers." In *War and Peace in the Ancient World.* K. Raaflaub, ed. Oxford: Blackwell: 98–120.

Bell, M. R. 1985. Gurob Tomb 605 and Mycenaean Chronology. In *Mélanges Gamal Eddin Mokhtar.* Cairo: L'Institute français d'archéologie orientale: 61–86, pls. I–VI.

———. 1991. The Tutankhamun Burnt Groups from Gurob: Basis for Absolute Chronology of LH III A and B. (PhD diss., University of Pennsylvania.)

Bennet, E. L. J., and J.-P. Olivier, eds. 1973. *The Pylos Tablets Transcribed,* vol. 1, *Texts and Notes.* Rome: Edizioni dell'Ateneo.

Benson, J. L. 1961. A Problem in Orientalizing Cretan Birds: Mycenaean or Philistine Prototypes? *JNES* 20: 73–84, pls. 3–6.

———. 1975. Birds on Cypro-Geometric Pottery. In *The Archaeology of Cyprus: Recent Developments.* N. Robertson, ed. Park Ridge, N.J.: Noyes: 129–150.

Betts, J. H. 1968. Trees in the Wind on Cretan Sealing. *AJA* 72: 149–150, pl. 61.

———. 1973. Ships on Minoan Seals. In *Marine Archaeology. Proceedings of the Twenty-Third Symposium of the Colston Research Society Held in the University of Bristol, April 4th to 8th, 1971.* D. J. Blackman, ed. London: Archon: 325–338.

Bianchi, R. S. 1997. The Theban Landscape of Rameses II. In *AEANE:* 87–94.

Bieńkowski, P. A. 1982. Some Remarks on the Practice of Cremation in the Levant. *Levant* 14: 80–89.

Biers, J. C., and J. Terry, eds. 2004. *Testament of Time: Selected Objects from the Collection of Palestinian Antiquities in the Museum of Art and Archaeology, University of Missouri–Columbia.* Madison, Wis.: Fairleigh Dickinson University Press.

Bietak, M. 1992. Minoan Wall-Paintings Unearthed at Ancient Avaris. *EgyArch* 2: 26–38.

———. 1993. The Sea Peoples and the End of the Egyptian Administration in Canaan. In *Biblical Archaeology Today, 1990, Proceedings of the Second International Congress on Biblical Archaeology.* A. Biran and J. Aviram, eds. Jerusalem: Israel Exploration Society: 292–306.

———. 1995. Connections between Egypt and the Minoan World: New Results from Tell el-Daʿba/Avaris. In *EAL:* 19–28, pls. 1–4, 14–17.

———. 1996. *Avaris the Capital of the Hyksos: Recent Excavations at Tell el-*Daʿba. The First Raymond and Beverly Sackler Foundation Distinguished Lecture in Egyptology. London: British Museum Press.

———. 2000. "Rich beyond the Dreams of Avaris; Tell el-Daʿba and the Aegean World—A Guide for the Perplexed": A Response to Eric H. Cline. *BSA* 95: 185–205, pls. A–C.

———. 2005. The Thutmoside Stronghold of Perunefer. *EgyArch* 26: 13–17.

———, N. Marinatos, and C. Palyvou. 2000. The Maze Tableau from Tell el Daʿba. In *WPT* 1: 77–90.

———, and A. Brysbaert. 2007. *Taureador Scenes in Tell el-Daʿba (Avaris) and Knossos. Denkschriften der Gesamtakademie,* Bd. 43. Vienna: Verlag der Österreichischen Akademie der Wissenschaften.

Blackman, A. M. 1923. *Luxor and Its Temples.* New York: Macmillan.

———. 1925. Oracles in Ancient Egypt. *JEA* 11: 249–255, pls. XXXV–XXXVIII.

———. 1926. Oracles in Ancient Egypt II. *JEA* 12: 176–185, pls. XXXIV–XLII.

———. 1941. The Stela of Shoshenḳ, Great Chief of the Meshwesh. *JEA* 27: 83–95, pls. 10–12.

Blackman, D. J. 2008. Reflections on the Importance of Shipsheds. In *Tropis X Abstracts.*

———, and M. C. Lentini. 2006. An Ancient Greek Dockyard in Sicily. In *The New View: Underwater Archaeology and the Historical Picture. Proceedings of the 2nd International Congress on Underwater Archaeology (IKUWA2) Zurich 2004. Antiqua* 40. A. Hafner, U. Niffeler, and U. Ruoff, eds. Basel: Archaeologie Schweiz: 193–197.

Blakolmer, F. 2000. Zum Charakter der frühägäischen Farben: Linear B und Homer. In *Österreichische Forschungen zur ägäischen Bronzezeit 1998. Akten der Tagung am Institut für Klassische Archäologie der Universität, Wien (2.–3. Mai 1998). Wiener Forschungen zur Archäologie* 3. F. Blakolmer, ed. Vienna: Phoibos: 225–239, Abb. 82–89.

Blegen, C. W., M. Rawson, L. W. Taylor, and W. P. Donovan. 1973. *The Palace of Nestor at Pylos in Western Messenia* III: *Acropolis and Lower Town Tholoi, Grave Circle, and Chamber Tombs Discoveries outside the Citadel.* Princeton, N.J.: University Press (University of Cincinnati).

Boardman, J. 1958. A Greek Vase from Egypt. *JHS* 78: 4–12, pls. I–II.

———. 1967. The Khaniale Tekke Tombs, II. *BSA* 62: 57–75, pls. 7–17.

———. 1999. *The Greeks Overseas: Their Early Colonies and Trade.* Fourth edition. Baltimore: Thames and Hudson.

Bobart, H. 1936. *Basketwork through the Ages.* London: Oxford University Press.

Bodenheimer, F. S. 1949. *Animal Life in Biblical Lands: From the Stone Age to the End of the Nineteenth Century* I. Jerusalem: Bialik Institute. (In Hebrew).

———. 1960. *Animal and Man in Bible Lands: Text.* (*Collection de travaux de l'Académie international d'histoire des sciences* 10). Leiden: Brill.

———. 1972. *Animal and Man in Bible Lands: Figures and Plates.* (*Collection de travaux de l'Académie international d'histoire des sciences* 10). Leiden: Brill.

Boegehold, A. L. 1996. Group and Single Competitions at the Panathenaia. In *Worshipping Athena:* 95–105.

Boetticher, C. 1863. *Bericht über die Untersuchungen auf der Akropolis von Athen im Frühjahre 1862.* Berlin: Von Ernst und Korn.

Borchardt, L. 1911. *Statuen und Statuetten von Königen und Privatleuten im Museum von Kairo,* vol. 1. *(Cat. Gén.).* Berlin: Reichsdr.

Bothe, C. I. 2007. Asphalt. In *Artists' Pigments: A Handbook of Their History and Characteristics,* vol. 4. B. H. Berrie, ed. Washington, D.C.: National Gallery of Art: 111–149.

Bouzek, J. 1985. *The Aegean, Anatolia, and Europe: Cultural Interrelations in the Second Millennium* B.C. SIMA 29. Gothenburg: Åström.

———. 1994. Late Bronze Age Greece and the Balkans: A Review of the Present Picture. *BSA* 89: 217–234.

———. 1996. The Problem of Migration in Mycenaean Greece. In *Atti e memorie del secondo Congresso internazionale di micenologia, Roma-Napoli, 14–20 ottobre 1991,* vol. 1, *Filologia.* Incunabula Graeca 98. E. De Miro, L. Godart and A. Sacconi, eds. Rome: Gruppo editoriale internazional: 685–692.

———. 2000. Review: *Seagoing Ships and Seamanship in the Bronze Age Levant.* By Shelley Wachsmann. *Eirene* 36: 199–200.

———. 2005. *Thracians and Their Neighbours: Their Destiny, Art, and Heritage* (*Studia Hercynia*). Prague: Institute of Classical Archaeology, Charles University and the Czech Society of Archaeology.

Brady, T. A. 1938. Review: A Festival of Isis in Rome under the Christian Emperors of the Fourth Century by Andrew Alföldi. *JRS* 28: 88–90.

Breasted, J. H. 1906. First Preliminary Report of the Egyptian Expedition. *AJSL* 23: 1–64.

Bremmer, J. 1994. *Greek Religion.* Oxford: Oxford University Press.

Brinkmann, V. 2008. The Polychromy of Ancient Greek Sculpture. In *Color of Life:* 18–39.

———, R. Wünsche, U. Koch-Brinkmann, S. Kellner, J. Köttl, O. Herzog, R. Batstone, S. Ebbinghaus, and A. Brauer. 2007. *Gods in Color: Painted Sculpture of Classical Antiquity.* Munich: Stiftung Archäologie, Glypothek Munich.

British Columbia Ministry of Culture and the Arts. 2010. British Columbia Shipwreck Recording Guide Appendix 1. Glossary. (June 20). http://www.tca.gov.bc.ca/archaeology/docs/shipwreck_recording_guide/glossary.htm.

Brommer, F. 1979. *The Sculptures of the Parthenon.* M. Whittall, trans. London: Thames and Hudson.

Brug, J. F. 1985. *A Literary and Archaeological Study of the Philistines.* BARIS 265. Oxford: B.A.R.

Bruneau, P. 1961. Isis Pélagia à Délos. *BCH* 85: 435–446.

———. 1974. Existe-t-il des statues d'Isis Pélagia? *BCH* 98: 333–381.

Brunton, G. 1934. Modern Painting on Predynastic Pots. *ASAE* 34:149–157.

Brunton, G., and R. Engelbach. 1927. *Gurob.* London: British School of Archaeology in Egypt and Bernard Quaritch.

Bryk, F. 1934. *Circumcision in Man and Woman.* D. Berger, trans. New York: American Ethnological Press.

Bryson, R. A., H. H. Lamb, and D. L. Dowley. 1974. Drought and the Decline of Mycenea. *Antiquity* 48: 46–50.

Buchholz, H.-G., and V. Karageorghis. 1973. *Prehistoric Greece and Cyprus: An Archaeological Handbook.* London: Phaidon.

Burckhardt, J., and E. Hohl. 1924. *Die Zeit Konstantins des Grossen.* Leipzig: Kröner (1st ed. 1852).

Burkert, W. 1983. *Homo Necans: The Anthropology of Ancient Greek Sacrificial Ritual and Myth.* Berkeley: University of California Press.

Caminos, R. A., and A. H. Gardiner. 1954. *Late-Egyptian Miscellanies.* Brown Egyptological Studies 1. London: Oxford University Press.

Cancik, H., and H. Schneider, eds. 2002–. *Brill's New Pauly: Encyclopedia of the Ancient World.* Leiden: Brill.

Canney, M. A. 1936. Boats and Ships in Temples and Tombs. In *Occident and Orient: Gaster Anniversary Volume.* B. Schindler and A. Marmorstein, eds. London: Taylor's Foreign Press: 50–57.

———. 1938. Boats and Ships in Processions. *Folklore* 49: 132–147.

Caprino, C., A. M. Colini, G. Gatti, M. Pallottino, and P. Romanelli. 1955. *La Colonna di Marco Aurelio.* Rome: "L'Erma" di Bretschneider.

Cariolou, G. A. 1997. KYRENIA II: The Return from Cyprus to Greece of the Replica of a Hellenic Merchant Ship. In *Res Maritimae:* 83–97.

Carlson, D. N. 2009. Seeing the Sea: Ships' Eyes in Classical Greece. *Hesperia* 78: 347–365.

Carpenter, T. 1986. *Dionysian Imagery in Archaic Art.* Oxford: Clarendon.

Carter, H. 2000. *The Tomb of Tut.ankh.amen: The Annexe and Treasury.* London: Duckworth (reprint).

———. 2001. *The Tomb of Tut.ankh.amen* II. London: Duckworth (reprint).

Cartwright, C. R. 2008A. The Plant Remains. In *The Painted Tomb-Chapel of Nebamun.* R. Parkinson, ed. London: British Museum Press: 47–48.

———. 2008B. Plant Remains from the Mud Plaster. In *The Nebamun Wall Paintings: Conservation, Scientific Analysis and Redisplay.* A. Middleton and K. Uprichard, eds. London: Archetype Publications and the British Museum: 25–29.

Caskey, J. L. 1964. Investigations in Keos, 1963. *Hesperia* 33: 314–335.

Casson, L. 1958. Hemiolia and Triemiolia. *JHS* 78: 14–18, pls. V–VI.

———. 1970. Phoenician Oared Ships. *MM* 56: 31.

———. 1975. Bronze Age Ships. The Evidence of the Thera Wall Paintings. *IJNA* 4: 3–10.

———. 1992. Odysseus' Boat (Od. 5.244–53). *IJNA* 21: 73–74.

———. 1995. *Ships and Seamanship in the Ancient World.* Baltimore: Johns Hopkins University Press (reprint with addenda and corrigenda).

———, and E. Linder. 1991. The Evolution in Shape of the Ancient Ram. In *The Athlit Ram.* Nautical Archaeology Series 3. L. Casson and J. R. Steffy, eds. College Station: Texas A&M University Press, 67–75.

Casson, L., and J. R. Steffy, eds. 1991. *The Athlit Ram.* College Station: Texas A&M University Press, 1991.

Catling, H. W. 1964. *Cypriot Bronzework in the Mycenaean World.* Oxford: Clarendon.

Cavanagh, W. and C. Mee. 1998. *A Private Place: Death in Prehistoric Greece.* SIMA 125. Jonsered: Paul Åstroms Förleg.

Cavillier, G. 2005. *Gli Shardana nell'Egitto ramesside.* BARIS 1438. Oxford: Archaeopress.

Černý, J. 1930. Une expression designant la réponse négative d'un oracle. *BIFAO* 30: 491–496.

———. 1935. Questions adressées aux oracles. *BIFAO* 35: 41–58.

———. 1941. Nouvelle série de questions adressées aux oracles. *BIFAO* 41: 13–24, pls. I–III.

———. 1962. Egyptian Oracles. In *A Saite Oracle Papyrus from Thebes in the Brooklyn Museum (Papyrus Brooklyn 47.218.3).* R. A. Parker, ed. Providence, R.I.: Brown University Press: 35–48.

Červiček, P. 1986. *Rock Pictures of Upper Egypt and Nubia.* Annales de l'institut universitaire oriental de Naples XLVI, I, suppl. 46. Naples: Istituto Universitario Orientale.

Chabas, F. J. 1873. *Études sur l'antiquité historique: D'après les sources égyptiennes et les monuments réputés préhistoriques.* Paris: Maisonncuvc.

Chadwick, J. 1976. *The Mycenaean World.* Cambridge: Cambridge University Press.

———. 1988. The Women of Pylos. In *Texts, Tablets, and Scribes: Studies in Mycenaean Epigraphy and Economy Offered to Emmett L. Bennett, Jr.* Supplement to Minos 10. J.-P. Olivier, E. L. Bennett, and T. G. Palaima, eds. Salamanca: Ediciones Universidad de Salamanca Servicio editorial: 43–95.

Chantraine, P. 1983. *Dictionnaire étymologique de la langue grecque.* Paris: Éditions Klincksieck.

Charlton, W. H. 1996. Rope and the Art of Knot-Tying in the Seafaring of the Ancient Eastern Mediterranean. (Master's thesis, Texas A&M University.)

Cherpion, N., J. Corteggiani, and F. Gout. 2007. *Le tombeau de Pétosisris à Touna el-Gebel.* Bibliothèque générale 27. Cairo: Institute française d'archéologie orientale.

Cincinnati Art Museum. 1970. *Sculpture Collection of the Cincinnati Art Museum.*

Cline, E. H. 1998. Rich beyond the Dreams of Avaris: Tell el-Daʿba and the Aegean World—A Guide for the Perplexed. *BSA* 93: 199–219.

Cochavi-Rainey, Z., and C. Lilyquist. 1999. *Royal Gifts in the Late Bronze Age, Fourteenth to Thirteenth Centuries B.C.E.: Selected Texts Recording Gifts to Royal Personages.* Beer-Sheva 13. Beersheva: Ben-Gurion University of the Negev.

Coldstream, J. N. 2006. *Geometric Greece: 900–700 BC.* London: Routledge.

Coleman, C., and M. Morrow. Forthcoming. *The Rock Drawings of Southern Upper Egypt* I–II (rev. ed. of H. Winkler's report).

Courbin, P. 1957. Une tombe géometrique d'Argos. *BCH* 81: 322–386.

Creasman, P. P., and N. Doyle. 2010. Overland Boat Transportation during the Pharaonic Period: Archaeology and Iconography. *JAEI* 2(3): 14–30.

Crouwel, J. 1973. Appendix: The Parasol Krater. *BSA* 68: 343–347.

Crowfeet Payne, J. A. Kaczmeveyk, and S. J. Fleming. 1977. Forged Decoration on Predynastic Pots. *JEA* 63: 5–12.

Cruz-Uribe, E. 1988. A New Look at the Adoption Papyrus. *JEA* 74: 220–223.

Csapo, E. 1997. Riding the Phallus for Dionysus: Iconology, Ritual, and Gender-Role De/Construction. *Phoenix* 51: 253–295.

Cunliffe, R. J. 1988. *A Lexicon of the Homeric Dialect.* Norman: University of Oklahoma Press.

Dakoronia, F. 1990. War-Ships on Sherds of LH IIIC Kraters from Kynos. In *Tropis* II: 117–122.

———. 1993. Homeric Towns in East Lokris: Problems of Identification. *Hesperia* 62: 115–127.

———. 1995. Editor's Note (War-Ships on Sherds of LH IIIC Kraters from Kynos?) In *Tropis* III: 147–148.

———. 1996A. Kynos . . . Fleet. In *Tropis* IV: 159–171.

———. 1996B. Mycenaean East Lokris. In *ASCIM:* 1167–1171, pls. 1–5.

———. 1999. Representations of Sea-Battles on Mycenaean Sherds from Kynos. In *Tropis* V: 119–128.

———. 2001. Further Finds from Kynos. In *Tropis* VI: 171.

———. 2002. Further Finds from Kynos. In *Tropis* VII: 283–290.

———. 2006. Mycenaean Pictorial Style at Kynos, East Lokris. In *Pictorial Pursuits:* 23–29.

Danelius, E., and H. Steinitz. 1967. The Fishes and Other Aquatic Animals on the Punt-Reliefs at Deir El-Bahri. *JEA* 53: 15–24.

Daressy, G. 1895. Une flottille phénicienne d'après une peinture récente. *RA:* 286–292, pls. XIV–XV.

———. 1915. Trois stèle de la période Bubastite. *ASAE* 26: 23–29.

Dattari, G. 1901. *Numi Augg. Alexandrini.* 2 vols. Cairo: Istituto Francese di Archeologia Orientale.

David, A. R. 1986. *The Pyramid Builders of Ancient Egypt: A Modern Investigation of Pharaoh's Workforce.* London: Routledge and Kegan Paul.

Davies, J. K. 1967. Demosthenes on Liturgies: A Note. *JHS* 87: 33–40.

Davies, M. I. 1978. Sailing, Rowing, and Sporting in One's Cups on the Wine Dark Sea. In *Athens Comes of Age: From Solon to Salamis. Papers of a Symposium Sponsored by the Archaeological Institute of America, Princeton Society, and the Department of Art and Archaeology.* Princeton, N.J.: Princeton University Press: 72–95.

Davies, N. d. G. 1920. *The Tomb of Antefoker, Vizier of Sesostris I, and of his Wife, Senet (no. 60).* London: Allen and Unwin.

———. 1925. The Tomb of Tetaky at Thebes (no. 15). *JEA* 11(1925): 10–18, pls. II–V.

———. 1926. An Apparent Perspectival Drawing. *JEA* 12: 110–111.

———. 1927. *Two Ramesside Tombs at Thebes.* New York: Metropolitan Museum of Art.

———. 1930. The Egyptian Expedition 1929–1930: The Work of the Graphic Branch of the Expedition. *BMMA* 25(12) (Part 2, December): 29–42.

———. 1932. The Egyptian Expedition: The Work of the Graphic Branch of the Expedition. *BMMA* 27 (Part 2, March): 51–62.

———. 1948. *Seven Private Tombs at Kurnah.* London: Egyptian Exploration Society.

———. 1973. *The Tomb of Rech-mi-rēʿ at Thebes.* 2 vols. Metropolitan Museum of Art Egyptian Expedition Publications. New York: Metropolitan Museum of Art (reprint).

———, and N. M. Davies. 1973. *The Tomb of Nefer-Hotep at Thebes.* New York: Arno.

Davies, N. d. G., and R. O. Faulkner. 1947. A Syrian Trading Venture in Egypt. *JEA* 33: 40–46, pl. VIII.

Davis, D. 2000. Navigation in the Ancient Eastern Mediterranean. (Master's thesis, Texas A&M University.)

———. 2007. Ancient Navigation. In *Oxford Encyclopedia of Maritime History.* I. J. B. Hattendorf, ed. New York: Oxford University Press: 85–93.

Davison, J. A. 1958. Notes on the Panathenaea. *JHS* 78: 23–42.

Dawson, W. R. 1925. An Oracle Papyrus, B.M. 10335. *JEA* 11: 247–248, pls. XXXV–XXVIII.

De Graeve, M.-C. 1981. *The Ships of the Ancient Near East (c. 2000–500 B.C.).* OLA 7. Leuven: Departement Oriëntalistiek.

de Ridder, A. 1896. *Catalogue des bronzes trouvés sur l'Acropole d'Athènes; publié sous les auspices de l'Académie des inscriptions et belles-lettres (Fondation Piot). Bibliothèque des Écoles françaises d'Athènes et de Rome,* facs. 74). Paris: Librairie Thorin et Fils.

de Rougé, E. 1867. Extraits d'un mémoire sur les attaques dirigées contre l'égypte par les peuples de la méditerranée. *RA:* 35–45.

Delcourt, M. 1965. *Pyrrhos et Pyrrha.* (*Bibliothèque de la Faculté de philosophie et lettres de l'Université de Liège,* 174).

Delgado, J. P., ed. 2008. *Nautical Archaeology, 2006–2007 Seasons. AJA* 112: 307–335.

Demangel, R. 1944–1945. La frise de socle du palais de Tirynthe. *BCH* 68–69: 404–410.

Desborough, V. d. A. 1964. *The Last Mycenaeans and Their Successors: An Archaeological Survey, c. 1200–c. 1000 B.C.* Oxford: Clarendon.

———. 1972. *The Greek Dark Ages.* London: Benn.

Detienne, M. 1970. Le navire d'Athéna. *Revue de l'histoire des religions* 178: 135–177.

Deubner, L. 1956. *Attische Feste.* Berlin: Akademie-Verlag.

DeVries, K., and M. Katzev. 1972. Greek, Etruscan, and Phoenician Ships and Shipping. In *HSBUA:* 37–64.

Dickinson, O. 2006. *The Aegean from Bronze Age to Iron Age: Continuity and Change between the Twelfth and Eighth Centuries B.C.* London: Routledge.

Dietrich, M., and O. Loretz. 1972. Die Schardana in den Texten von Ugarit. In *Antike und Universalgeschichte: Festschrift Hans Erich Stier zum 70. Geburtstag am 25. Mai 1972.* G. A. Lehmann and R. Stiehl, eds. Münster: Aschendorff: 39–42.

Dikaios, P. 1969. *Enkomi: Excavations 1948–1958,* vol. 3a, pls. 1–239. Mainz: von Zabern.

———. 1971. *Enkomi: Excavations 1948–1958,* vol. 2, *Chronology, Summary and Conclusions, Catalogue, Appendices.* Mainz: von Zabern.

Dittman, K. H. 1941. Der Segelwagen von Medinet Madi. *MDIK* 10: 60–78, pl. 16.

Dothan, M. 1986. Sardinia at Akko? In *Studies in Sardinian Archaeology,* vol. 2, *Sardinia in the Mediterranean.* M. S. Balmuth, ed. Ann Arbor: University of Michigan Press: 105–115.

Dothan, T. 1982A. *The Philistines and Their Material Culture.* New Haven, Conn.: Yale University Press and Israel Exploration Society.

———. 1982B. What We Know about the Philistines. *BibAR* 8(4): 20–44.

———, and M. Dothan. 1992. *People of the Sea: The Search for the Philistines.* New York: Macmillan.

Dothan, T., and A. Zuckerman. 2004. A Preliminary Study of the Mycenaean IIIC:1 Pottery Assemblages from Tel Miqne-Ekron and Ashdod. *BASOR* 333: 1–54.

Doumas, C. 1983. *Thera: Pompeii of the Ancient Aegean.* London: Thames and Hudson.

———. 1992. *The Wall-Paintings of Thera.* A. Doumas, trans. Athens: Thera Foundation.

———. 1999. *The Wall-Paintings of Thera.* A. Doumas, trans. 2nd ed. Athens: Thera Foundation.

———. 2008. The Wall Paintings of Thera and the Eastern Mediterranean. In *Beyond Babylon:* 124–125.

Doyle, N. 1998. Iconography and the Interpretation of Ancient Egyptian Watercraft. (Master's thesis, Texas A&M University.)

Duff Gordon, L. 1969. *Letters from Egypt (1862–1869) by Lady Duff Gordon.* G. Waterfield, ed. New York: Praeger.

Dugas, C. 1936. Céramique. Peinture. Mosaïque. *RÉG* 49: 222–234.

Dunand, F. 1973. *Le culte d'Isis dans le bassin oriental de la Méditerranée,* vol. 1, *Le culte d'Isis et les Ptolémées.* Études préliminaires aux religions orientales dans l'Empire romain, T. 26. Leiden: Brill.

———, and C. Zivie-Coche. 2004. *Gods and Men in Egypt: 3000 BCE to 395 CE.* Ithaca, N.Y.: Cornell University Press.

Dunand, M. 1950. *Fouilles de Byblos,* vol. 2, *1933–1938,* Atlas. Paris: Geuthner.

———. 1954. *Fouilles de Byblos,* vol. 2, *1933–1938,* Text. Paris: Geuthner.

Dyer, L. 1891. *Studies of the Gods in Greece at Certain Sanctuaries Recently Excavated.* London: Macmillan.

Eastaugh, N., V. Walsh, T. Chaplin, and R. Siddal. 2004A. *The Pigment Compendium: A Dictionary of Historical Pigments.* Amsterdam: Elsevier Butterworth-Heinemann.

———. 2004B. *The Pigment Compendium: Optical Microscopy of Historical Pigments.* Amsterdam: Elsevier Butterworth-Heinemann.

Edgerton, W. F., and J. A. Wilson. 1936. *Historical Records of Ramses III: The Texts in Medinet Habu,* 2 vols. Translated with explanatory notes. SAOC 12. Chicago: Oriental Institute of the University of Chicago.

Egberts, A. 1991. The Chronology of *The Report of Wenamun. JEA* 77: 57–67.

Eisenberg, J. M. 2007. News from Egypt: Egypt's Largest Fortress Uncovered in North Sinai. *Minerva* 18(6): 7.

El Goresy, A. 2000. Polychromatic Wall Painting Decorations in Monuments of Pharaonic Egypt: Compositions, Chronology, and Painting Techniques. In *WPT* 2: 49–76.

Ellmo, A., and D. Lidén. 2005. *Sexualitet i Japan.* Linköping: Linköpings Universitet.

Evelyn-White, H. G., ed. 1998. *Hesiod, Homeric Hymns, Epic Cycle Homerica.* Cambridge, Mass.: Harvard University Press.

Fabre, D. 2004. *Seafaring in Ancient Egypt.* London: Periplus.

Fakhri, A. 1944. *Siwa Oasis: Its History and Antiquities.* Cairo: Government Press.

Fattovich, R. 2005. Marsā Gawāsīs: A Pharaonic Coastal Settlement by the Red Sea in Egypt. In *People of the Red Sea: Proceedings of Red Sea Project II Held in the British Museum, October 2004.* BAR S1395, Series for Arabian Studies 3. J. C. M. Starkey, ed. Oxford: Archaeopress: 15–22.

———, and K. A. Bard. 2006. *Joint Archaeological Expedition at Mersa/Wadi Gawasis (Red Sea, Egypt) of the University of Naples "L'Orientale" (Naples, Italy), Instituto Italiano per l'Africa e l'Oriente (Rome, Italy), and Boston University (Boston, USA): 2005–2006 Field Season.* http://www.archaeogate.org/egittologia/article/441/9/joint-archaeological-expedition-at-mersawadi-gawasis-re.html.

Faulkner, R. O. 1941. Egyptian Military Standards. *JEA* 27: 12–18.

———. 1947. The Wars of Setos I. *JEA* 33: 34–40.

———. 1953. Egyptian Military Organization. *JEA* 39: 32–47.

Fay, B. 1986. *Egyptian Museum, Berlin.* Berlin: Ägyptisches Museum der Staatlichen Museen Preussischer Kulturbesitz.

Finkelstein, I. 1995. The Date of the Settlement of the Philistines in Israel. *Tel Aviv* 22: 213–239.

———. 2000. The Philistine Settlements: When, Where, and How Many? In *Sea Peoples:* 159–180.

Fitzgerald, M. A. 1994. The Ship. In *The Harbours of Caesarea Maritima: Results of the Caesarea Ancient Harbour Excavation Project 1980–1985,* vol. 2, *The Finds and the Ship.* BARIS 594. J. P. Oleson, ed. Oxford: Tempus Reparatum: 163–223, pls. 29–44.

Foucart, G. 1921–1922. Un temple flottant: Le vaisseau d'or d'Amun-Ra. *MonPiot* 25: 143–169.

Freed, R. A., L. M. Berman, and D. M. Doxey. 2009. *The Secrets of Tomb 10A: Egypt 2000 BC.* Boston: MFA Publications.

French, E. B. 1973. Note: A Possible Chariot Figure with Parasol. *BSA* 68: 347–348.

Frickenhaus, A. 1912. Der Schiffskarren des Dionysos in Athen. *JDI* 27: 61–79, Tafs. I–IV.

Frost, H. 1979. Egypt and Stone Anchors: Some Recent Discoveries. *MM* 65: 137–161.

Furumark, A. 1941. *The Mycenaean Pottery: Analysis and Classification.* Stockholm: Kungl.

Gaballa, G. A. 1972. Some Nineteenth-Dynasty Monuments in Cairo Museum. *BIFAO* 71: 129–137.

———. 1973. Three Documents from the Reign of Ramesses III. *JEA* 59: 109–113.

———, and K. A. Kitchen. 1969. The Festival of Sokar. *Orientalia* 38: 1–76, Tab. I–II.

Gardiner, A. H. 1920. The Ancient Military Road between Egypt and Palestine. *JEA* 6: 99–116.

———. 1933. The Dakhleh Stela. *JEA* 19: 19–30, pls. 5–7.

———. 1940. Adoption Extraordinary. *JEA* 26: 23–29.

———. 1941. Ramesside Texts relating to the Taxation and Transport of Corn. *JEA* 27: 19–73.

———. 1947. *Ancient Egyptian Onomastica,* 3 vols. London: Oxford University Press.

———. 1953. The Harem at Miwer. *JNES* 12: 145–149.

———. 1968. *Ramesside Administrative Documents.* Oxford: Griffith Institute, Ashmolean Museum.

———, and H. I. Bell. 1943. The Name of Lake Moeris. *JEA* 29: 37–46.

Gardner, E. A. 1914. The Panathenaic Ship of Herodes Atticus. (Philostratus, *Vitae Sophistarum,* II. v.). *CR* 28: 225–226.

Gardner, P. 1881A. Boat-Races among the Greeks. *JHS* 2: 90–97.

———. 1881B. Boat-Races at Athens. *JHS* 2: 315–317.

Gesell, G. C. 1985. *Town, Palace, and House Cult in Minoan Crete.* Gothenburg: Åström.

Gilboa, A. 2005. Sea Peoples and Phoenicians along the Southern Phoenician Coast: A Reconciliation: An Interpretation of Šikila (SKL) Material Culture. *BASOR* 337: 47–78.

Gitin, S., A. Mazar, and E. Stern, eds. 1998. *Mediterranean Peoples in Transition: Thirteenth to Early Tenth Centuries* BCE*. In Honor of Trude Dothan.* Jerusalem: Israel Exploration Society.

Gjerstad, E., J. Lindros, E. Sjöqvist, A. Westholm, and the Svenska Cypernexpeditionen. 1934A. *The Swedish Cyprus Expedition: Finds and Results of the Excavations in Cyprus, 1927–1931,* vol. 1, *Text.* Stockholm: Swedish Cyprus Expedition.

———. 1934B. *The Swedish Cyprus Expedition: Finds and Results of the Excavations in Cyprus, 1927–1931,* vol. 2, *Plates.* Stockholm: Swedish Cyprus Expedition.

Gladstone, W. E. 1877. The Colour Sense. *Nineteenth Century* 2: 366–388.

Glanville, S. R. K. 1931. Records of a Royal Dockyard of the Time of Thutmose III: Papyrus British Museum 10056: Part I. *ZÄSA* 66: 105–121.

———. 1932. Records of a Royal Dockyard of the Time of Thutmose III: Papyrus British Museum 10056: Part II. Commentary. *ZÄSA* 68: 7–41.

Gollaher, D. 2000. *Circumcision: A History of the World's Most Controversial Surgery.* New York: Basic Books.

Goneim, Z. 1957. *Excavations at Saqqara: Horus Sekhem-khet, The Unfinished Step Pyramid at Saqqarra.* Cairo: Service des antiquités [et] l'Inst. français d'archéologie orientale.

Goren, Y., I. Finkelstein, and N. Na'aman. 2002. Petrographic Investigation of the Amarna Tablets. *NEA* 65(3): 196–205.

———. 2004. *Inscribed in Clay: Provenance Study of the Amarna Tablets and Other Ancient Near Eastern Texts.* Tel Aviv: Sonia and Marco Nadler Institute of Archaeology, Tel Aviv University.

Göttlicher, A. 1978. *Materialien für ein Korpus der Schiffsmodelle im Altertum.* Mainz: von Zabern.

———. 1992. *Kultschiffe und Schiffskulte im Altertum.* Berlin: Mann.

———. 2004. A Newly Acquired Ancient Ship-Model in Kassel, Germany. *IJNA* 33: 154–157.

———. 2008. The Archaic Shipmodel Kassel T832 in the Context of 6th/5th B.C. Naval Architecture. In *Tropis X Abstracts.*

Götze, A. 1957. *Kleinasien. Kulturgeschichte des alten Orients* 3. (Abschnitt, 1). Munich: Beck.

Grant, E. 1932. *Ain Shems Excavations (Palestine): 1928–1929–1930–1931.* Part II. Biblical and Kindred Studies no. 4. Haverford, Penn.: Haverford College.

Gray, D. *Seewessen.* Archaeologia Homerica, Bd. 1, Kapitel G. Göttingen: Vandenhoeck und Ruprecht.

Green, J. R. 2007. Let's Hear It for the Fat Man: Padded Dancers and the Prehistory of Drama. In *The Origins of Theater in Ancient Greece and Beyond: From Ritual to Drama.* E. Csapo and M. C. Miller, eds. Cambridge: Cambridge University Press: 96–117.

Griffith, F. L. 1898A. *The Petrie Papyri: Hieratic Papyri from Kahun and Gurob (Principally of the Middle Kingdom).* Text. London: Quaritch.

———. 1898B. *The Petrie Papyri: Hieratic Papyri from Kahun and Gurob (Principally of the Middle Kingdom).* Plates. London: Quaritch.

Griffiths, J. G. 1975. *Apuleius of Madauros: The Isis Book (Metamorphoses, Book XI).* Leiden: Brill.

Grinsell, L. V. 1961. The Breaking of Objects as a Funerary Rite. *Folklore* 72: 475–491.

Güntner, W. 2000. *Figürlich bemalte mykenische Keramik aus Tiryns.* (*Tiryns* XII). Mainz: von Zabern.

———. 2006. Same Motif, Another Context. In *Pictorial Pursuits:* 177–182.

Habachi, L. 1972. *The Second Stela of Kamose and His Struggle against the Hyksos Ruler and His Capital.* Abhandlungen des Deutschen Archäologischen Institutes Kairo, Ägyptologische Reihe 8. Glückstadt: Augustin.

———. 1980. The Military Posts of Ramesses II on the Coastal Road and the Western Part of the Delta. *BIFAO* 80: 13–30.

Hale, J. R. 2009. *Lords of the Sea: The Epic Story of the Athenian Navy and the Birth of Democracy.* New York: Viking.

Hallager, E., M. Vlasakis, and B. P. Hallager. 1992. New Linear B Tablets from Khania. *Kadmos* 31: 61–87.

Halleux, R. 1969. Lapis-Lazuli, azurite ou pâte de verre? A propos de *kuwano* et *kuwanowoko* dans les tablettes mycéniennes. *SMEA* 39: 47–66.

Hamilton, R. 1978. A New Interpretation of the Anavysos Chous. *AJA* 82: 385–387.

Hammacher, A. M. 1985. *René Magritte.* New York: Abrams.

Hankey, V. 1981. The Aegean Interest in El-Amarna. *JMAA* 1: 38–49.

———, and P. Warren. 1974. The Absolute Chronology of the Aegean Late Bronze Age. *BICS* 21: 142–152.

Harding, A. F. 2000. *European Societies in the Bronze Age.* Cambridge: Cambridge University Press.

Harrison, J. E. 1885. Odysseus and the Sirens-Dionysian Boat Races: A Cylix by Nikosthenes. *JHS* 6: 19–29.

Hassler, A. 2011. Mycenaeans at Tell Abu Gurob? *In Intercultural Contacts in the Ancient Mediterranean. Proceedings of the International Conference at the Netherlands–Flemish Institute in Cairo, 25th to 29th October 2008.* OLA 202. K. Duistermaat and I. Regulski, eds. Leuven, Vitgeverij Peters, 125–135.

Hayes, W. C. 1955. *A Papyrus of the Late Middle Kingdom in the Brooklyn Museum, (Papyrus Brooklyn 35.1446).* Wilbour Monographs 5. Brooklyn, N.Y.: Brooklyn Museum.

———. 1959. *The Scepter of Egypt. A Background for the Study of the Egyptian Antiquities in the Metropolitan Museum of Art.* Part II, *The Hyksos Period and the New Kingdom (1675–1080 B.C.).* Cambridge, Mass.: Published for the Metropolitan Museum of Art by Harvard University Press.

———. 1990. *The Scepter of Egypt: A Background for the Study of the Egyptian Antiquities in the Metropolitan Museum of Art.* Part I, *From the Earliest Times to the End of the Middle Kingdom.* New York: Metropolitan Museum of Art.

Hellbing, L. 1979. *Alasia Problems.* SIMA 57. Gothenburg: Åström.

Heltzer, M. 1979. Some Questions regarding the Sherdana in Ugarit. *IOS* 9: 9–16.

———. 1982. *The Internal Organization of the Kingdom of Ugarit.* Wiesbaden: Reichert.

———. 1988. Sinaranu, Son of Siginu, and the Trade Relations between Ugarit and Crete. *Minos* 23: 7–13.

Hencken, H. 1968A. *Tarquinia and Etruscan Origins.* New York: Praeger.

———. 1968B. *Tarquinia, Villanovans, and Early Etruscans,* 2 vols. American School of Prehistoric Research, Peabody Museum, Harvard University Bulletin 23. Cambridge, Mass.: Peabody Museum.

———. 1971. *The Earliest European Helmets: Bronze Age and Early Iron Age.* Cambridge, Mass.: Peabody Museum of Archaeology and Ethnology, Harvard University.

Hennessy, J. B. 1966. Excavation of a Bronze Age Temple at Amman. *PEQ:* 155–162, pls. XXXII–XXXVI.

———. 1985. Thirteenth-Century B.C. Temple of Human Sacrifice at Amman. In *Studia Phoenicia,* vol. 3, *Phoenicia and Its Neighbours. Proceedings of the Colloquium Held on the 9th and 10th of December 1983 at the Vrije Universiteit Brussel.* E. Lipinski, ed. Leuven: Uitgeverij Peeters: 85–104.

Henschel-Simon, E. 1937. The "Toggle-Pins" in the Palestine Archaeological Museum. *QDAP* 6: 169–209.

Herodotus. 1997. *The Histories.* G. Rawlinson, trans. New York: Knopf.

———, R. B. Strassler, and A. L. Purvis. 2007. *The Landmark Herodotus: The Histories.* New York: Pantheon.

Herr, L. G. 1983. The Amman Airport Structure and the Geopolitics of Ancient Transjordan. *BibArch* 46(4): 223–229.

Heubeck, A., and A. Hoekstra. 1989. *A Commentary on Homer's Odyssey,* vol. 2, *Books* IX–XVI. Oxford: Clarendon.

Heubeck, A., S. West, and J. B. Hainsworth. 1988. *A Commentary on Homer's Odyssey,* vol. 1, *Introduction and Books* I–VIII. Oxford: Clarendon.

Heywood, A. 2001. The Use of Huntite as a White Pigment in Ancient Egypt. In *Colour and Painting in Ancient Egypt.* W. V. Davies, ed. London: British Museum Press: 5–9.

Hillman, A., and N. Liphschitz. 2004. The Wood. In *The Ma'agan Mikhael Ship: The Recovery of a 2400-Year-Old Merchantman,* vol. 2. Y. Kahanov and E. Linder, eds. Jerusalem: Israel Exploration Society and University of Haifa: 145–155.

Hocker, F. 1998. Glossary of Nautical Terms. In Wachsmann 1998: 377–381.

Hoekstra, A. 1969. *Homeric Modifications of Formulaic Prototypes: Studies in the Development of Greek Epic Diction.*[2] Amsterdam: Noord-Hollandsche Uitg.

Hoftijzer, J., and W. H. van Soldt. 1998. *Texts from Ugarit Pertaining to Seafaring.* In Wachsmann 1998: 333–344.

Holland, L. B. 1929. Mycenaean Plumes. *AJA* 33: 173–205.

Hollander, S. 2011. Ancient Greek Warship to NY? *Wall Street Journal Online* (January 28). http://online.wsj.com/video/ancient-greek-warship-to-ny/A726C787-76A5-4DE3-A92C-BE689C200247.html.

Holmes, Y. L. 1971. The Location of Alashiya. *JAOS* 91: 426–429.

———. 1973. Egypt and Cyprus: Late Bronze Age Trade and Diplomacy. *AOAT* 22: 91–98.

Hölscher, U. 1954. *Medinet Habu* V: *Post-Ramesside Remains.* OIP 66. Chicago: University of Chicago Press.

Hölscher, W. 1955. *Libyer und Ägypter: Beiträge zur Ethnologie und Geschichte libyscher Völkerschaften nach den altägyptischen Quellen.* Ägyptologische Forschungen 4. Glückstadt: Augustin.

Homan, M. M. 2010. Did the Ancient Israelites Drink Beer? *BibAR* 36(5): 48–56, 78.

Hood, S. 1993. Amber in Egypt. In *Amber in Archaeology: Proceedings of the Second International Conference on Amber in Archaeology, Liblice 1990.* C. W. Beck, J. Bouzek, and D. Dreslerova, eds. Praha, Czech Republic: Institute of Archaeology, Czech Academy of Sciences: 230–235, p. XVI.

Hornell, J. 1938. Boat Processions in Egypt. *Man* 38: 145–146, pl. I–J.

———. 1943. The Prow of the Ship: Sanctuary of the Tutelary Deity. *Man* 43: 121–128, pl. B.

———. 1970. *Water Transport.* Newton Abbot, UK: David and Charles. (First published 1946).

Hornig, K. 2002. Relikte altägyptischer Segalfahrzeuge. *Skyllis* 5(1): 4–18.

Howard, D. M., Jr. 1994. Philistines. In *Peoples of the Old Testament World.* A. J. Hoerth, G. L. Mattingly, and E. M. Yamauchi, eds., 231–250.

Hulin, L. 2011. Pragmatic Technology: Issues in the Interpretation of Libyan Material Culture. In *Intercultural Contacts in the Ancient Mediterranean. Proceedings of the International Conference at the Netherlands-Flemish Institute in Cairo 25th to 29th October 2008.* OLA 202. K. Duistermaat and I. Regulsk, eds. Leuven, Uitgeverij Peters: 101–114.

Hurwit, J. M. 1999. *The Athenian Acropolis: History, Mythology, and Archaeology from the Neolithic Era to the Present.* Cambridge: Cambridge University Press.

———. 2004. *The Acropolis in the Age of Pericles.* Cambridge: Cambridge University Press.

Hutchinson, R. W. 1962. *Prehistoric Crete.* Baltimore: Penguin.

Huxley, G. L. 1960. *Achaeans and Hittites.* Belfast: Queen's University.

Ingholt, H. 1940. *Rapport préliminaire sur sept campagnes de fouilles à Hama en Syrie (1932–1938).* Copenhagen: Munksgaard.

Irwin, M. E. 1974. *Colour Terms in Greek Poetry.* Toronto: Hakkert.

Isler-Kerényi, C. 2007. *Dionysos in Archaic Greece: An Understanding through Images.* Leiden: Brill.

Israel Antiquities Authority. 2008. Rare 2,500-Year-Old Marble Discus Found at Yavne-Yam. *Israel Ministry of Foreign Affairs* (July 13), http://www.mfa.gov.il/MFA/History/Early+History+-+Archaeology/Rare+marble+discus+found+at+Yavne-Yam+13-Jul-2008.htm.

Isserlin, B. S. J. 1991. The Canal of Xerxes: Facts and Problems. *BSA* 86: 83–91, pl. 4.

———, R. E. Jones, S. Papamarinopoulos, G. E. Syrides, Y. Maniatis, G. Facorellis, and J. Uren. 1996. The Canal of Xerxes: Preliminary Investigations in 1993–1994. *BSA* 91: 329–340.

———, R. E. Jones, S. Papamarinopoulos, and J. Uren. 1994. The Canal of Xerxes: Preliminary Investigations of 1991–2. *BSA* 89: 277–284, pls. 43–44.

Janko, R., in preparation. Two Seals Inscribed in Linear B from Bernstorf in Bavaria: Diplomatic Gifts from Mycenaean Pylos along the Amber Route?

Janssen, J. J. 1996. The Elusive Date of the Papyrus Amiens. *GM:* 61–63.

———. 2004. *Grain Transport in the Ramesside Period: Papyrus Baldwin (BM EA 10061) and Papyrus Amiens.* Hieratic Papyri in the British Museum 8. London: British Museum Press.

Johnston, P. F. 1985. *Ship and Boat Models in Ancient Greece.* Annapolis, MD.: Naval Institute Press.

Jones, D. 1990. *Model Boats from the Tomb of Tut'ankhamūn.* Tut'ankhamūn's Tomb Series 9. Oxford: Griffith Institute.

———. 1995. *Egyptian Bookshelf: Boats.* Austin: University of Texas Press.

Jones, R. E., B. S. J. Isserlin, V. K. Karastathis, S. P. Papamarinopoulos, G. E. Syrides, J. Uren, I. Balatsas, Ch. Kapopoulos, Y. Maniatis, and Y. Facorellis. 2000. Exploration of the Canal of Xerxes, Northern Greece: The Role of Geophysical and Other Techniques. *AP* 7: 147–170.

Jones, S. 1997. *The Archaeology of Ethnicity: Constructing Identities in the Past and Present.* London: Routledge.

Jordan, B. 1972. *The Athenian Navy in the Classical Period.* Berkeley: University of California Press.

Kahanov, Y. 1998. "Sewing System" in the Hull Construction of the Ma'agan Mikhael Shipwreck: A Comparative Study with Mediterranean Parallels. (PhD diss., Haifa University [in Hebrew]).

———. 2004. The Sewing of the Ship. In *The Ma'agan Mikhael Ship: The Recovery of a 2400-Year-Old Merchantman,* vol. 2. Y. Kahanov and E. Linder, eds. Jerusalem: Israel Exploration Society and University of Haifa: 1–79.

———, and P. Pomey. 2004. The Greek WN Shipbuilding Tradition and the *Ma'agan Mikhael* Ship: A Comparison with Mediterranean Parallels from the Sixth to the Fourth Centuries BC. *MM* 90: 6–28.

———, J. Royal, and J. Hall. 2004. The Tantura Wrecks and Ancient Mediterranean Shipbuilding. In *The Philosophy of Shipbuilding: Conceptual Approaches to the Study of Wooden Ships.* F. M. Hocker

BIBLIOGRAPHY 298

and C. A. Ward, eds. College Station: Texas A&M University Press: 113–127.

Kahl, J. 1995. Les Šerdanū et la fin d'Ougarit: A propos des documents d'Égypte, de Byblos et d'Ougarit relatifs aux Shardana (by O. Loretz): Annexe: Les témoignage textuels égyptiens sur les Shardana. In *Pays d'Ougarit:* 137–140.

Kamil, J. 1996. *Luxor: Ancient Thebes and the Necropolis.* Rev. ed. Cairo: Egyptian International Publishing–Longman.

Kantor, H. J. 1947. The Aegean and the Orient in the Second Millennium B.C. *AJA* 51:1–103.

Karageorghis, V. 1960. Supplementary Notes on the Mycenaean Vases from the Swedish Tombs at Enkomi. *OpAth* 3: 135–153.

———. 1963. Une tomb de guerrier à Palaepaphos. *BCH* 87: 265–300.

———. 1967. Chronique des fouilles à Chypre en 1966. *BCH* 91: 275–370.

———. 1982. *Cyprus: From the Stone Age to the Romans.* London: Thames and Hudson.

———, and J. Des Gagniers. 1974A. *La céramique chypriote de style figuré. Âge du fer (1050–500 av. J.-C.): Texte.* Rome: Edizioni dell'Ateneo.

———. 1974B. *La céramique chypriote de style figuré. Âge du fer (1050–500 av. J.-C.): Illustrations et descriptions des vases.* Rome: Edizioni dell'Ateneo.

Karastathis, V. K., and S. P. Papamarinopoulos. 1994. Preliminary Results of the Implementation of the Shallow Seismic Techniques in Order to Detect the King Xerxes' Canal: Extended Abstract. *NEGS* 53 (December): 8–9.

———. 1997. The Detection of the Xerxes Canal by the Use of Shallow Reflection and Refraction Seismics: Preliminary Results. *GP* 45: 389–401.

———, S. P. Papamarinopoulos, and R. E. Jones. 2001. 2-D Velocity Structure of the Buried Ancient Canal of Xerxes: An Application of Seismic Methods in Archaeology. *JAG* 47: 29–43.

Katary, S. L. P. 1976. *Land Tenure in the Ramesside Period: A Reconsideration of the Wilbour Papyrus and Related Documents.* Toronto: Kegan Paul International.

Keel, O. 1994. Philistine "Anchor" Seals. *IEJ* 44: 21–35.

———, M. Shuval, and C. Uehlinger. 1990. *Studien zu den Stempelsiegeln aus Palästina/Israel.* Band 3, *Die Frühe Eisenzeit: Ein Workshop.* Freiburg/Göttingen: Universitätsverlag/Vandenhoeck und Ruprecht.

Kelder, J. M. 2009. Royal Gift Exchange between Mycenae and Egypt: Olives as "Greeting Gifts" in the Late Bronze Age Easter Mediterranean. *AJA* 113: 339–353.

Kemp, B. J. 1978. The Harim-Palace at Medinet el-Ghurab. *ZÄSA* 15: 122–133.

———. 1991. *Ancient Egypt: Anatomy of a Civilization.* London: Routledge.

———, and G. Vogelsang-Eastwood. 2001. *The Ancient Textile Industry at Amarna.* London: Egypt Exploration Society.

Kennedy, D. H. 1976. Cable Reinforcement of the Athenian Trireme. *MM* 62: 159–168.

Kerényi, C. 1976. *Dionysos: Archetypal Image of Indestructible Life.* Bollingen Series LXV: 2. R. Manheim, trans. Princeton, N.J.: Princeton University Press.

Kilian, K. 1988. Ausgrabungen in Tiryns 1982/1983: Bericht zu den Grabungen. *AA:* 105–151.

Killebrew, A. E. 2005. *Biblical Peoples and Ethnicity: An Archaeological Study of Egyptians, Canaanites, Philistines, and Early Israel, 1300–1100 B.C.E.* Atlanta: Society of Biblical Literature.

———. 2006–2007. The Philistines in Context: The Transmission and Appropriation of Mycenaean-Style Culture in the East Aegean, Southeastern Coastal Anatolia, and the Levant. *SM* 27–28: 245–266.

———. 2008. Aegean-Style Pottery and Associated Assemblages in the Southern Levant: Chronological Implications regarding the Transition from the Late Bronze II to Iron I and the Appearance of the Philistines. In *Israel in Transition: From Late Bronze II to Iron IIA (c. 1250–850 BCE): The Archaeology.* L. L. Grabbe, ed. London: T&T Clark: 54–71.

Killen, J. T. 1983. PY AN 1. *Minos* 18: 71–79.

Kimmig, W. 1964. Seevölkerbewegung und Urnenfelderkultur. In *Studien aus Alteuropa,* vol. 1. R. von Uslar and K. Narr, eds. Cologne: Böhlau: 220–283.

Kirk, G. S. 1949. Ships on Geometric Vases. *BSA* 44: 93–153, pls. 38–40.

———. 1985. *The Iliad, a Commentary,* vol. 1, *Books* 1–4. Cambridge: Cambridge University Press.

Kitchen, K. A. 1973. *The Third Intermediate Period in Egypt (1100–650 B.C.).* Warminster: Aris and Phillips.

———. 1974. Nakht-Thuty: Servitor of Sacred Barques and Golden Portals. *JEA* 60: 168–174.

———. 1978. Review: *The Sea Peoples and Egypt.* By Alessandra Nibbi. *JEA* 64: 169–171.

———. 1982. *Pharaoh Triumphant: The Life and Times of Ramses II.* Warminster: Aris and Phillips.

———. 1983. Review: *Ancient Egypt and Some Eastern Neighbours.* By Alessandra Nibbi. *PEQ* 64: 169–171.

———. 1987. The Basics of Egyptian Chronology in Relation to the Bronze Age. In *High, Middle, or Low? (Acts of an International Colloquium on Absolute Chronology Held at the University of Gothenburg 20th–22nd August 1987).* P. Åström, ed. Gothenburg: Åström: 37–55.

———. 1990. The Arrival of the Libyans in Late New Kingdom Egypt. In *Libya and Egypt:* 15–27.

Kober, A. E. 1932. *The Use of Color Terms in the Greek Poets, Including All the Poets from Homer to 146 B.C. except the Epigrammatist.* Geneva: Humphrey.

Kopcke, G. 1967. Neue Holzfunde aus dem Heraion von Samos. *AM* 82: 100–148, Beilagen 46–83.

Korrés, G. S. 1989. Representation of a Late Mycenaean Ship on the Pyxis from Tragana, Pylos. In *Tropis* I: 177–202 (English summary, 199).

Kossack, G. 1954. *Studien zum Symbolgut der Urnenfelder- und Hallstattzeit Mitteleuropas.* Berlin: de Gruyter.

Kozloff, A. P., B. M. Bryan, L. M. Berman, and E. Delange. 1992. *Egypt's Dazzling Sun: Amenhotep III and His World.* Cleveland: Cleveland Museum of Art in cooperation with Indiana University Press.

Krutchten, J.-M. 1986. *Le grand texte oraculaire de Djéhoutymose: Intendant du domaine d'Amon sous le pontificat de Pinedjem II.* Monographies Reine Élisabeth 5. Brussels: Fondation égyptologique Reine Élisabeth.

Kurt, C. 1979. *Seemännische Fachausdrücke bei Homer: Unter Berücksichtigung Hesiods und der Lyriker bis Bakchylides.* Göttingen: Vandenhoeck and Ruprecht.

Kyle, D. G. 1992. The Panathenaic Games: Sacred and Civic Athletics. In *Goddess and Polis:* 76–101.

———. 1996. Gifts and Glory: Panathenaic and Other Greek Athletic Prizes. In *Worshipping Athena:* 106–136.

Kyrieleis, H. 1980. Archaische Holzfunde aus Samos. *AM* 95: 87–147, Taf. 18–40, Beilage 7.

Lackenbacher, S., and F. Malbran-Labat. 2005. Ugarit et les Hittites dans les archives de la "maison d'Urtenu." *SMEA* 47: 227–240.

Lacovara, P. 1997. Gurob and the New Kingdom *"Harim"* Palace. In *AEANE:* 297–306.

Laing, G. 1918. The Church Fathers and the Oriental Cults. *CJ* 13: 246–257.

Landström, B. 1970. *Ships of the Pharaohs.* Garden City, N.Y.: Doubleday.

Lane, E. W. 1973. *The Modern Egyptians.* New York: Dover. (Unabridged republication of the 5th ed., as published by John Murray, London, in 1860. First published 1836.)

Languri, G. M. 2004. *Molecular Studies of Asphalt, Mummy, and Kassel Earth Pigments: Their Characterisation, Identification, and Effect on the Drying of Traditional Oil Paint.* MolArt Report 9. London: Archetype Publications.

Laviosa, C. 1972. La Marina Micenea. *ASAtene* 47–48 (1969–1970; n.s. 31–32). Rome: 7–40.

Leahy, A. 2000. Ethnic Diversity in Ancient Egypt. In *Civilizations of the Ancient Near East,* vol. 1. Part 3: *Population* J. M. Sasson, ed. Peabody, MA: Henderickson: 225–234.

Le Baron Bowen, R. 1960A. Egypt's Earliest Sailing Ships. *Antiquity* 34: 117–131.

———. 1960B. Man's Earliest Sails. *MM* 49: 144–146.

Lefebvre, M. G. 1923. *Le tombeau de Petosiris,* vol. 3, *Vocabulaire et planches.* Cairo: Institute française d'archéologie orientale.

———. 1924. *Le tombeau de Petosiris,* vol. 1, *Description.* Cairo: Institute française d'archéologie orientale.

Legrain, G. 1914. *Louqsor sans les pharaons: Légendes et chansons populaires de la Haute Égypte, recueillies.* Brussels: Vromant.

———. 1917. Le logement et transport des barques sacrées et des statues des dieux dans quelques temples égyptiens. *BIFAO* 13: 1–76, pls. I–VII.

Lehmann, G. A. 1979. Die Šikalāyū—Ein neues Zeugnis zu den "Seevölker"-Heerfahrten im späten 13. Jh. v. Chr. (RS 34.129). *UF* 11: 481–494.

Lehmann, P. W., and K. Lehmann. 1973. *Samothracian Reflections.* Princeton, N.J.: Princeton University Press.

Lentini, M. C., and D. J. Blackman. 2009. L'arsenale navale di Naxos di Sicilia: La ricerca 2003–2006. In M. C. Lentini with D. J. Blackman, *Naxos di Sicilia: l'abitato e l'arsenale navale, scavi 2003–2006.* Messina: Sicania40–79.

———. 2010. Ultima ricerca nell'arsenale di Naxos di Sicilia. In *Ricoveri per navi militari nei porti del Mediterraneo antico e medievale.* Collana CUEBC Ravello: Archeologia, storia, cultura 5. D. J. Blackman and M. C. Lentini, eds. Bari: Edipuglia: 37–54.

Lenz, J. R. 1998. Homer's *nhysì koronísin.* In Wachsmann 1998: 199–200.

Lesko, L. H. 1980. The Wars of Ramses III. *Serapis* 6: 83–86.

———. 1992. Egypt in the 12th Century B.C. In *Crisis Years:* 151–156.

———. 2004. *A Dictionary of Late Egyptian,* vol. 2. 2nd ed. Providence, R.I.: Scribe.

Levi, D. 1942. Mors Voluntaria: Mystery Cults on Mosaics from Antioch. *Berytus* 7: 19–55, pls. I–VII.

———. 1947. *Antioch Mosaic Pavements,* 2 vols. Princeton, N.J.: Princeton University Press.

Lilliu, G. 1966. *Sculture della Sardegna nuragica.* Cagliari: Edizioni "La Zattera."

Lilyquist, C. 1999. On the Amenhotep III Inscribed Faience Fragments from Mycenaea. *JAOS* 119: 303–308, fig. 1.

Littauer, M. A., and J. H. Crouwel. 1979A. An Egyptian Wheel in Brooklyn. *JEA* 65: 107–120, pls. XII–XVI.

———. 1979B. *Wheeled Vehicles and Ridden Animals in the Ancient Near East.* Leiden: Brill.

Liverani, M. 1977. Le Chêne de Sherdanu. *VT* 27: 212–216.

Lloyd, A. B. 1975. Were Necho's Triremes Phoenician? *JHS* 95: 45–61.

Lohmann, H. 2008. Diolkos, Diholkos. In *Brill's New Pauly: Encyclopedia of the Ancient World* (electronic version). Leiden: Brill. Brill Online. Texas A&M University. July 17. http://www.brillonline.nl.ezproxy.tamu.edu:2048/subscriber/entry?entry=bnp_e12221250.

Loretz, O. 1995. Les Šerdanū et la fin d'Ougarit: À propos des documents d'Égypte, de Byblos et d'Ougarit relatifs aux Shardana. In *Pays d'Ougarit:* 125–140.

Lucas, A., and J. R. Harris. 1962. *Ancient Egyptian Materials and Industries.* London: Arnold.

MacDonald, B. R. 1986. The Diolkos. *JHS* 106: 191–195.

Maeir, A. M. 2007. A New Interpretation of the Term *'opalim* (עפלים) in Light of Recent Archaeological Finds from Philistia. *JSOT* 32: 23–40.

———. 2008. Did Captured Ark Afflict Philistines with E.D.? *BibAR* 34(3): 46–51.

Malamat, A. 1971. Syro-Palestinian Destinations in a Mari Tin Inventory. *IEJ* 21: 31–38.

———. 1998. Mari and the Bible. *Studies in the History and Culture of the Ancient Near East*, vol. 12. Leiden: Brill: 33–40.

Mandel-Elzinga, U. 1988. Ptolemaische Relief Keramik. *JDI* 103: 247–307.

Mansfield, J. M. 1985. The Robe of Athena and the Panathenaic Peplos. (PhD diss., University of California, Berkeley.)

Marazzi, M., 2009 [2010]. Il corpus delle iscrizioni in lineare B oggi: organizzazione e provenienze. *Pasiphae* 3: 121–154.

Mariette, A. 1981. *Monuments divers recueillis en Égypte et en Nubie.* Wiesbaden: LTR (reprint 1872).

Marinatos, N. 1984. *Art and Religion in Thera: Reconstructing a Bronze Age Society.* Athens: Mathioulakis.

———. 1993. *Minoan Religion: Ritual, Image, and Symbol.* Columbia: University of South Carolina Press.

Marinatos, S. 1933. La marine créto-mycénienne. *BCH* 57: 170–235, pls. XIII–XVII.

———. 1974. A Note about Minoan Wooden Columns. *AAA* 7: 95–96.

Mark, S. 1991. Odyssey 5.234–53 and Homeric Ship Construction: A Reappraisal. *AJA* 95: 441–445.

———. 1996. Odyssey (5.234–53) and Homeric Ship Construction: A Clarification. *IJNA* 25: 46–48.

———. 2005. *Homeric Seafaring.* College Station: Texas A&M University Press.

———. 2008. The Earliest Naval Ram. *IJNA* 37: 253–272.

Martin, H. G. 1999. Ein Schiffsmodell vom Kerameikos und der Panathenaienzug. *Skyllis:* 2(2): 138–152.

Marucchi, O. 1927. *Guide du musée Egýptien du Vatican.* Rome: Imprimerie Polyglotte Vaticane.

Maspero, G. 1892. *Egyptian Archaeology.* A. A. Edwards, trans. 2nd ed. New York: Putnam.

———. 1915. *Guide du Visiteur au Musée du Caire.* Cairo: Institute française d'archéologie orientale.

Maxwell-Stuart, P. G. 1981. *Studies in Greek Colour Terminology,* vol. 1, *Glaukos.* Leiden: Brill.

Mazar, A. 1985. The Emergence of the Philistine Material Culture. *IEJ* 35: 95–107.

———. 1990. *Archaeology of the Land of the Bible: 10,000–586 B.C.E.* The Anchor Bible Reference Library. New York: Doubleday.

McLeod, W. E. 1958. An Unpublished Egyptian Composite Bow in the Brooklyn Museum. *AJA* 62: 397–401.

McPherson, J. W. 1941. *The Moulids of Egypt (Egyptian Saints-Days).* Cairo: Nile Mission Press.

Meiberg, L. G., 2011. Figural Motifs on Philistine Pottery and Their Connections to the Aegean World, Cyprus and Coastal Anatolia. (PhD diss., University of Pennsylvania.)

Melena, J. 1975. *Studies on Some Mycenaean Inscriptions for Knossos Dealing with Textiles. Minos Suppl.* 5: 50–63, pls. II–III.

Mellink, M. J. 1966. Review: *Cypriot Bronzework in the Mycenaean World,* by H. W. Catling. *Mnemosyne:* 456–458.

Meriç, R., and P. A. Mountjoy. 2002. Mycenaean Pottery from Bademgediği Tepe (Puranda) in Ionia: A Preliminary Report. *IstMitt* 52: 79–98.

Merrillees, R. S. 1968A. *The Cypriote Bronze Age Pottery Found in Egypt.* SIMA 18. Lund: Åström.

———. 1968B. Review: Studies on the Arts and Crafts of the Late Cypriote Bronze Age. *PEQ* 100: 64–67.

———. 1973. Mycenaean Pottery from the Time of Akhenaten in Egypt. In *MEM:* 175–186.

———. 1987. *Alashia Revisited.* Paris: Gabalda.

———. 2006. Alashia Redux, Was It Cyprus: No. *AO* 9(1): 27.

———, and J. Winter. 1972. Bronze Age Trade between the Aegean and Egypt: Minoan and Mycenaean Pottery in the Brooklyn Museum. *MW* 1: 105–133.

Merriman, A. 2011. *Egyptian Watercraft Models from the Predynastic to Third Intermediate Periods.* BARIS 2263. Oxford, Archaeopress.

Mertens, P. 1960. Les peuples de la mer. *Cd'É* 35: 65–88.

Meskell, L. 2002. *Private Life in New Kingdom Egypt.* Princeton, N.J.: Princeton University Press.

Michaud, J. P. 1974. Chronique des fouilles et découvertes archéologiques en Grèce en 1973. *BCH* 98: 579–722.

Mikalson, J. D. 1975. *The Sacred and Civil Calendar of the Athenian Year.* Princeton, N.J.: Princeton University Press.

Miles, M. M. 1998. *The City Eleusinion.* Agora 31. Princeton, American School of Classical Studies in Athens.

Mohen, J.-P., and C. Eluère. 2000. *The Bronze Age in Europe.* New York: Abrams.

Moll, I. F. 1929. *Das Schiff in der bildenden Kunst: Vom Altertum bis zum Ausgang des Mittelalters.* Bonn: Schroeder.

Monloup, T. 1984. *Salamine de Chypre* XII: *Les figurines de terre cuite de tradition archaïque.* Paris: Diffusion de Boccard.

Monroe, C. M. 1990. The Boatbuilding Industry of New Kingdom Egypt. (Master's thesis, Texas A&M University, rev. ed.)

Moonwomon, B. 1994. Color Categorization in Early Greek. *JIES* 22: 37–65.

Moran, W. L. 1987. *The Amarna Letters.* Baltimore: Johns Hopkins University Press.

Moretti, M. 1961. *Tarquinia: La tomba della nave.* Milan: Lerici.

Morgan, L. 1995. Minoan Painting and Egypt: The Case of Tell el Da'ba. In *EAL:* 29–53, pl. 18.

Morgan Brown, L. 1978. The Ship Procession in the Miniature Frieze. In *Thera* I: 629–644.

Morricone, L. 1975. Coo-Scavi e scoperte nel "Serraglio" in località minori (1935–1943). *ASAtene* 50–51 (1972–1973, n.s. 34–35): 139–396.

Morrison, J. S., J. F. Coates, and N. B. Rankov. 2000. *The Athenian Trireme: The History and Reconstruction of an Ancient Greek Warship.* 2nd ed. Cambridge: Cambridge University Press.

Morrison, J. S., and R. T. Williams. 1968. *Greek Oared Ships, 900–322 B.C.* Cambridge: Cambridge University Press.

Mountjoy, P. A. 2005. Mycenaean Connections with the Near East in LH IIIC: Ships and Sea Peoples. In *Emporia:* 423–427, pls. XCV–XCVIII.

———. 2006A. Mycenaean Pictorial Pottery from Anatolia in the Transitional LH IIIB2–LH IIIC Early and the LH IIIC Phases. In *Pictorial Pursuits:* 107–121.

———. 2006B. Reconstruction and Description of a Philistine Ship: A Pictorial Motif Represented on a Krater. In *Tel Miqne-Ekron Excavations 1995–96: Field INE East Slope Report: End of the Late Bronze Age and the Iron I Early Philistine City.* M. W. Meehl, T. Dothan, and S. Gitin, eds. Jerusalem: W. F. Albright Institute of Archaeological Research and Institute of Archaeology, Hebrew University of Jerusalem: 85 (156, 157, fig. 3.27: 2 for pottery description and pottery figure).

———. 2011. A Bronze Age Ship from Ashkelon with Particular Reference to the Bronze Age Ship from Bademgediği Tepe. *AJA* 115: 483–488.

Muhly, J. D. 1980. The Bronze Age Setting. In *Age of Iron:* 25–67.

———. 2006. Alashia Redux, Was It Cyprus: Yes. *AO* 9(1): 25–26.

Muhly, J. D. and V. Kassianidou. 2012 Parallels and Diversities in the Production Trade and Use of Copper and Iron in Crete and Cyprus from the Bronze Age to the Iron Age. In *Parallel Lives: Ancient Island Societies in Crete and Cyprus. Papers Arising from the Conference in Nicosia Organized by the British School at Athens, the University of Crete, and the University of Cyprus, in November–December 2006.* British School at Athens Studies 20. G. Cadogan, M. Iacovou, K. Kopaka, and J. Whitley, eds. Exeter: British School at Athens: 119–140.

Murnane, W. J. 1980. *United with Eternity: A Concise Guide to the Monuments of Medinet Habu.* Chicago: Oriental Institute of the University of Chicago and the American University in Cairo Press.

Myśliwiec, K. 2000. *The Twilight of Ancient Egypt: First Millennium B.C.E.* Ithaca, N.Y.: Cornell University Press.

Nagy, B. 1980. A Late Panathenaic Document. *Ancient World* 3: 107–111.

Naville, E. 1898. *The Temple of Deir el Bahri.* Part III, *End of the Northern Half and Southern Half of the Middle Platform.* London: Egypt Exploration Fund.

———. 1901. *The Temple of Deir el Bahri.* Part IV, *Plates LXXXVII–CXVIII. The Shrine of Hathor and the Southern Hall of Offerings.* London: Egypt Exploration Fund.

Neils, J. 1992. The Panathenaia: An Introduction. In *Goddess and Polis:* 13–27.

———. 1996. Pride, Pomp, and Circumstance: The Iconography of Procession. In *Worshipping Athena:* 177–197.

———. 2001. *The Parthenon Frieze.* Cambridge: Cambridge University Press.

———. 2005. "With Noblest Images on All Sides": The Ionic Frieze of the Parthenon. In *The Parthenon: From Antiquity to the Present.* J. Neils, ed. Cambridge: Cambridge University Press: 199–223.

———. 2007. Replicating Tradition: The First Celebrations of the Greater Panathenaia. In *Panathenaic Games:* 41–51.

———, and S. V. Tracy. 2003. *The Games at Athens.* Princeton, N.J.: American School of Classical Studies at Athens.

Nelson, H. H. 1929. The Epigraphic Survey of the Great Temple of Medinet Habu (Seasons 1924–1925 to 1927–1928). *Medinet Habu 1924–1928. OIC* 5: 1–36.

———. 1943. The Naval Battle Pictured at Medinet Habu. *JNES* 2: 40–45.

Newberry, P. E. 1893. *Beni Hasan,* vol. 1. Archeological Survey of Egypt. London: Kegan Paul, Trench, Trübner.

———. 1942. Notes on Seagoing Ships. *JEA* 28: 64–66.

Niemeier, B., and W.-D. Niemeier. 2000. Aegean Frescoes in Syria-Palestine: Alalakh and Tel Kabri. In *WPT* 2: 763–802.

Niemeier, W.-D. 1991. Minoan Artisans Traveling Overseas: The Alalakh Frescoes and the Painted Plaster Floor at Tel Kabri (Western Galilee). In *Thalassa:* 189–201, pls. 46–51.

———. 1998. The Mycenaeans in Western Anatolia and the Problem of the Origins of the Sea Peoples. In *Peoples in Transition:* 17–65.

Nilsson, M. P. 1923. Fire-Festivals in Ancient Greece *JHS* 43: 144–148.

———. 1955. *Geschichte der griechischen Religion.* 2 vols. Munich: Beck'sche Verlagsbuchhandlung.

———. 1963. *A History of Greek Religion.* F. J. Fielden, trans. 2nd ed. Oxford: Clarendon.

Nims, C. F. 1976. Ramaesseum Sources of Medinet Habu Reliefs. In *Studies in Honor of George R. Hughes: January 12, 1977.* SAOC, no. 39. Chicago: Oriental Institute of the University of Chicago: 169–175.

Norman, N. J. 1983. The Panathenaic Ship. *ArchNews* 12: 41–46.

Nowak, T. J. 1999. A Marble *Ophthalmos* from Tektas, Burnu. *INAQ* 26(4): 10–11.

———. 2001. A Preliminary Report on *ophthalmoi* from the Tektaş Burnu Shipwreck. *IJNA* 30: 86–94.

———. 2006. Archaeological Evidence for Ship Eyes: An Analysis of Their Form and Function. (Master's thesis, Texas A&M University.)

O'Conner, D. 1990. The Nature of Tjemhu (Libyan) Society in the Later New Kingdom. In *Libya and Egypt:* 29–113.

———. 2000. The Sea Peoples and the Egyptian Sources. In *Sea Peoples:* 85–101.

Ohly, D. 1953. Holz. *MA* 68: 77–126.

Onassoglou, A. 1985. *Die "talismanischen" Siegel.* Berlin: Mann.

Oren, E. D. 1984. Migdol: A New Fortress on the Edge of the Nile Delta. *BASOR* 256: 7–44.

———. 1987. The "Ways of Horus" in North Sinai. In *Egypt, Israel, Sinai: Archaeological and Historical Relationships in the Biblical Period.* A. F. Rainey, ed. Jerusalem: Tel Aviv University: 69–119.

Østergaard, J. S. 2008. Emerging Colors: Roman Sculpture Polychromy Revived. In *Color of Life:* 40–61.

Otto, E. 1967. *Ancient Egyptian Art: The Cults of Osiris and Amon.* New York: Abrams.

Palagia, O. 2008. The Date and Iconography of the Calendar Frieze on the Little Metropolis, Athens. *JDI* 123: 215–237.

Palaima, T. G. 1991. Maritime Matters in the Linear B Tablets. In *Thalassa:* 273–310, pl. LXIII.

Palmer, L. R. 1955. A Mycenaean Calendar of Offerings (PY Kn 02). *Eranos* 53: 1–13.

———. 1998. *The Interpretation of Mycenaean Texts.* Oxford: Clarendon (reprint).

Panagiotopoulos, D. 2006. Foreigners in Egypt in the Time of Hatshepsut and Thutmose III. In *Thutmose III: A New Biography.* E. H. Cline and D. O'Connor, eds. Ann Arbor: University of Michigan Press: 370–412.

Pare, C. 1989. From Dupliaja to Delphi: The Ceremonial Use of the Wagon in Later Prehistory. *Antiquity* 63: 80–100.

Parke, H. W. 1967. *The Oracles of Zeus: Dodona, Olympia, Ammon.* Cambridge, Mass.: Harvard University Press.

———. 1977. *Festivals of the Athenians.* London: Thames and Hudson.

Parker, R. 1996. *Athenian Religion.* Oxford: Clarendon.

———. 2005. *Polytheism and Society at Athens.* Oxford: Oxford University Press.

———, ed. 1962. *A Saite Oracle Papyrus from Thebes in the Brooklyn Museum (Papyrus Brooklyn 47.218.3).* Providence, R.I.: Brown University Press.

Parkinson, R., and L. Schofield. 1993A. Akhenaten's Army? *EgyArch* 3: 34–35.

———, 1993B. Mycenaeans Meet the Egyptians at Last. *Art Newspaper* 24 (January): 10.

———. 1995. Images of Mycenaeans: A Recently Acquired Painted Papyrus from El-Amarna. In *EAL:* 125–126, pl. 8.

Parry, A., ed. 1971. *The Making of Homeric Verse: The Collected Writings of Milman Parry.* Oxford: Clarendon.

Parry, M. 1928. *L'épithète traditionelle dans Homère: Essai sur un problème de style homérique.* Paris: Belles Lettres.

———. 1930. Studies in the Epic Technique of Oral Verse-Making I: Homer and Homeric Style. *HSCP* 41: 73–147.

Partridge, R. B. 1996. *Transport in Ancient Egypt.* London: Rubicon.

Paton, J. M., G. P. Stevens, L. D. Caskey, H. N. Fowler, and T. W. Heermance. 1927. *The Erechtheum.* Cambridge, Mass.: Harvard University Press.

Pausanias. 1969–1985. *Pausanias: Description of Greece,* 5 vols. W. H. S. Jones, trans. Cambridge, Mass.: Harvard University Press and Heinemann.

Peck, W. H. 1994–1995. Tales of Golden Boats and Beautiful Ladies. *KMT* 5(4): 70–75.

Pellecchia, M., R. Negrini, L. Colli, M. Patrini, E. Milanesi, A. Achilli, G. Bertorelle, et al. 2007. The Mystery of Etruscan Origins: Novel Clues from *Bos taurus* Mitochondrial DNA. *Proceedings of the Royal Society B:* 1175–1179.

Pendlebury, H. W., J. D. S. Pendlebury, and M. B. Money-Coutts. 1937–1938 (1940). Excavations in the Plain of Lasithi. III. Karphi: A City of Refuge of the Early Iron Age in Crete. *ABSA* 38: 57–145.

Petrie, H. F. 1952. *Seven Memphite Tomb Chapels.* London: British School of Egyptian Archaeology.

Petrie, W. M. F. 1888. *Tanis,* part 2, *Nebesheh (AM) and Defenneh (Tahpanhes),* 1886. Memoirs of the Egypt Exploration Fund 4. London: Trübner.

———. 1890. *Kahun, Gurob, and Hawara.* London: Kegan Paul, Trench, Trübner.

———. 1891. *Illahun, Kahun, and Gurob 1889–1990.* London: Nutt.

———. 1905. *Ehnasya 1904.* London: Egypt Exploration Fund.

———. 1921. *Corpus of Prehistoric Pottery and Palettes.* London: British School of Egyptian Archaeology.

———. 1931. *Seventy Years in Archaeology.* London: Sampson Low, Marston.

———. 1933A. Egyptian Shipping: Outlines and Notes. *Ancient Egypt and the East* (parts I–II, March–June): 1–14.

———. 1933B. Egyptian Shipping (Continued). *AncEg* (Parts III–IV, September–December): 65–75.

———. 1974A. *Objects of Daily Use.* Warminster: Aris and Phillips (reprint).

———. 1974B. *Prehistoric Egypt: Corpus of Prehistoric Pottery and Palettes.* Warminster: Aris and Phillips (reprint).

Pettegrew, D., 2011. The *Diolkos* of Corinth. *AJA* 115: 549–574.

Philostratus and Eunapius. 1921. *Lives of the Sophists and Lives of the Philosophers and Sophists.* Loeb Classical Library. W. C. Wright, trans. Cambridge, Mass.: Harvard University Press.

Picard, C. 1931. Les influences étrangères au tombeau de Petosiris. *BIFAO* 30: 201–207.

Pickard-Cambridge, A. W., J. Gould, and D. M. Lewis. 1968. *The Dramatic Festivals of Athens.* London: Oxford University Press.

Pinney, G. F. 1988. Pallas and Panathenaea. In *Proceedings of the 3rd Symposium on Ancient Greek and Related Pottery (Copenhagen, August 31–September 4, 1987).* J. Christiansen and T. Melander, eds. Copenhagen: Nationalmuseet, Ny Carlsberg Glyptotek and Thorvaldsens Museum: 465–477.

Pittakis, K. S. 1862. To en to Erechtheio eurethen ploion. *ArchEph,* second series, 1: 91–94.

Plutarch. 1914. *Plutarch's Lives.* I. B. Perrin, trans. Cambridge, Mass.: Harvard University Press and Heinemann.

Plutschow, H. E., and P. G. O'Neill. 1996. *Matsuri: The Festivals of Japan.* Surrey: Japan Library.

Politi, J. 2001. The Papyri and the "Burnt Groups." *GM* 182: 107–111.

Polzer, M. E. 2004. An Archaic Laced Hull in the Aegean: The 2003 Excavation and Study of the Pabuç Burnu Ship Remains. *INAQ* 31(3) (Fall): 3–11.

———. 2005. Hull Remains from the Archaic Shipwreck at Pabuç Burnu, Turkey: Signs of Shifting Technology? In *AIA 106th Annual Meeting Abstracts* 28. Boston: Archaeological Institute of America: 165.

———. In press. Laced Hull Remains from the 6th-Century B.C. Shipwreck at Pabuç Burnu, Turkey. In *Tropis* IX.

Pomey, P. 1981. L'épave de Bon-Porté et les bateaux cousus de Méditerranée. *MM* 67: 225–243.

———. 1982. Le navire romain de la Madrague de Giens. *CRAI:* 133–154.

———. 1985. Mediterranean Sewn Boats in Antiquity. In *Sewn Plank Boats. Archaeological and Ethnographic Papers Presented to a Conference at Greenwich (November 1984).* BARIS 276. S. McGrail and E. Kentley, eds. Oxford: British Archaeological Reports: 35–47.

———. 1996. Navigation and Ships in the Age of Greek Colonization. In *The Greek World: Art and Civilization in Magna Graeca and Sicily.* G. P. Carratelli, ed. New York: Rizzoli: 133–140.

———, ed. 1997A. Un exemple d'évolution des techniques de construction navale antique: De l'assemblage par ligatures à l'assemblage par tenons et mortaises. In *Techniques et économie antiques et médiévales: le temps de l'innovation. Colloque International (CNRS) (21–23 Mai 1996).* D. Meeks and D. Garcia, eds. Aix-en-Provence: Errance, 195–203.

———, 1997B. *La navigation dans l'antiquité.* Aix-en-Provence: Édisud.

Pomey, P., Y. Kahanov, and E. Rieth. 2012. Transition from Shell to Skeleton in Ancient Mediterranean Ship-Construction: Analysis, Problems, and Future Research. *IJNA* 41: 235–314.

Popham, M. R. 1986. A Late Minoan IIIC Pyxis in the Andreades Collection. *OJA* 5(2): 157–164.

———. 2001. The Collapse of Aegean Civilization at the End of the Late Bronze Age. In *The Oxford Illustrated History of Prehistoric Europe.* B. Cunliffe, ed. Oxford: Oxford University Press: 277–303.

Porten, B., and A. Yardeni. 1993. *Textbook of Aramaic Documents from Ancient Egypt: Literature, Accounts, Lists,* vol. 3. Jerusalem: Hebrew University.

Porter, B., and R. L. B. Moss. 1937. *Topographical Bibliography of Ancient Egyptian Hieroglyphic Texts, Reliefs, and Paintings,* vol. 5, *Upper Egypt: Sites (Deir Rîfa to Aswân, Excluding Thebes and the Temples of Abydos, Dendera, Esna, Edfu, Kôm Ombo, and Philae).* Oxford: Clarendon.

———. 1960. *Topographical Bibliography of Ancient Egyptian Hieroglyphic Texts, Reliefs, and Paintings,* vol. 1, *The Theban Necropolis:* Part 1, *Private Tombs.* Oxford: Clarendon.

Press, L. 1973. On the Origin of Minoan Palatial Architecture. *ArcheologiaWar* 24: 1–11.

Pritchard, J. B. 1951. Syrians as Pictured in the Paintings of the Theban Tombs. *BASOR* 122: 36–41.

———, ed. 1969A. *The Ancient Near East in Pictures Relating to the Old Testament.* Princeton, N.J.: Princeton University Press.

———, ed. 1969B. *Ancient Near Eastern Texts Relating to the Old Testament.* Princeton, N.J.: Princeton University Press.

Pryor, J. H. 1995. From Dromon to Galea: Mediterranean Bireme Galleys AD 500–1300. In *Age of the Galley:* 101–116.

Pulak, C. 1988A. The Bronze Age Shipwreck at Ulu Burun, Turkey: 1985 Campaign. *AJA* 92: 1–37.

———. 1988B. Excavations in Turkey: 1988 Campaign. *INAN* 15(4): 13–17.

———. 1989. Ulu Burun: 1989 Excavation Campaign. *INAN* 16(4): cover, 4–10.

———. 1990. Ulu Burun: 1989 Excavation Season. *INAN* 17(4): cover, 8–13.

———. 1991. The Late Bronze Age Shipwreck at Ulu Burun, 1991 Field Season: "Ingot Summer." *INAN* 18(4): cover, 4–10.

———. 1992. The Shipwreck at Ulu Burun, Turkey: 1992 Excavation Season. *INAQ* 19(4): cover, 4–11, 21.

———. 1993. The Shipwreck at Ulu Burun, Turkey: 1992 Excavation Campaign. *INAN* 19(4): cover, 4–12.

———. 1994. 1994 Excavation at Uluburun: The Final Campaign. *INAN* 21(4): 8–16.

———. 1995. Das Schiffswrack von Uluburun. In *In Poseidons Reich: Archäologie unter Wasser.* Zaberns Bildbände zur Archäologie 23. Mainz: von Zabern: 43–58.

———. 1996. Analysis of the Weight Assemblages from the Late Bronze Age Shipwrecks at Uluburun and Cape Gelidonya, Turkey. (PhD diss., Texas A&M University.)

———. 1997. The Uluburun Shipwreck. In *Res Maritimae:* 188–224.

———. 2000. The Balance Weights from the Late Bronze Age Shipwreck at Uluburun. In *Metals Make the World Go Round: The Supply and Circulation of Metals in Bronze Age Europe: Proceedings of a Conference Held at the University of Birmingham in June 1997.* C. F. E. Pare, ed. Oxford: Oxbow: 247–266.

———. 2001. Cedar for Ships. *AHL* 14: 24–36.

———. 2005A. Das Schiffswrack von Uluburun. In *Das Schiff von Uluburun: Welthandel vor 3000 Jahren; Katalog der Ausstellung im Deutschen Bergbau-Museum Bochum vom 15. Juli 2005 bis 16. Juli 2006.* U. Içın, Ö. Ipek, and O. Medenbach, eds. Bochum: Deutsches Bergbau-Museum: 55–102.

———. 2005B. Who Were the Mycenaeans aboard the Uluburun Ship? In *Emporia:* 295–310, pls. LXX–LXXI.

———. 2008. The Uluburun Shipwreck and Late Bronze Age Trade. In *Beyond Babylon:* 288–305, artifact catalogue: 306–310, 313–321, 324–333, 336–348, 350–358, 366–370, 372–378, 382–385.

Raban, A. 1989. The Medinet Habu Ships: Another Interpretation. *IJNA* 18: 163–171.

Rackham, O. 1978. The Flora and Vegetation of Thera and Crete before and after the Great Eruption. In *Thera* I: 755–764.

Rainey, A. F. 1965. The Military Personnel of Ugarit. *JNES* 24: 17–27.

———. 1967. *A Social Structure of Ugarit.* Jerusalem: Mosad Byalik (in Hebrew).

———. 1982. Toponymic Problems (Cont.). *Tel Aviv* 9: 130–136.

———. 2003. Some Amarna Collations. *Eretz Israel* 27: 192*–202*.

———, and R. S. Notley. 2006. *The Sacred Bridge: Carta's Atlas of the Biblical World.* Jerusalem: Carta.

Redford, D. B. 1983. Review: *Die Beziehungen Ägyptens und Vorderasiens zur Ägäis bis ins 7. Jahrhundert v. Chr.* by Wolfgang Helck. *JAOS* 103: 481–483.

———. 1992. *Egypt, Canaan, and Israel in Ancient Times.* Princeton, N.J.: Princeton University Press.

———. 1997. Textual Sources for the Hyksos Period. In *Hyksos:* 1–44.

———. 2000. Egypt and Western Asia in the Late New Kingdom: An Overview. In *Sea Peoples:* 1–20.

———. 2010. *City of the Ram-Man: The Story of Ancient Mendes.* Princeton: Princeton University Press.

Redmount, C. A. 1994. The Wadi Tumilat and the Canal of the Pharaohs. *JNES* 54: 127–135.

Reeves, C. N. 1990. *The Complete Tutankhamun: The King, the Tomb, the Royal Treasure.* New York: Thames and Hudson.

Rehak, P., and J. G. Younger. 2001. Review of Aegean Prehistory VII: Neopalatial, Final Palatial, and Postpalatial Crete. In *Aegean Prehistory: A Review.* T. Cullen, ed. Boston: Archeological Institute of America: 383–465.

Reich, R. 1991. A Note on the Roman Mosaic at Magdala on the Sea of Galilee. *LA* 41: 455–458.

Reimer, P. J., M. G. L. Baillie, E. Bard, A. Bayliss, J. W. Beck, P. G. Blackwell, C. B. Ramsey, et al. 2009. IntCa109 and Marine09 Radiocarbon Age Calibration Curves, 0–50,000 Years cal BP. *Radiocarbon* 51: 1111–1150.

Reinach, S. 1909. *Répertoire de reliefs grecs et romains,* vol. 1. Paris: Leroux.

Reisner, G. A. 1913. *Models of Ships and Boats.* Cairo: Institut français d'archéologie orientale.

Renfrew, C. 1967. Cycladic Metallurgy and the Aegean Early Bronze Age. *AJA* 71: 1–20, pls. 1–10.

Rice, E. E., and Kallixeinos. 1983. *The Grand Procession of Ptolemy Philadelphus. (OCPM).* Oxford: Oxford University Press.

Richard, H. 1992. *Choes and Anthesteria: Athenian Iconography and Ritual.* Ann Arbor: University of Michigan Press.

Riederer, J. 1997. Egyptian Blue. In *Artists' Pigments: A Handbook of Their History and Characteristics,* vol. 3. E. W. FitzHugh, ed. Washington: National Gallery of Art/New York: Oxford University Press: 23–45.

Rife, J. L. 2008. The Burial of Herodes Atticus: Elite Identity, Urban Society, and Public Memory in Roman Greece. *JHS* 128: 92–127.

Riis, P. J. 1948. *Hama: Fouilles et recherches 1931–1938,* vol. 2: 3, *Les cimetières à crémation.* Copenhagen: Gyldendakske Boghandel, Nordisk Forlag.

Ritner, R. K. 2009. *The Libyan Anarchy: Inscriptions from Egypt's Third Intermediate Period.* Writings from the Ancient World 21. Atlanta: Society of Biblical Literature.

Rival, M. 1991. *La charpenterie navale romaine: Matériaux, méthodes, moyens.* Travaux du Centre Camille Jullian, 4. Paris: Éditions du Centre national de la recherche scientifique.

Roberts, R. G. 2009. Identity, Choice, and the Year 8 Reliefs of Ramesses III at Medinet Habu. In *Forces of Transformation: The End of the Bronze Age in the Mediterranean. Proceedings of an International Symposium Held at St. John's College, University of Oxford 25–6th March 2006.* C. Bachhuber and R. G. Roberts, eds. Oxford: Oxbow: 60–68.

Robertson, N. 1985. The Origin of the Panathenaea. *RhM* 128: 231–295.

———. 1992. *Festivals and Legends: The Formation of Greek Cities in the Light of Public Ritual.* Toronto: University of Toronto Press.

———. 1993. Athens' Festival of the New Wine. *HSCP* 95: 197–250.

———. 1996. Athena's Shrines and Festivals. In *Worshipping Athena,* 27–77.

Roccos, L. J. 1995. The Kanephoros and Her Festival Mantle in Greek Art. *AJA* 99: 641–666.

Romey, K. 2003. The Vogelbarke of Medinet Habu. (Master's thesis, Texas A&M University.)

Rowe, S., R. Siddall, and R. Stacey. 2010. Romano-Egyptian Gilded Cartonnage: Technical Study and Conservation of a Mummy Mask from Hawara. In *Decorated Surfaces on Ancient Egyptian Objects: Technology, Deterioration, and Conservation.* J. Dawson, C. Rozeik, and M. M. Wright, eds. London: Archetype Publications: 106–121.

Ruipérez, M. S. 1983. The Mycenaean Name of Dionysos. In *Res Mycenaeae. Akten de VII. International Mykenologischen Colloquiums in Nürnberg vom 6–10 April 1981.* A Heubeck and G. Neumann eds. Göttingen Vandenhoeck and Ruprect, 408–412.

Runciman, S. 1993. *A History of the Crusades,* vol. 2, *The Kingdom of Jerusalem.* Cambridge: Cambridge University Press.

Russman, E. R. 2008. Stele of an Asiatic Soldier. In *Beyond Babylon:* 264–265.

Saglio, E., E. Pottier, and G. Lafaye. 1877. *Dictionnaire des antiquités grecques et romaines d'après les textes et les monuments,* vol. 3. Paris: Librairie Hachette.

Sakellarakis, J. A. 2006. *Heraklion Museum: Illustrated Guide.* Athens: Ekdotike Athenon S.A.

Sandars, N. K. 1985. *The Sea Peoples: Warriors of the Ancient Mediterranean 1250–1150 B.C.* London: Thames and Hudson (rev. ed.).

Säve-Söderbergh, T. 1946. *The Navy of the Eighteenth Dynasty.* Uppsala: Lundequistska Bokhandeln.

Sayed, A. M. A. H. 1977. Discovery of the Site of the Twelfth Dynasty Port at Wadi Gawasis on the Red Sea Shore (Preliminary Report on the Excavations of the Faculty of Arts, University of Alexandria, in the Eastern Desert of Egypt, March 1976). *RÉg:* 140–178.

Scandurra, E. 1972. The Maritime Republics: Medieval and Renaissance Ships in Italy. In *HSBUA:* 205–224.

Schaden, O. J. 1979. Some Observations on the Sea Peoples. In *Studies in Honor of Tom B. Jones.* T. B. Jones, M. A. Powell, and R. H. Sack, eds. Kevelaer, Germany: Butzon und Bercker: 143–155.

Schaeffer, C. F. A. 1952. *Enkomi-Alashiya: nouvelles missions en Chypre, 1946–1950.* Paris: Klincksieck.

Schäfer, H. 1974. *Principles of Egyptian Art.* J. Baines, trans. Oxford: Clarendon.

Schauer, P., 1987. Der vierrädige Wagen in Zeremonialgeschehen und Bestattungsbrauch der orientalisch-ägäischen Hochkulturen und ihrer Randgebiete. In *Vierrädrige Wagen der Hallstattzeit: Unterrsuchungen zu Geschichte und Technik.* RGMZM Monographien, Bd. 12. F. E. Barth, ed. Mainz: Röminisch-Germanisches Zentralmuseum: 1–23.

Scheid, J., and V. Huet, eds. 2000. *La colonne aurélienne: autour de la colonne Aurélienne: geste et image sur la colonne de Marc Aurèle à Rome.* Bibliothèque de l'École des hautes études. Turnhout: Brepols.

Scherf, J. 2008. Phallus. In *Brill's New Pauly: Encyclopedia of the Ancient World* (electronic version). H. Cancik and H. Schneider, eds. Leiden: Brill. (Brill Online). Texas A&M University. July 13. http://www.brillonline.nl.ezproxy.tamu.edu:2048/subscriber/entry?entry=bnp_e917960.

Schofield, L., and R. Parkinson. 1994. Of Helmets and Heretics: A Possible Egyptian Representation of Mycenaean Warriors on a Papyrus from El-Amarna. *BSA* 89: frontispiece, 157–170, pls. 21–22.

Schulman, A. R. 1976. The Royal Butler Ramessesemperrē.' *JARCE* 13: 117–130.

Scott, D. A. 2002. *Copper and Bronze in Art: Corrosion, Colorants, Conservation.* Los Angeles: Getty Conservation Institute.

Seiradaki, M. 1960. Pottery from Karphi. *BSA* 55: 1-37.

Seligmann, C. G. 1966. Ancient Egyptian Beliefs in Modern Egypt. In *Essays and Studies Presented to William Ridgeway on His Sixtieth Birthday, 6 August 1913.* W. Ridgeway and E. C. Quiggin, eds. Freeport, N.Y.: Books for Libraries Press: 448–457.

Serpico, M., and R. White. 2000. Oil, Fat, and Wax. In *Ancient Egyptian Materials and Technology.* P. T. Nicholson and I. Shaw, eds. Cambridge: Cambridge University Press: 390–429.

———. 2001. The Use and Identification of Varnish on New Kingdom Funerary Equipment. In *Colour and Painting in Ancient Egypt.* W. V. Davies, ed. London: British Museum Press: 33–42.

Seymour, T. D. 1914. *Life in the Homeric Age.* New York: Macmillan.

Shapiro, H. A. 1996. Democracy and Imperialism: The Panathenaia in the Age of Perikles. In *Worshipping Athena:* 215–225.

Shaw, I., 2007. Gurob: The Key to Unlocking an Egyptian Harem? *Current World Archaeology Magazine* 23(June/July): 12–19.

———, 2008a. Gurob Harim Palace Survey. University of Liverpool, School of Archaeology, Classics and Egyptology. November 23. http://www.liv.ac.uk/sace/research/projects/gurob.htm.

———, 2008b. A Royal Harem Town of the New Kingdom: New Fieldwork at Medinet el-Gurob. In *Queens of Egypt: From Hetepheres to Cleopatra.* C. Ziegler, ed. Paris, Somogy Art Publishers: 104–115.

———, 2010. The Royal Harim at Medinet el-Ghurob: New Fieldwork (2005–7). In *The Second Young Egyptologists' Conference, Lisbon, 2006,.* L. M. de Araujo and J. das Candeas Sales, eds. Lisbon: 256–264. [published as a CD]

———, 2011a. New Fieldwork at the Medinet el-Gurob New Kingdom Settlement: Investigating a Harem Palace Town in the Faiyum (Summary of the 2009–10 Seasons). In *Achievements and Problems of Modern Egyptology: Proceedings of the International Conference Held in Moscow on September 29-October 2, 2009.* G. A. Belova, ed. Moscow, Russian Academy of Sciences: 348–364.

———, 2011b. Seeking the Ramesside Royal Harem: New Fieldwork at Medinet el-Gurob. In *Ramesside Studies in Honour of Ken Kitchen,* M. Collier and S. Snape, eds. Bolton, Rutherford Press: 453–463.

———, and P. Nicholson, eds. 1995. *The Dictionary of Ancient Egypt.* New York: Abrams.

Shaw, J. W. 1973. *Minoan Architecture: Materials and Techniques.* Annuario 49, n.s. 33. Rome: Instituto Poligrafico dello Stato.

———, and M. C. Shaw. 2006. *Kommos,* vol. 5, *The Monumental Minoan Buildings at Kommos.* Princeton, N.J.: Princeton University Press.

Shear, J. L. 2001. Polis and Panathenaia: The History and Development of Athena's Festival. (PhD diss., University of Pennsylvania.)

Sheffer, A., and A. Tidhar. 1988. Identification of Textile Remains. In *The Egyptian Mining Temple at Timna.* B. Rothenberg, ed. London: Institute for Archaeo-Metallurgical Studies: 223–230.

Sherratt, S., 2005. "Ethnicities," "Ethnonyms," and Archaeological Labels. Whose Ideologies and Whose Identities? In *Archaeolgical Perspectives on the Transmission and Transformation of Culture in the Eastern Mediterranean.* Levant Supplementay Series 2. J. Clarke, ed. Oxford: Oxbow Books: 25–38.

Shisha-Halevy, A. 1978. An Early North-West Semitic Text in the Egyptian Hieratic Script. *Orientalia* 47: 145–162.

Simon, E. 1983. *Festivals of Attica: An Archaeological Commentary.* Madison: University of Wisconsin Press.

———. 1996. Theseus and Athenian Festivals. In *Worshipping Athena:* 9–26.

Simpson, W. K. 1953. New Light on the God Reshef. *JAOS* 73: 86–89.

———, ed. 1972. *The Literature of Ancient Egypt: An Anthology of Stories, Instructions, and Poetry.* New Haven, Conn.: Yale University Press.

Singer, I. 1985. The Beginning of Philistine Settlement in Canaan and the Northern Boundary of Philistia. *Tel Aviv* 12: 109–122.

———. 1992. How Did the Philistines Enter Canaan? A Rejoinder. *BibAR* 18(6): 44–46.

———. 2006. Ships Bound for Lukka: A New Interpretation of the

Companion Letters RS 94.2530 and RS 94.2523. *Altorientalische Forschungen* 33: 242–262.

Sjöqvist, E. 1940. *Problems of the Late Cypriote Bronze Age.* Stockholm: Swedish Cyprus Expedition.

Snodgrass, A. M. 1980. Iron and Early Metallurgy in the Mediterranean. In *Age of Iron:* 335–374.

———. 2000. *The Dark Age of Greece.* New York: Routledge (reprint).

Sparks, R. S. 2004. Canaan in Egypt: Archaeological Evidence for a Social Phenomenon. In *Invention and Innovation: The Social Context of Technological Change.* vol. 2, *Egypt, the Aegean, and the Near East, 1650–1150 BC.* J. Bourriau and J. S. Phillip, eds. Oxford: Oxbow: 25–54.

Spathari, E. 1995. *Sailing through Time: The Ship in Greek Art.* D. Hardy, trans. Athens: Kapon.

Sperber, D. 1986. *Nautica Talmudica.* Bar-Ilan Studies in Near Eastern Languages and Culture. Ramat Gan: Bar-Ilan University Press/ Leiden: Brill.

Spetsieri-Choremi, A. 2000. Thrávsma anathimatikoù Anaglífou apo tin periochí tou Athinaikoù Elefsiníou. *ArchEph:* 1–18.

Spiegelberg, W., and A. Erman. 1898. Grabstein eines syrischen Söldners aus Tell Amarna. *ZÄSA* 36: 126–129, pl. XVII.

Stager, L. 1991. When Canaanites and Philistines Ruled Ashkelon. *BibAR* 17(2): 24–37, 40–43.

———. 1995. The Impact of the Sea Peoples (1185–1050 BCE). In *The Archaeology of Society in the Holy Land.* T. E. Levy, ed. New York: Facts on File: 332–348, 583–585.

———. 1996. Ashkelon and the Archaeology of Destruction: Kislev 604 BCE. *Eretz Israel* 25: 61*–74*.

Stahl, C. 2006. *Mitteleuropäische Bernsteinfunde von der Frühbonze-bis zur Frühlatènezeit ihre Verbreitung, Formgebung, Zeitstellang, urd Herkunft.* Dettelbach: J. H. Roll.

Staïs, V. 1910. *Marbres et bronzes du Musée National.* Athens: Sakellarios.

Steffy, J. R. 1990. The Boat: A Preliminary Study of Its Construction. In *EABSG:* 29–47.

———. 1991. The Ram and Bow Timbers. In *The Athlit Ram:* 6–39.

———. 1994. *Wooden Ship Building and the Interpretation of Shipwrecks.* College Station: Texas A&M University Press.

Stern, H. 1981. Les calendriers romains illustrés. In *Aufstieg und Niedergang der römischen Welt II Principat:* Bd. 12.2. H. Temporini and W. Haase, eds. Berlin: de Gruyter: 431–475, pls. I–XXXIX, tables I–II.

Stevenson Smith, W. 1949. Three Late Egyptian Reliefs. *BMFA* 47(268): 21–29.

Stieglitz, R. R. 1972–1975. An Ancient Terra-Cotta Ship from Cyprus. *Sefunim* 4: 44–46.

Stone, B. J. 1995. The Philistines and Acculturation: Culture Change and Ethnic Continuity in the Iron Age. *BASOR* 298: 7–32.

Strange, J. 1980. *Caphtor/Keftiu.* Leiden: Brill.

Strassler, R. B., ed. 1996. *The Landmark Thucydides: A Comprehensive Guide to the Peloponnesian War.* New York: Touchstone.

Stuiver, M., and H. A. Polach. 1977. Discussion: Reporting of ^{14}C Data. *Radiocarbon* 19: 355–363.

Suys, É. 1927. *Vie de Petosisris: Grand prêtre de Thot à Hermopolis-la-Grande.* Brussels: Fondation égyptologique Reine Élisabeth.

Svoronos, J. N. 1899. Der athenische Volkskalender. *JIAN* 2: 21–78.

———. 1914. Stylides, ancres, hierae, aphlasta, stoloi, akrostolia, embola, proembola et totems marins. *JIAN* 16: 81–152.

Sweeney, D., and A. Yasur-Landau. 1999. Following the Path of the Sea Persons: The Women in the Medinet Habu Reliefs. *Tel Aviv* 26: 116–145.

Sylvester, D. 1992. *Magritte: The Silence of the World.* New York: Menil Foundation, Abrams.

Symeonoglou, S. 1973. *Kadmeia.* SIMA 35. Gothenburg: Åström (S. vägen 61).

Tandy, D. W. 2000. *Warriors into Traders: The Power of the Market in Early Greece.* Berkeley: University of California Press.

Taylor, J. H. 1996. *Egyptian Bookshelf: Unwrapping a Mummy.* Austin: University of Texas Press.

———. 2001. *Death and the Afterlife in Ancient Egypt.* Chicago: University of Chicago Press.

Terrace, E. L. B. 1963. Ancient Egyptian Jewelry in the Horace L. Mayer Collection. *AJA* 67: 269–274.

Thomas, A. P. 1981A. *Gurob: A New Kingdom Town: Introduction and Catalogue of Objects in the Petrie Collection.* Warminster: Aris and Phillips.

———. 1981B. *Gurob: A New Kingdom Town: Plates.* Warminster: Aris and Phillips.

Thomas, C. G., and C. Conant. 1999. *Citadel to City-State: The Transformation of Greece 1200–700 B.C.E.* Bloomington: Indiana University Press.

Thompson, H., and R. E. Wycherley. 1972. *The Agora of Athens: The History, Shape, and Uses of an Ancient City Center.* Agora 14 Princeton, : American School of Classical Studies in Athens.

Thureau-Dangin, F., M. Dunand, L. Cavro, and G. Dossin. 1936. *Til-Barsib.* Paris: Geuthner.

Tite, M. S., M. Bimson, and M. R. Cowell. 1987. The Technology of

Egyptian Blue. British Museum Occasional Paper 56. M. Bimson and I. C. Freestone, eds. London: British Museum Press: 39–46.

Tobin, J. 1991. The Panathenaic Ship of Herodes Atticus. *American School of Classical Studies at Athens Newsletter* (Spring): 5.

———. 1993. Some New Thoughts on Herodes Atticus's Tomb, His Stadium of 143/4, and Philostratus VS 2.550. *AJA* 97: 81–89.

Tomkinson, J. L. 2003. *Festive Greece: A Calendar of Tradition.* Athens: Anagnosis.

Tooley, A. M. J. 1995. *Egyptian Models and Scenes.* Buckinghamshire: Shire Publications.

Torczyner, H. 1979. *Magritte: The True Art of Painting.* New York: Abradale Press and Abrams.

Torr, C. 1892. Review: *Illahun, Kahun, and Gurob*, by W. M. Flinders Petrie. *CR* 6: 127–131.

———. 1964. *Ancient Ships.* Chicago: Argonaut.

Trakadas, A. 2008. Athenian Naumachiai: Hellenistic and Imperial Naval Festivals and Their Contexts. In *Tropis X Abstracts.*

Trigger, B. G., B. J. Kemp, D. O'Conner, and A. B. Lloyd. 1996. *Ancient Egypt: A Social History.* Cambridge: Cambridge University Press.

Trope, B. T., S. Quirk, and P. Lacovara. 2005. *Excavating Egypt: Great Discoveries from the Petrie Museum of Egyptian Archaeology.* Atlanta: Michael C. Carlos Museum.

Tykot, R. H. 1994. Sea Peoples in Etruria? Italian Contacts with the Eastern Mediterranean in the Late Bronze Age. *EtrStud* 1: 59–83.

Tylor, J. J. 1896. *Wall Drawings and Monuments of El Kab: The Tomb of Sebeknekht.* London: Egypt Exploration Fund.

Tzachili, I. 2005. Anthodokoi Talaroi: The Baskets of the Crocus Gatherers from Xesté 3, Akrotiri, Thera. In *Aegean Wall Painting: A Tribute to Mark Cameron.* British School at Athens Studies 13. L. Morgan, ed. London: British School at Athens: 113–118.

Tzalas, H. 2008. The Ship Model no. 17420 of the Distomo Archaeological Museum. In *Tropis X Abstracts.*

Ucko, P. J. 1969. Penis Sheaths: A Comparative Study. *PRAI* 1969: 24A–67.

van Doorninck, F. H., Jr. 1982. Protogeometric Longships and the Introduction of the Ram. *IJNA* 11: 277–286.

van Hoorn, G. 1951. *Choes and Anthesteria.* Leiden: Brill.

Van Siclen, C. C. 1986. Amenhotep II's Barque Chapel for Amun at North Karnak. *Extrait du Bulletin de l'Institut français d'archéologie orientale* 86: 353–359.

van Walsem, R. 1997. *The Coffin of Djedmonthuiufankh in the National Museum of Antiquities at Leiden.* Leiden: Nederlands Instituut voor het Nabije Oosten.

Vandersleyen, C. 1971. *Les guerres d'Amosis, fondateur de la XVIIIe dynastie.* Monographies de la Fondation égyptologique Reine Élisabeth I. Brussels: Fondation égyptologique Reine Élisabeth.

Vandkilde, H. 2007. *Culture and Change in Central European Prehistory: 6th to 1st Millennium BC.* Aarhus: Aarhus University Press.

Vasić, R. 2004. The Dupljaja Cart Again. In *Festschrift für Florin Medeleţ.* Bibliotheca Historica et Archaeologica Banatica 32. P. Rogozea and V. Cedică, eds. Timişoara: Editura Mirton Timişoara: 155–162.

Veenhof, K. R. 1995. Kanesh: An Assyrian Colony in Anatolia. In *Civilizations of the Ancient Near East,* vol. 2. Part 5: *History and Culture.* J. M. Sasson, ed. New York: Scribner: 859–871.

Veldmeijer, A. J., and C. Zazzaro. 2008. The "Rope Cave" at Mersa/Wadi Gawasis (poster). In *Abstracts of Papers. Tenth International Congress of Egyptologists (Rhodes, Greece, 22–29 May 2008).* P. Kousoulis, ed. Rhodes: International Association of Egyptologists and University of the Aegean: 260 and handout.

Ventris, M., and J. Chadwick. 1973. *Documents in Mycenaean Greek.* Cambridge: Cambridge University Press.

Vercoutter, E. 1954. *Essai sur les relations entre Égyptiens et Préhellènes. L'Orient illustré* 6. Paris: Librairie Maisonneuve.

———. 1956. *L'Égypt et le monde égéen préhellénique.* Bibliothèque d'étude 22. Cairo: l'Institut français d'archéologie orientale.

———. 1964. Desert Highway Built for Ships. *UNESCO Courier* (December): 36–37.

———. 1965. Excavations at Mirgissa II (Oct. 1963–March 1964). *Kush* 13: 62–73.

Vermeule, E. 1964. *Greece in the Bronze Age.* Chicago: University of Chicago Press.

Vernier, É. 1927A. *Bijoux et Orfèvreries.* Musée des antiquités égyptiennes, Catalogue général nos. 52640–53171, vol. 1, *Texte.* Cairo: l'Institut français d'archéologie orientale.

———. 1927B. *Bijoux et Orfèvreries.* Musée des antiquités égyptiennes, Catalogue général nos. 52640–53171, vol. 2, *Index et Planches.* Cairo: l'Institut français d'archéologie orientale.

Vigneau, A., and E. Drioton. 1949. *Le Musée du Caire: Encyclopédie photographique de l'art.* Paris: Éditions "Tel."

Vinson, S. M. 1993. The Earliest Representations of Brailed Sails. *JARCE* 30: 133–150.

———. 1994. *Egyptian Boats and Ships.* Shire Egyptology Series no. 20. Buckinghamshire: Shire Publications.

von Bissing, F. W. 1900. *Ein the banischer Grabfunde, aus dem Anfang des neuen Reichs.* Berlin: Duncker.

Von Bothmer, D. 1985. *The Amasis Painter and His World: Vase-Painting in Sixth-Century B.C. Athens.* Malibu: Getty Museum and Thames and Hudson.

Vos, R. L. 1993. *The Apis Embalming Ritual: P. Vindob. 3873*. OLA 50. Leuven: Uitgeverij Peeters en Departement Oriëntalistiek.

Wachsmann, S. 1981. The Ships of the Sea Peoples. *IJNA* 10: 187–220.

———. 1982. The Ships of the Sea Peoples (*IJNA* 10[3]: 187–220): Additional notes. *IJNA* 11: 297–304.

———. 1987. *Aegeans in the Theban Tombs*. OLA 20. Leuven: Uitgeverij Peters.

———. 1990. First Century CE Kinneret Boat Classes. In *EABSG*: 119–124.

———. 1995. The 1994 INA/CMS Joint Expedition to Tantura Lagoon. *INAQ* 22(2): cover, 3–8.

———. 1996. Bird-Head Devices on Mediterranean Ships. In *Tropis* IV: 539–572.

———. 1997. Were the Sea Peoples Mycenaeans? The Evidence of Ship Iconography. In *Res Maritimae*: 339–356.

———. 1998. *Seagoing Ships and Seamanship in the Bronze Age Levant*. College Station: Texas A&M University Press.

———. 1999. The Pylos Rower Tablets Reconsidered. In *Tropis* V: 491–504.

———. 2000. To the Sea of the Philistines. In *Sea Peoples*: 103–143.

———. 2002A. Sailing into Egypt's Past: Does a Celebration of Luxor's Patron Saint Echo Ancient Pharaonic Traditions? *Archaeology* 55(4): 36–39.

———. 2002B. The Moulid of Abu el Haggag: A Contemporary Boat Festival in Egypt. In *Tropis* VII: 821–835.

———. 2003. The Forum: Hatshepsut's Journey to Punt. *AO* 6(3): 10–11.

———. 2006. Alashia Redux, Was It Cyprus: Yes. *AO* 9(1): 26–27.

———. 2008. Underwater Survey, 1996–1997. In *Ashkelon Excavation Report*, vol. 1, *Introduction and Overview (1985–2006)*. L. E. Stager, J. D. Schloen, and D. M. Master, eds. Winona Lake, Ind.: Eisenbrauns: 97–100.

———. 2009. *The Sea of Galilee Boat: A 2000 Year Old Discovery from the Sea of Legends*. 3rd ed. College Station: Texas A&M University Press.

———. 2010. Ahhotep's Silver Ship Model: The Minoan Context. *JAEI* 2(3): 31–41.

———. 2012. The Panathenaic Ships: The Iconographic Evidence. *Hesperia 81:237–266.*

———, K. Raveh, and E. Amos. 1990. The Excavation. In *EABSG*: 9–14.

———, and Y. Kahanov. 1997. Shipwreck Fall: The 1995 INA/CMS Joint Expedition to Tantura Lagoon, Israel. *INAQ* 24(1): cover, 3–18.

Wainwright, G. A. 1939. Some Sea Peoples and Others in the Hittite Archives. *JEA* 25: 148–153.

———. 1959. The Teresh, the Etruscans, and Asia Minor. *AnatSt* 9: 197–213.

———. 1961. Some Sea Peoples. *JEA* 47: 71–90.

———. 1962. The Meshwesh. *JEA* 48: 89–99.

Wallace, F. E. 1927. Colour in Homer and in Ancient Art: Preliminary Studies. *Smith College Classical Studies* 9: 1–83.

Walters, H. B. 1914. *Catalogue of the Greek and Roman Lamps in the British Museum*. London: British Museum.

Ward, C., and C. Zazzaro. 2010. Evidence for Pharaonic Seagoing Ships at Mersa/Wadi Gawasis, Egypt. *IJNA* 39: 27–43.

Ward, W. A. 1989. Some Foreign Personal Names and Loan Words from the Deir el-Medineh Ostraca. In *Essays in Ancient Civilization Presented to Helene J. Kantor*. Studies in Ancient Oriental Civilization 47. A. Leonard and B. B. Williams, eds. Chicago: Oriental Institute of the University of Chicago: 287–303.

———. 1994. Foreigners Living in the Village. In *Pharaoh's Workers: The Villagers of Deir El Medina*. L. H. Lesko, ed. Ithaca, N.Y.: Cornell University Press: 62–66, 164–167.

Wardle, K. A. 1973. A Group of Late Helladic IIIB2 Pottery from within the Citadel at Mycenae. *BSA* 68: 297–342, pls. 57–62.

Warren, P. 1995. Minoan Crete and Pharaonic Egypt. In *EAL*: 1–18.

Watzinger, C. 1924. *Griechische Vasen in Tübingen*. Tübingen: Tübingen Universität.

Weber, W. 1914. *Die ägyptisch-grieshischen Terrakotten*. Berlin: Curtius.

Webster, T. B. L. 1953–1954. Greek Comic Costume. *BJRL* 36: 563–587.

———. 1970. *Greek Theatre Production*. London: Methuen.

Wedde, M. 1999. War at Sea: The Mycenaean and Iron Age Oared Galley. In *Polemos*: 465–476, pls. LXXVII–XCII.

———. 2000. *Towards a Hermeneutics of Aegean Bronze Age Ship Imagery*. Mannheim: Bibliopolis.

———. 2002. Birdshead Revisited: The Bow Morphology of the Early Greek Galley. In *Tropis* VII: 837–856.

———. 2006. Pictorial Evidence for Partial System Survival in the Greek Bronze to Iron Age Transition. In *Pictorial Pursuits*: 255–269.

Weeks, K. R. 2005. *The Treasures of Luxor and the Valley of the Kings*. Vercelli: White Star.

Welch, K. 1998. Greek Stadia and Roman Spectacles: Asia, Athens, and the Tomb of Herodos Atticus. *JRA* 11: 117–145.

Wente, E. 1967. *Late Ramesside Letters*. SAOC 33. Chicago: University of Chicago Press.

———. 1990. *Letters from Ancient Egypt.* Society of Biblical Literature Writings from the Ancient World Series 1. Atlanta: Scholars Press.

Werker, E. 1990. Identification of the Wood. In *EABSG:* 65–75.

———. 2005. Identification of the Wood in the Ancient Boat from the Sea of Galilee. *'Atiqot* 50: 233–236.

Werner, W. 1997. The Largest Ship Trackway in Ancient Times: The Diolkos of the Isthmus of Corinth, Greece, and Early Attempts to Build a Canal. *IJNA* 26: 98–117.

West, M. L. 1997. Homer's Meter. In *A New Companion to Homer.* I. Morris and B. Powell, eds. Leiden: Brill: 218–237.

Westerberg, K. 1983. *Cypriote Ships from the Bronze Age to c. 500 B.C.* Gothenburg: Åström.

White, R. 1990. Analysis of Resinous Material. In *EABSG:* 81–88.

Wickett, E. 1990. *For Those Who Sail to Heaven.* Documentary film (48:31 minutes).

———. 2009. Archaeological Memory, the Leitmotifs of Ancient Egyptian Festival Tradition, and Cultural Legacy in the Festival Tradition of Luxor: The *Moulid* of Sidi Abu'l Hajjaj al Uqsori and the Ancient Egyptian "Feast of Opet." JARCE 45: 403–426.

Widmer, W. 1975. Zur Darstellung der Seevölker am Grossen Tempel von Medinet Habu. *ZÄSA* 102: 67–77.

Wiener, M. 2009. Locating Ahhiyawa. In *Doron: Timitikos Tomos Gia Ton Kathgiti Spyro Iakovidi.* D. Daniilidou, ed. Athens: Academy of Athens: 701–715.

Wilbour, C. E. 1936. *Travels in Egypt (December 1880 to May 1891): Letters of Charles Edwin Wilbour.* J. Capart, ed. New York: Brooklyn Museum.

Wilkinson, J. G. 1837. *The Manners and Customs of the Ancient Egyptians,* vol. 2. London: Murray.

Williams, E. R. 1985. Isis Pelagia and a Roman Marble Matrix from the Athenian Agora. *Hesperia* 54: 109–119, pls. 21–23.

Williams, R. 1986. Nestor's War Effort (Stat. Ach. 1.422). *CQ,* n.s., 36: 280–283.

Wilson, J. A. 1935. The Libyans and the End of the Egyptian Empire. *AJSL* 51: 73–82.

———. 1944. Funeral Services of the Egyptian Old Kingdom. *JNES* 3: 201–218.

Wilson, L. M. 1938. *The Clothing of the Ancient Romans.* Baltimore: Johns Hopkins Press.

Wilson, R. J. A. 1981–1982. Archaeology in Sicily, 1977–1981. *AR* 28: 84–105.

———. 1987–1988. Archaeology in Sicily, 1982–87. *AR* 34: 105–150.

Winkler, H. A. 1939. *Rock-Drawings of Southern Upper Egypt,* vol. 2. London: Egyptian Exploration Society, H. Milford, Oxford University Press.

Winlock, H. E. 1924. The Tombs of the Kings of the Seventeenth Dynasty at Thebes. *JEA* 10: 217–277.

———. 1947. *The Rise and the Fall of the Middle Kingdom in Thebes.* New York: Macmillan.

———. 1955. *Models of Daily Life in Ancient Egypt from the Tomb of Meket-Re at Thebes.* Publications of the Metropolitan Museum of Art Egyptian Expedition 18. Cambridge, Mass.: Published for the Metropolitan Museum of Art by Harvard University Press.

Winnicki, J. K. 2009. *Late Egypt and Her Neighbours: Foreign Population in Egypt in the First Millennium BC.* Journal of Juristic Papyrology Supplements 12. Warsaw: Warsaw University, Faculty of Law and Administration, Chair of Roman and Antique Law.

Winton, A. J. 1906. Saffron, Safflower. In *The Microscopy of Vegetable Foods.* New York: Wiley: 674–678.

Wiseman, J. 1978. *The Land of the Ancient Corinthians.* SIMA 50. Gothenburg: Åström.

Witt, R. E. 1997. *Isis in the Ancient World.* Baltimore: Johns Hopkins University Press.

Wittmann, W. 1938. *Das Isisbuch des Apuleius. Untersuchungen zur Geistesgeschichte des 1. Jahrhunderts.* Forschungen zur Kirchen- und Geistesgeschichte 12. Stuttgart: Kohlhammer.

Wood, B. 1991. The Philistines Enter Canaan: Were They Egyptian Lackeys or Invading Conquerors? *BibAR* 17(6): 44–52, 89–90.

Woudhuizen, F. C. 2006. The Ethnicity of the Sea Peoples. (PhD diss., Erasmus University.)

Wreschner, E. 1971. Prehistoric Rock-Engravings in Nahal ha-Me'erot, Mount Carmel. *IEJ* 21: 217–218, pls. 46–47.

Wreszinski, W. 1988. *Atlas zur altaegyptischen Kulturgeschichte.* 2 vols. Geneva: Slatkine Reprints.

Yacoub, M. 2007. *The Splendours of Tunisian Mosaics.* Tunis: Ministry of Culture and Heritage Conservation and the Agency for the Development of Heritage and Cultural Promotion.

Yadin, Y. 1963. *The Art of Warfare in Biblical Lands, in the Light of Archaeological Study.* M. Pearlman, trans. London: Weidenfeld and Nicolson.

———. 1968. "And Dan, Why Did He Remain in Ships?" (Judges 5:17). *AJBL* 1: 9–23.

Yardeni, A. 1994. Maritime Trade and Royal Accountancy in an Erased Customs Account from 475 B.C.E. on the Ahiqar Scroll from Elephantine. *BASOR* 293: 67–78.

Yasur-Landau, A. 2010. *The Philistines and Aegean Migration at the End of the Late Bronze Age.* Cambridge: Cambridge University Press.

Yon, M. 1971. *Salamine de Chypre,* vol. 2, *La Tombe T. 1 de XIe s. av. J.-C.* Paris: de Boccard.

———. 1992. The End of the Kingdom of Ugarit. In *Crisis Years:* 111–122.

Yoyotte, J. 1949. Les stèles de Ramses II à Tanis. *Kêmi* 10: 58–74.

———. 1951. Un document relatif aux rapports de la Libye et de la Nubie. *BSFE* 5: 9–14.

Yurco, F. J. 1997. Merneptah's Wars, the "Sea Peoples," and Israel's Origins. In *AEANE:* 497–506.

Index

Note: Page numbers in *italics* indicate illustrative figures and tables; page numbers in **bold** indicate maps.